Axial Flow Fans and Ducts

AXIAL FLOW FANS AND DUCTS

R. ALLAN WALLIS
M.E., A.S.T.C., F.I.E. Aust., M.R.Ae.S.

formerly of
Aeronautical Research Laboratories (Melbourne)
and
CSIRO Division of Mechanical Engineering (Melbourne)

KRIEGER PUBLISHING COMPANY
MALABAR, FLORIDA
1993

Original Edition 1983
Reprint Edition 1993 with new material

Printed and Published by
**KRIEGER PUBLISHING COMPANY
KRIEGER DRIVE
MALABAR, FLORIDA 32950**

Copyright © 1983 by John Wiley and Sons, Inc.
Reprinted by Arrangement.

All rights reserved. No part of this book may be reproduced in any form or by any means, electronic or mechanical, including information storage and retrieval systems without permission in writing from the publisher.
No liability is assumed with respect to the use of the information contained herein.
Printed in the United States of America.

Library of Congress Cataloging-In-Publication Data
Wallis, R. Allan.
    Axial flow fans and ducts / R. Allan Wallis.
      p.   cm.
    Reprint. Originally published: New York : Wiley c1983.
    Includes bibliographical references and index.
    ISBN 0-89464-644-3
    1. Fans (Machinery)   2. Air ducts.   I. Title.
  [TJ960.W34   1991]
    621.6′1—dc20   91-22967
                                                                    CIP

10 9 8 7 6 5 4 3 2

Foreword

With the publication of "Axial Flow Fans and Ducts" Mr. Wallis has made a significant contribution to axial flow fan design and to the proper application of axial flow fans. This book, which is a modern replacement for an earlier book by the author, will be of importance to fan designers, engineers who are applying axial flow fans to air handling systems, and engineers responsible for the operation of these systems.

The fan designer will find design procedures presented in a logical sequence and the background of the design equations carefully developed. This book will serve as one of the important elements in planning an axial flow fan design development program. Mr. Wallis is careful to point out the limitations of the design method and to call attention to the importance of laboratory testing for new fan designs. However, an understanding and application of this design procedure will allow the designer to arrive at a theoretical design that will be an optimum starting point for the test program. In this way, much of the "cut and try" part of the test program can be eliminated and a new design can be developed more efficiently. The results of the test program can be incorporated into the design procedure and the effect of changes in the design variables can be evaluated in an orderly and logical manner.

A knowledge of the design principles of axial flow fans is of importance to the engineer applying these fans to air moving systems since many of the shortcomings of faulty air handling systems can be traced to a lack of understanding of what makes a fan work efficiently. Mr. Wallis explains the principles of axial flow fan design in a way that is quite understandable to engineers engaged in the design and operation of air moving systems. An understanding of these principles is necessary for the engineer to avoid those duct configurations that will be detrimental to the fan operation. The emphasis on energy conservation requires that fans be operated efficiently. Mr. Wallis does an excellent job of combining fan design principles with duct design principles so that a unified picture is presented to the application engineer.

This book will also be of value to engineers charged with the operation of air moving systems. Many of their operating problems can be more easily solved if they have an understanding of the principles of fan design, the principles of duct design, and the interrelationship between the two.

Although Mr. Wallis has based his design procedures on a solid theoretical basis, his presentation is a very practical one and shows the results of his own extensive practical engineering experience. He has written this book for qualified engineers who want to use current, valid engineering design procedures to improve their work.

For those engineers who wish to investigate any particular phase of the material, there are extensive references cited which will be most helpful.

J. BARRIE GRAHAM

Santa Fe, New Mexico

Preface to the Reprint Edition

This edition of the book contains new information in relation to arbitrary vortex flow design. Experimental research work carried out since 1983 at the CSIRO Division of Building, Construction, and Engineering, has answered most of the important questions regarding design limitations. It was the lack of definitive experimental data that was responsible for the cautionary advice expressed in the previous edition. I am extremely grateful to Mr. Martin Welsh for giving permission for the work to proceed, and to Mr. Ron Downie for the truly excellent manner in which these tests were conducted. My sincere thanks go to both these colleagues.

In collaboration with Mr. Graham Kipp, a comprehensive suite of nine user-friendly programs has been prepared. A minimum of aerodynamic knowledge is required in their use. The question of the spanwise distribution of lift coefficient did not present a serious programming problem, and the other matters mentioned in my original Preface have now been successfully handled. This edition of the book provides an ideal backup for the information crystallized in these computer programs.*

Without the co-operation of the CSIRO Establishment, and the assistance of the above personnel, these major and important developments would have been impossible.

In view of the continuing performance, efficiency, and noise problems that are frequently associated with installed axial flow fans, Chapter 23 has been expanded to highlight some of the common installation errors associated with this fan type. The role of flow features in precipitating these undesirable situations has been emphasized.

*Information regarding the programs can be obtained from
> Wallis and Kipp
> 19 Corby Street
> North Balwyn 3104
> Victoria, Australia

Preface

It is my belief that uncertainty is a major factor inhibiting the problem-solving capability of many engineers; lack of confidence is usually associated with an inadequate physical understanding of basic airflow phenomena.

The necessary background for engineers handling problems in these areas is presented herein. An understanding of the boundary layer and its characteristics provides the key to design progress. Likewise, a familiarity with aerofoil design theory, particularly the interrelationship between shape changes and the boundary layer, is important.

The formulation and proof of many of the basic theorems and concepts on which aerodynamics is based have been omitted as they are adequately dealt with in the text book literature. In particular, the reader is required to accept the non-dimensional concept as essential to progress in relation to all design, analysis, and test exercises.

The standards adopted in this book promote the attainment of proficiency in respect to all theoretical and practical aspects of fan and duct technology.

This book would not be complete without some mention of computer programs. The uncomplicated expressions used in "free vortex flow" designs are, in my opinion, more suited to the hand-held calculator. However, the tedium of repetitative calculations such as those involved in deriving blade section coordinates, analyzing probable fan performance, or reducing fan test data will be reduced by computerization.

A unique program cannot be written for fan design since the spanwise distribution of lift is not a fixed entity. A series of alternative, plausible assumptions should therefore be made for insertion into the computer program at the appropriate juncture. A fully computerized design method must be programmed to include additional criteria relating to wall stall, blade load and swirl limitations, blade aspect ratio, and other fan properties.

Mechanical features, strength, materials, noise, and dust and water erosion are matters which require consideration before completing the aerodynamic design; their interactions with the latter are outlined in the text.

I am particularly indebted to Mount Isa Mines Ltd. for the opportunity to work closely with their engineers in putting theory into practice on a large and satisfying scale. Much of the stimulus for the applied research studies undertaken owes its origin to equipment problems encountered by the company. Research contracts funded by the Australian Mineral Industries Research Association Ltd. were an important factor in widening the activity spectrum.

It would be very remiss of me if I failed to record my thanks and appreciation for the support and encouragement received from the managements of both the Aeronautical Research Laboratories and the CSIRO Division of Energy Technology (formerly the Division of Mechanical Engineering) during my time of duty. I am especially indebted to the secretarial staff of the latter for the manuscript typing, an arrangement approved by the Divisional Chiefs, Dr. B. Rawlings and Dr. D. C. Gibson.

I am grateful to the University of New South Wales for allowing me to use material from a Master of Engineering thesis submitted in 1954. The Institution of Engineers in Australia has kindly given permission for the use of selected data from seven research papers published on my behalf in their Transactions; these are referenced within.

Acknowledgment is also made to Dr. G. N. Patterson who initiated the fan design approach developed herein. Messrs. M. C. Welsh, I. C. Shepherd, V. J. Smith, and Dr. R. Parker all contributed some input to the finished work. My special thanks are extended to the above and to Mr. G. W. Kipp for assistance in proofreading the galleys.

Finally the typing assistance, support, and tolerance displayed by my wife during the manuscript preparation and publication periods is applauded.

R. ALLAN WALLIS

Melbourne, Australia
June

Contents

1 Introduction 1
2 Introduction to the Fluid Dynamics of Fans and Ducts 8
3 Boundary Layer and Skin Friction Relations 30
4 Duct Component Design and Losses 61
5 Duct System Design and Development 137
6 Airfoil Data for Blade Design 143
7 Introduction to Fan Design Methods 177
8 Rotor: Momentum Considerations, Free Vortex Flow 184
9 Rotor Blade Design 197
10 Rotor Losses 217
11 Stators: Design Considerations 230
12 Stator and Strut Support Losses 241
13 Ancillary Component Design 248
14 Overall Efficiency, Torque, Thrust, and Power 260
15 Design Optimization 264
16 Fan Materials, Mechanics, and Noise 277
17 Fan Applications 302
18 Design Examples 312
19 Rotor and Stator Analysis 324
20 Examples of Fan Analysis 336
21 Fan Testing—Commercial 344
22 Fan Testing—Developmental Research 361
23 Review of Design Assumptions and Limitations 390

Appendix A Approximate Design Procedures 399
Appendix B Arbitrary Vortex Flow Design 409
Appendix C Diaphragm Mounted and Shrouded Fans 425

Appendix D Air Circulating Fans 427
Appendix E Airfoil Section Data 433

Notation 451
Index 457

Axial Flow Fans and Ducts

CHAPTER 1
Introduction

Propeller, or axial flow, fans are finding greater acceptance in industrial applications, as alternative equipment to the radial flow variety. However, the full potential of the axial flow fan will only be realized when modern design techniques and the latest information are utilized to the fullest extent.

At present the percentage of ineffectual axial flow fan installations is considerably higher than for its radial flow counterpart. The problem arises because of an unrefined aerodynamic design approach adopted by many duct and fan engineers. When duct resistance exceeds the pressure duty capability of the fan, stalling occurs, causing a rapid falloff in the volume flow rate. Hence before designing or installing a fan it is essential to ascertain with reasonable accuracy the operational duct resistance. Good fan inlet flow conditions are also of great importance in achieving a satisfactory degree of fan installation effectiveness.

In addition to outlining an up-to-date design approach to axial flow fans, a compendium of duct design information and approach methods that have proved adequate in practice is presented. The interrelated nature of fan and duct design is thereby acknowledged.

When adequate attention is paid to aerodynamic detail, fan efficiency can be high and noise level low.

1.1 GENERAL TYPES OF AXIAL FLOW FAN

Axial flow fans can be placed in three main categories, namely:

1. *Air circulator, or free fan.* A free fan is one that rotates in a common unrestricted air space. Desk, wall, pedestal, and ceiling fans fall into this category.
2. *Diaphragm-mounted fan.* This type of fan transfers air from one relatively large air space to another.
3. *Ducted fan.* A fan is ducted when the air is constrained by an enclosing duct to enter and leave the fan blading in an axial direction.

The minimum duct length required to satisfy this condition will be in excess of the distance between inlet to and outlet from the blading. When additional fan stages are fitted in series, the attainable pressure rise increases. In the extreme case, a multistage unit becomes a compressor.

The first two types are relatively long-established, being currently in common usage. The task that they perform, generally speaking, cannot be economically performed by radial flow equipment. Although improvements have been made in the design of these unducted types, there are still many of a relatively crude type in existence. Exhaust fans that operate in short, compact ducts constitute a worthwhile improvement over the diaphragm-mounted variety and could eventually supersede the latter. Higher efficiencies and lower noise levels will, of course, be associated with the better class of air circulator and exhaust units.

The major portion of the technical material presented in this book is concerned, however, with ducted fan design, analysis, and tests. This fan type lends itself to a refined aerodynamic treatment that has been developed in response to a demand for higher pressure capability and hence increased power requirements. Design guidance on the first two fan types is presented in Appendixes C and D.

1.2 ELEMENTS OF A DUCTED FAN UNIT

The various elements that go to make up a ducted fan unit are illustrated in Fig. 1.1. Rotor blades are a series of airfoils that by virtue of relative motion with respect to the air mass add total pressure to the airstream. It is desirable that this function should be discharged with minimum friction, secondary flow, and flow separation losses.

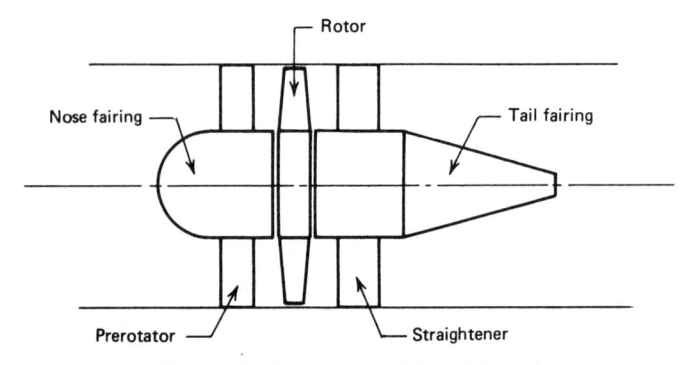

Figure 1.1 Components of ducted fan unit.

1.3 DUCTED FAN DUTY

Stationary vanes, termed *stators,* are normally located upstream and/or downstream of the rotor.

In well-designed units the rotor boss diameter is usually 40 to 70% of the rotor diameter. On approaching the rotor axis both blade velocity and swept area become increasingly small; as a result the potential work output and volume flow rate in this region are limited. The larger boss ratios are usually associated with high-pressure-rise units.

Suitably shaped fairings upstream and downstream of the rotor boss are an essential part of good design. In a multistage unit of co-rotating rotors, a stator row is required between rotor stages. Contra-rotating rotors do not normally require prerotator or straightener vanes.

1.3 DUCTED FAN DUTY

The duty of a ducted fan is specified by the volumetric flow rate of air that is forced through a duct system against resistance comprised of static and dynamic pressure losses that are more conveniently expressed in terms of one variable, namely, total pressure loss. These losses are due to skin friction, flow separation, secondary flows, and energy dissipation at system discharge.

In recent years there has been a growing realization that fan performance cannot be considered in isolation from the associated duct assembly. This is now acknowledged in the revised and new fan test codes.

Fans may be designed either as standard "off-the-shelf" units, covering a wide range of possible duties, or as equipment meeting some uniquely specified duty requirement. In each instance, matching the fan and duct characteristics (Fig. 1.2) in the installation design stage can only be achieved with reasonable and acceptable accuracy when both duct and fan components are subject to careful aerodynamic design. Hence it follows that a certain standard must be adopted in relation to the design developments presented here.

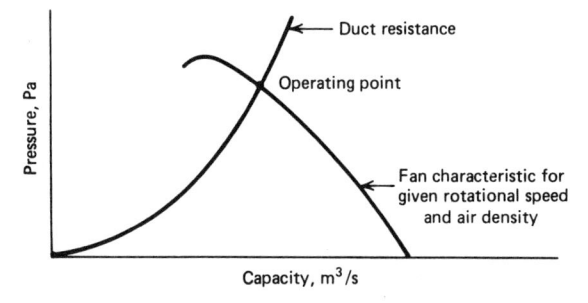

Figure 1.2 Matching of fan and duct system.

The relevant standard requires a fan inlet flow that is steady, axisymmetric, and free of significant swirl. The fan outlet flow must also be of good quality, particularly when the fan is connected to downstream ducting.

Denoting the mean total pressure (gauge) at fan inlet and outlet by the symbols ITP and OTP, the fan total pressure (FTP) is given by

$$\text{FTP} = \text{OTP} - \text{ITP} \tag{1.1}$$

In the case of exhaust units the useful total pressure load on the fan is represented by ITP, since OTP, which consists of the fan exit velocity pressure, is dissipated at discharge. Therefore ITP is numerically equal to fan static pressure (FSP) when the latter term is reserved exclusively for exhaust fan usage. Replacing FSP by the more meaningful term *fan inlet total pressure* (FITP), we eliminate the negative sign of ITP by expressing the latter as the useful total pressure rise across the fan unit. Therefore,

$$\text{FITP} = 0 - \text{ITP} \tag{1.2}$$

where ITP is a negative quantity. When discharging to a plenum chamber in which the static pressure differs from atmospheric, the zero term must be replaced by the appropriate gauge static pressure of the chamber, namely,

$$\text{FITP} = \text{chamber static pressure} - \text{ITP} \tag{1.3}$$

where the chamber pressure is either negative or positive with respect to atmospheric pressure.

From a useful total pressure load viewpoint there are therefore two pressures defining fan duty. When useful work is done downstream of the fan, FTP is the correct quantity; this includes all in-line and blowing-type installations. In the case of exhaust fans that have no useful downstream work component, FITP is the desired value. The elimination of the term *fan static pressure* is believed fully justified in view of the serious interpretation difficulties associated with its use in the past; it is clearly a misnomer.

When giving joint consideration to duct and fan duty matters, it is important to define clearly the upstream and downstream fan boundaries. For present purposes, these limit planes shall be fixed, respectively, just ahead of the nose fairing and at exit from the downstream diffusing passage or exhaust fan exit extremity.

The fan pressure rises normally associated with axial flow fans are not such as to necessitate the treatment of air as a compressible fluid. Hence fan performance can be expressed, for standard atmospheric conditions, as volume flow rate versus fan pressure at a given rotational speed.

When the specified duties call for changes in the volume flow rate, provision has to be made either for resistance-type control devices in the ductwork or for speed and/or blade pitch changes in the fan unit. The latter may

1.4 TYPES OF DUCTED FAN: AERODYNAMIC CLASSIFICATION

be in the form of either adjustable-pitch rotor blading or variable-incidence inlet vanes.

1.4 TYPES OF DUCTED FAN: AERODYNAMIC CLASSIFICATION

When a fan rotor adds total pressure to the air flowing through it, the angular momentum of the stream is altered. For example, a rotor receiving air that is approaching in an axial direction discharges it with a tangential component of velocity, resulting in the appearance of a phenomenon that in the theory of aircraft propellers is known as *slipstream rotation*. The change of angular momentum in the airstream is related to the torque on the rotor shaft.

The efficiency of the fan unit is influenced by the amount of swirl left in the air after it has passed the last stage of blading in the unit. The swirl momentum can play no part in overcoming the resistance of the duct system unless the associated tangential component is removed and its velocity pressure converted into static pressure.

Five main design possibilities arise as a result of the previously mentioned aerodynamic phenomenon of slipstream rotation.

1. *Rotor unit.* The swirl passes downstream of the rotor and the associated momentum is lost.
2. *Rotor-straightener unit.* The swirl is removed by stators placed downstream of the rotor and the associated dynamic pressure is recovered in the form of a static pressure rise.
3. *Prerotator-rotor unit.* Stators are used to impart a preswirl in the opposite sense to the rotor motion and the rotor then removes the swirl.
4. *Prerotator-rotor-straightener unit.* A combination of the preceding two configurations.
5. *Contrarotating rotors.* The second rotor removes the swirl introduced by the first.

The preceding provides a very convenient method for the classification of ducted fans.

The rotors in a contrarotating fan assembly will differ, with the rear rotor possessing blading with lower solidity and pitch settings than the leading one.

Multistage co-rotating units may be designed with identical rotors and stators, except for instances where air compressibility becomes an important factor.

Fans provided with pitch change capability are known as adjustable- or variable-pitch units, where the latter term refers to equipment in which the pitch can be changed without stopping the fan.

1.5 ON STATIC PRESSURE RISE IN A FAN UNIT

The magnitude of the pressure rise required produces detailed design differences between fans that are nominally of a similar type. For example, low-pressure-rise fans possess a smaller number of blades than the high-pressure-rise variety; the relative boss diameter will be greater in the latter case.

The static pressure rise through the rotor blading is related to the retardation of the air relative to the blade as it passes from inlet to outlet of the rotor stage.

The air flowing through prerotating stators is accelerated and, in accordance with Bernoulli's equation, the static pressure falls. However, the reverse occurs in straighteners, where a static pressure rise accompanies the removal of the tangential velocity component.

Unfortunately, when the static pressure rises along a solid surface in the direction in which the air is flowing, separation of the flow from the surface tends to occur, with undesirable consequences. It is this phenomenon, known as stalling, that restricts the static pressure rise obtainable from a given fan unit.

Static pressure rise considerations have inspired the design of unusual types of fan. In an impulse type of rotor the static pressure rise through the blading is zero. It is designed to have equal inlet and outlet velocities of the air relative to the blades; the direction of the vectors has, however, been changed. In other words, the complete increase in total pressure exists as swirl velocity pressure. The straighteners are then charged with the responsibility of converting this dynamic pressure into static pressure.

An alternative method that has been evolved for the purpose of restricting the static pressure rise through the rotor consists in increasing the axial velocity component from inlet to outlet. This is achieved by a flow contraction due to conical surfaces on either the enclosing duct or rotor boss. A portion of the excess axial velocity pressure is then recovered as static pressure in a downstream diffuser. Because of radial flows within the blade passages, the fan is theoretically of a minor mixed-flow variety. However, as will be illustrated in Appendix B for conical hubs, arbitrary vortex flow design procedures can produce results of acceptable accuracy, within specified limits.

In the majority of design cases the static pressure rise requirement is moderate and can be met by employing normal axial flow fan design techniques. Increases in the design pressure rise can be achieved by greater rotor speed and/or larger boss ratios; the latter is preferable from a noise control viewpoint.

The static pressure recovery in the downstream annular diffuser also constitutes an important component of overall fan pressure rise capability.

Good design involves the attainment of the optimum degree of static

1.6 SCOPE OF TREATMENT

pressure recovery in each and every component of the selected fan assembly.

1.6 SCOPE OF TREATMENT

A comprehensive aerodynamic treatment of ducted axial flow fans of the "free vortex flow" type is presented in the present work. This covers all aspects of design, analysis, and testing. Special attention is given to optimized fan design techniques. Although the highest efficiency is obtained for fans of the preceding type, design difficulties may occur at the blade root, necessitating a local "arbitrary vortex flow" design approach. An approximate design method is appended. As a result of recent experimental validation, the latter method for blade design can now be used with confidence for swirl-free inlet flows, within the limits outlined in Appendix B. However, because of the experimentally established unimportance of the design forces associated with the radial flows, no need is felt for a more general radial flow theory, with its added complexity.

Basic fluid phenomena are briefly reviewed as an introduction to fan aerodynamics and as a framework structure in support of the duct design sections. Flow imcompressibility is assumed throughout, in line with the normal definition of a fan.

CHAPTER 2

Introduction to the Fluid Dynamics of Fans and Ducts

2.1 THE ATMOSPHERE

Air, as the working medium, has a number of physical properties that are of importance in the design and testing of axial flow fans. These are pressure, temperature, density, viscosity, and humidity, with the latter, however, being of relatively minor importance. (For gases other than air, the duct and fan design techniques apply when the appropriate density, pressure, and viscosity are inserted.)

2.1.1 Temperatures and Pressures at Mean Sea Level (MSL)

Data on temperature and pressure are presented in Table 2.1.

2.1.2 Density

The air density ρ is obtained from the equation of state, namely,

$$\frac{p_{ab}}{\rho} = RT_{ab} \qquad (2.1)$$

where p_{ab} = absolute pressure (Pa)
ρ = density (kg/m³)
T_{ab} = absolute temperature (K)   (absolute zero = -273°C)
R = universal constant for an air mass of 1 kg (287.0 J/K)

2.1 THE ATMOSPHERE

Table 2.1

Condition	T (°C)	p_{ab} (kPa)	Ref.
"Normal" air	0	101.325 (760 mmHg)	[2.1]
"Standard" air, for fans	20	101.325	[2.2]

For dry air at a standard temperature of 20°C the air density is given in [2.1] as 1.2054 kg/m³ at MSL. However, in [2.2] a standard humidity of 65% is assumed and hence it can be shown that the standard density for fans becomes 1.2 kg/m³. Therefore for general use

$$\rho = 1.2 \left(\frac{293B}{760(273 + T)} \right) \tag{2.2}$$

where B is barometric pressure in mm of Hg, and T is in °C.

2.1.3 Humidity

The effect of water vapor on air density [2.1] can be established from the relation

$$\frac{\rho}{\rho_{dry}} = \frac{p_{ab} - (0.378 p'_v \times \text{RH})}{p_{ab}} \tag{2.3}$$

where the relative humidity RH (p_v/p'_v) is expressed as a fraction (e.g., 0.60) and p'_v is the saturated vapor pressure of water in air [2.1] for a wet bulb temperature identical to the measured dry bulb temperature, the latter being used in Table 2.2 for determining p'_v.

2.1.4 Variation of Temperature with Altitude

The "lapse rate" specified by the International Commission of Air Navigation (ICAN) is −6.5°C per 1000 m of height up to an altitude of 11,000 m [2.1].

Table 2.2 Saturated Vapor Pressure of Water

Air temperature, °C	0	10	20	30	40	50
p'_v, Pa	611	1228	2335	4228	7350	12,305

10   INTRODUCTION TO THE FLUID DYNAMICS OF FANS AND DUCTS

2.1.5 Variation of Pressure with Altitude

The pressure at altitude, using the preceding temperature relation, is given from a development of Eq. (2.1) in [2.1] as

$$p_h = p_0 \left(\frac{T_h}{T_0}\right)^{5.256} \quad \text{(Pa)} \tag{2.4}$$

where the subscripts h and 0 refer to altitude and sea level conditions, respectively, and p and T are absolute quantities.

2.1.6 Viscosity

The viscosity of air is solely a function of temperature being independent of air pressure. Sutherland's equation [2.1] expresses the relationship as

$$\mu = 1.484 \frac{T^{1.5}}{T + 117} \quad \text{(Pa} \cdot \text{s)} \tag{2.5}$$

where the coefficient is adjusted to give a viscosity of 18.30×10^{-6} Pa · s at 23°C [2.1]. For standard air at 20°C the value is 18.16×10^{-6} Pa · s.

2.1.7 Kinematic Viscosity

The quantity is given by the viscosity to density ratio ν, having a value for standard air conditions of 15.13×10^{-6} m²/s.

2.1.8 Compressibility

In axial flow fan design practice, pressure changes in the fluid are too small to change the density to any significant extent. The desire to restrict tip speed, for noise limitation reasons, also ensures a relative air Mach number below the value at which compressibility becomes an important factor. Hence, in line with the usual practice, the concept of incompressibility is adopted throughout this book.

2.1.9 Thermodynamic Matters

The slight increase in air temperature that accompanies the transfer of energy from the rotor to the air is also ignored. The transfer of heat to and from the duct system, by means of heat exchangers, is briefly considered in Section 4.1.1.

2.2 BERNOULLI'S EQUATION

When we consider the flow path of a particle in a fluid in which viscosity can be neglected, changes in the inertia forces must equal the changes in the pressure forces. When the flow at a point is steady with respect to time, the equation of motion for flow along a streamline is

$$\rho U \frac{dU}{dS} = -\frac{dp}{dS} \tag{2.6}$$

where U = velocity (m/s)
p = static pressure (Pa)
ρ = density (kg/m^3)

Integrating the preceding for constant density,

$$\tfrac{1}{2}\rho U^2 = -p + \text{const}$$

or

$$\tfrac{1}{2}\rho U^2 + p = \text{const} = H \tag{2.7}$$

The constant of integration H is called the total pressure of the fluid and the quantity $\tfrac{1}{2}\rho U^2$ is known as the dynamic or velocity pressure. The implications of the preceding equation can be illustrated in Fig. 2.1.

As a particle moves from A to B, it is accelerated, and hence it follows, from the foregoing equation, that the static pressure is progressively reduced.

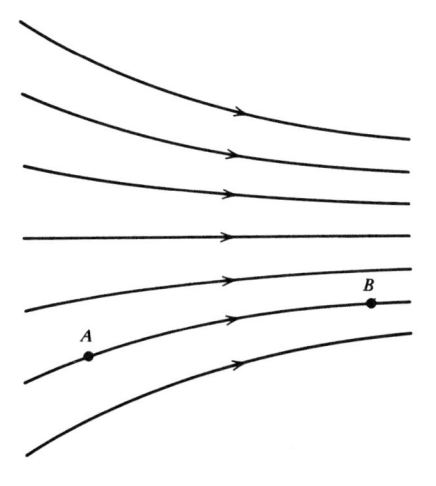

Figure 2.1 Streamline motion.

In the reverse process, the flow is retarded and velocity pressure is converted into static pressure; this process is called pressure recovery or diffusion. The static pressure rise produced in a fan unit can be attributed to pressure recoveries occurring in various components of the unit (Section 1.5).

Bernoulli's equation applies strictly to flow paths in which viscosity may be neglected, but qualitatively the principle of the interchangeability of static pressure and velocity pressure remains unaltered even where viscosity has an appreciable influence.

2.3 VISCOSITY AND BOUNDARY LAYERS

2.3.1 Skin Friction

In developing Bernoulli's equation, viscosity was neglected. This assumption is, however, not valid near a solid surface. Air in direct contact with the surface is at rest and as a result there is a large rate of fluid shear as the velocity u increases rapidly with distance from the surface until it reaches a constant value U in the "free" stream (Fig. 2.2). This region of flow retardation, called the *boundary layer,* represents a loss of fluid momentum that is related to the "skin friction" force acting on the surface.

2.3.2 Boundary Layer Thickness

A boundary layer starts to grow at the point where the airstream first contacts the body. As more and more momentum is taken from the outer flow in order to maintain the forward motion of the air near the surface, the boundary layer thickness, in general, increases with distance x (Fig. 2.3).

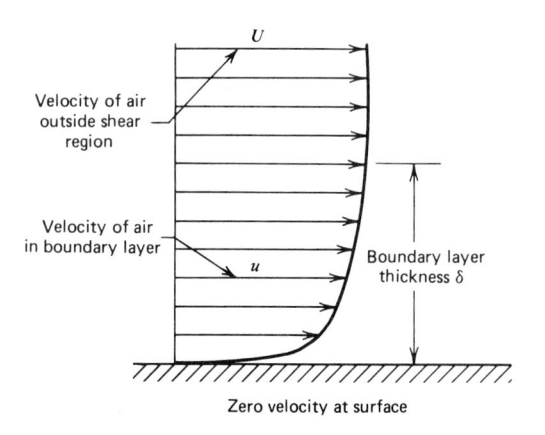

Figure 2.2 Boundary layer flow.

2.3 VISCOSITY AND BOUNDARY LAYERS

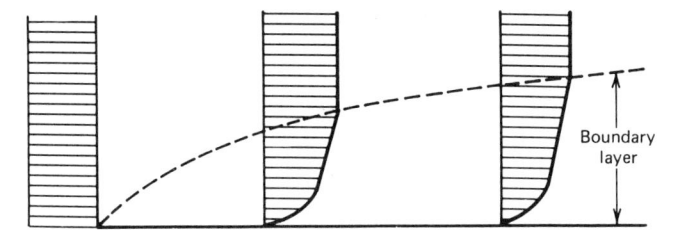

Figure 2.3 Growth of boundary layer on flat plate.

This thickness δ is relatively hard to define as the velocity in the boundary layer approaches the free stream velocity asymptotically. The point at which the velocity is 99.5% of its value in the free stream is frequently taken as the edge of the boundary layer. Somewhat more useful quantities are the "displacement" and "momentum" thicknesses.

2.3.3 Displacement Thickness

Because of the retarded flow in the boundary layer there is a displacement of the main body of fluid away from the surface. When the flow deficiency $(U - u)$ is integrated across the boundary layer in the y direction and equated to the product of the free stream velocity U and a thickness δ^*, one obtains

$$U\delta^* = \int_0^\delta (U - u)\, dy$$

or

$$\delta^* = \int_0^\delta \left(1 - \frac{u}{U}\right) dy \qquad (2.8)$$

where δ^* is called the displacement thickness. No difficulty is experienced in determining this thickness, as the flow deficiency near the outside edge of the boundary layer is negligible and hence the choice of the limit δ is not critical.

The displacement thickness represents the amount by which the main flow has been displaced from the surface and is equivalent to an increased body thickness.

2.3.4 Momentum Thickness

Probably the most useful thickness is the one that defines the loss of momentum in the boundary layer. In a manner similar to the foregoing, the loss of momentum is integrated across the layer and equated to the hypothetical case where the whole free stream momentum contained in a width θ is lost. This gives

$$\rho U^2 \theta = \rho \int_0^\delta u(U - u)\, dy$$

or

$$\theta = \int_0^\delta \frac{u}{U}\left(1 - \frac{u}{U}\right) dy \qquad (2.9)$$

where θ is the momentum thickness. This thickness is of fundamental importance when assessing fluid friction losses.

2.3.5 Transfer of Momentum by Laminar and Turbulent Means

To maintain the forward motion of the air particles near the surface, against the retarding action of skin friction, momentum must be transferred inward from a region possessing higher momentum. The transfer mechanism can be either laminar or turbulent.

The action of a laminar boundary layer can be illustrated from the simple model of an infinite number of thin fluid strata, parallel to each other and the surface; each stratum possesses relative motion with respect to its neighbor and as a consequence exerts a viscous pull on it. In this manner the desired momentum transfer toward the solid surface is effected through the medium of fluid shear.

No simple concept can be advanced for a turbulent boundary layer. However, it is known that momentum is transferred in a direct and speedy manner by particles moving in an eddying fashion. The eddy motions passing through a given point are exceedingly numerous and are continually changing their characteristics as they are swept downstream. At any such point, however, the turbulence properties have definite mean statistical values with time; this implies that the mean motion is a steady one and that the turbulence obeys definite laws. This type of turbulence should not be confused with the violent large-scale eddies that exist downstream of bluff bodies, for example, cylinders and plates normal to the stream.

The intense mixing that takes place in a turbulent boundary layer implies large fluid shear stresses. As a consequence, skin friction is considerably greater for turbulent boundary layers than for laminar ones of comparable thickness. An excellent simplified physical model of turbulent shear flows is available in [2.3].

2.3.6 Boundary Layer Transition

Laminar boundary layers are normally associated with very low fluid velocities or with the newly formed layers downstream of the point at which the

2.3 VISCOSITY AND BOUNDARY LAYERS

flow meets a solid surface. The leading edge of an airfoil is such a region. With increasing speed, or increasing distance downstream of the initial contact point, the laminar layer eventually gives way to turbulent flow; the change from laminar to turbulent flow is called *transition*.

Depending on its cause, transition may occur abruptly or may be prolonged over a finite length of surface. During this period the skin friction will progressively increase from a laminar to a turbulent value.

2.3.7 Boundary Layer Growth

As indicated previously, the boundary layer grows as the flow progresses along a solid surface. In a pipe this growth continues until the wall boundary layers meet the axis. This latter condition is termed "fully developed" pipe flow.

Since skin friction is greater in a turbulent layer than in a laminar one, the turbulent boundary layer grows at a faster rate. This is illustrated in Fig. 2.4 for flow along a flat plate; the thickness scale has been greatly magnified.

Other factors influencing skin friction and boundary layer growth are surface roughness and the static pressure gradient in the stream direction.

2.3.8 Effects of Streamwise Static Pressure Gradients

In an accelerating flow, static pressure is converted into velocity pressure; as a result, dynamic pressure is added to the fluid near the surface in addition to that transferred by the shear. Moreover, since the flow as a whole is speeded up, appreciable thinning of the layer in the flow direction may occur in highly accelerated flows.

Conversely, when the fluid is decelerating and the pressure rising, the momentum transfer mechanism has to contend with the loss of momentum because of skin friction together with that associated with the conversion of dynamic pressure into static pressure. Because of its effective mixing mechanism, a turbulent boundary layer can meet these demands more readily than can a laminar one.

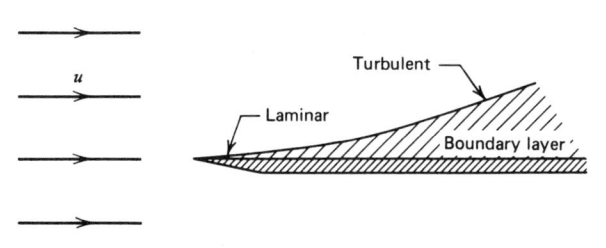

Figure 2.4 Comparative rates of growth.

2.3.9 Boundary Layer Flow Separation

When in a decelerating flow the loss of momentum in the flow immediately adjacent to a surface exceeds the rate at which momentum is being transferred inward to the surface, the boundary layer approaches the condition where steady local flow is impossible. The flow then leaves the surface in the manner shown in Fig. 2.5.

In other words, the particles near the surface no longer have a downstream velocity component and as a consequence the main flow is forced away from the surface. When this occurs, unsteady large-scale eddies are initiated. Large energy losses and drag are associated with such a phenomenon.

The region downstream of a boundary layer separation point is often referred to as a "stagnant" zone, because of the relatively low air velocities therein. Although this is not a strictly accurate description, it is nevertheless for some purposes convenient to think of it in this manner. The "stagnant" air can be considered to be an extension to the solid surface insofar as the main flow is concerned. Such an extension represents an alleviation of the adverse pressure gradient.

The "stagnant" region is, in effect, a highly inefficient mixing zone, as the losses will imply. However, when the adverse gradients have been reduced in the preceding manner, it is possible for the eddy mixing to bring about a reattachment of the main flow to the surface at some downstream location.

This phenomenon of flow separation constitutes one of the major problems in fluid dynamics, as it restricts the lift that can be obtained from airfoils, increases resistance to flow, and produces unsatisfactory conditions where uniformity of flow velocity is essential. Various devices exist for controlling this separation, but the additional complication that their installation involves has, in most cases, exerted a deterrent effect.

Separation avoidance by the skillful and informed design of flow surfaces, with each individual surface accepting its optimum share of the overall pressure recovery load, represents a sounder design approach.

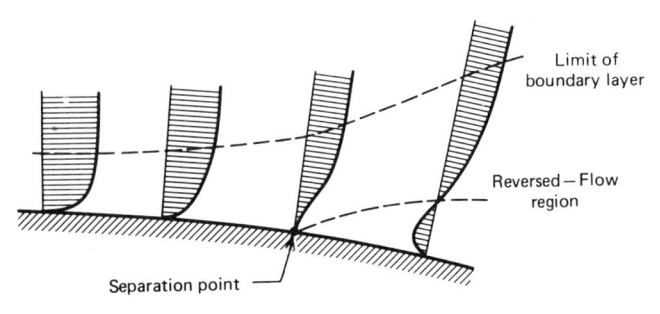

Figure 2.5 Growth of layer in adverse pressure gradient.

2.4 SIMILARITY AND NONDIMENSIONAL NUMBERS

The subjects of similarity and dimensional analysis are adequately covered by many fluid dynamics textbooks and therefore the discussion in these pages will be limited to the proper use of such data as they affect duct and fan design.

When the results of tests carried out on an item of aerodynamic equipment are expressed in a suitable nondimensional manner, the data so obtained can be used in predicting the performance of similar items. Full use must be made of nondimensional parameters in design methods.

Although the beginner may have initial difficulty in understanding the physical significance of certain parameters, a mastery of their use will facilitate the execution of successful fan designs.

2.4.1 Similarity

There are three complementary types of similarity.

1. *Geometric similarity.* Two units are geometrically similar if the corresponding length dimensions have a constant ratio throughout. In practice, each dimension of a unit is expressed as a ratio of some reference length (e.g., the diameter of a pipe). A unit can then be specified by a number of such ratios and two systems will be similar when the corresponding ratios are identical throughout.

2. *Kinematic similarity.* The basic dimension of time has now been added to that of length, and this implies that the flow velocity at any point in the system should be in constant ratio to the velocity at a corresponding part of the similar unit. As before, it is more convenient to express the velocities in one unit as a ratio of a reference velocity, for example, the mean velocity of flow through a pipe at the selected station.

3. *Dynamic similarity.* Dynamic similarity requires the ratio of the forces acting to be similar for both systems. The three basic dimensions of length, time, and mass are therefore included.

Three different aspects of dynamic similarity are discussed in the following subsections.

2.4.2 Reynolds Number

This is probably the best known of all fluid motion parameters but is unfortunately not always properly understood. When a dimensional analysis is applied, it is found that the forces acting on a solid surface and the flow phenomena associated with them are greatly dependent on the ratio of the inertia to the viscous forces, namely,

$$\frac{\rho U l}{\mu}$$

where U and l are characteristic values of velocity and length, respectively, and ρ and μ the relevant values of density and viscosity.

It is in the choice of U and l that confusion arises and to clarify this point a more basic approach than usual will be made. Osborne Reynolds, after whom the ratio is named, demonstrated experimentally in a pipe that the type of boundary layer flow, i.e., laminar or turbulent, is dependent on a critical value of the relation

$$R_d = \frac{\rho \bar{U} d}{\mu} \qquad (2.10)$$

where $\bar{U}$ is the mean velocity and d is the internal diameter of the pipe [2.4]. For Reynolds' experiments, the flow in the pipe was "fully developed." Hence it follows that an equally suitable number is

$$R_\delta = \frac{\rho U \delta}{\mu} \qquad (2.11)$$

where U is the velocity on the axis, that is, at the edge of the boundary layer, and δ is the boundary layer thickness, which in this case equals $d/2$. The characteristic velocity and length are therefore related to the boundary layer. This is very reasonable when it is remembered that the viscous forces are of no practical importance outside the region of the boundary layer. The Reynolds number given by Eq. (2.11) is the one with real meaning in relation to pipe inlet flow.

In the general case of flow along a plate, the boundary layer thickness δ is difficult to define accurately, and hence it is more convenient to use θ, the momentum thickness. This gives

$$R_\theta = \frac{\rho U \theta}{\mu} \qquad (2.12)$$

which is the most basic Reynolds number to be discussed here.

It is obvious, however, that for flow over airfoils and along plates, the preceding Reynolds number will vary with distance and thus some Reynolds number capable of expressing the integrated effect is required. Such numbers will be discussed in subsequent subsections, but it must be remembered that they can only be justified if they satisfy the requirements of boundary layer theory.

2.4 SIMILARITY AND NONDIMENSIONAL NUMBERS

As suggested earlier, the Reynolds number was first used in the prediction of transition, i.e., the change from laminar to turbulent flow. Reynolds suggested a certain critical value of

$$\frac{\rho \bar{U} d}{\mu} \simeq 2000$$

but added the qualification that initial disturbances in the fluid could markedly influence its value. The truth of this has since been strikingly demonstrated by both Ekman and Taylor, who by exercising great care in ensuring a very steady initial flow, obtained critical values of 5×10^4 and 3.2×10^4, respectively [2.5].

Since the Reynolds number is the ratio of the inertia to the viscous forces, two systems cannot be dynamically similar unless the component Reynolds numbers of one system are identical with those in the geometrically similar unit. It is in this connection that the Reynolds number finds its greatest field of application.

2.4.3 Force Coefficients

For a body immersed in a fluid and having a relative motion with respect to that fluid, it can be shown by dimensional analysis that the forces on the body can be expressed by the relation

$$F = m\rho l^2 U^2 f \left(\frac{lU\rho}{\mu}\right) \tag{2.13}$$

where m is a constant and $f(lU\rho/\mu)$ is a function of Reynolds number. As before, l and U are characteristic values of length and velocity.

Dimensional analysis is useful for indicating the correct grouping and presentation of results but actual experience and test data must be employed in choosing the appropriate characteristic dimensions. Although l appears three times in the preceding equation, it is not essential to substitute the same dimension in each case.

Some important applications of the preceding equation will now be considered in detail.

1. *Skin frictional force in zero pressure gradient.* The element of force dF acting on an elementary area dA can be written

$$dF = c_f \tfrac{1}{2} \rho U^2 \, dA \tag{2.14}$$

where U is the local free stream velocity and $c_f = 2mf(R_\theta)$.

For two-dimensional flow, dA can be replaced by dx, as unit length is assumed in the direction normal to the flow path. Hence

$$dF \text{ per unit width} = c_f \tfrac{1}{2}\rho U^2 \, dx \tag{2.15}$$

In the preceding it will be seen that the characteristic lengths have been replaced by unity, dx and θ, respectively. The nondimensional coefficient c_f is called the local skin friction coefficient.

Assuming a smooth surface, the process of determining the total force F acting on a surface can be simplified. On integrating the preceding expression

$$F \text{ per unit width} = \tfrac{1}{2}\rho U^2 \int_0^{x_e} c_f \, dx$$

$$= C_f \tfrac{1}{2}\rho U^2 x_e \tag{2.16}$$

where

$$C_f = \frac{1}{x_e} \int_0^{x_e} c_f \, dx \tag{2.17}$$

and x_e is the effective length of surface.

It follows that C_f is a mean value of the skin friction coefficient, and since the boundary layer will grow with x in a unique manner, the length θ in the Reynolds number can be replaced by x_e. In other words, C_f is a function of $U x_e \rho / \mu$ that is unique in the same manner as the c_f versus R_θ relation.

To illustrate the preceding development, the case of a sharp-edged, two-dimensional plate in a uniform stream will be considered in Fig. 2.6.

For simplicity, transition is assumed to occur abruptly at point T. The effective length in the laminar flow region is obvious, but for the turbulent

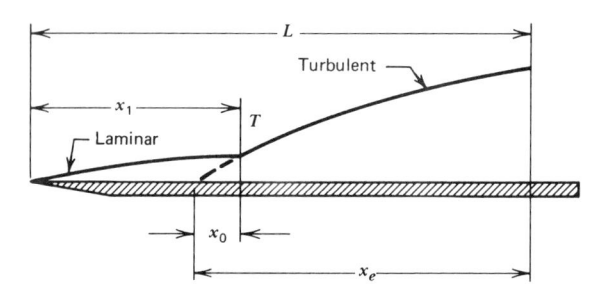

Figure 2.6 Lengths used in skin friction calculations.

2.4 SIMILARITY AND NONDIMENSIONAL NUMBERS

flow the boundary layer thickness must be extrapolated forward to zero to obtain the appropriate effective length x_e.

The skin frictional force acting on one surface of the preceding plate, between the limits $x = 0$ and $x = L$, is

$$F \text{ per unit width} = C_f \tfrac{1}{2}\rho U^2 x_1 + C_f \tfrac{1}{2}\rho U^2 x_e - C_f \tfrac{1}{2}\rho U^2 x_0 \quad (2.18)$$

where the values of C_f are obtained from the appropriate curves of C_f versus $Ux\rho/\mu$.

When the dimension L is large and the dimension x_1 small, the force F can be approximated by

$$F \text{ per unit width} = C_f \tfrac{1}{2}\rho U^2 L \quad (2.19)$$

This latter expression is the one usually employed, but the assumptions made in obtaining this simple expression should not be forgotten and each configuration should be examined to determine whether the expression may safely be used.

In cases when the velocity is variable with x, where nonuniform roughness exists, or where other complications are present, good accuracy can be obtained only by boundary layer growth calculations and by subsequent integration of the local elementary skin friction forces. Sufficient data to facilitate such calculations will be given later.

The foregoing paragraphs have illustrated the manner in which nondimensional coefficients can be used in the calculation of skin friction forces. Flow over various types of solid surface will be considered in the relevant subsections.

2. *Airfoil Forces.* The force experienced by an airfoil, when situated in an airstream, is usually resolved into two components: the lift force, which acts perpendicular to the direction of the oncoming stream, and the drag force, which acts in the stream direction. By rewriting Eq. (2.13), these forces are given by

$$L = C_L \tfrac{1}{2}\rho U_0^2 A \quad (2.20)$$

and

$$D = C_D \tfrac{1}{2}\rho U_0^2 A \quad (2.21)$$

where U_0 is the velocity of the oncoming stream, A is the planform area of the airfoil, and C_L and C_D are the lift and drag coefficients, respectively.

The force coefficients are usually obtained experimentally in the following form:

$$C_L = f_1\left(\frac{U_0 c \rho}{\mu}, \alpha\right) \quad (2.22)$$

$$C_D = f_2\left(\frac{U_0 c \rho}{\mu}, \alpha\right) \quad (2.23)$$

where c and α represent the airfoil chord and incidence, respectively.

For two-dimensional flow,

$$L \text{ per unit span} = C_L \tfrac{1}{2}\rho U_0^2 c \quad (2.24)$$

$$D \text{ per unit span} = C_D \tfrac{1}{2}\rho U_0^2 c \quad (2.25)$$

In the preceding development, the chord Reynolds number R_c rather than the boundary layer one R_θ has been used. It can readily be shown, however, that for a given Reynolds number and incidence, and for complete geometric similarity, the boundary layer thickness at any point on the surface is a fixed proportion of the chord dimension. Hence it follows that the chord is a suitable dimension for expressing the integrated effect of the boundary layers on the lift and drag forces. When the boundary layers are influenced by inadvertent surface roughness or inaccuracy in airfoil profile or by appreciable changes in free stream turbulence, inaccuracies must be expected in calculating the forces from chord Reynolds number data.

The local velocity at any point on a particular type of airfoil section is usually determined by the quantities U_0 and α. Hence kinematic similarity is achieved, provided there are no irregular boundary layer effects similar to those just mentioned.

2.4.4 Pressure Coefficients

When dynamic similarity is achieved, it follows that the pressures that are responsible for the forces must also be capable of nondimensional treatment. In Eq. (2.13), ρU^2 has the dimensions of pressure, hence the choice of a dynamic pressure, $\tfrac{1}{2}\rho U^2$, in reducing quantities to a nondimensional form.

The absolute value of pressure is of limited interest in incompressible aerodynamics; as a result, pressures are usually measured with respect to some reference value such as atmospheric pressure, normally referred to as gauge pressures.

As an example of the preceding, the static pressure p at some point on an airfoil surface is expressed by the nondimensional coefficient

$$C_P = \frac{p - p_0}{\tfrac{1}{2}\rho U_0} \quad (2.25)$$

2.5 FREE VORTEX FLOW

where p_0 and U_0 are the undisturbed values of static pressure and velocity sufficiently far upstream of the airfoil.

In general, therefore, it can be said that a pressure coefficient consists of three parts, namely, the pressure concerned, a reference pressure level, and a reference dynamic pressure.

2.5 FREE VORTEX FLOW

A vortex can be qualitatively described as a circulatory flow about an axis OZ (Fig. 2.7). When the fluid has a velocity component in the direction OZ, the air particles trace out helical flow paths.

When these paths maintain a constant radius, a condition for radial equilibrium exists, namely, a balance is maintained between the centrifugal and pressure forces acting on the particle. The equilibrium requirements for this two-dimensional balance, termed *free vortex flow*, are now established.

Assume that an element of fluid at radius r with unit length in the direction OZ rotates with angular velocity ω about the axis OZ. The centrifugal force acting on the element is, from Fig. 2.8, given by

$$F_c = s \cdot dr \cdot \rho \frac{(\omega r)^2}{r}$$

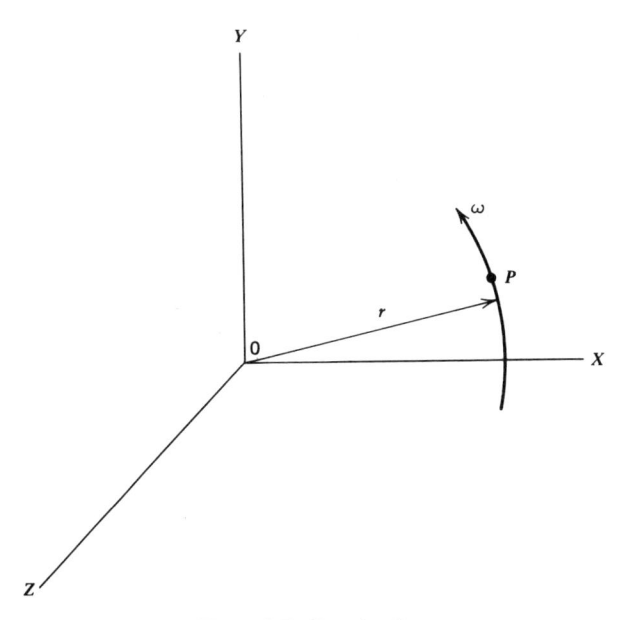

Figure 2.7 Rotating flow.

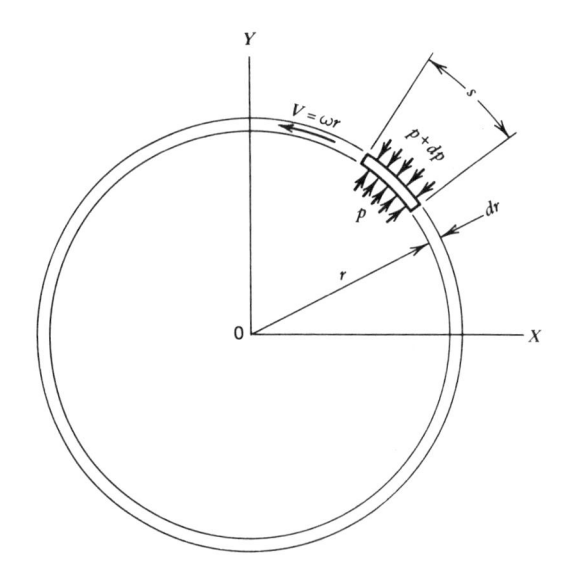

Figure 2.8 Rotating element.

and the pressure force by

$$F_p = dp \cdot s$$

where dp is the pressure difference between the two faces of the element. Equating the two forces gives

$$\frac{dp}{dr} = \rho \omega^2 r \tag{2.27}$$

which is a universal requirement for radial equilibrium.

The total pressure of a particle in equilibrium is

$$H = p + \tfrac{1}{2}\rho V_a^2 + \tfrac{1}{2}\rho (\omega r)^2 \tag{2.28}$$

where V_a and ωr are the axial and tangential velocity components, respectively. Differentiating with respect to r,

$$\frac{dH}{dr} = \frac{dp}{dr} + \tfrac{1}{2}\rho \frac{dV_a^2}{dr} + \tfrac{1}{2}\rho \frac{d(\omega r)^2}{dr} \tag{2.29}$$

When H and V_a are constant with radius, Eq. (2.29) reduces to

$$\frac{dp}{dr} = -\tfrac{1}{2}\rho \frac{d(\omega r)^2}{dr} \tag{2.30}$$

2.7 SECONDARY FLOWS

Combining Eqs. (2.27) and (2.30), it follows that

$$\omega r^2 = \text{constant} \tag{2.31}$$

that is, ωr is inversely proportional to r.

The preceding constitutes the framework on which the free vortex design technique for axial flow fans is based.

2.6 ARBITRARY VORTEX FLOW

In some instances it is necessary to design axial flow fan equipment with radial gradients of axial velocity and total pressure. Low-pressure-rise cooling tower fans, of small boss ratio, provide one example of this design requirement brought about by practical blade construction considerations. Although approximate design methods are available for use with tangential velocity distributions of the form

$$\omega r = a + br \tag{2.32}$$

where a and b are constants, the computerized "streamline curvature" and "matrix" techniques would appear to provide more accurate design procedures. However, these theoretical methods must be adjusted to conform to practical situations such as those arising from boundary layer growth. However, recent experimental work based on the approximate method has shown good agreement between fan performance and the theory.

2.7 SECONDARY FLOWS

2.7.1 Unvaned Corner

Adjacent fluid particles of different total pressure, following a curved path with a relatively constant radial static pressure gradient, will trace out different trajectories, as in Fig. 2.9. The particle possessing the lesser centrifugal force, namely p', will move with a component toward the center of rotation, because of noncompliance with the foregoing equilibrium requirement. The air movement in the side wall boundary layer toward the inner wall creates the situation illustrated in Fig. 2.10, leading to the appearance of secondary flow vortices on this curved surface. For fully developed pipe flow conditions, the strength of these vortices is at a maximum.

Since the dynamic energies associated with the tangential velocity components of secondary flow are seldom recoverable, the resulting loss is usually unavoidable. These losses are at a minimum when the side wall boundary layers are thin, thus restricting the secondary flow air quantities. Flow sepa-

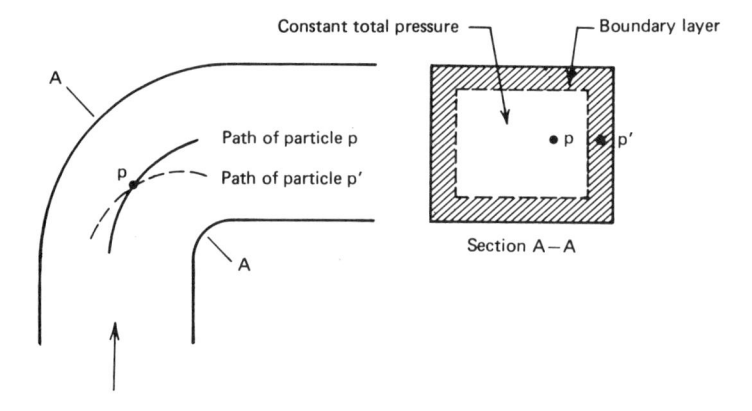

Figure 2.9 Origin of secondary flow.

ration with its large attendant total pressure gradients can, through interrelated flow phenomena, lead to losses that are a multiple of the minimum.

2.7.2 Vaned Corner

The individual magnitude of the preceding corner vortices, in a right-angled corner, can be reduced and controlled by the insertion of turning vanes (Fig. 2.11). In addition, with less low-energy air being transported toward the inner curved duct wall, the tendency toward flow separation is reduced. These features are of most benefit for compact bends of small turning radii.

2.7.3 Turning Flow Through Fan Blading

The function of a rotor or stator blade is similar to that of corner turning vanes. In the first place, the blades produce an air deflection; in the second, the extremities of the blades operate in the wall boundary layers of both the

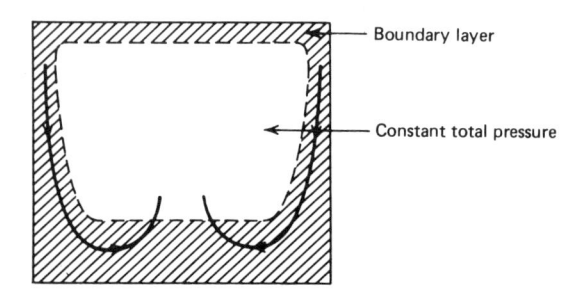

Figure 2.10 Schematic presentation of secondary flow condition at A–A (Fig 2.9).

2.8 INTERACTION AMONG VARIOUS FLOW TYPES

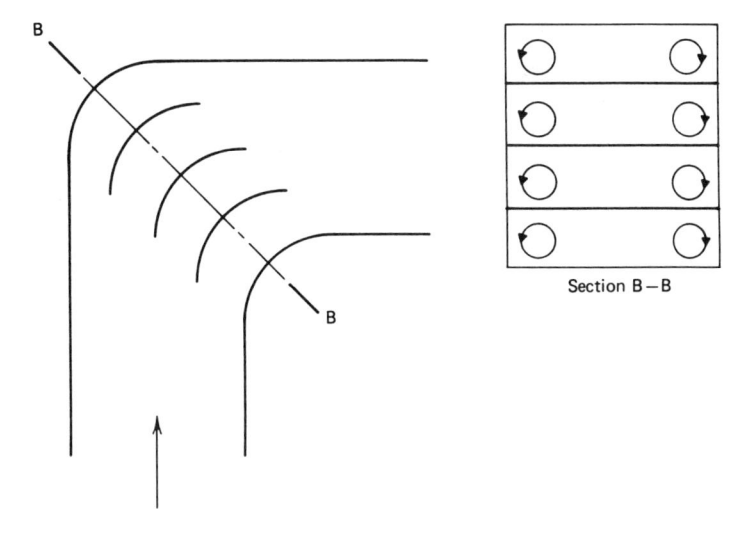

Figure 2.11 Secondary flows in vaned corner.

encasing duct and the rotor boss. Hence secondary flows will appear on the blade surfaces at their extremities; the accompanying losses will be minimized for streamlined in-flow conditions and for well-designed and efficient blading. The local leading-edge flow separation characteristic of constant thickness blade sections, at other than design conditions, will be attended by increased losses.

The axes of these flows will be of a curved helical nature when main vortex flows of the types discussed in the preceding subsection are present. The complexity of the flow at the rotor blade extremities is further increased by tip clearance effects and by the centrifugal influences of the boundary layer particles in contact with the rotating blades. These design problems are reviewed in [2.6].

The problems of accurately analyzing and assessing these complex fluid momentum losses, however, remain unsolved, forcing the designer to continue using ad hoc relations of limited applicability.

2.8 INTERACTION AMONG VARIOUS FLOW TYPES

The main types of flow encountered in duct and fan design are summarized in Table 2.3. When these flows interact with each other, the degree of flow complexity is increased, making the design task a difficult one. In seeking reliable data the designer must, on occasion, resort to model test programs. However, in the majority of cases design guidance is adequate, although unfortunately there are far too many instances where these guidelines are

Table 2.3 Simplified Classification of Flow Types

Type of Flow	Quasi-Two-Dimensional Flows		Three-Dimensional Vortex Flows	
Steady shear-free potential flow	Parallel flow Curved flow	(a)	"Free" vortex flow ($V_t r = $ const) "Forced" vortex flow ($V_t r \neq $ const)	(d)
Steady shear flow	Boundary layers Wake flow	(b)	Secondary flow, resulting from the turning of shear flows	(e)
Unsteady separated flow	Detached from continuous surface Detached from bluff termination	(c)	Detached from three-dimensional body	(f)

N.B.: The degrees of freedom available in turbulent fluids allow the preceding flows to interact with each other, in an extremely complex manner.

either ignored or poorly understood. Adherence to the principles and data contained here is therefore strongly recommended.

The most recent guide to engineering style turbulent flow calculations is available in [2.7].

REFERENCES

2.1 G. W. C. Kaye and T. H. Laby, *Physical and Chemical Constants,* Longmans, London, 10th ed. 1948.

2.2 International Standards Organization, Air performance test methods of industrial fans using standardized airways (in preparation).

2.3 P. Bradshaw, *An Introduction to Turbulence and Its Measurement,* Chap. 1, Pergamon, Oxford, 1971.

2.4 O. Reynolds, On experimental investigation of the circumstances which determine whether the motion of water shall be direct or sinuous and of the law of resistance in parallel channels, *Phil. Trans. Part III,* 939–982, 1883.

2.5 S. Goldstein (Editor) *Modern Developments in Fluid Dynamics,* Vol. 1, Oxford University Press, 1938, p. 321.

2.6 J. H. Horlock and H. J. Perkins, Annulus wall boundary layers in turbomachines, *AGARDograph No. 185,* May 1974.

2.7 P. Bradshaw, T. Cebeci, and J. H. Whitelaw, *Engineering Calculation Methods for Turbulent Flow,* Academic Press, London, 1981.

CHAPTER 3

Boundary Layer and Skin Friction Relations

In this chapter are presented selected boundary layer data, of both a quantitative and qualitative nature, for use in design calculations of engineering accuracy. Provided boundary layer flow separation can be avoided to a major degree on all ducting and blading surfaces, these data will permit the reader to approach the design problem in a confident and proficient manner.

Because of the increasing concern about matters relating to conservation of energy and the environment, optimum design solutions must be sought as a prime objective. Hence separated flows that are wasteful of energy and are the root cause of excessive aerodynamic and related mechanical noise and vibration must be avoided, wherever possible, by employing established and proved design techniques.

With a view to keeping the presentation simple, the ensuing boundary layer developments are essentially two-dimensional; when applied correctly these lead to an adequate level of design accuracy. Despite recent good progress with respect to three-dimensional boundary layer problems (e.g., with relevance to the highly complex blade tip and root flows), much remains to be done. At present a high degree of empiricism is required in obtaining approximate design solutions to these problems. For this reason there is advantage in having some understanding of the way in which these data were derived.

3.1 EQUATIONS OF MOTION FOR BOUNDARY LAYER FLOW

For two-dimensional steady *laminar* flow, the Prandtl boundary layer equations are

$$u \frac{\partial u}{\partial x} + \frac{\partial u}{\partial y} = -\frac{1}{\rho} \frac{\partial p}{\partial x} + \nu \frac{\partial^2 u}{\partial y^2} \tag{3.1}$$

3.1 EQUATIONS OF MOTION FOR BOUNDARY LAYER FLOW

$$-\frac{1}{\rho}\frac{\partial p}{\partial y} = 0 \tag{3.2}$$

where $\nu = \mu/\rho$, or kinematic viscosity, x and y are coordinates parallel and normal to the surface, respectively, and u and v are the corresponding velocity components in these directions.

These equations apply to the flow at a point (x, y) in the boundary layer; when integrated they relate to the boundary layer as a whole. The first relation equates the changes occurring in the inertia, pressure, and viscous forces. From the second equation it follows that the static pressure for a given x coordinate remains constant throughout the boundary layer. Experimental aerodynamics makes wide use of this fact.

The preceding equations contain too many unknowns to be of universal use, although isolated solutions have been obtained. One of these is due to Blasius, who obtained a solution for the laminar layer in a zero pressure gradient, i.e., $\partial p/\partial x = 0$. Solutions exist for fully developed flow in a pipe and for other flow cases.

In a *turbulent* boundary layer, the turbulent stresses that provide the turbulent mixing are much greater than the viscous ones. In an attempt to represent the turbulent stresses, the first equation of motion is empirically modified to include a term ϵ, called the "apparent" kinematic viscosity due to turbulent shear. Hence

$$u\frac{\partial u}{\partial x} + v\frac{\partial u}{\partial y} = -\frac{1}{\rho}\frac{\partial p}{\partial x} + (\nu + \epsilon)\frac{\partial^2 u}{\partial y^2} \tag{3.3}$$

Despite many attempts, no successful solution of this equation has been obtained. This is not surprising in view of the complexity of the actual boundary layer flow.

The preceding equations, although of little direct use, have been included by virtue of their fundamental importance and general interest.

The equation of continuity for two-dimensional flow is usually grouped with the preceding equations and used in conjunction with them. This states

$$\frac{\partial u}{\partial x} + \frac{\partial v}{\partial y} = 0 \tag{3.4}$$

In other words, a small loss of fluid in the x direction is exactly balanced by a gain of fluid in the y direction, since there can be no loss or gain from the fluid as a whole.

Von Karman by considering the momentum flow into and out of a "control box" developed the well-known two-dimensional momentum integral equation, namely,

$$\frac{d\theta}{dx} + \left(\frac{\delta^*}{\theta} + 2\right)\frac{\theta}{U} \cdot \frac{dU}{dx} = \frac{\tau_0}{\rho U^2} \tag{3.5}$$

When $dU/dx = 0$,

$$\frac{d\theta}{dx} = \frac{\tau_0}{\rho U^2} \tag{3.6}$$

or

$$\theta = \int \frac{\tau_0}{\rho U^2} dx \tag{3.7}$$

In accord with normal engineering practice, the contribution of the Reynolds normal stresses to the momentum equation has been ignored since, with the exception of flows in the vicinity of separation, the skin friction term is the dominant one [3.1]. This equation is not valid for thick boundary layers on longitudinally curved wall ducting, as discussed later.

3.2 LAMINAR BOUNDARY LAYERS

As illustrated in Fig. 2.3, the velocity in the boundary layer changes continuously from zero at the surface to the free stream value at the outer edge of the layer. Also, for reasons given in Section 2.3.8, the velocity profile will be influenced by the pressure gradient in the stream direction. It has been established by Pohlhausen that similarity of profile is achieved when two boundary layers have the same value of the nondimensional parameter

$$\Lambda = \frac{\delta^2}{\nu} \cdot \frac{dU}{dx} \tag{3.8}$$

Hence it follows that Λ defines a family of velocity profiles; these are illustrated in Fig. 3.1. The approach of boundary layer separation for retarded flow, that is, dU/dx negative, is self-evident.

The velocity at any point in the layer is given in [3.2] by

$$\frac{u}{U} = 2\frac{y}{\delta} - 2\left(\frac{y}{\delta}\right)^3 + \left(\frac{y}{\delta}\right)^4 + \frac{\Lambda}{6}\frac{y}{\delta}\left(1 - \frac{y}{\delta}\right)^3 \tag{3.9}$$

from which it can be shown that the displacement and momentum thicknesses, as defined in Sections 2.3.3 and 2.3.4, are given by

$$\delta^* = \frac{\delta}{120}(36 - \Lambda) \tag{3.10}$$

3.2 LAMINAR BOUNDARY LAYERS

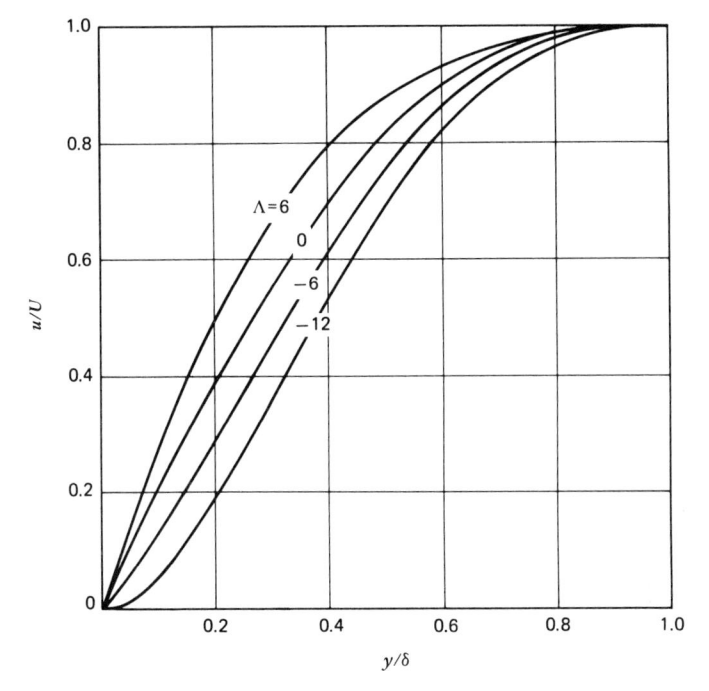

Figure 3.1 Laminar boundary layer profiles.

and

$$\theta = \frac{\delta}{315}\left(37 - \frac{\Lambda}{3} - \frac{5\Lambda^2}{144}\right) \qquad (3.11)$$

The shape parameter defined by

$$\lambda = \frac{\theta^2}{\nu} \cdot \frac{dU}{dx} \qquad (3.12)$$

is now favored, because of the greater accuracy with which θ can be calculated, as compared with δ.

The growth of the laminar layer can be expressed in the following simple equation due to Thwaites [3.3]:

$$\theta^2 = \frac{0.45\nu}{U^6}\int_0^x U^5\,dx \qquad (3.13)$$

34 BOUNDARY LAYER AND SKIN FRICTION RELATIONS

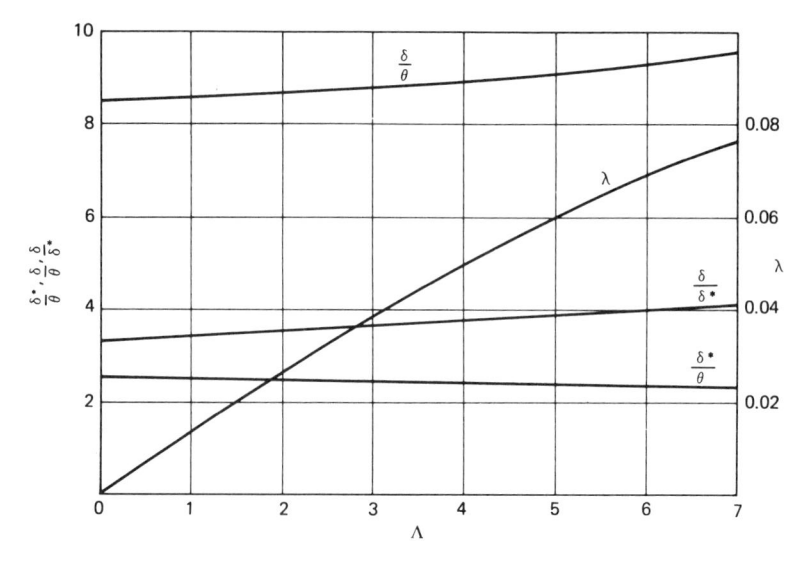

Figure 3.2 Laminar boundary layer relations.

This integral, which is based on the von Karman relation, Eq. (3.5), can be solved by simple graphical means.

Because of the inability of a laminar layer to persist far into a region of increasing pressure without separating, it is usually sufficient, for engineering purposes, to assume that separation occurs just downstream of the point where the pressure begins to rise. Hence the use of the preceding relationships will be limited to positive values of λ and Λ; some useful relations are given graphically in Fig. 3.2.

An exception to the preceding statement occurs in the case of small chord blading operating at relatively low Reynolds numbers. In this instance, however, overall airfoil drag data are available and hence detailed calculations are not normally required.

No simple solution is available for an arbitrary three-dimensional flow but, for the case of an axisymmetric body, the boundary layer grows [3.4] according to the expression

$$\theta^2 = \frac{0.45\nu}{U^6 r^2} \int_0^x r^2 U^5 \, dx \qquad (3.14)$$

where r is the local body radius and x is the distance along the surface.

3.3 LAMINAR SKIN FRICTION

There are at least six main simple configurations for which estimates of skin friction are required: two-dimensional flat plate flow; fully developed flows in pipes, annuli, and channels; initial flow in a pipe prior to fully developed conditions; and flow over an axisymmetric body. Although small differences are indicated by mathematical studies, these are not sufficient to warrant a separate treatment for each condition.

The shear stress at any point in the boundary layer is given by

$$\tau = \mu \frac{\partial u}{\partial y} \tag{3.15}$$

and hence the surface value, that is, the skin friction, is

$$\tau_0 = \mu \left(\frac{\partial u}{\partial y} \right)_{y=0} \tag{3.16}$$

From the various two-dimensional boundary layer solutions reviewed by Thwaites [3.3], it can be shown that for favorable pressure gradients (i.e., λ positive), the parameter T, defined as

$$T = \left(\frac{\partial u}{\partial y} \right)_{y=0} \cdot \frac{\theta}{U} \tag{3.17}$$

can be approximated by

$$T = 1.3\lambda + 0.23 \tag{3.18}$$

It follows from a rearrangement of Eq. (2.13) that

$$c_f = \frac{\tau_0}{\frac{1}{2}\rho U^2} \tag{3.19}$$

and by using Eqs. (3.16) to (3.19)

$$\boxed{c_f = \frac{2.6\lambda + 0.46}{R_\theta}} \tag{3.20}$$

This is the basic skin friction relationship that is here assumed to hold in a universal manner, provided λ is not negative.

3.3.1 Zero Velocity Gradients

For this condition of flow, certain developments are possible. Since λ is zero,

$$c_f = \frac{0.46}{R_\theta} \tag{3.21}$$

$$\frac{\delta}{\theta} = 8.51 \tag{3.22}$$

and hence

$$c_f = \frac{3.92}{R_\delta} \tag{3.23}$$

where

$$R_\delta = \frac{\delta U}{\nu}$$

Equation (3.21) is presented in Fig. 3.3.

1. *Flow along a two-dimensional flat plate at zero incidence.* Integrating Eq. (3.13) for $U = $ constant,

$$\theta = 0.671 x R_x^{-1/2} \tag{3.24}$$

where

$$R_x = \frac{xU}{\nu}$$

and using Eqs. (3.10) and (3.11),

$$\delta^* = 1.71 x R_x^{-1/2} \tag{3.25}$$

and

$$\delta = 5.71 x R_x^{-1/2} \tag{3.26}$$

Using the relationship between θ and x given in Eq. (3.24) and substituting in Eq. (3.21),

$$\boxed{c_f = 0.686 R_x^{-1/2}} \tag{3.27}$$

3.3 LAMINAR SKIN FRICTION

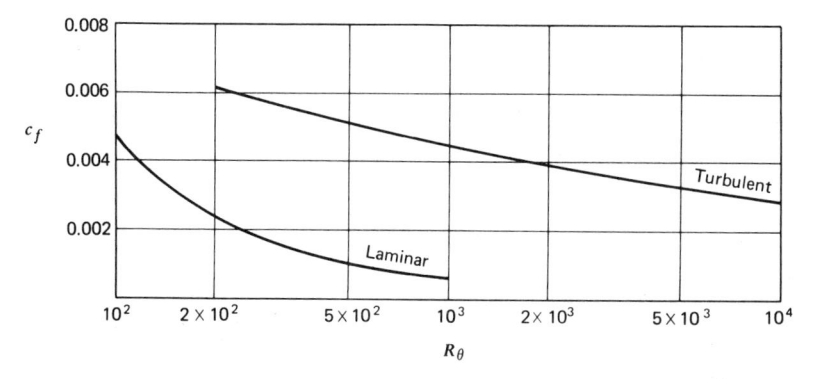

Figure 3.3 Local skin friction coefficient, zero pressure gradient [Eqs. (3.21), (3.45)].

From Eq. (2.16),

$$C_f = \frac{1}{x} \int_0^x 0.686 R_x^{-1/2} \, dx \qquad (3.28)$$

therefore

$$\boxed{C_f = 1.37 R_x^{-1/2}} \qquad (3.29)$$

This equation is illustrated in Fig. 3.4.

2. *Fully developed flow in a pipe.* In this instance the velocity at a given radius remains constant along the pipe and the static pressure falls, owing to the viscous losses. The ratio of maximum to mean velocities is given in [3.5] by,

$$\frac{U_{max}}{\bar{U}} = 2$$

and hence by employing Eqs. (3.23) and (3.19) and remembering that $\delta = d/2$

$$\boxed{\gamma = \frac{\tau_0}{\tfrac{1}{2}\rho \bar{U}^2} = 15.7 \left(\frac{\bar{U}d}{\nu}\right)^{-1}} \qquad (3.30)$$

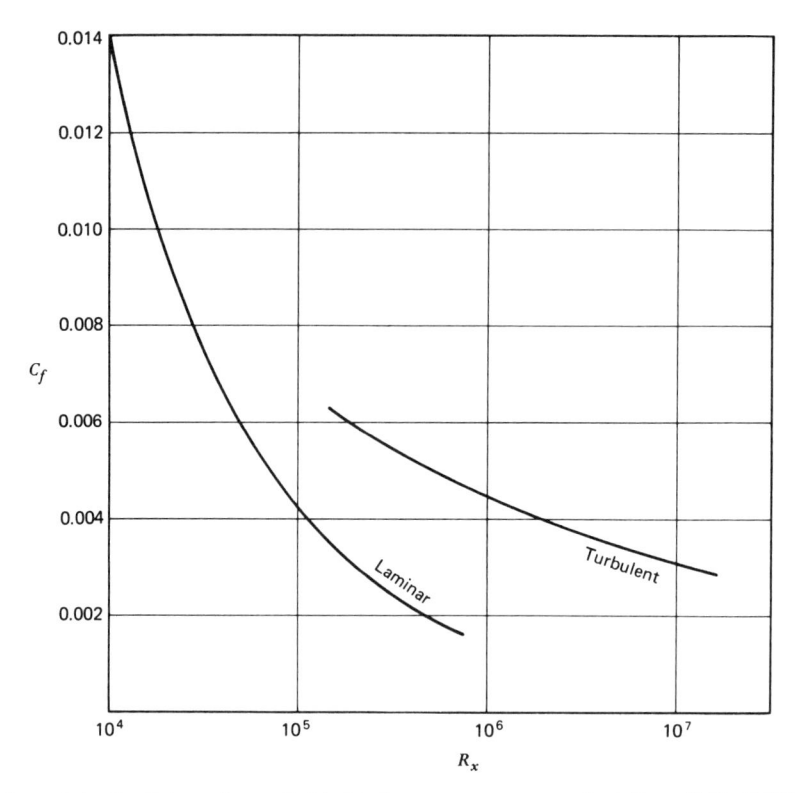

Figure 3.4 Coefficient of overall skin friction, zero pressure gradient [Eqs. (3.29, (3.56)].

3. *Fully developed flow in an annulus.* Relationships for this ducting configuration are given in [3.6], namely,

$$\frac{r_{max}}{r_0} = \sqrt{\frac{1 - (r_i/r_0)^2}{2 \ln r_0/r_i}} \tag{3.31}$$

$$\frac{U_{max}}{\overline{U}} = 2\left[\frac{1 - (r_{max}/r_0)^2[1 + 2 \ln r_0/r_{max}]}{1 + (r_i/r_0)^2 - 2(r_{max}/r_0)^2}\right] \tag{3.32}$$

where r_{max}, r_o, and r_i represent the maximum velocity location and the outer and inner wall positions, respectively, and ln signifies the natural logarithm.

The skin friction coefficient is given by

$$\boxed{\gamma = 16\left[\frac{\overline{U}(d_o - d_i)}{\nu}\right]^{-1} f\left(\frac{d_i}{d_o}\right)} \tag{3.33}$$

3.4 TURBULENT BOUNDARY LAYERS

where

$$f\left(\frac{d_i}{d_o}\right) = \frac{(1 - d_i/d_o)^2}{1 + (d_i/d_o)^2 + [(1 - d_i^2/d_o^2)/\ln(d_i/d_o)]}$$

This function has values of 1 and 1.5 for the diameter ratio limits of zero and unity, respectively.

4. *Fully developed flow in a two-dimensional channel.* Similarly, for a channel of height h, where the maximum/mean velocity ratio is

$$\frac{U_{max}}{\bar{U}} = 1.5$$

it follows from Eq. (3.23) that

$$\gamma = 23.6 \left(\frac{2\bar{U}h}{\nu}\right)^{-1} \tag{3.34}$$

5. *Initial pipe flow.* In view of the relatively small differences between the coefficients presented in Eqs. (3.29), (3.30), and (3.34), as derived from Eq. (3.23), and the respective theoretical values of 1.33, 16, and 24, flat plate conditions may be assumed when carrying out engineering-type calculations in relation to passage entry flow.

3.3.2 Arbitrary Velocity Gradients

Within the qualifications outlined in Section 3.2, only favorable pressure gradient conditions need be considered. When the distribution of U with respect to x is known, the distribution of θ can be obtained from Eq. (3.13). The variation of λ with x follows from Eq. (3.12), which then permits the computation of the local skin friction coefficients c_f from Eq. (3.20).

The total skin friction force is obtained by integrating Eq. (2.14), namely,

$$F \text{ per unit width} = \tfrac{1}{2}\rho \int_0^x U^2 c_f \, dx \tag{3.35}$$

This integration can be carried out by simple graphical methods.

3.4 TURBULENT BOUNDARY LAYERS

Because of the mathematical difficulties associated with calculating turbulent layer characteristics from the basic equations of motion, most of the formulas available are empirical. There is, however, a marked similarity between the laminar and turbulent equation forms, and full use will be made of this feature.

A widely used parameter in turbulent boundary layer theory is the shape parameter H, expressed by

$$H = \frac{\delta^*}{\theta} \qquad (3.36)$$

It has been shown experimentally that, for a given value of H, the boundary layer profile is approximately unique and can be represented [3.5] by an expression of the form

$$\frac{u}{U} = \left(\frac{y}{\delta}\right)^n \qquad (3.37)$$

where n is related to H as follows:

$$H = 2n + 1 \qquad (3.38)$$

For zero velocity gradient and moderate Reynolds numbers, $n \approx 1/7$; at high Reynolds numbers $n \approx 1/9$. With severe adverse pressure gradients, n increases to a value approaching 4/5 at the point of flow separation. The representation given by such a relationship is not particularly accurate in the immediate vicinity of the surface (especially in the case of flows approaching separation), but for the purposes of this book the preceding expressions are acceptable. A family of boundary layer profiles derived from experimental data [3.7] is illustrated in Fig. 3.5.

The value of n decreases in the case of accelerated flows. [Relaminarization can occur when the flow is strongly accelerated [3.8], for example, through a smooth contraction of large area ratio.]

When the power law is adopted, the relationships between the boundary layer thicknesses are given by

$$\delta^* = \delta\left(\frac{H-1}{H+1}\right) \qquad (3.39)$$

$$\theta = \delta\left[\frac{H-1}{H(H+1)}\right] \qquad (3.40)$$

Considerable use is made of the empirically established "universal law of the wall" when assessing the effects of certain variables on the velocity profile and on the skin friction. Defining a "friction" velocity, which is characteristic of the turbulent fluctuating motion, as

$$u_\tau = \sqrt{\frac{\tau_o}{\rho}} \qquad (3.41)$$

3.4 TURBULENT BOUNDARY LAYERS

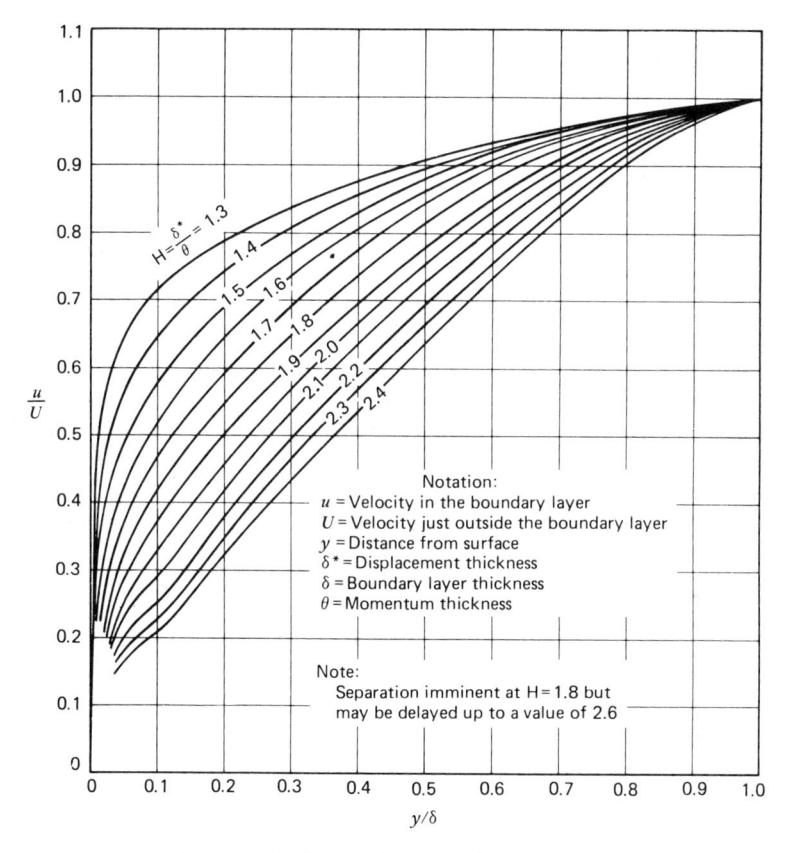

Figure 3.5 Turbulent boundary layer profiles.

and nondimensionalizing the boundary layer velocity by means of this quantity, the universal law for the inner layer is given by

$$\frac{u}{u_\tau} = A \log_{10} \frac{y u_\tau}{\nu} + B \tag{3.42}$$

where A and B, the universal constants, are approximately 5.75 and 5.2, respectively. This relation applies outside the thin surface layer within which the laminar law applies, namely,

$$\frac{u}{u_\tau} = \frac{y u_\tau}{\nu} \tag{3.43}$$

The thickness of this "laminar" sublayer is Reynolds number dependent.

The universal law was confirmed by Ludwieg and Tillman [3.9] for adverse and favorable pressure gradients. They developed a widely accepted skin friction relation in terms of both R_θ, and H, namely,

$$c_f = 0.246 \times 10^{-0.678H} R_\theta^{-0.268} \tag{3.44}$$

The effect of high free stream turbulence on boundary layer characteristics was investigated by Kline et al. [3.10], who concluded that when the free stream turbulence exceeds about 4%, the layers fail to conform to the preceding universal law, both constants now being functions of the turbulence level; the boundary layer growth rate is noticeably increased. (The insertion of rods in the free stream is usually required in attaining this high turbulence level, which is relevant to multistage compressor flows.)

The universal law was also selected by Ellis and Joubert [3.11] in their study of turbulent boundary layers within a curved duct with zero longitudinal pressure gradient. The results show that the law of the wall takes on a modified form with the usual logarithmic portion applying over a very small range of yu_τ/ν for the convex surface and a slightly larger range for the concave surface. The normal plane flow relations for boundary layer growth and skin friction do not apply; it is concluded that turbulence is suppressed on the convex wall and amplified on the concave one in accordance with the classical Rayleigh stability criteria. A growing volume of research literature supports these conclusions.

The high turbulence and wall curvature exceptions are discussed later for diffusers and corners, respectively, to which they are more specifically related.

For flat plate conditions, simple power laws that relate c_f to R_θ over a wide range of Reynolds number are equally acceptable to logarithmic equations based on the law of the wall. The one suggested by Spence [3.12], namely,

$$\boxed{c_f = 0.01766 R_\theta^{-1/5}} \tag{3.45}$$

is adopted in the present instance. In applying this equation to longitudinal pressure gradient cases, Spence decided to neglect the relationship between c_f and H, expressed in Eq. (3.44) and obtained the simple equation

$$\theta = \frac{0.0106 R_\theta^{-1/5}}{U^4} \int_{x_i}^{x} U^4 \, dx + \text{constant} \tag{3.46}$$

where the constant is θ_i, the value of θ at x_i. This initial quantity is normally the calculated laminar value of θ at the estimated transition point x_i. Reasons for assuming c_f to be independent of H are related in part to the discarding of turbulence terms from the momentum equation (see Section 3.1) used in

3.5 TURBULENT SKIN FRICTION, SMOOTH SURFACES

establishing Eq. (3.46); the acknowledged error in skin friction appears to compensate for their omission [3.12].

For moderate adverse pressure gradients, it is possible to represent the boundary layer growth with the equation [3.13]

$$\frac{\theta}{\theta_i} = \left(\frac{U_i}{U}\right)^{2+G} \tag{3.47}$$

where G is a function of R_θ but has an average value of 2.8.

In the special case of zero velocity gradient [see Eq. (3.7)],

$$\theta = \frac{1}{2}\int_{x_i}^{x} c_f \, dx + \theta_i \tag{3.48}$$

When c_f is known as a function of x, this expression can be evaluated.

Following a procedure similar to that adopted in [3.4], it can be shown that Eq. (3.46) can be written, for the axisymmetric case, as

$$\theta = \frac{0.0106 R_\theta^{-1/5}}{U^4 r} \int_{x_i}^{x} U^4 r \, dx + \theta_i \tag{3.49}$$

Except for the outer edge of the boundary layer there is a large degree of similarity in the turbulence characteristics of pipe and flat plate boundary layer flows [3.14]. As a consequence, the universal law of the wall appears valid for channel, pipe, and boundary layer flow, with and without external pressure gradient, as affirmed in [3.15]. For present purposes therefore a single approach treatment to these flows is justified. This is in line with the foregoing laminar flow treatment. Reference should be made to [3.16] when increased accuracy is required for special duct flow purposes.

3.5 TURBULENT SKIN FRICTION, SMOOTH SURFACES

The skin friction relation based on the law of the wall can be adequately represented by the simple power law of Eq. (3.45) over a wide range of Reynolds numbers; hence this latter expression has been adopted here. It can be shown [3.5] that this relationship leads to a constant value of 1/8 for the power n [Eq. (3.37)]. This is the average figure for moderate to high Reynolds number flow conditions. However, the commonly used value of 1/7 is selected in the current design equation development. This gives, from Eqs. (3.38) to (3.40),

$$H = \frac{\delta^*}{\theta} = \frac{9}{7}; \quad \frac{\delta}{\theta} = \frac{72}{7}; \quad \frac{\delta}{\delta^*} = 8 \tag{3.50}$$

Until recently, fully developed flows along straight ducts with constant cross-sectional area, of all passage shapes, were assumed to obey a unique turbulent skin friction law in terms of the Reynolds number

$$R_{d_h} = \frac{\bar{U} d_h}{\nu}$$

where d_h is the hydraulic, or equivalent, mean diameter expressed as

$$d_h = 4 \frac{\text{cross-sectional area}}{\text{wetted perimeter}} \quad (3.51)$$

This generalization has been superseded by estimation methods that introduce geometry factors that either correct the skin friction coefficient for a given R_{d_h} or introduce a corrected R_{d_h}. Both methods require a knowledge of the geometry factor for laminar flow through a passage identical to the one in question. Further details are given in Chapter 4.

For the present, however, in parallel with the laminar flow treatment, the same four cases will be developed in terms of Eq. (3.45). The selection of a constant value of n (which should vary with R_d), is a small source of error, as the boundary layer thickness and mean/maximum velocity ratios are all dependent on n.

1. *Flow along a two-dimensional flat plate at zero incidence.* Integrating Eq. (3.46) for $U = $ constant,

$$\theta = 0.0227 x R_x^{-1/6} \quad (3.52)$$

and using the relations in Eq. (3.50)

$$\delta = 0.233 x R_x^{-1/6} \quad (3.53)$$

$$\delta^* = 0.0292 x R_x^{-1/6} \quad (3.54)$$

Equation (3.45) can now be rewritten as

$$\boxed{c_f = 0.0377 R_x^{-1/6}} \quad (3.55)$$

and, by substituting in Eq. (2.16),

$$\boxed{C_f = 0.0452 R_x^{-1/6}} \quad (3.56)$$

3.5 TURBULENT SKIN FRICTION, SMOOTH SURFACES

In developing Eqs. (3.52) to (3.56), it is assumed that x is equal to x_e (see Section 2.4.3) and hence θ_i is zero. When R_x is large, the difference between x_e and x (see Fig. 2.6) will usually be small. Before applying the foregoing equations, it should be established whether the relations are suitable for the task in hand.

2. Fully developed flow in a pipe. The relation between maximum and mean velocities is readily established, for $n = 1/7$, as

$$\frac{U_{max}}{\overline{U}} = \frac{(n+1)(n+2)}{2} = 1.22 \tag{3.57}$$

After substitution Eq. (3.45) gives

$$\gamma = \frac{\tau_0}{\frac{1}{2}\rho \overline{U}^2} = 0.0465 \left(\frac{\overline{U}d}{\nu}\right)^{-1/5} \tag{3.58}$$

For Reynolds numbers below 10^5 but above the critical, the Blasius relation, namely,

$$\gamma = 0.079 \left(\frac{\overline{U}d}{\nu}\right)^{-1/4} \tag{3.59}$$

gives slightly better accuracy.

3. Fully developed flow in an annulus. The radius of maximum velocity is given in [3.18] by

$$\left(\frac{r_{max}}{r_o}\right)^3 = \frac{r_i/r_o(1 + r_i/r_o)}{2} \tag{3.60}$$

At low Reynolds numbers this equation leads to significant error.

The ratio of maximum to mean velocities is strictly a function of r_i/r_o, but experimental studies [3.19] have established a value of

$$\frac{U_{max}}{\overline{U}} = 1.13$$

for the usual range of radius ratios.

46 BOUNDARY LAYER AND SKIN FRICTION RELATIONS

The skin friction coefficient is given in [3.20] as

$$\gamma = 0.087\left[\frac{\bar{U}(d_o - d_i)}{\nu}\right]^{-1/4} \tag{3.61}$$

where the hydraulic mean diameter, from Eq. (3.51), is $(d_o - d_i)$; radius ratio is not a significant variable [3.20].

This expression gives γ values that are 10% higher than those obtained from Eq. (3.59) for a cylindrical pipe, at identical hydraulic diameter Reynolds numbers (see also Section 4.2.2).

4. *Fully developed flow in a two-dimensional channel.* The hydraulic mean diameter in this instance is equal to twice the channel width h. The practice of establishing skin friction, in terms of $2\bar{U}h/\nu$, from pipe flow data will result, according to [3.16], in a slight underestimate of resistance. However, Patel and Head [3.17] suggest that closer agreement with pipe flow data is achieved when channel flow friction is plotted against $\bar{U}h/\nu$. This matter is more fully discussed in Section 4.2.2 in relation to large aspect ratio rectangular ducts; in the limit, these are equivalent to two-dimensional channels.

Pursuing the procedures adopted earlier in this chapter, the following relationships are obtained:

$$\frac{U_{max}}{\bar{U}} = 1.4$$

for a 1/7 power velocity profile law, and

$$\gamma = 0.0475\left(\frac{2\bar{U}h}{\nu}\right)^{-1/5} \tag{3.62}$$

when the relevant substitutions are made in Eq. (3.45).

This expression results in a 2% higher skin friction than that for pipe flow, for equal Reynolds numbers based on d_h.

3.6 TURBULENT BOUNDARY LAYERS IN PRESSURE GRADIENTS

The growth of a turbulent layer in a pressure gradient, for example, around an isolated airfoil, can be approximately calculated from Eq. (3.46), pro-

3.7 BOUNDARY LAYER TRANSITION

vided there is no flow separation. The science of predicting flow separation in adverse pressure gradients is, at the present time, not an exact one, but a method such as that outlined by Spence [3.12] can be very useful in assessing the deterioration of the layer as it enters a region of rising pressure. Spence's method, like the efforts of other workers, attempts to calculate the distribution of H, that is, δ^*/θ, along the surface. The parameter H increases as flow separation is approached and hence an upper limit is usually set above which flow separation is considered to be imminent. This problem of separation normally takes precedence over that of skin friction, and hence accurate estimates of the latter are seldom required.

An indication of the complexity of the art can be obtained from [3.21], where a direct comparison between alternative methods is available.

The Spence method, however, is inappropriate for most internal duct flow problems where the boundary layer either fills the duct or constitutes a large proportion of the internal flow, thus eliminating the "potential flow" component.

A favorable pressure gradient, such as occurs in a contraction, can lead to a thinning in the turbulent layer. First, there is the general contraction of the stream that reduces the boundary layer dimensions. Second, there is a large conversion of static pressure into dynamic pressure. Since there is no variation in static pressure across the duct just downstream of entry to the contraction, the conversion will result in a fairly uniform stream at outlet to the contraction. Wind tunnel designers apply this principle in achieving a satisfactory airstream in the working or test section of the tunnel. In cases of extreme acceleration the boundary layer may revert to a quasi-laminar state before reestablishing itself as a turbulent layer [3.8].

3.7 BOUNDARY LAYER TRANSITION

3.7.1 General Considerations

Transition is the process that causes a changeover from a laminar flow regime to one in which turbulent stresses are predominant. This transformation can be initiated by a variety of circumstances; in no case, however, is the mechanism of transition fully understood [3.22].

At very low Reynolds numbers the viscous forces dominate the boundary layer, with the result that turbulent flow cannot exist. If, for instance, turbulence were artificially introduced into such a laminar layer, it would quickly die out under the influence of viscosity.

With increasing Reynolds number, conditions more favorable to the establishment and maintenance of turbulent flow are approached. Beyond a specific Reynolds number, which is a function of flow characteristics, disturbances within the boundary layer grow rapidly as Reynolds number is further increased. The laminar layer is then replaced by a normal turbulent layer, and once this has taken place, additional increases in Reynolds number produce only minor changes in the properties of the turbulent flow within this layer.

The seed of the transition process is an initial flow disturbance. Since this perturbation need be of only infinitesimal magnitude in the first instance, it can be seen that transition is inevitable in a real fluid at high enough Reynolds numbers. The wide range of flow disturbances responsible for initiating transition is well exemplified in [3.22].

Transition is brought about by one or more of the following circumstances:

1. Adverse pressure gradients.
2. Instability with respect to internal flow disturbances or surface curvature.
3. Surface irregularities.

3.7.2 Adverse Pressure Gradients

When the static pressure rises in the direction of flow, the laminar layer rapidly approaches a condition of separation. This occurs when the stream velocity has been reduced to approximately 90% of its value at the start of the diffusion process. The laminar layer leaves the surface and becomes a "free" shear layer. Except for small Reynolds numbers, transition commences in the "free" layer; this may be followed by a reattachment of the flow to the surface, provided the downstream static pressure gradient is not excessively adverse. The transition process is completed immediately downstream of reattachment.

The position of laminar separation is independent of Reynolds number, but increasing Reynolds number speeds up the transition process, thus facilitating flow reattachment. The losses involved in this transition phenomenon are a function of the size of the separated region and may be considerable. For this reason "trip" wires are often used upstream of a possible danger region in order to ensure that the flow is turbulent at the commencement of diffusion.

Adverse pressure gradients also have an influence on the growth of boundary layer disturbances, as discussed in the following subsection.

3.7.3 Boundary Layer Instabilities

The subject of boundary layer instabilities is a very complex one; hence only the salient features will be outlined here.

The onset of transition is accompanied by the appearance of turbulence "spots" or "spikes" that are swept downstream in wedge-shaped regions with apex angles of approximately 14°. These spots occur at irregular intervals of time and at randomly distributed points within the boundary layer

3.7 BOUNDARY LAYER TRANSITION

flow. Hence at any point along a surface the transition phenomenon can be related to an intermittency factor that is the ratio of total time that the flow is turbulent to the whole elapsed time. Transition is therefore not an instantaneous event but one that is spread over a finite length of surface. When the intermittency factor reaches unity, the transition process can be considered complete.

Random vorticity with streamwise axes is an essential and dominant feature of turbulent boundary layer flow. Hence the transition mechanism must encompass the creation of longitudinal vortices. A physical study of the interrelationship between "horseshoe" vortices, velocity profile inflection points, the turbulence "spikes," and the intermittency factor has been carried out by Thomson [3.23]; as a result, we now have a clearer understanding of the transition mechanism.

An increase in free stream turbulence intensity will, of course, result in an earlier onset of transition and a substantial reduction in the length of surface required to complete transition. According to [3.22] a lower onset limit of

$$R_\theta = 190$$

exists for high turbulence levels and zero pressure gradient. This is slightly above the critical stability limit calculated by Tollmien for zero pressure gradient and quoted in [3.24], namely,

$$R_\theta = 165 \quad \left(\text{for } \frac{\delta^*}{\theta} = 2.55\right)$$

Hence between the stability limit and the first turbulent burst the Reynolds number margin is approximately 25, which was accepted in [3.24] as being reasonable. This point of neutral stability results from calculations that consider the effect of viscous forces acting on hypothetical wave motions within the laminar layer. The existence of such waves was verified experimentally by Schubauer and Skramstad in [3.25]. Turbulent spots, in some circumstances, are considered to result from a breakup of the amplified wave motion. Hence the rate of amplification is also of great practical importance. As free stream turbulence is lowered in intensity, the amplification rate is also presumably reduced, resulting in higher values of $R_{\theta_{tr}}$, the Reynolds number at which transition is complete.

In the case of fully developed pipe flow with strong disturbances at inlet, the critical onset Reynolds number is in the vicinity of 2000. This can be related to R_θ by the identity

$$R_\theta \approx 0.12 \left(\frac{\bar{U}d}{\nu}\right)$$

after making the appropriate substitutions for laminar flow. The resulting value of R_θ, namely, 240, is greater than the suggested onset limit figure of 190 for flat plate boundary layers. On the other hand, the critical onset Reynolds number $(\bar{U}h/\nu)$ for fully developed channel flow has been established in the vicinity of 1400 [3.17], which when converted to R_θ by the expression

$$R_\theta \approx 0.09\left(\frac{\bar{U}h}{\nu}\right)$$

gives a value of 125. This experimental finding confirms earlier transition work carried out by Davies and White, as discussed in [3.17]. No quantitative explanation is offered for the disparsity of these figures, but the increased sensitivity of laminar flow to two-dimensional rather than three-dimensional disturbances is a well-established fact. The preceding data satisfy this qualitative argument.

For the preceding two strongly disturbed inlet flow situations, transition is complete at Reynolds numbers of 3000 and 2500 to 3000, for the pipe and channel flow cases, respectively; these convert to $R_{\theta_{tr}}$ values of 360 and 225 to 270. The suggested figure for the flat plate case is 320 [3.22].

Transition data for extremely low levels of turbulence are restricted to the pipe flow case only. A number of workers have achieved critical values in excess of 20,000 ($R_\theta > 2400$), but for normal engineering purposes such low levels of turbulence are seldom achieved and hence the preceding data appear acceptable. However, for carefully contoured flared inlets, with a low turbulence entry flow, higher $R_{\theta_{tr}}$ must be expected.

The stability theory of transition predicts increased neutral stability limits and lower rates of amplification with increasingly favorable pressure gradients; the reverse is true for adverse gradients. Some appreciation of the "state of the art" can be obtained from [3.22], where an attempt has been made to present the available but relatively scarce experimental data in a combined pressure gradient/turbulence level plot. For further information this latter reference should be consulted, as the context in which the tentative recommendations are forwarded is important.

A wavy surface initiates early transition by virtue of the changing surface pressure gradients created. Downstream of the crests the gradients are normally adverse, reducing boundary layer stability. With increasing crest height, for a given wavelength, laminar separation is eventually precipitated.

Görtler-Taylor type vortices, which eventually trigger transition, are present on concave surfaces. The centrifugal forces acting on the boundary layer are a function of both the velocity distribution and surface curvature. When the centrifugal, pressure, and viscous forces are out of balance, a number of vortices on the scale of the boundary layer, and with streamwise axes, are produced. The convex surface acts as a stabilizing influence and hence such surfaces can be considered to be flat, when assessing the likelihood of transition.

3.7 BOUNDARY LAYER TRANSITION

Liepmann [3.27] suggested a transition criterion of

$$\left[R_\theta \sqrt{\frac{\theta}{r}}\right]$$

where r is the radius of surface curvature; the values of this parameter varied between 9.5 and 6 for the turbulence range 0.06 to 0.3%. These data are only applicable to smooth, discontinuity-free surfaces.

3.7.4 Surface Irregularities

When a foreign body becomes attached to a surface, transition may be precipitated. These excrescences usually produce a laminar separation in their vicinity, which has the dual effect of increasing the thickness of the downstream layer and introducing large disturbances into the boundary layer flow. Transition usually occurs close to the obstacle; downstream of pimples and similar bodies, transition spreads in a cross-stream direction, resulting in a turbulent wedge of approximately 14° included angle.

When the roughness is less than its critical height, the disturbance dies out and no discernible effects on skin friction or boundary layer thickness occur. When the critical size is present, the initiation of transition increases skin friction because of the resulting turbulent layer. Heights greater than the critical progressively increase the losses by reason of the increased thickness of the initial turbulent boundary layer.

Simple relations [3.28] are presented in Fig. 3.6 for two common types of irregularity, the first being similar to sand particles and the second to transition trip wires.

It was found experimentally that these curves, which were obtained from tests on isolated obstacles, could also be applied to arrangements of sparsely distributed roughnesses; the most critical irregularity triggers transition.

The curves suggest that the boundary layer is insensitive to roughnesses that are less than 0.1 of the boundary layer thickness, provided R_θ does not exceed 600. At these high R_θ, however, transition will probably occur because of other causes.

When roughness is present, $R_{\theta_{tr}}$ is considerably reduced below 190, the value previously linked to a high turbulence level. When the pimple height equals the boundary layer thickness, transition occurs at

$$R_\delta \approx 670, \quad \text{i.e.,} \, R_\theta \approx 80$$

This aspect of the transition prediction problem thus becomes paramount when appreciable roughness is present.

The laminar boundary layer thickness at the disturbance location can be calculated from Eqs. (3.13) and (3.11), or from Eq. (3.26) for $dU/dx = 0$.

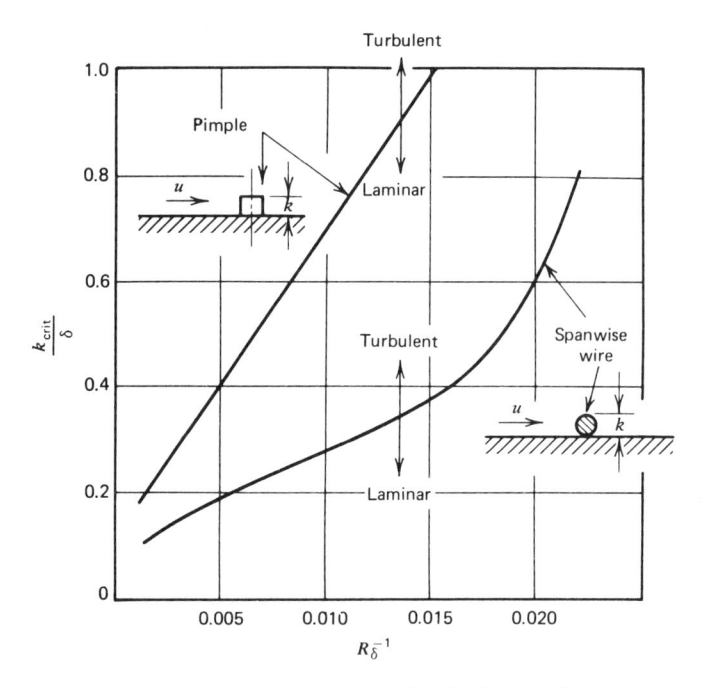

Figure 3.6 Critical roughness height for laminar flow.

3.8 SURFACE ROUGHNESS IN TURBULENT BOUNDARY LAYERS

3.8.1 General Considerations

A surface is "aerodynamically smooth" when the roughness height does not exceed the thickness of the "laminar sublayer." Viscous forces are predominant in this sublayer, which is that part of the turbulent boundary layer in the immediate vicinity of the surface. The laminar sublayer has a thickness that is given approximately in [3.29] by

$$\delta_s = \frac{4\nu}{\sqrt{\tau_0/\rho}} \tag{3.63}$$

The type of roughness encountered in practice rarely shows uniformity of height or shape, and hence it is difficult, from uniform sand roughness test data, to assess the overall increase in skin friction. Roughness density too can play an important part; an increase in "population" may not always lead to an increase in skin friction [3.30]. It is fairly obvious, therefore, that test

3.8 SURFACE ROUGHNESS IN TURBULENT BOUNDARY LAYERS

data relating to the actual surface are most desirable. The finished products of certain manufacturing processes often possess a similarity in their surface condition. Hence test results from representative specimens of these products can provide useful data in estimating skin friction. Nevertheless, care must always be exercised when applying this information, since, for example, cast steel produced by one foundry may have a surface significantly different from that produced by another.

With increasing roughness height, a stage is reached when the eddies produced by the roughness elements virtually control the fluid losses occurring in the boundary layer. Viscous stresses become unimportant and the skin friction in consequence is independent of Reynolds number; the surface is then considered to be "fully rough."

An "equivalent roughness height" can now be defined for surfaces possessing nonuniform roughness. It is the height of a uniform sand roughness that will give the surface a skin friction similar to that experienced by the nonuniform specimen, both being for fully rough conditions.

3.8.2 Smooth and Fully Rough Limits

These will naturally be a function of the nonuniformity in roughness height on a given surface. This information is not available in most cases and parameters that are a function of the equivalent roughness height k must be adopted. Averaging out the results of numerous tests gives the following approximate relationships [3.31].

1. Aerodynamically smooth,

$$\frac{k\sqrt{\tau_0/\rho}}{\nu} < 3 \qquad (3.64a)$$

2. Fully rough,

$$\frac{k\sqrt{\tau_0/\rho}}{\nu} > 60 \qquad (3.64b)$$

For values between 3 and 60, the skin friction is a function of both Reynolds number and roughness distribution. In this transitional region between smooth and fully rough conditions, the equivalent roughness rule tends to be inaccurate when the individual heights of the roughness present vary by an appreciable amount. Hence experimental data for a similar class of surface roughness must be available before skin friction on the actual specimen can be assessed accurately in this range.

54   BOUNDARY LAYER AND SKIN FRICTION RELATIONS

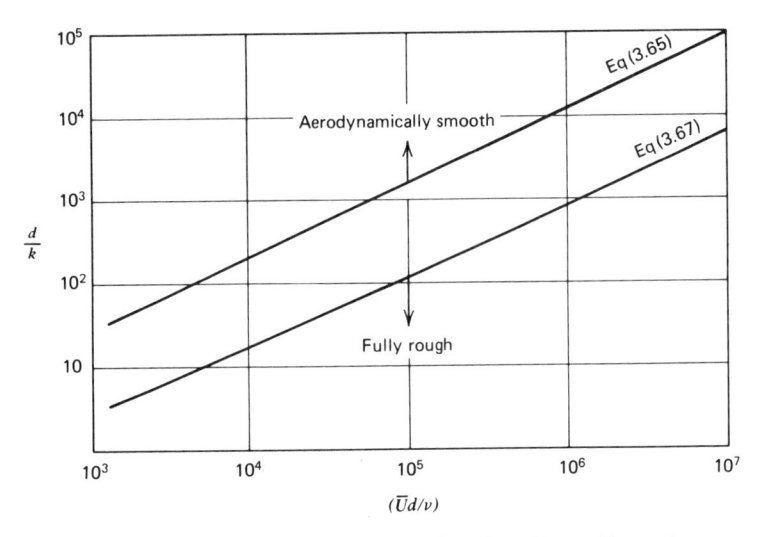

Figure 3.7 Smooth and rough limits as function of Reynolds number.

3.8.3 Roughness in Cylindrical Ducts, Fully Developed Flow

The skin friction for flow along a smooth surface is obtained from Eq. (3.58); when substitution is made in Eq. (3.64), the Reynolds number above which a certain roughness size has an influence can be calculated from

$$\frac{k}{d} \approx 20 \left(\frac{\bar{U}d}{\nu}\right)^{-0.9} \tag{3.65}$$

When the surface is fully rough, the skin friction can be expressed empirically by [3.29]

$$\gamma^{-1/2} = \sqrt{\frac{\tfrac{1}{2}\rho \bar{U}^2}{\tau_0}} = 3.46 + 4.00 \log_{10} \frac{d}{2k} \tag{3.66}$$

Substituting in Eq. (3.64), the relation between roughness height and the Reynolds number above which the surface is fully rough is given by

$$\frac{k}{d} \approx 85 \left(\frac{\bar{U}d}{\nu}\right)^{-1} \left(3.46 + 4.00 \log_{10} \frac{d}{2k}\right) \tag{3.67}$$

Equations (3.65) and (3.67) are graphically presented and the zones indicated in Fig. 3.7.

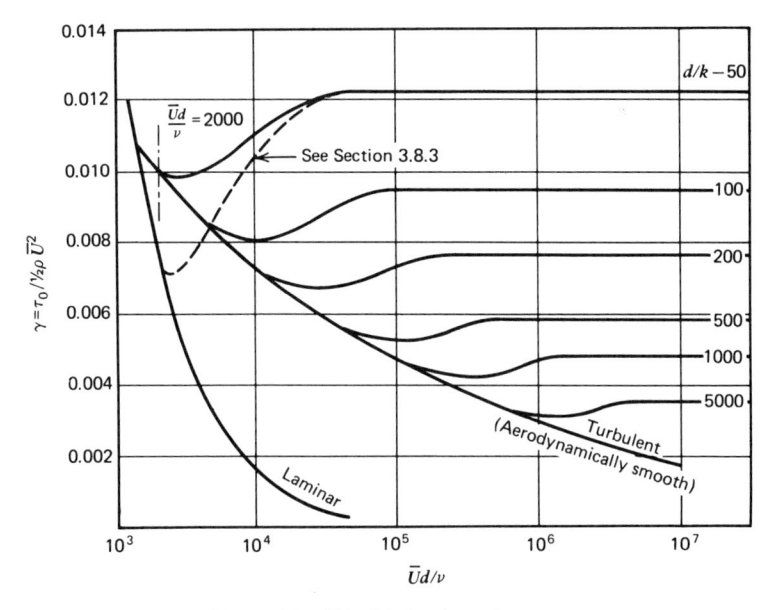

3.8 SURFACE ROUGHNESS IN TURBULENT BOUNDARY LAYERS

Figure 3.8 Skin friction in a pipe.

Using the foregoing data, the skin friction coefficient has been given as a function of Re and equivalent roughness height in Fig. 3.8. The intermediate roughness curves, which join the fully rough and aerodynamically smooth lines, are somewhat arbitrary but should give estimates of sufficient accuracy for engineering design purposes. Figure 3.9 presents γ as a function of d/k for fully rough surfaces.

A design consideration that should receive serious study is the likely

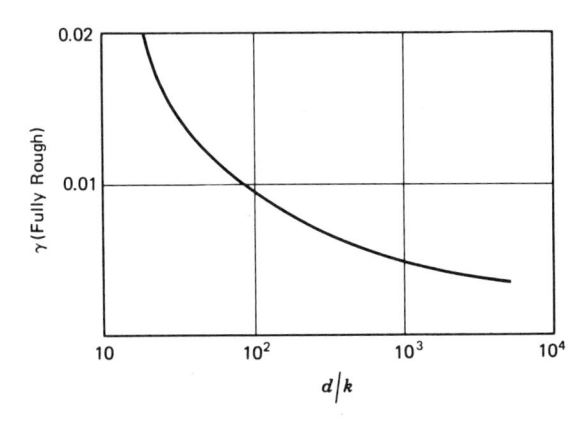

Figure 3.9 Pipe skin friction for fully rough conditions.

deterioration of the surface under operational conditions; rust, surface flaking, dust, and mud buildups comprise a few of the various possibilities.

In long pipe or duct lines where high accuracy of loss estimation is essential, the results of relevant tests must be employed.

The total pressure loss in a pipe of length l in which the flow is fully developed is given by

$$\Delta H_x = \frac{\tau_0 A_w}{A} \qquad (3.68)$$

where A_w and A are the wetted and cross-sectional areas, respectively. Since the velocity pressure remains constant, the preceding quantity represents the static pressure differential Δp. Expanding,

$$\Delta p = \gamma \frac{\frac{1}{2}\rho \bar{U}_x^2 \pi dl}{\pi d^2/4}$$

$$\frac{\Delta p}{\frac{1}{2}\rho \bar{U}_x^2} = \frac{4\gamma l}{d} \qquad (3.69)$$

The required γ value can be obtained from Eq. (3.69), on substitution of the relevant experimental pressure drop coefficient.

3.8.4 Roughness in Other Types of Flow Passages, Fully Developed Flow

Extension of roughness effects from the cylindrical duct to other flow passages demands certain assumptions. Because of the lack of published experimental data, this expedient is necessarily taken.

First, the relations in Eq. (3.64) are adopted. Second, in establishing the aerodynamically smooth R_{d_h} limits for various effective roughness heights, the appropriate γ and hence τ_0 should be substituted in Eq. (3.64) to give an equation of the form

$$\frac{k}{d_h} \approx a \left(\frac{\bar{U} d_h}{\nu}\right)^{-p} \qquad (3.70)$$

Third, fully rough conditions exist above a R_{d_h} obtained from Eq. (3.67), where $\bar{\delta}$ is substituted for $d/2$, namely,

$$\frac{k}{d_h} \approx 85 \left(\frac{\bar{U} d_h}{\nu}\right)^{-1} \left(3.46 + 4.00 \log_{10} \frac{\bar{\delta}}{k}\right) \qquad (3.71)$$

where $\bar{\delta}$ is an average boundary layer thickness. For example, in the case of

3.8 SURFACE ROUGHNESS IN TURBULENT BOUNDARY LAYERS

an annular passage

$$\bar{\delta} = \frac{d_o - d_i}{4}$$

Fourth, the value of γ for fully rough conditions is approximately given by

$$\gamma^{-1/2} = \sqrt{\frac{\tfrac{1}{2}\rho \bar{U}}{\tau_0}} = 3.46 + 4.00 \log_{10} \frac{\bar{\delta}}{k} \tag{3.72}$$

The extent to which these expressions can be used in estimating roughness effects in a general manner must be a matter for personal judgment. However, their application to annular and rectangular ducts appears reasonable, particularly in view of the difficulty in selecting the appropriate equivalent roughness height.

1. *Annular ducts.* The aerodynamically smooth condition is related to Eq. (3.70), which in this case is given by

$$\frac{k}{d_h} \approx 15 \left(\frac{\bar{U} d_h}{\nu} \right)^{-0.875} \tag{3.73}$$

when Eq. (3.61) is substituted in Eq. (3.64). The fully rough state is reached at a R_{d_h} as given by

$$\frac{k}{d_h} \approx 85 \left(\frac{\bar{U} d_h}{\nu} \right)^{-1} \left(3.46 + 4.00 \log_{10} \frac{d_h}{4k} \right) \tag{3.74}$$

The skin friction for fully rough conditions is

$$\gamma^{-1/2} = 3.46 + 4.00 \log_{10} \frac{d_h}{4k} \tag{3.75}$$

2. *Rectangular ducts.* For ducts with aspect ratios up to 3, the cylindrical duct data can be directly applied, in terms of hydraulic mean diameter. However, in the two-dimensional channel case (infinite aspect ratio) the substitution of Eq. (3.62) for Eq. (3.58) will produce no significant change from Eq. (3.65) when d_h is introduced, namely,

$$\frac{k}{d_h} \approx 20 \left(\frac{\bar{U} d_h}{\nu} \right)^{-0.9} \tag{3.76}$$

Since the average boundary layer thickness is given by $d_h/4$, as in the previous case, Eqs. (3.74) and (3.75) remain unchanged.

For aspect ratios of intermediate value, interpolation procedures should be adopted.

3. *Other passage shapes.* It is suggested that other passage shapes be treated on their merits, in terms of the preceding discussion.

3.8.5 Roughness on Flat Plates and Airfoils

A flat plate in a zero pressure gradient has, for fully rough conditions, a local skin friction coefficient given by [3.30]

$$c_f = \left(2.87 + 1.58 \log_{10} \frac{x}{k}\right)^{-2.5} \tag{3.77}$$

The overall skin friction coefficient C_f can be obtained by a solution of Eq. (2.16). Equation (3.77) is approximately valid for $10^2 < l/k < 10^6$, where l is the total x dimension. With a similar procedure to that followed in the previous subsection, the expression for smooth and fully rough limits is as follows:

$$\frac{k}{x} \approx 22 R_x^{-11/12} \tag{3.78}$$

$$\frac{k}{x} \approx 85 R_x^{-1}\left(2.87 + 1.58 \log_{10} \frac{x}{k}\right)^{1.25} \tag{3.79}$$

with the usual reminder that x has been assumed to be approximately equal to x_e.

Since boundary layer thickness grows with x, the effect of a fixed roughness height will diminish with increasing x, in contrast to fully developed pipe flow.

The roughness level on airfoils can usually be kept very small and fully rough conditions are seldom experienced. Owing to the presence of adverse pressure gradients and other variables, the effect of roughness on drag can best be determined by experiment on the actual airfoil or a model of it. By combining Eqs. (3.45) and (3.64), however, the following is obtained

$$\frac{k}{\theta} \approx 32 R_\theta^{-0.9} \tag{3.80}$$

as the roughness limit for aerodynamic smoothness.

On fan blades, an increase in drag causes a decrease in efficiency and a possible early onset of the stall; the latter restricts the maximum lift that the blade can supply. If the blades were very rough, as the result of rough casting, corrosion or the presence of dirt, the fan characteristics might, in addition, be affected, because of a reduced deflection of the stream. An explanation of this feature can be found in the increase of the boundary layer thickness, which is particularly large for leading-edge crudities.

3.8.6 Effect of Roughness in Adverse Pressure Gradients

There are no quantitative data available on the effect of roughness on skin friction and flow separation in an adverse gradient, and here generalities must suffice.

In adverse gradients, roughness usually increases the tendency toward separation. This is due to the greater skin friction losses and hence increased boundary layer thickness.

(It has been shown, however, that coarse roughness can be used as a means of increasing the mixing rate of the turbulent layer in wide-angle diffusers with the result that separation is prevented. The total pressure loss associated with the roughness is usually very high.)

REFERENCES

3.1 B. Thwaites (Ed.), *Incompressible Aerodynamics Fluid Motion Memoirs*, Chap. II, Clarendon, Oxford, 1960.

3.2 H. Schlichting, *Boundary Layer Theory*, Chap. XII, McGraw-Hill, New York, 4th ed., 1960.

3.3 B. Thwaites, Approximate boundary layer equations, *Aeronaut. Quart.*, **1** (3), 245, 1949.

3.4 N. Rott and L. F. Crabtree, Simplified laminar boundary layer calculations for bodies of revolution and for yawed wings, *J. Aero. Sci.*, **19** (8), 553, 1952.

3.5 H. Schlichting, op. cit., Chap. XX.

3.6 J. G. Knudsen and D. L. Katz, *Fluid Dynamics and Heat Transfer*, McGraw-Hill, New York, 1958, pp. 92–97.

3.7 A. E. Von Doenhoff and N. Tetervin, Determination of general relations for the behaviour of turbulent boundary layers, *NACA Tech. Report 772*, 1943.

3.8 R. Narasimha and K. R. Sreenivasan, Relaminarization in highly accelerated turbulent boundary layers, *J. Fluid Mech.*, **61**, 417–447, 1973.

3.9 H. Ludwieg and W. Tillmann, Investigations of the wall shearing stress in turbulent layers, *NACA Tech. Memo. 1285*, 1950.

3.10 S. J. Kline, A. V. Lisin, and B. A. Waitman, Preliminary experimental investigation of effect of freestream turbulence on turbulent boundary layer growth, *NASA Tech. Note D-368*, 1960.

3.11 L. B. Ellis and P. N. Joubert, Turbulent shear flow in a curved duct, *J. Fluid Mech.*, **62**, 65–84, 1974.

3.12 D. A. Spence, The development of turbulent boundary layers, *J. Aero. Sci.* **23** (1), 3, 1956.

3.13 D. Ross and J. M. Robertson, An empirical method for calculation of the growth of a turbulent boundary layer, *J. Aero. Sci.*, **21** (5), 355, 1954.

3.14 B. Thwaites, op. cit., Chap. I.

3.15 J. C. Rotta, *Turbulent Boundary Layers in Incompressible Flow*, Progress in Aeronautical Sciences, Vol. 2, Pergamon, Oxford, 1962, p. 55.

3.16 A. Quarmby, Improved application of the von Karman similarity hypothesis to turbulent flow in ducts, *J. Mech. Eng. Sci.*, **11**, 14–21, 1969.

3.17 V. C. Patel and M. R. Head, Some observations on skin friction and velocity profiles in fully developed pipe and channel flows, *J. Fluid Mech.*, **38**, 181–201, 1969.

3.18 M. R. Doshi and W. N. Gill, Fully developed turbulent pipe flow in an annulus: radius of maximum velocity, ASME *J. Appl. Mech.*, **38** (4), 1090–1091, 1971.

3.19 K. Sridhar et al. Settling length for turbulent flow of air in smooth annuli with square-edged or bellmouth entrances, *ASME J. Appl. Mechs.*, **37** (1), 25–28, 1970.

3.20 J. A. Brighton and J. B. Jones, Fully developed turbulent flow in annuli, *ASME J. Basic Eng.*, **86** (4), 835–844, 1964.

3.21 S. J. Kline et al. (Eds.) *Computation of Turbulent Boundary Layers*, Stanford Conf., Vol. 1, Stanford Univ., 1969.

3.22 J. E. Ffowcs Williams et al. Transition from laminar to turbulent flow, *J. Fluid Mech.*, **39**, 547–559, 1969.

3.23 K. D. Thomson, The prediction of inflexional velocity profiles and breakdown in boundary layer transition, *WRE-Rep. 1052*, Weapons Research Establishment, Dept. of Supply, Australia, 1973.

3.24 D. J. Hall and J. C. Gibbings, Influence of stream turbulence and pressure gradient upon boundary layer transition, *J. Mech. Eng. Sci.*, **14**, 134–146, 1972.

3.25 G. B. Schubauer and H. K. Skramstad, Laminar boundary layer oscillations and transition on a flat plate, *NACA Report 909*, 1948.

3.26 H. Schlichting, op. cit., Chap. XVI.

3.27 H. W. Liepmann, Investigation of boundary layer transition on concave walls, *NACA Adv. Conf. Report 4J28*, 1945.

3.28 E. H. Cowled, The effects of surface irregularities on transition in the laminar boundary layer, Aust. Dept. of Supply, Aero. Research Labs, ARL *Report A86*, 1954.

3.29 S. Goldstein (Ed.). *Modern Developments in Fluid Dynmics*, Vol. II, Oxford University Press, Oxford, p. 380, 1938.

3.30 S. Goldstein, op. cit., p. 382.

3.31 L. E. Prosser and R. C. Worster, *The Hydro-mechanics of Fluid Flow, in some Aspects of Fluid Flow*, Edward Arnold, London, p. 65, 1951.

CHAPTER 4
Duct Component Design and Losses

4.1 DUCTS IN GENERAL

The purpose of a fan is to increase the total pressure of the air to a value equal to the total resistance losses in any given duct system. Loss calculation from available published data is the accepted procedure, but for complex duct assemblies model testing may be necessary in order to obtain the desired level of certainty. The latter procedure eliminates the guesswork associated with additional losses arising from the effect of swirling, asymmetric flows discharging from one duct component into its neighbor.

In general, published data apply to a limited range of inlet flow conditions, where symmetry exists and the swirl is usually less than 10°. The art of duct design is to approach these conditions as closely as possible, thus avoiding the need for model testing; this is not always feasible. Test philosophies are discussed in [4.1]. The results are always reduced to nondimensional terms before being applied to the full-scale system.

As the fan stall is approached there is a rapid increase in the nondimensional fan pressure rise coefficient and hence modest inaccuracies in duct loss prediction are not necessarily serious. However, the error margin for satisfactory and stable flow operation is more restricted than that for the centrifugal fan case. As a result axial flow fans are rejected by many users who fail to apply realistic interference factors in estimating system losses. For complete safety and operational satisfaction, accuracy in duct loss determination is vital; the fan should possess a 15 to 20% total pressure (dimensional) stall margin over the required operational value. Axial flow turbomachinery can be designed to meet any given pressure duty condition.

Resistance methods of volume flow rate control, such as damper vanes and butterfly valves, are potentially dangerous for this fan type. However, when correctly placed in the system and allowed for in the loss estimate, the difficulties should be minimal.

62 DUCT COMPONENT DESIGN AND LOSSES

A comparable level of proficiency should apply to duct and fan design alike; the former is possibly the most demanding.

Heat transfer by conductive, convective, and radiation mechanisms is adequately covered in the published literature. The main influences of heat transfer on fan and duct design are in modifying the density and Reynolds number in accordance with the gas and viscosity laws given in Chapter 2.

Heat addition can alter the boundary layer properties and hence affect loss estimates. However, these considerations are usually ignored as the degree of uncertainty associated with ordinary published loss data is of greater importance.

In the case of exhaust flue gases, buoyancy forces must, of course, be taken into consideration when assessing the fan pressure requirements.

4.1.1 Duct Components

Any duct system can be subdivided into a number of elementary components such as:

1. Ducts of constant cross-sectional area and shape.
2. Contractions, of which the inlet is a special case.
3. Diffusers, with the cross-sectional area increasing in the flow direction.
4. Corners, in which the general flow direction is changed.
5. Flow junctions, where two streams either meet or diverge.
6. Internal bodies such as struts, fairings, air straighteners, screens, and obstructions.

The preceding list embraces ducts of all cross-sectional shape and ducting whose shape is changing with distance along the duct. Complex arrangements of these elements require special design treatment, often requiring model or full-scale developmental test studies.

Special-purpose assemblies of these components, such as ejectors and cyclone separators, have been the subject of extensive research effort. A "black box" treatment of these devices is normal practice in system design and loss estimation exercises.

4.1.2 Definition of Duct Loss

A duct loss may be defined as the mean loss of total pressure sustained by the stream in passing through the specific duct element. The losses are due to the following causes:

1. *Skin friction.* Mean total pressure is lost within the attached wall boundary layers, because of the action of fluid shear forces.

4.1 DUCTS IN GENERAL

2. *Eddying or separated flow.* Losses from eddying or separated flow have their origin in the random large-scale turbulence that is fed from the main stream per the medium of the free shear layer.
3. *Secondary flow.* Streamwise total pressure is lost to the cross-flow whose energy is progressively dissipated with downstream distance.
4. *Discharge loss.* In an open-circuit duct system, the mean velocity pressure of the air leaving the discharge duct is lost.

Since the unrecovered velocity pressure associated with the cross-flows has been defined as a loss, we need only consider the streamwise component in the general total pressure loss expression for a given component. Hence for Fig. 4.1,

$$\Delta H_x = \frac{1}{ab} \int_0^a \int_0^b (\tfrac{1}{2}\rho u_a^2 + p)_A \, dy \, dx$$

$$- \frac{1}{cd} \int_0^c \int_0^d (\tfrac{1}{2}\rho u_a^2 + p)_B \, dy \, dx \quad (4.1)$$

where A and B denote upstream and downstream stations, respectively, u_a is the axial velocity component, and p is the static pressure.

When the duct is of circular cross section, the general equation is

$$\Delta H_x = \frac{1}{\pi R_A^2} \int_0^{R_A} \int_0^{2\pi} (\tfrac{1}{2}\rho u_a^2 + p)_A r \, dr \, d\theta$$

$$- \frac{1}{\pi R_B^2} \int_0^{R_B} \int_0^{2\pi} (\tfrac{1}{2}\rho u_a^2 + p)_B r \, dr \, d\theta \quad (4.2)$$

where R_A and R_B are the respective radii and θ is the circular measure in terms of radians.

When the flow is axisymmetric and the static pressure is constant across the inlet and outlet planes, Eq. (4.2) can be written

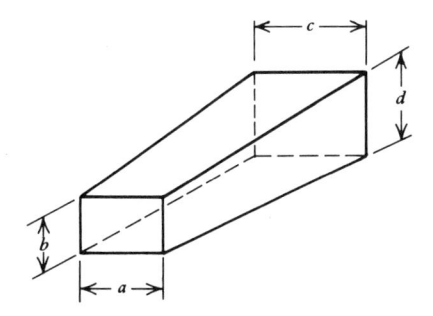

Figure 4.1 Rectangular duct of changing cross section.

$$\Delta H_x = (P + \alpha \tfrac{1}{2}\rho U_a^2)_A - (P + \alpha \tfrac{1}{2}\rho U_a^2)_B \qquad (4.3)$$

where U_a represents the mean axial velocity and α is the dynamic energy coefficient given by

$$\alpha = \frac{1}{U_a^3 A} \int_0^R u_a^3 \pi r \, dr \qquad (4.4)$$

where A is the cross-sectional area at a given station.

The preceding expressions employ "area weighting" as a means of determining mean total pressure. However, Livesey and Hugh [4.2] consider the "mass weighted" method of data reduction as the more accurate one.

In any case, further simplification for design purposes is required, since the cross-sectional velocity distributions at both stations are seldom known. Hence the preceding expressions are mainly useful in relation to research studies.

The general expression

$$\Delta H_x = \tfrac{1}{2}\rho U_A^2 \left[1 - \left(\frac{A_A}{A_B}\right)^2 \right] + (P_A - P_B) \qquad (4.5)$$

which is centered on mean static and velocity pressure values, represents a practical approximation to the actual total pressure losses. When experimental data for a specified inlet flow condition are reduced and presented in terms of the preceding variables, this information can be used with confidence in assessing the loss in an equivalent duct element with similar entry flow properties.

In the case of fully developed pipe flow in a straight duct of constant area, the dynamic energy coefficients are equal and cancel each other in Eq. (4.3); Eqs. (4.3) and (4.5) are then identical and reduce to

$$\Delta H_x = P_A - P_B \qquad (4.6)$$

For developing pipe flows, the preceding expression will normally provide adequate accuracy for loss estimation purposes.

Equation (4.5) must be interpreted correctly in relation to bends and diffusers. In the former instance, transverse static pressure gradients are present in planes marking the start and finish of flow turning. Therefore the quotation for loss should be for a combination of bend with specified lengths of straight inlet and outlet ducting. The accepted practice with respect to diffusers is discussed in Section 4.4

4.1.3 Nondimensional Treatment

When the total pressure loss in a ducting element is expressed in terms of a reference velocity pressure, the parameter so obtained is a function of

4.1 DUCTS IN GENERAL

Reynolds number, namely,

$$K_x = \frac{\Delta H_x}{\frac{1}{2}\rho U_A^2} = f(\text{Re}) \tag{4.7}$$

where U_A is the mean duct entry velocity. The most fundamental Re is the one based on boundary layer thickness at entry, but practicalities dictate the use of

$$R_d = \frac{U_A d}{\nu} \tag{4.8}$$

where d is the entry diameter, or equivalent. However, the duct designer should consider each item of ducting on its merits before reaching an estimate of loss. For example, a flared inlet will possess an extremely thin boundary layer, and hence the use of R_d for this component is inappropriate and can be misleading, particularly when assessing whether the flow is laminar or turbulent.

Duct inlet flow conditions are known to have a marked influence on pressure loss. In addition, inadvertent surface roughness, and manufacturing flaws and excrescences are often present. However, the actual flow and duct properties often remain unknown in the design stage, and hence the designer, on the basis of past experience, is forced to make allowance for possible contingencies.

The flow resistance of many duct components is quoted for "turbulent flow," the dependence on Reynolds number being ignored. Greater loss uncertainties associated with other relevant parameters can become a more important consideration in such cases.

The losses in long, straight pipelines of known roughness can be predicted with reasonable accuracy in terms of Reynolds number, providing one of few exceptions to the preceding general situation.

The component loss data are normally presented with reference to the mean inlet velocity pressure.

4.1.4 Prediction of Duct System Resistance

In most assemblies, one or more of the duct components will be subject to flow interference and hence increased losses. Compound bends with two or more directional changes, and diffuser/bend combinations, might result in additional losses of 100% or more for each duct element. When these higher losses are not provided for in the system estimate, the fan and duct characteristic curves may intersect to the left of the stall point (Fig. 1.2).

The maximum and minimum nondimensional fan total pressure coefficients embracing the projected operational range of a system can be expressed, for constant air density, by

$$K_T = \frac{\Delta H_T}{\frac{1}{2}\rho \overline{V_a^2}} = K_1\left(\frac{A_f}{A_1}\right)^2 + K_2\left(\frac{A_f}{A_2}\right)^2 + \cdots + K_n\left(\frac{A_f}{A_n}\right)^2 \qquad (4.9)$$

where $AV = A_f \overline{V}_a$, $\overline{V}_a$ is the mean axial fan annulus velocity, A_f is the fan annulus area, and each term represents a duct component loss [as in Eq. (4.7)] corrected by a multiplication factor for interactive and interference effects. Hence, in tabulating the component losses, a column for this factor should be provided.

The highest K_T evaluated must be given prime consideration when choosing the design value of fan total pressure coefficient.

When the flow system possesses parallel flow paths, the total pressure loss in each leg must be equal, for the individual flow quantities. Since it is difficult to design the required volume-flow split into a duct system proposal, dampers or other resistance-type controls are normally fitted in one or more of the duct branches. The unboosted flow path with the highest undampered nondimensional resistance will then determine the fan total pressure coefficient requirement, namely,

$$K_T = \frac{\Delta H_T}{\frac{1}{2}\rho_f \overline{V_a^2}} = K_1 \frac{\rho_1}{\rho_f}\left(\frac{U_1}{\overline{V}_a}\right)^2 + K_2 \frac{\rho_2}{\rho_f}\left(\frac{U_2}{\overline{V}_a}\right)^2 + \cdots + K_n \frac{\rho_n}{\rho_f}\left(\frac{U_n}{\overline{V}_a}\right)^2 \qquad (4.10)$$

where ρ_f is the air density in the fan annulus and the mean velocities (U_1, U_2, etc.) are those for individual duct components. Air flow balance computations based on the assumption that resistance is proportional to velocity squared will normally identify the preceding flow path, but on occasions this path will be self-evident.

The required fan volume flow rate is, of course, obtained from the addition of the separate branch flows.

4.2 STRAIGHT DUCTS OF CONSTANT CROSS-SECTIONAL AREA

A wide variety of cross-sectional shapes is not unusual in industrial practice. The most common of these is the circular duct or pipe that incidentally lends itself to rigorous theoretical and experimental treatment. Skin friction constitutes the sole resistance component as opposed to rectangular and other kindred ducts that also possess small secondary corner flows. However, experiment has shown that these flows make no significant contribution to duct loss, thereby facilitating direct comparison with circular pipe data. These data are also useful in loss estimation procedures relating to complex flow passages, for example, between tube arrays in heat exchangers.

Since turbulent flow is present in most industrial installations, skin friction data are normally available for the fully developed pipe flow case. However,

4.2 STRAIGHT DUCTS OF CONSTANT CROSS-SECTIONAL AREA

more attention is presently being given to studying the developing pipe flow features [4.3, 4.4]. In addition to the circular duct, investigations now cover annular ducts. A study of the developing laminar boundary layer flow for circular and other duct shapes is available in [4.5]. A major objective of the investigations is the establishment of an entrance length defined as the distance from entry to the plane of fully developed turbulent pipe flow.

4.2.1 Entrance Lengths

The majority of studies either assume, or experimentally promote, transition at the duct inlet. If the laminar layer were allowed to develop naturally, the variation in the transition point with different turbulence levels would make for unacceptable complication. By fixing transition at entry the entrance lengths can be substantially reduced. The promotion of flow separation at a sharp-edged duct entry will, of course, greatly shorten this length to 10 diameters or less.

Mathematical treatment of the entrance length problem is hampered by the downstream reduction in the "potential flow" core. However, a satisfactory treatment and confirmatory experimental data are given in [4.3] for a wide range of Reynolds numbers. The entrance length finding of 40 to 50 pipe diameters is supported by [4.4], which identifies inlet turbulence and disturbances as important variables.

The entrance length for annular ducts was experimentally investigated in [4.5] for a bellmouth outer wall entry and a hemispherically capped center body. The experiments were repeated with the bellmouth removed, leaving a square-edged outer wall entry condition. In both cases the test duct was attached to one wall of an upstream box. The tests covered the radius ratio range from 0.31 to 0.84 for naturally occurring transition in a stream possessing, in all probability, an appreciable level of turbulence at duct inlet. The entrance length for the square-edged entry varied between 25 and 35 hydraulic mean diameters; adding the bellmouth increased these values by 10 to 15. As indicated in Section 3.5, the hydraulic mean diameter is equal to twice the annulus width.

The other matter of interest in regard to developing pipe flow in annular ducts is the skin frictional resistance. By reason of the smaller boundary layer Reynolds number, the pressure loss will exceed that for fully developed pipe flow. The effect of a given size and distribution of roughness will also be greater owing to the reduced boundary layer thickness. Therefore in accepting the usual engineering practice of assuming fully developed flow conditions, the user should be aware of these higher resistance features and, where appropriate, make some allowance for them. Flat-plate calculations will fix the order of the resultant increments, in instances where concern exists.

4.2.2 Estimation of Skin Friction, Fully Developed Flow

The recent development of empirical factors relating friction data for a variety of cross-sectional shapes to those pertaining to the circular duct has greatly simplified the task of accurately establishing loss estimates.

Once established, the skin friction coefficient can be translated into a dimensional quantity by means of Eq. (3.68) and the relation for γ, namely,

$$\gamma = \tau_0 / \tfrac{1}{2}\rho \overline{U}_x^2 \tag{4.11}$$

The length of duct in terms of hydraulic mean diameter, in which one mean velocity pressure is lost because of skin friction, represents a convenient way of expressing resistance. The development of Eq. (3.68) results in the substitution of d_h for d in Eq. (3.69); equating the left-hand side to unity gives

$$l' = \frac{0.25}{\gamma} d_h \tag{4.12}$$

where l' is the required length.

1. *Circular ducts.* For smooth surfaces, the fully developed turbulent pipe flow has a skin friction coefficient given by Eq. (3.58). Substituting this value in Eq. (4.11) gives resistance in the desired form (Fig. 4.2).

When the roughness height to duct diameter ratio and the Reynolds number both combine to produce the fully rough condition, as defined in Fig. 3.7, the resulting resistance data can be obtained from Fig. 4.3.

Roughness ratios that lie between the fully rough and smooth limits will result in intermediate values of l'/d. The transitionary curves of Fig. 3.8 can vary in shape, depending on turbulence and other relevant factors. Hence for long duct lines where accuracy is essential, the loss estimates must be obtained from test work. A degree of conservatism should be exercised when experimental data are unavailable.

Choosing the correct equivalent roughness height can also be a problem, especially when rust or coarse grain particles are present. In such cases the roughness ratio selected should be on the conservative side unless actual test data are available for the surface in question. For surfaces produced by normal industrial processes, the data given in Table 4.1 are reasonably reliable.

A substantial degree of conservatism is recommended when dealing with flexible ducts. The loss data quoted by the maker are for the most favorable duct conditions, a situation seldom achieved in practice. The user must apply a factor based on experience.

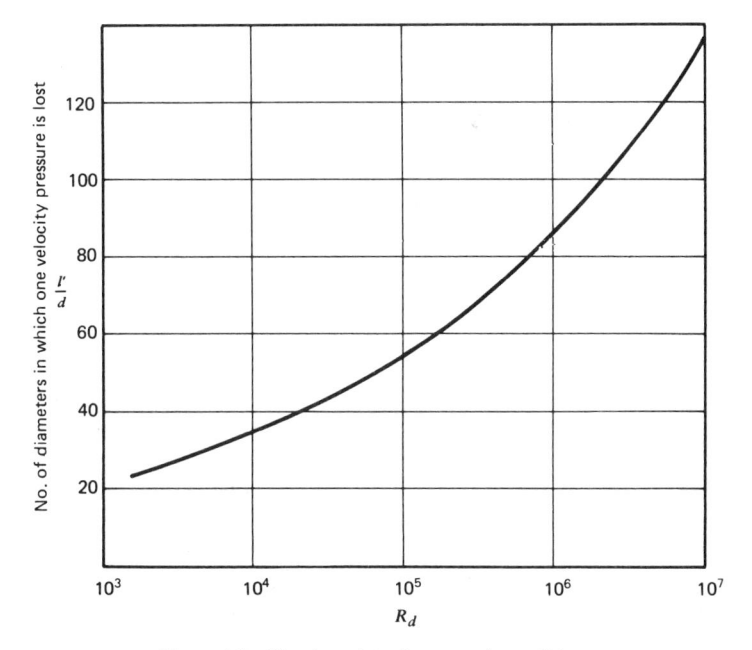

4.2 STRAIGHT DUCTS OF CONSTANT CROSS-SECTIONAL AREA 69

Figure 4.2 Pipe loss data for smooth conditions.

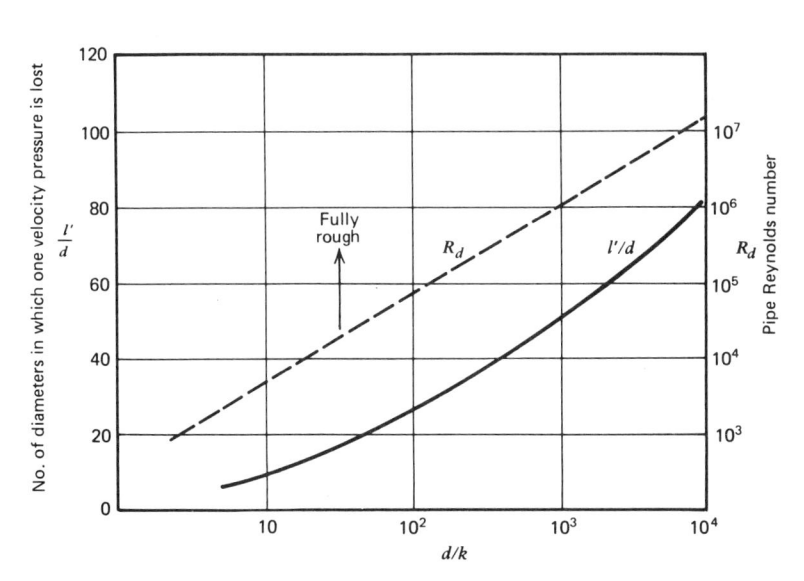

Figure 4.3 Pipe loss data for fully rough conditions.

Table 4.1 Roughness of Common Surfaces

Material	Equivalent Roughness (mm)
Drawn nonferrous metals and glass tubing	0.0015 to 0.0025
Rolled steel aluminum tubing	0.025 to 0.05
Galvanized iron (normal finish)	0.125 to 0.15
Cast metals	0.15 to 0.25

The data presented in Table 4.1 were extracted from five different references and represent the normal range of values but not the possible limits. An illustration of encountered extremes is presented in [4.7], where equivalent roughness heights for uncoated steel range between 0.01 mm for the best surface and 25 mm for heavy surface scaling.

Relevant practical experience is a valuable aid to the engineer seeking to apply the published information. Surface deterioration during the operational life of the ductwork, such as deposition or corrosion, should be anticipated by increased design loss coefficients.

2. *Annular ducts.* As indicated in Section 3.5, the skin friction coefficient in this instance exceeds that for circular pipe flows by approximately 10%, for equal Reynolds numbers based on hydraulic mean diameter. This quantity, which is based on the widely acknowledged work of [4.8], is therefore recommended for use, although [4.9] nominates the increase as 5%.* The procedure of [4.7] is to take skin friction as being equal to that in the pipe flow case.

In the absence of published roughness data relating to annular ducts, the approximations suggested in Section 3.8.4 are recommended for use along the procedural lines outlined for circular ducts.

3. *Rectangular ducts.* Experimental work on ducts possessing aspect ratios of 1 and 2 (as in [4.9]) has produced a common skin friction curve to that for circular ducts, in terms of R_{d_h}. In the former case d_h is given by $2ab/(a+b)$. However, it is now known that this is not the case for large aspect ratio ducts.

In the two-dimensional channel (infinite aspect ratio) flow studies, reported in Section 3.5, the skin friction coefficient lies somewhere between R_d and $R_{d_h}/2$ on the γ versus R_{d_h} curve for circular ducts.

---

* The recommendation of O. C. Jones, Jr., and J. C. M. Leung in ASME J. Fluids Eng. Vol. 103, pp. 615–623, 1981, is to apply the same Reynolds number correction procedure as that for rectangular ducts [4.11] where ϕ^* is given by Fig. 4.7. The increase in skin friction is approximately 10%.

4.2 STRAIGHT DUCTS OF CONSTANT CROSS-SECTIONAL AREA

Two later publications, [4.10] and [4.11], report developments in which this matter is quantified in terms of geometry factors, as determined from laminar flow friction studies. The latter reference specifically investigates large aspect ratio ducts and concludes that the findings for laminar flow can be carried over into turbulent flow. At a given R_{d_h}, the laminar skin friction coefficient for channel flow is 23.6/15.7 times the circular pipe value [Eqs. (3.30) and (3.34)]. Hence the magnitude of the geometry parameter ϕ^* at infinite aspect ratios is $\tfrac{2}{3}$ (Fig. 4.4). Since γ_{lam} is inversely proportional to R_{d_h}, then

$$R_{d_h}^* = \phi^* R_{d_h} \qquad (4.13)$$

This enables the γ_{lam} for all cross-sectional shapes to be obtained by adjusting R_{d_h} in the preceding manner and by inserting this number in Eq. (3.30).

When applied to turbulent boundary layers, the duct loss can be obtained from Fig. 4.2, for the corrected Reynolds number (Eq. 4.13). According to [4.11] the loss estimate will be within $\pm 5\%$, of experimentally established data. The value of ϕ^* at a duct aspect ratio of 2.25, namely, unity, vindicates the traditional use of circular duct data in the usual aspect ratio range.

Instead of applying the geometry factor directly in the preceding fashion, [4.10] assumes similarity between the laminar and turbulent expressions. A geometry factor A and a turbulent flow factor G^* are recommended, being presented in Fig. 4.5. These are then employed in calculating the skin

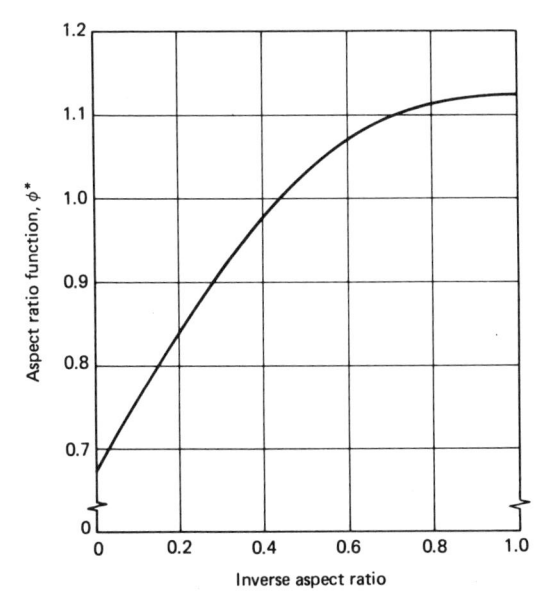

Figure 4.4 Geometry parameter, laminar equivalent. (Reproduced with permission of ASME from O. C. Jones [4.11]).

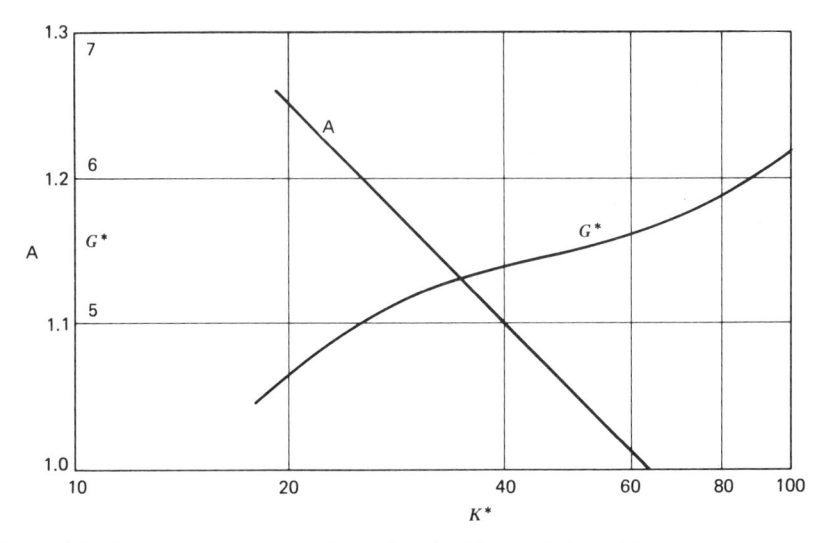

Figure 4.5 Geometry parameters. (Reproduced with permission of Pergamon Press from K. Rehme [4.10].

friction coefficient for the required R_{d_h} from

$$\sqrt{2}\,\gamma^{-1/2} = A\left[2.5\ln\left(R_{d_h}\sqrt{\frac{\gamma}{2}}\right) + 5.5\right] - G^* \quad (4.14)$$

where trial and error methods are used in obtaining γ. The laminar geometry factor [4.10] is expressed as $K^* = [4\gamma R_{d_h}]_{\text{lam}}$ and is related to ϕ^* by the ratio

$$\phi^* = \frac{K_c^*}{K^*} = \left[\frac{\gamma_c}{\gamma}\right]_{\text{lam}}$$

where the suffix c refers to the circular duct case and $K_c^* = 64$.

In assessing the effect of roughness for normal duct aspect ratios, the data of Fig. 4.3 can be used. Guidance in other cases can be obtained from Section 3.8.4.

4. *Ducts of arbitrary shape.* Ducting shapes for heat exchangers or other specific applications often require segmental or multitube cross-sectional arrangements. When the laminar flow geometry factor is known, use of Fig. 4.5 enables the solution of Eq. (4.14). The ϕ^* factors given in [4.7] for triangular, sector, and annular ducts are presented in Figs. 4.6 and 4.7, along with the hydraulic mean diameter relationships. Additional data are presented in Table 4.2.

4.2 STRAIGHT DUCTS OF CONSTANT CROSS-SECTIONAL AREA 73

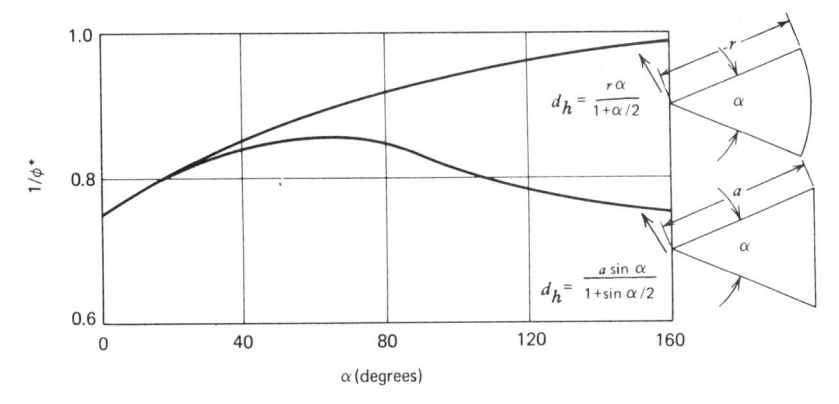

Figure 4.6 Geometry parameters for triangular and sector ducts. (Adapted with permission of ESDU International Ltd. from [4.7].)

$$d_h = \frac{r\alpha}{1+\alpha/2}$$

$$d_h = \frac{a \sin \alpha}{1+\sin \alpha/2}$$

When estimating the resistance of tube bundles or more complex duct shapes, reference should be made to [4.10]. However, the preceding information should be adequate in most practical instances.

In the majority of cases coming under the preceding heading, reliable roughness effects can only be obtained from actual experiment, as outlined in Section 3.8.3.

4.2.3 Other Losses

When duct construction and installation are below standard, disturbed flow caused by duct protrusions and leaking joints can substantially increase duct resistance. Examples of the former are illustrated in Fig. 4.8.

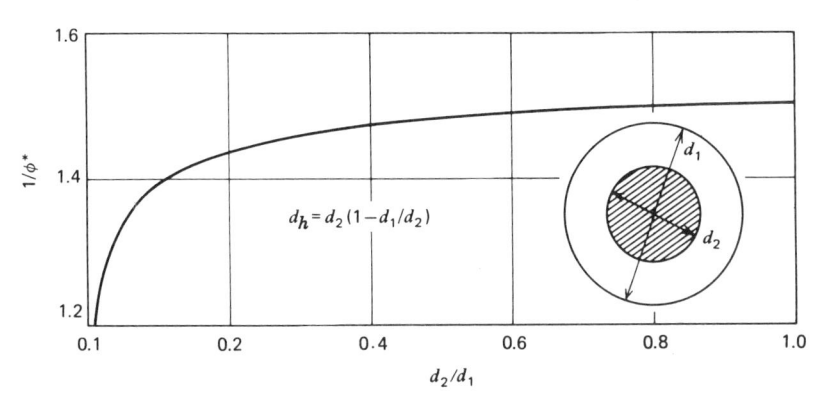

$$d_h = d_2(1 - d_1/d_2)$$

Figure 4.7 Geometry parameter for annular ducts. (Adapted with permission of ESDU International Ltd. from [4.7].)

DUCT COMPONENT DESIGN AND LOSSES

Table 4.2 Data for Calculating Interstice Friction

Air Passage	$1/\phi^*$	d_h
![]	0.441	$0.240r$
![]	0.406	$0.334r$
![]	0.406	$0.546r$

When the internal static pressure is in excess of the external value, air will leak from the duct. The momentum lost in this manner must therefore be a debit against overall system performance. When the leakage flow is reversed, the duct boundary layer is distributed, thus increasing the momentum losses. There is also the effect of an increased mass flow on duct losses and system performance.

Rectangular ducting can also sustain increased losses because of inadequately stiffened walls, thus giving rise to local contractions and expansions.

Together the preceding features resulting from a poor standard of workmanship can reduce the l'/d_h ratio by up to 20%. This fact illustrates the importance of overall perspective when estimating system resistance.

The resistance of constant cross-sectional area noise silencers must be established experimentally with respect to each particular type, because of differing surface conditions.

In conclusion, it should be emphasized that the preceding loss estimates apply to fully developed flow conditions. Adjustment to other flow states should be made on the basis of boundary layer considerations (Chapter 3).

4.3 CONTRACTIONS

Contractions are as a rule comparatively short, and hence skin friction losses are small.

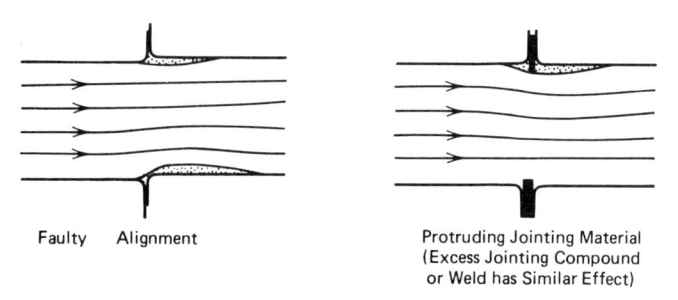

Faulty Alignment

Protruding Jointing Material (Excess Jointing Compound or Weld has Similar Effect)

Figure 4.8 Ducting faults.

4.3 CONTRACTIONS

In accelerating flows pressure energy is converted into velocity pressure; this process is usually free of undesirable features. However, with abrupt changes in cross section, the air is overspeeded near the surface and the consequent retardation produces regions of eddying flow; the larger these regions, the greater the losses.

A large amount of information is available on the design of good contractions (see [4.12] for listing), but for general purposes a smooth contour constructed with two radii is adequate. For circular units Chmielewski [4.12] chose a particular acceleration rate, in terms of two power functions. The values selected determine the contour, which in turn can be approximated by an inlet and an outlet radius. A minimum length for avoiding the preceding separated flow regions is given in Fig. 4.9, for a preferred choice of powers. The contour for a contraction ratio of 5 is presented in Fig. 4.10; the inlet curvature can be approximated by a radius that is roughly 60% of the outgoing one. This ratio restricts the surface length over which an adverse pressure gradient exists in the contraction inlet. The length of the contraction is taken from the point at which centerline flow acceleration commences, which is approximately 30% upstream of the change in duct geometry.

Two-dimensional contractions of similar area ratio to the circular variety require a length increase of approximately 25% to allow for a linear rather than an inverse square root reduction in the throat dimension with area ratio.

The loss data tabulated in [4.13] for conical contractions have been rationalized and presented in Fig. 4.11, for an unknown Re. However, provided the wall angle is approximately 10°, skin friction presumably is a substantial proportion of the ensuing loss.

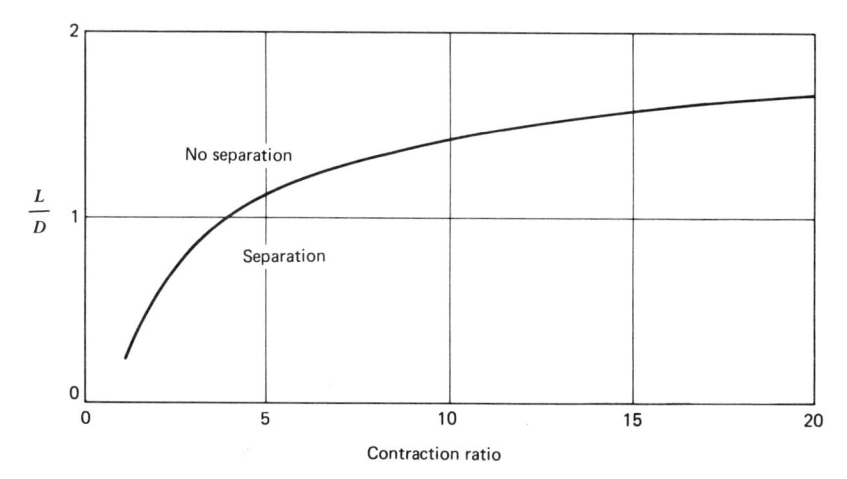

Figure 4.9 Minimum length for Fig. 4.10-type contractions, without separation. (Reproduced with permission of American Institute of Aeronautics and Astronautics from G. E. Chmielewski [4.12].)

76 DUCT COMPONENT DESIGN AND LOSSES

When the wall angle is 90°, the variation of loss coefficient with area ratio, from two sources, is given in Fig. 4.12. At ratios above 4 or 5 the upstream duct is essentially a plenum, from the inlet loss viewpoint. A difference in the undisclosed test Re values could provide some explanation for the displacement of the two curves in Fig. 4.12.

A plenum chamber can be defined as one in which the velocity and associated dynamic pressure are extremely small and hence negligible. For example, a contraction ratio of 10 results in an upstream velocity pressure that is only 1% of the downstream value. Hence for all practical purposes the plenum static pressure equals total pressure. A flared inlet to the downstream duct with a radius one-quarter that of its diameter will ensure good inlet conditions; losses rise rapidly when the radius ratio is less than 0.15 (Fig. 4.13). Provided the flow disturbance in the flared inlet due to the local adverse pressure gradient is of restricted extent, the conversion of static pressure to velocity pressure on entry will result in a fairly uniform downstream distribution of velocity. Countersinking (Fig. 4.13) produces notable loss reductions and consequently flow improvements; the optimum angle is approximately 30°.

Inlet boxes are frequently used in relation to boiler fans for electric power generation. The duct unit combines an unvaned corner of up to 90° deflection and a plenum chamber of restricted dimensions. Model testing to determine the "optimal" design, defined as a relatively compact unit with moderate losses, has been carried out by Bernard [4.14]. The experiments were restricted to swirl-free, uniform inlet velocity conditions. The connection between the box and the fan unit features a similar flare to that associated with plenum chambers (Fig. 4.14).

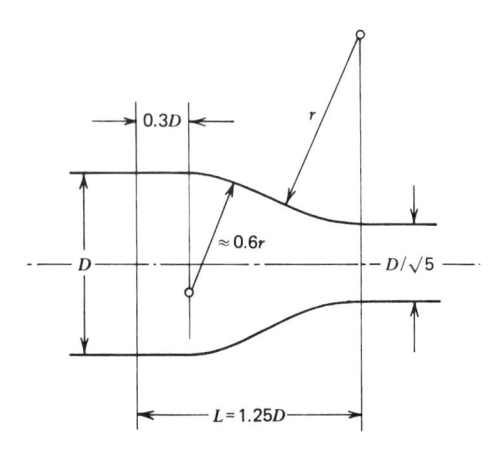

Figure 4.10 Area ratio 5 contraction. (Adapted with permission of American Institute of Aeronautics and Astronautics from G. E. Chmielewski [4.12]).

4.3 CONTRACTIONS

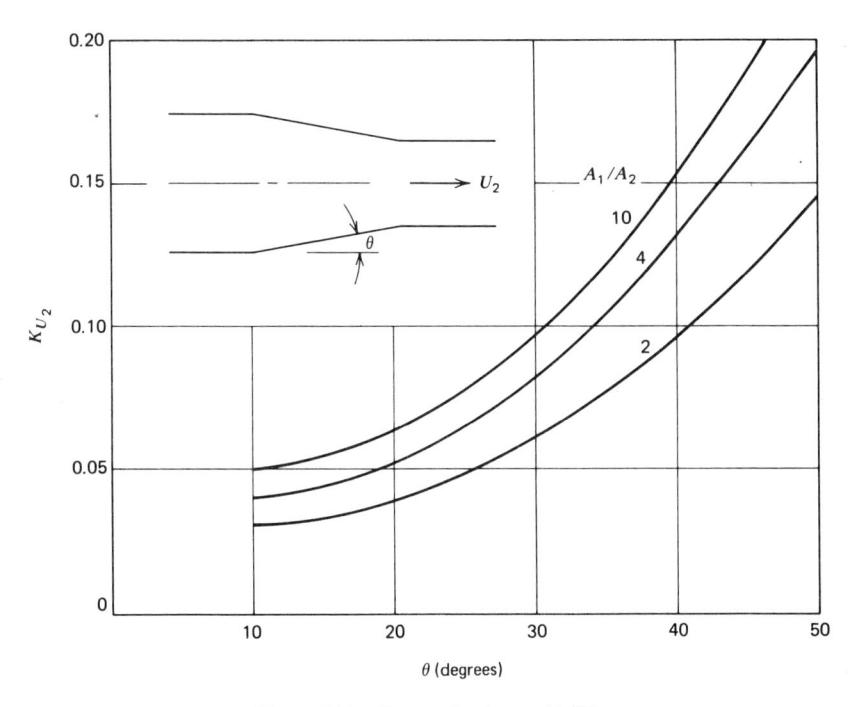

Figure 4.11 Contraction losses [4.13].

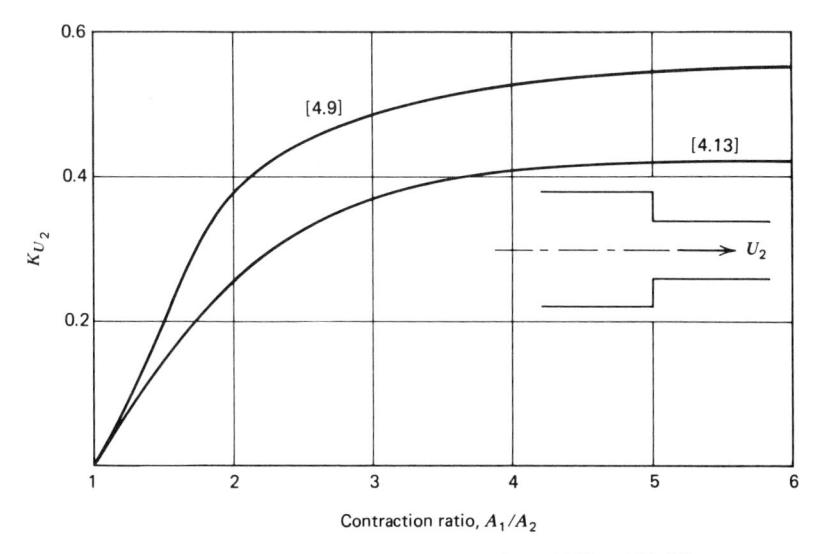

Figure 4.12 Sudden contraction losses from [4.9] and [4.13].

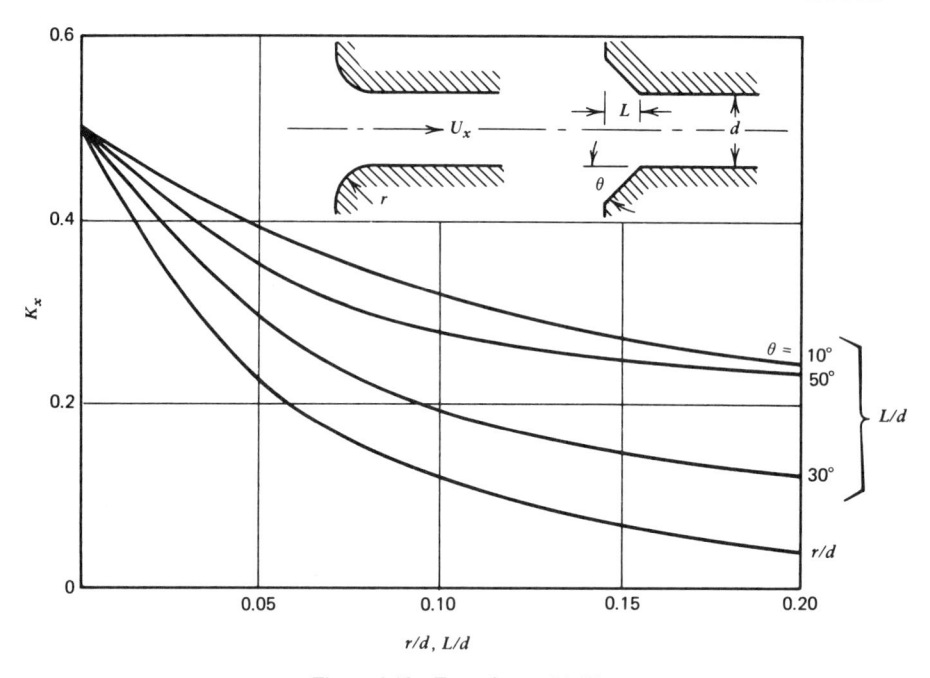

Figure 4.13 Entry losses [4.13].

However, axial flow fan performance is noticeably dependent on inadvertent swirl and inlet flow asymmetry. These latter properties will exist to some degree in all box/fan systems and hence some interference with the normal fan characteristic and efficiency curves can be expected. These fan inlet flow consequences will be minimized with a large contraction ratio, as produced by increased inlet box dimensions and a large fan boss ratio.

The overall inlet box/fan assembly, for an ideal box inlet flow, resulted in the performance data of Fig. 4.15 for the geometries of Fig. 4.16. For the optimal box geometry defined in Fig. 4.14 and a boss ratio of 0.488, the peak efficiency and fan pressure rise were both reduced by small amounts, the former by less than 2%.

When the rotor is mounted on an overhung bearing, the support struts across the annulus can be designed to remove the swirl components at fan entry. In almost all practical instances, swirl and possible flow instability will be present.

As the fan drive shaft shroud becomes increasingly large, the potential for flow instability is heightened, particularly for asymmetric inlet flow. Interaction between the cross-flows in the toe of the box, the main source of diffi-

4.3 CONTRACTIONS

culty, can be prevented by the addition of an axially aligned plate in the plane of symmetry extending from the shroud to the box boundary. Substantial swirl at the box inlet will also create stability problems, which can only be minimized by a flow-straightening device in the upstream duct.

Limited additional information on inlet box design and losses is available in [4.15] and reproduced in Fig. 4.17. The use of a dual box with a double-sided inlet arrangement is reported therein. Presumably this would require the addition of a plate to separate the opposing flows.

The wide range of inlet box geometries, as developed by the various fan manufacturers, would suggest that within reason shape is not all that critical. The contraction ratios also vary greatly. However, every detailed inlet box proposal should be subjected to careful model testing, including the expected inlet flow features. Failure to model test can result in unacceptable losses.

The term *inlet* is normally applied to contractions of infinite area ratio. Shapes vary in a similar manner to contractions from a contoured type of inlet to an abrupt, sharp-edged variety (Fig. 4.18). The ideal contour (Fig. 4.19) is one that possesses an increasing radius of curvature in the throat direction [4.16]. An inlet [4.17] specially developed in a wind tunnel to give unseparated flow up to a crosswind/duct velocity ratio of 1.25 is illustrated in Fig. 4.20. The experimental development was in connection with a 6.1-m-diameter mine ventilation unit of the downcast type. The combination of

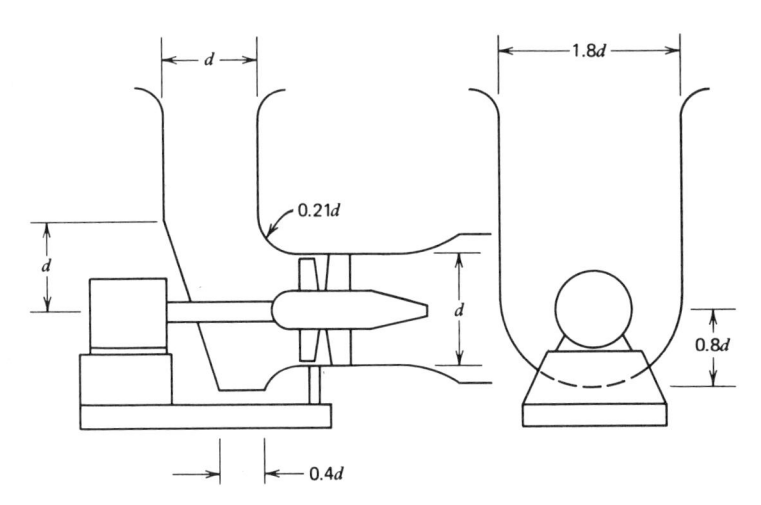

Figure 4.14 Preferred inlet box geometry. *Note:* Efficiency losses decrease with increase in boss ratio. (Reproduced with permission from T. Bernard [4.14] in Proc. 3rd Conf. on Fluid Mech. and Fluid Mach. Akadémiai Kiadó 1969.)

80   DUCT COMPONENT DESIGN AND LOSSES

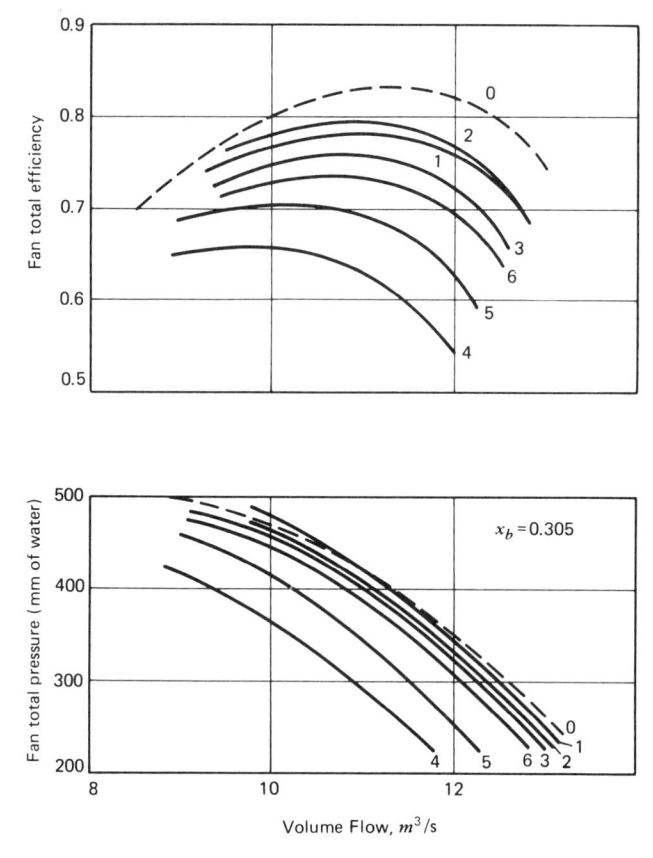

Figure 4.15 Performance data for Fig. 4.16 geometries. (Reproduced with permission from T. Bernard [4.14] in Proc. 3rd Conf. on Fluid Mech. and Fluid Mach. Akadémiai Kiadó 1969.)

duct contour, ring airfoil, and wire screen provided a low loss arrangement for delaying the onset of inlet flow separation, for increasing crosswind strengths. A good measure of unsteady-load relief for the adjacent rotor blades was thereby achieved.

In the absence of a general crosswind component, a semicircular contour of radius $0.15D$, where D is the duct diameter, provides the minimum bellmouth shape when seeking a low loss entry. Although losses increase for smaller radii, some rounding is always worthwhile. For larger inlet units where spinning or other three-dimensional shape-forming processes are impractical, the welding together of a number of semicylindrical surfaces, arranged in a segmental pattern, provides an acceptable alternative.

Two types of conical inlet are illustrated in Figs. 4.18 and 4.21. The latter is used as a flow-measuring device. The large static pressure change as the

4.3 CONTRACTIONS

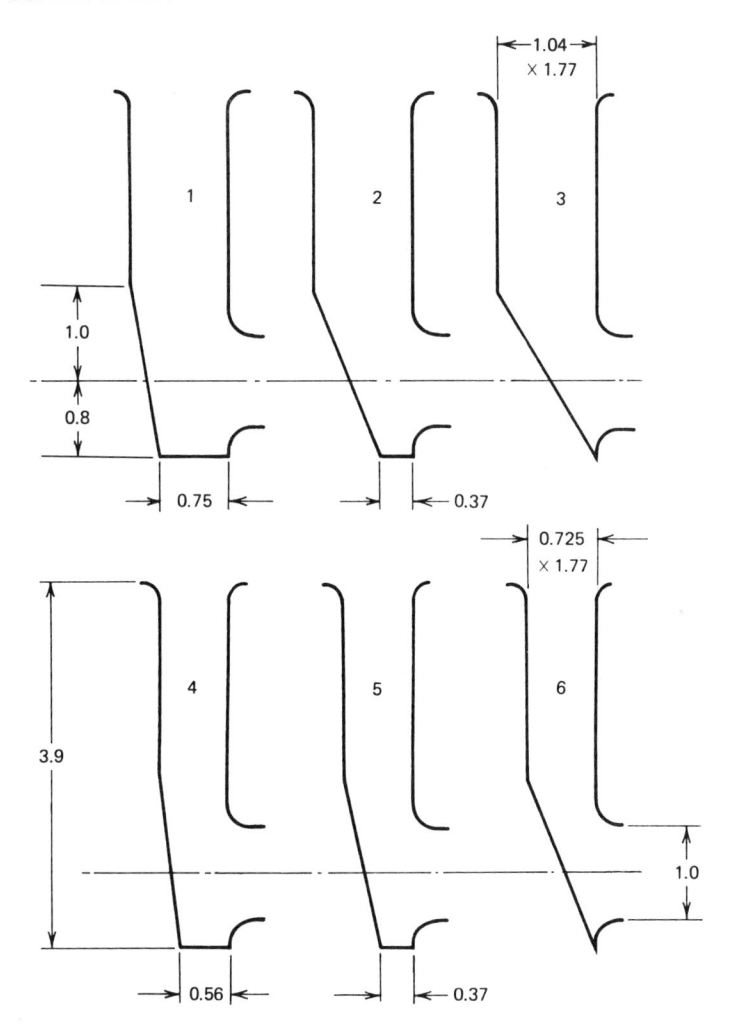

Figure 4.16 Inlet box geometries tested in [4.14]. (Reproduced with permission from T. Bernard [4.14] in Proc. 3rd Conf. on Fluid Mech. and Fluid Mach. Akadémiai Kiadó 1969.)

flow is accelerated from rest to duct velocity enables the latter, and hence the flow rate, to be established as a function of static pressure. The variant illustrated has been extensively tested at N.E.L. [4.18] and its discharge coefficient C_{b_c} established as a function of R_d. The loss coefficient is given by $K_x = (1 - C_{b_c}^2)/C_{b_c}^2$; the values presented in Fig. 4.21 are based on the relationship of Fig. 21.4, and show a marked dependence on R_d. Variations

82 DUCT COMPONENT DESIGN AND LOSSES

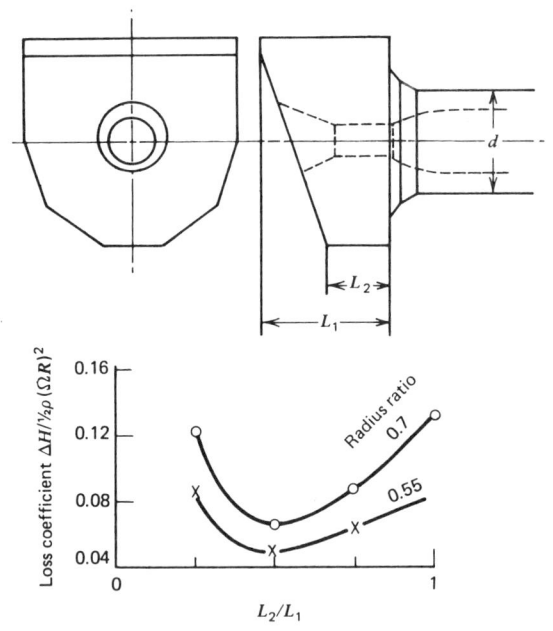

Figure 4.17 Inlet box geometries and test results. (Reproduced with permission of ASME from N. Yamaguchi [4.15].)

in loss coefficient will exist with respect to all contractions and entries and therefore the data presented here should be taken as representative rather than precise.

Sharp-edged Borda mouthpieces are now seldom used for flow measurement. At the other end of the scale, negligible losses and hence $C_{b_c} = 1$ can be assumed for contoured entries of the type illustrated in Fig. 4.19; the duct velocity pressure will approximately equal the static pressure differential, thus achieving independence of Re for flow determination.

Another loss-reducing, flow-improving device consists of two or three louvers arranged as in Fig. 4.22. The arrangement illustrated is attached to an engine test cell (Plate 1) for the purpose of eliminating separated flow conditions arising from strong crosswinds [4.19]. A feature of this device is the important part played by the leeward side louvers as distinct from the obvious role of the windward louvers. The cell is used in the testing of turboprop aircraft engines.

4.4 DIFFUSERS

An expanding duct component possesses an apparent incompatibility between aerodynamic performance and unit compactness. Since space limita-

4.4 DIFFUSERS

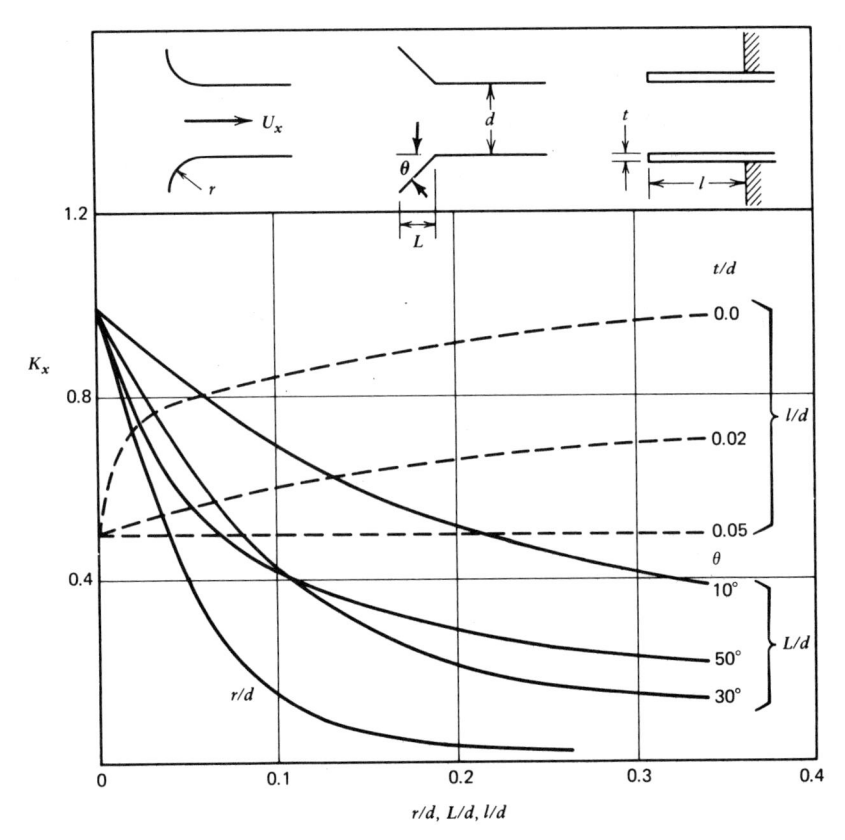

Figure 4.18 Inlet losses [4.13].

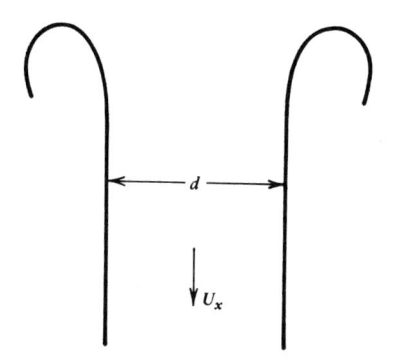

Figure 4.19 Preferred inlet geometry [4.16].

84   DUCT COMPONENT DESIGN AND LOSSES

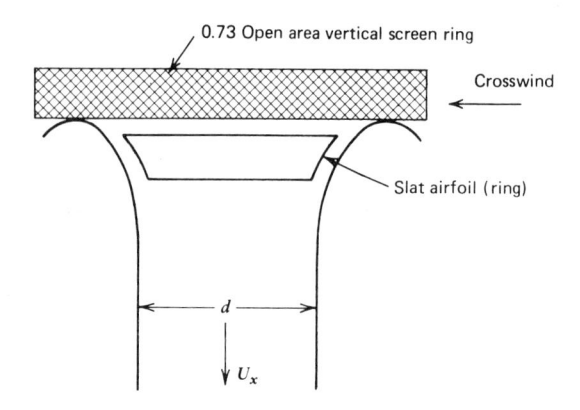

Figure 4.20 Inlet geometry for large mine fan [4.17].

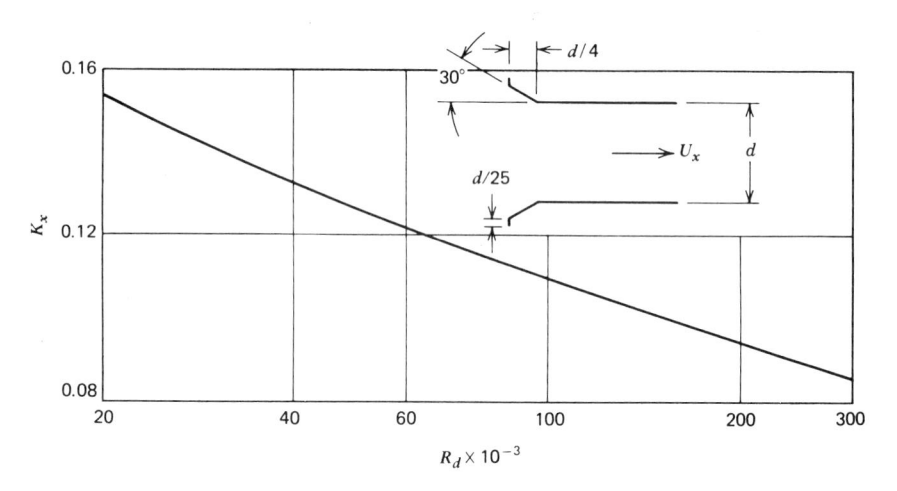

Figure 4.21 Inlet loss for conical flow measuring device.

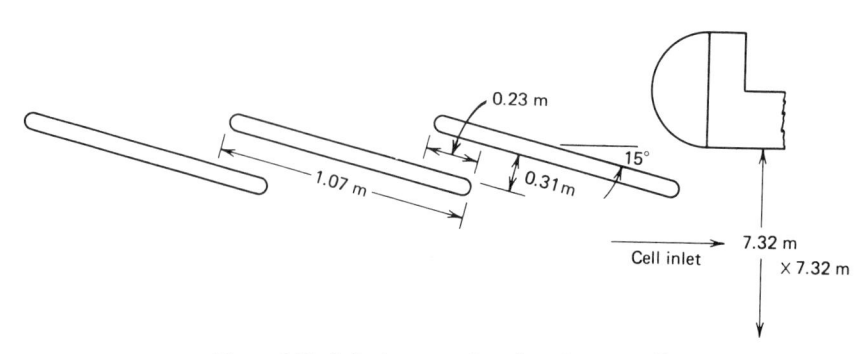

Figure 4.22 Inlet louvers, aircraft engine test cell.

4.4 DIFFUSERS

Plate 1. Wind protection louvers at engine test cell entry.

tion is often the deciding factor, increases in air resistance and aerodynamically induced noise are generally accepted as being unavoidable. However, these consequences can be minimized by adopting or developing a suitable arrangement from the various options discussed here.

The large number of variables that influence diffuser performance has necessitated the conduct of numerous research programs. To minimize confusion, however, it is important to discuss the subject in terms of the salient engineering design parameters.

Ideally, diffusers have a dual function to perform, namely, (1) to reduce the mean velocity in a steady manner and (2) to recover as static pressure a large proportion of the subsequent difference in dynamic pressure. For zero total pressure loss, Eq. (4.5) can be written

$$(P_B - P_A)_i = \tfrac{1}{2}\rho U_A^2 \left[1 - \left(\frac{A_A}{A_B}\right)^2 \right] \tag{4.16}$$

The diffuser efficiency (i.e., effectiveness) can be expressed as the ratio of the actual to the ideal pressure recovery, namely,

$$\eta_D = \frac{P_B - P_A}{\tfrac{1}{2}\rho U_A^2 [1 - (A_A/A_B)^2]} \tag{4.17}$$

or,

$$\eta_D = 1 - \frac{\Delta H_D}{\frac{1}{2}\rho U_A^2[1 - (A_A/A_B)^2]} \qquad (4.18)$$

Because Eqs. (4.17) and (4.18) ignore the inlet and outlet flow energy coefficients, as defined in Eq. (4.4), they are not precise measures of diffuser performance. However, they provide an adequate and practical engineering method for loss estimation.

The ideal static pressure rise coefficient is given by

$$C_{p_i} = 1 - \left(\frac{A_A}{A_B}\right)^2 \qquad (4.19)$$

and for real flow by

$$C_p = \frac{P_B - P_A}{\frac{1}{2}\rho U_A^2} \qquad (4.20)$$

Finally, the loss coefficient is

$$K_D = 1 - \eta_D \qquad (4.21)$$

The dual functions of a diffuser can lead to different design solutions. First, when a reduction in pipe flow velocity is the only requirement, with no length restriction, a small divergence angle is desirable. However, if the angle is too small for the required expansion ratio, an excess loss caused by skin friction will be present in the lengthy unit. On the other hand, a large angle will precipitate flow separation, introducing far greater losses. The optimum angle for least overall loss in the pipe system is a matter for judgment, since the loss remains constant over a small range of angles.

The second diffuser function, namely, maximum pressure recovery within a given length, is characterized by the presence of a small degree of transitory separation. Since the diffuser effectiveness for this condition is close to the peak value, and with most engineering tasks requiring this type of optimum solution, the ensuing presentation of data will be mainly centered around this objective.

In addition, the presentation will be for straight-walled configurations only, since these provide near-optimum design solutions. These arrangements possess relatively large initial adverse pressure gradients that reduce rapidly toward the diffuser outlet; the change in gradient matches the reducing capacity of the layer to withstand separation [4.21]. Experimental studies with curved walls have failed to produce significant gains [4.21].

4.4 DIFFUSERS

The effect of wall curvature on thick turbulent boundary layer properties could be a contributing factor. In the simplest terms, the adverse influence can be related to a trend toward the elimination of randomly appearing, longitudinal vorticity in turbulent shear flow. The importance of these vortices in the normal fluid-mixing process is stressed in [4.22]. Fundamentally the same principle applies, in reverse, as that which explains the presence of Görtler-Taylor vortices in laminar flow along concave walls (see Section 3.7.3). Bradshaw [4.23] comprehensively reviews these features in a liberally referenced monograph.

4.4.1 Flow Features

The various flow regimes in a two-dimensional diffuser have been exhaustively studied by Kline and others. These are illustrated in Fig. 4.23 [4.24].

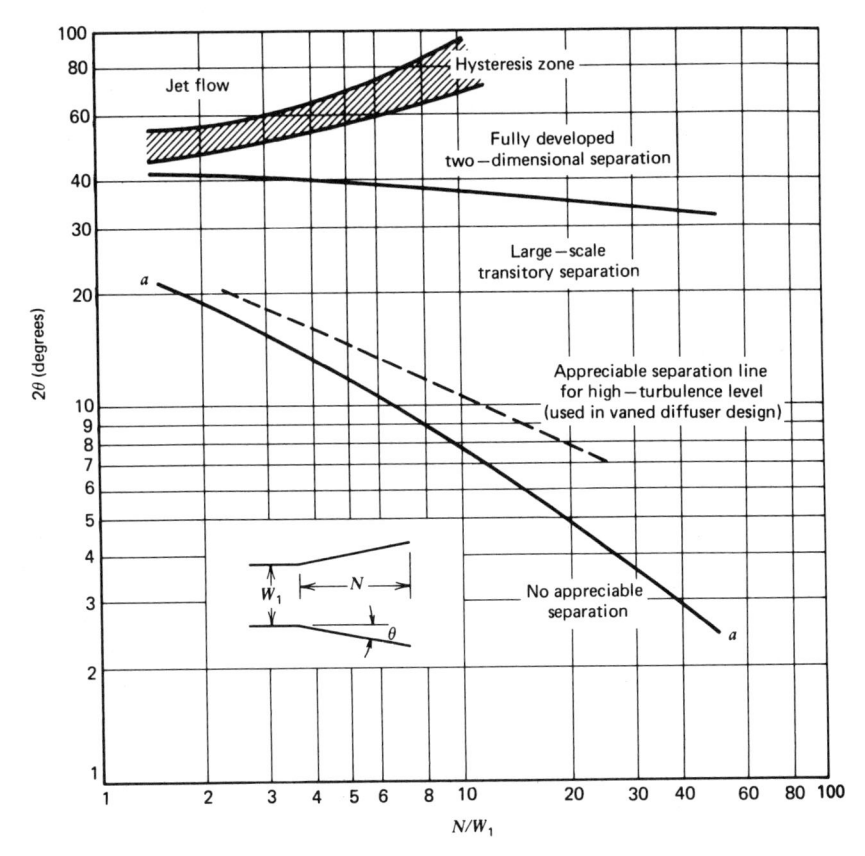

Figure 4.23 Flow regimes in two-dimensional diffusers, thin inlet boundary layers. (Reproduced with permission of ASME from R. W. Fox and S. J. Kline [4.24].)

The "transitory stall" regime is characterized by intermittent changes in separation patterns, and locations; unsteady flow results. "Two-dimensional stall," or separation from a relatively fixed location on one wall, is a more stable flow state. The forward movement of the separation phenomenon to the diffuser inlet, with angle increases, results in "jet flow," or separation from both walls. The latter two regimes are separated by a hysteresis zone in which the flow alternates between these two phenomena.

These regimes are specifically for thin inlet boundary layer conditions but thick layers only lower the a–a line by 1 to 2° without changing the position of the other lines. High inlet flow turbulence has the opposite tendency.

An important deduction that can be made as a result of the foregoing is that the optimum diffuser geometry is independent of inlet conditions, as established by experiment. However, the aerodynamic performance shows a marked sensitivity to changes in inlet flow quality.

It may be argued that a conservative choice of wall angle for increased boundary layer thickness may result in improved diffuser flow. However, one must balance this against reduced area ratio, for fixed length, and hence a further loss in C_p in addition to the reduction caused by entry conditions.

Hence for most practical engineering purposes the optimum wall angles, as established from thin inlet boundary layer conditions, can be considered independent of inlet boundary layer thickness. However, asymmetric and irregular inlet flows will produce design, loss estimation, and outlet flow distribution problems that can only be satisfactorily resolved by experimentation, for each and every set of actual circumstances.

4.4.2 Optimum Diffuser Geometry

Sufficient data are available for the establishment of design recommendations, based on the $C_{p_{max}}$ line for a given diffuser length, with respect to two-dimensional, conical, and annular configurations. Information on square diffusers is more limited. The performance of the first variety is not influenced by aspect ratio (inlet width/inlet depth), provided this number exceeds unity [4.25].

The results obtained by McDonald and Fox [4.26] have been selected to represent the conical diffuser case. These data have been successfully extended to a development covering annular diffuser design cases. It was established in [4.27] that the Sovran and Klomp optimum design solution [4.28] does not cover all annular diffuser geometries.

Annular diffusers can possess either converging or diverging centerbodies; the former are preferable for "in-line" duct arrangements, whereas the latter are frequently used in exhaust applications. The findings of experimental studies covering both types are available in [4.29] and [4.30]. The expanding centerbody studies were concentrated on optimum geometries that pos-

4.4 DIFFUSERS

sessed the shortest length for a given area ratio. The development was based on an analogy with conical diffusers, as postulated by Kmonicek [4.31]. On comparing the resulting design recommendations with those of [4.28], where testing was centered on this general geometric type, satisfactory agreement was attained.

Data for four common diffuser types are presented in Figs. 4.24 to 4.26. In all cases the experimental studies were for diffusers with no tailpipe. The optimum included angle for square section diffusers of area ratio 4 is given in [4.32] as 6° approximately. (*Equivalent angle* is defined as the included angle of a conical diffuser with identical inlet and outlet areas, and length, to that of the diffuser in question.)

4.4.3 Diffuser Performance

Performance may be expressed in terms of either C_p or η_D. The latter is often favored, since for optimum geometries its decline with increasing area ratio is not substantial and hence remains relatively constant.

The loss in peak effectiveness resulting from changing inlet conditions is the subject of extensive current research. From the engineering viewpoint, however, the average performance decrement can be assumed in design.

The addition of a tailpipe increases the performance of optimum diffusers by a small amount. This increment is related to the peak velocity reduction

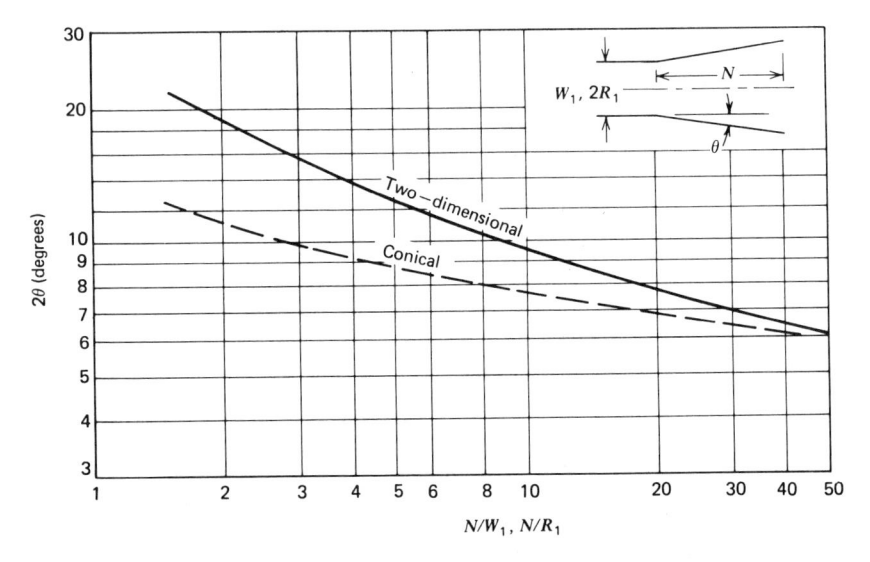

Figure 4.24 Recommended included angle for two-dimensional and conical diffusers.

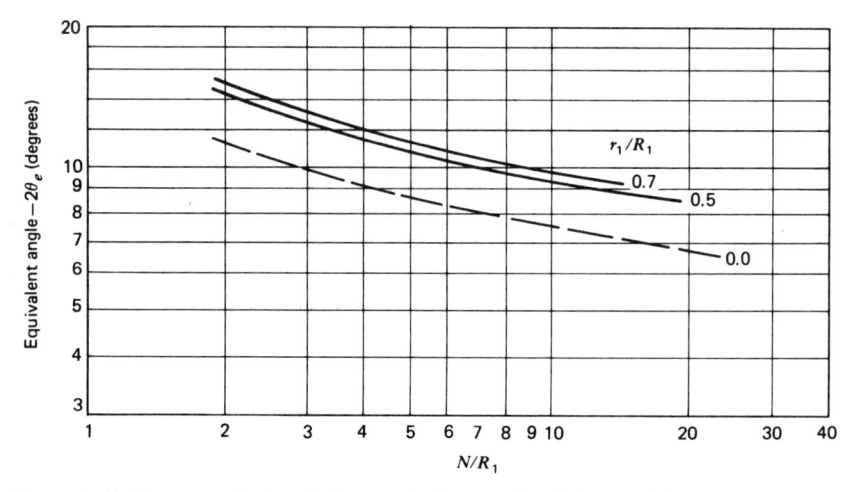

Figure 4.25 Recommended equivalent angle for annular diffusers with convergent centerbodies, exhaust case.

and subsequent downstream static pressure recovery. This gain is complete after a length of from two to six outlet duct diameters, depending on diffuser inlet conditions and area ratio.

Provided the Reynolds number, based on the inlet dimension and velocity, is above 2×10^5 approximately, diffuser performance remains virtually unchanged.

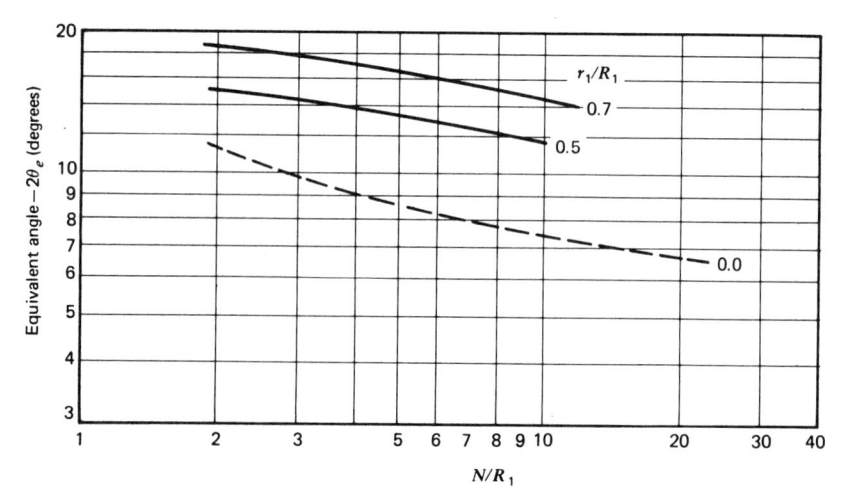

Figure 4.26 Recommended equivalent angle for annular diffusers with divergent centerbodies, exhaust case.

4.4 DIFFUSERS

The agreement between performance values as obtained by various workers for relatively comparable geometries and flow conditions can only be classed as fair. In an attempt to avoid confusion among engineers unfamiliar with flow phenomena, the number of variables is herein limited and the use of mean performance criteria is recommended.

With the appropriate choice of η_D, the related C_p can be estimated from Fig. 4.27. It will be apparent that for constant η_D, the higher area ratios have little significant influence on C_p. The small decline in peak η_D with area ratio actually levels off the maximum attainable C_p.

Published diffuser effectiveness data indicate peak values slightly in excess of 0.90 for two-dimensional, conical, and annular diffusers with thin inlet boundary layers but without tailpipe. Since the tailpipe can make a substantial improvement to performance for poor quality outlet flows, one is forced to assume that, in the preceding tests, a relatively good-quality outlet

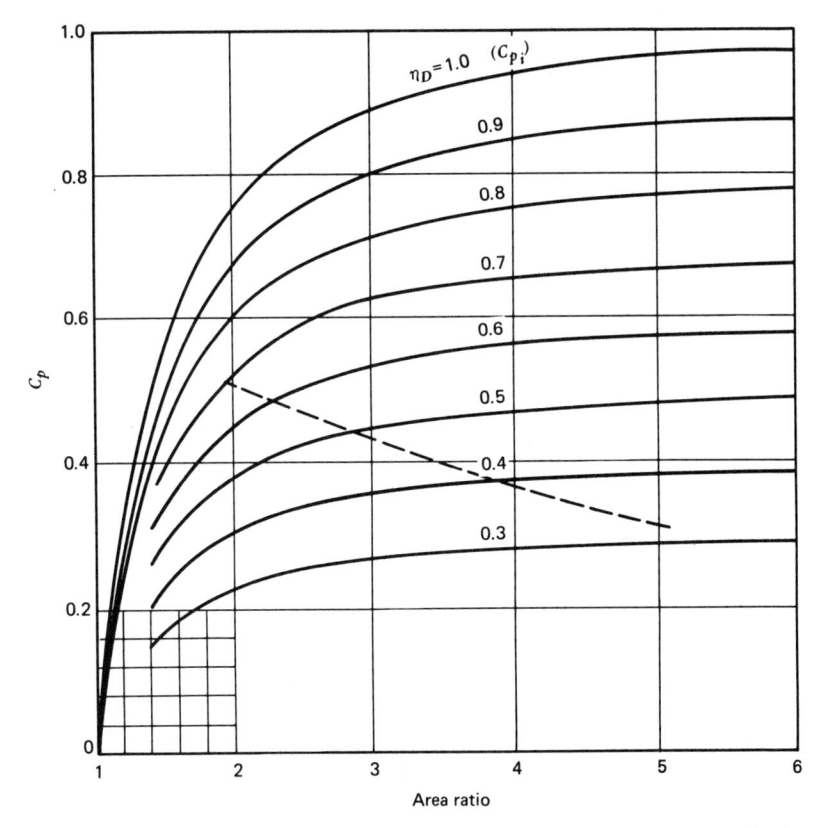

Figure 4.27 Pressure recovery coefficient versus area ratio and diffuser effectiveness. Dotted line represents C_p for "jet flow" (Fig. 4.23), wide-angle conical diffuser with tailpipe.

92 DUCT COMPONENT DESIGN AND LOSSES

flow was present. (Theoretically, η_D has a maximum attainable value of 0.94; skin friction is the limiting factor.)

With thick inlet boundary layers and no tailpipe, peak effectiveness is normally in excess of 0.80 for all three diffuser types. The presence of flow asymmetries or severe nonuniformities will result in greater performance reductions. A guide to their magnitude can be obtained from [4.33]. Flows with a linear velocity gradient in the freestream (as distinct from the boundary layer regions) suffer relatively large losses, particularly for moderate to large area ratios. This is particularly relevant to annular diffusers downstream of axial flow fans, as demonstrated in [4.29]. As a consequence, deviations from good free vortex flow design techniques for this fan type will result in diffuser performance penalties.

Other data relating to annular diffusers with distorted and nonuniform inlet flow are available in [4.34] and [4.35].

For normal engineering applications with symmetrical inlet flow conditions it is recommended that, for optimum geometries, η_D be taken as 0.80 ($K_D = 0.2$), a conservative estimate. For thin inlet boundary layers this can be increased by 7 to 8%. The addition of a tailpipe may increase η_D by a further 1 to 4%, depending on diffuser outlet flow features.

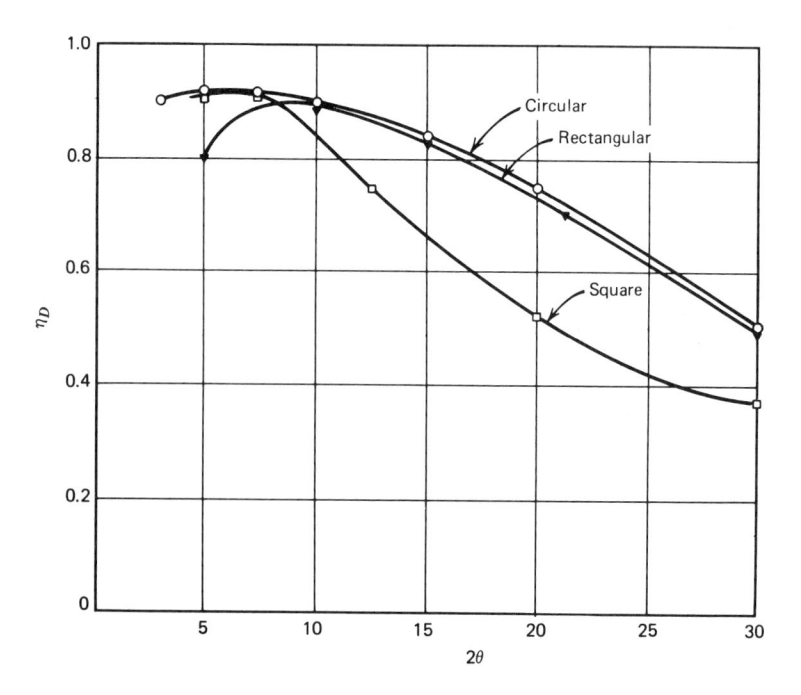

Figure 4.28 Diffuser efficiency as function of cross-sectional shape and included angle, area ratio of 4 [4.32].

4.4 DIFFUSERS

The design recommendations of Figs. 4.24 to 4.26 give maximum pressure recovery for a given length. Since the difference between the related η_D and the peak values discussed earlier is small, and within the normal limits of uncertainty, no distinction is necessary in practical design situations.

When the wall angles exceed the optimum values, a measure of the reduced performance can be obtained from Fig. 4.28.

The pressure recovery for "jet flow" in conical diffusers with tailpipe is superimposed on Fig. 4.27, representing the lower limit condition. Since the effectiveness exceeds 0.3 for area ratios less than 5 approximately, a reasonable proportion of the difference in dynamic pressure, between upstream and downstream, is recovered as static pressure; this is effected through the agency of turbulent mixing in the downstream duct.

Unless special precautions are taken in the design of engineering duct systems, with respect to providing regular, symmetrical inlet flow conditions, unknown performance penalties and gross nonuniformity of outlet flows must be accepted as obvious eventualities. Two courses of design action are available, namely, to guess the performance on the basis of published data or to establish by experiment the unknown data from model or similar installation tests. The latter has merit, since the experimental scope can be increased to include the development of a suitable boundary layer control system, thus minimizing the incurred disabilities.

The wide scatter of published effectiveness data for varying inlet flow irregularities does not permit out-of-context statements to be made herein; hence the reader must consult the literature for guidance.

4.4.4 Cropped Diffusers

Diffuser length for a required area ratio may be subject to practical restriction, necessitating the use of excessive wall angles. However, an improved pressure recovery performance is obtainable with diffusers of optimum angle, for the given permissible length, followed by a sudden expansion at the tailpipe attachment plane.

Limited published data on two-dimensional units suggest that appreciable truncation can be made with little deterioration in pressure recovery.

Studies of the conical variety [4.36] relative to a 5° wall angle unit indicate that truncations of up to 50% of the original diffuser produce typical C_p reductions of only 4 to 5%. Truncation did not appear to alter significantly the downstream tailpipe location at which the maximum pressure recovery was recorded.

The C_p reduction in conical diffuser performance can be calculated according to [4.36] from

$$\Delta C_p = \frac{(A_A/A_T - A_A/A_B)^2}{1 - (A_A/A_B)^2} \qquad (4.22)$$

where A_T is the downstream area of the truncated cone. Evaluation of this expression indicates a relative insensitivity to area ratio in the normal range of engineering application, namely, ratios of up to 5 or 6. For higher ratios the degree of truncation required will in all probability be large, resulting in poor downstream velocity distributions for restricted tailpipe lengths, for example, in heat exchanger installations. In these circumstances vanes can offer a preferred solution.

Comparison between a cropped conical diffuser ($\theta = 5°$) with varying truncation ratios (S/N), and diffusers of large wall angle ($\theta = 7.5°$, $10°$, and $12.5°$) and of length N' and untruncated, is made on Fig. 4.29. The curve derived from Eq. (4.22) approximated the mean experimental values of [4.36] for a thick inlet boundary layer and a wide range of area ratios. A cropped diffuser ($\theta = 5°$) is clearly preferable to a short, wider-angled one.

The similar values of C_p for all area ratios above approximately 3 would indicate a reducing effectiveness factor with this ratio. This could be due to the wall angle ($5°$) being in excess of the optimum indicated in Fig. 4.24. In addition, the pressure recovery coefficients will diverge, with area ratio, as the truncation ratio S/N approaches zero, the sudden expansion case (see Fig. 4.27). However, the data given in Fig. 4.29 can be accepted as being of sufficient design accuracy for most practical applications.

Truncation also improves flow stability when comparison is made with short diffusers of excessive wall angle.

The preceding principle has been applied to annular diffusers that discharge into the annular combustor chambers of aircraft jet engines [4.37]. The chief nonaeronautical application, however, occurs when either the tail fair-

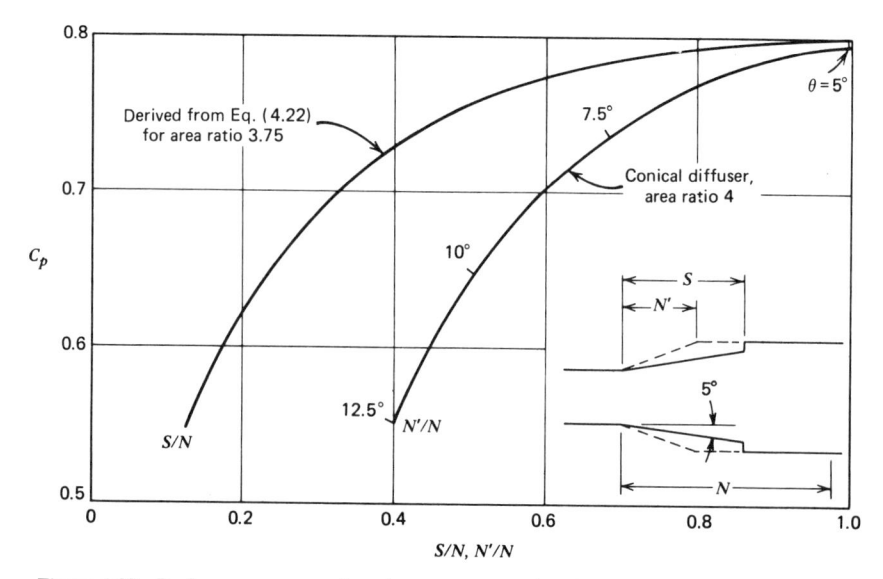

Figure 4.29 Performance comparison between cropped and wide-angle conical diffusers.

4.4 DIFFUSERS

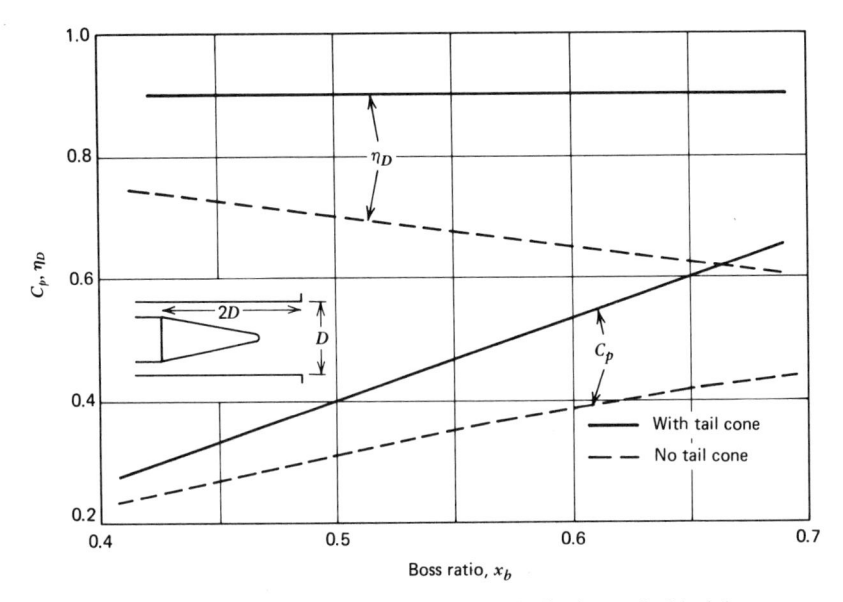

Figure 4.30 Performance data for conical tail bodies in a cylindrical duct.

ing is omitted from an axial flow fan assembly or the fairing possesses an abrupt termination. The latter is associated with a conical outer casing for both in-line and exhaust fan applications. Provided the desired area ratio is achieved, an exhaust fan suffers no loss penalty as a result of the preceding installed centerbody type.

The data presented in Fig. 4.30 for radius ratios of 0.5 and 0.69 were obtained in relation to well-designed, rotor-straightener exhaust fan units ([4.29] and unpublished work). For design purposes a linear variation of effectiveness with radius ratio has been assumed. These data are limited to the configurations listed in Fig. 4.30 and apply to good-quality flow conditions. Adjustments for changed circumstances must be made on the basis of personal experience.

Information on the performance of annular diffusers with a sudden centerbody termination, on a plane within the enclosing outer cone, is not available for in-line fan arrangements. However, limited unpublished work [4.38] on an expanding centerbody exhaust diffuser of area ratio, 3, has demonstrated that body termination at 85% diffuser length leaves the performance unchanged from the full-length case, where the peak η_D was 0.93; reducing the body length to 67% incurred a η_D loss of 0.03. The inner and outer wall angles were 3° and 6°, respectively, and the radius ratio was 0.5.

A similar exhaust diffuser of lesser outer wall angle and a short converging conical centerbody, with various amounts of abrupt shortening of the latter, exhibited no significant performance changes up to the point at which 55% of the body length had been removed [4.29]. The test unit, illustrated in Fig. 4.31, possessed a peak η_D of approximately 0.90. Removal of the tail cone, for this particular radius ratio, resulted in a peak η_D reduction of approximately 0.08. The pressure recovery resulting from the expanding outer wall tends to reduce the importance of centerbody truncation when comparison is made with the η_D loss of 0.2 for radius ratio 0.5, as presented in Fig. 4.30.

Although the preceding test data are limited in scope, they provide some guidance for other design configurations. However, since the preceding figures relate to a moderately high Re,

$$\left[\frac{\bar{U}(R_1 - r_1)}{\nu} \simeq 3.5 \times 10^5\right]$$

and are for good-quality inlet flow conditions, amendments based on practical experience are required for other flow eventualities.

4.4.5 Vaned Diffusers

In designing short diffusers of moderately large area ratio, larger wall angles than those recommended previously are frequently sought.

The insertion of vanes into a wide-angled unit reduces the included angles of individual passages to predetermined values. Design guidance with respect to rectangular units is provided in [4.39] and [4.40], which supplement one another. Constructional difficulties have restricted the use of conical vanes in circular diffusers. One notable exception can be observed in injection-molded units employed as exhaust registers in air distribution systems. An alternative device consists of longitudinal plates arranged along diametrical planes [4.41, 4.42].

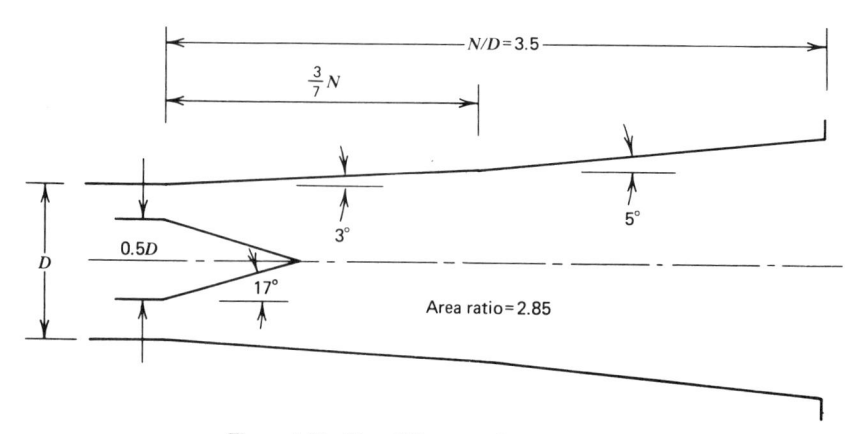

Figure 4.31 Test diffuser configuration [4.29].

4.4 DIFFUSERS

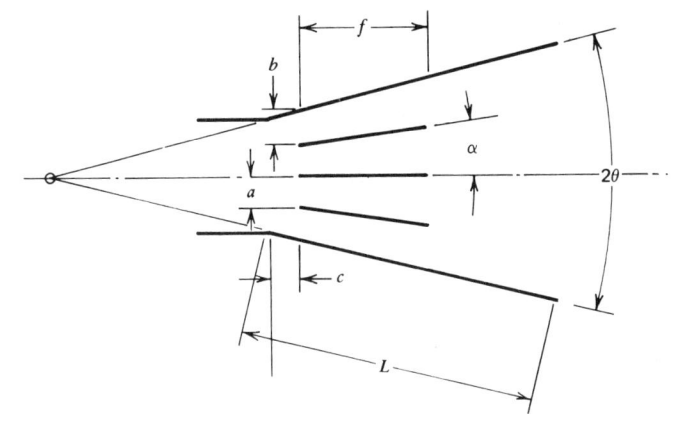

Figure 4.32 Two-dimensional vaned diffuser [4.39].

The vaned two-dimensional diffuser design recommendations of [4.39] and [4.40] are similar in regard to the selection of the vane numbers n and the vane length f (Figs. 4.32 and 4.33). The difference is related to the wall passage design. The experimental work of [4.39] established an optimum b/a of 1.2, which incidentally determines the length c. However, the diffuser flow is closer to critical conditions, and a lesser value of b/a is here advised.

Area Ratio	$\Delta\theta$
3.0	4.5°
3.5	6.0°
4.0	6.8°
5.0	7.5°

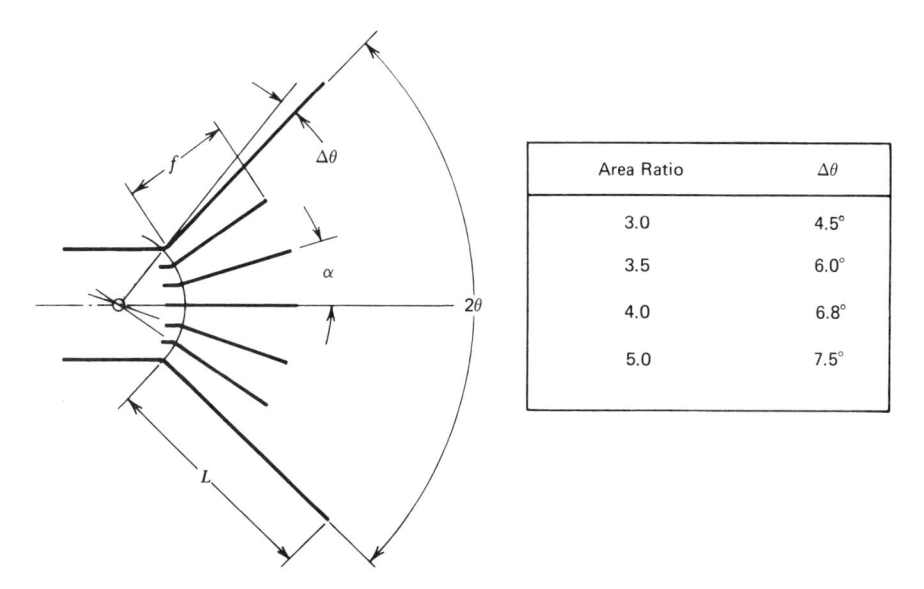

Figure 4.33 Two-dimensional vaned diffuser. (Reproduced with permission of ASME from O. G. Feil [4.40].)

When the wall passage included angle is marginally less than that of the vane passages, the mean crest of the flat-topped optimum curve is reached. The work of [4.40] highlights the need for the vane leading edges to approach the diffuser throat and for the wall passage angle to be less than that of the vane passages. The introduction of the parameter $\Delta\theta$ results in a translation of the virtual point source (Fig. 4.33), which makes the wall passage angle a function of area ratio. For the test range of divergence angles, 40 to 80°, no point source correlation with diffuser angle was apparent; however, with increasing inlet boundary layer thickness, a trend toward a large $\Delta\theta$ was apparent.

The vane divergence angle is given in [4.39] by

$$\alpha = \frac{2\theta}{n + 1} \qquad (4.23)$$

and in [4.40] by

$$\alpha = \frac{2(\theta + \Delta\theta)}{n + 1} \qquad (4.24)$$

The value of n is adjusted to keep θ within the limits, 3.5° and 5°, with the larger angle applying to short diffusers; the vanes should terminate at the diffuser outlet.

The vane length f is determined from the line of appreciable separation, given in Fig. 4.23. For the preceding angle limits, the ratio f/a (i.e., N/W_1) has values between 25 and 12, respectively. The best overall performance appears to be governed by this separation condition. A leading-edge extension that is curved to assume an axial direction in the throat is required, in the avoidance of local separation, when the diffuser divergence angles are large [4.40].

Diffuser effectiveness for optimum vane arrangements is presented in Fig. 4.34, for the exhaust configuration. It will be noted that thick inlet turbulent boundary layers, a normal feature of exhaust diffusers, severely limit the potential gains. However, a more uniform outlet flow is achieved in all instances.

Details of design and performance modifications for tailpipe addition are not available, although the differences from the exhaust case can be expected to increase with divergence angle. An experimental approach to any specific design requirement may therefore be advisable.

The good agreement between performances at the common point ($2\theta \simeq 40°$), for diffusers possessing significant differences in inlet geometry, suggests a relative insensitivity to such matters. Experience has shown that the vane leading edges should not be far from the throat, for flow stabilization, and should be tangential to the local flow direction.

In the case of circular wide-angle diffusers, outlet flow uniformity can be improved by inserting conical or radial vanes, or both. The latter is an

4.4 DIFFUSERS

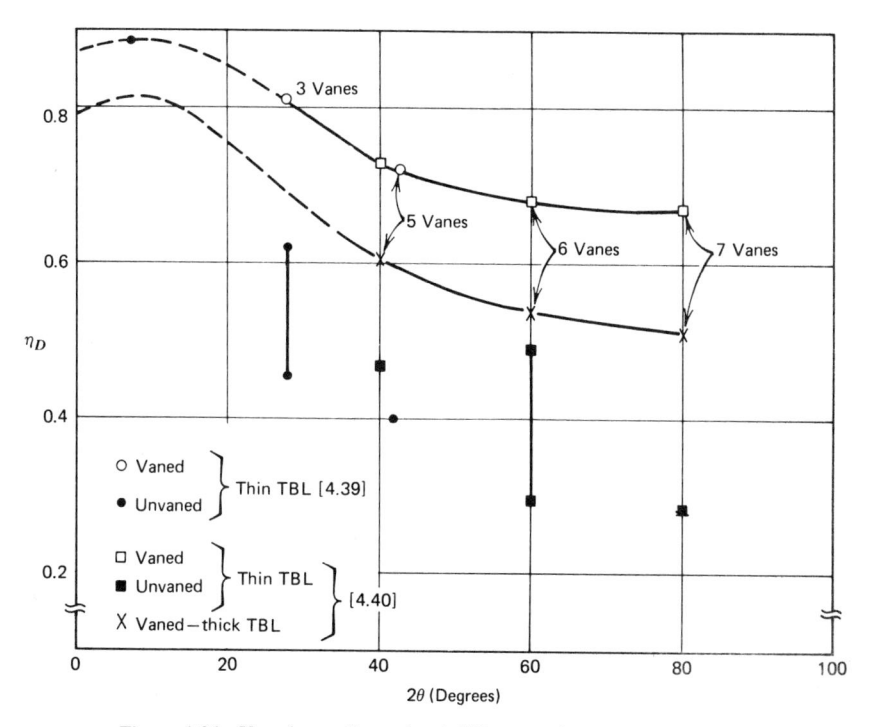

Figure 4.34 Vaned two-dimensional diffuser performance [4.39, 4.40].

extension of the "eggbox" type flow straightener principle, as used in cylindrical ducts. However, when poor-quality inlet flows are present, such a device restrains the airflow paths and does not permit the radial movements required for a more uniform outlet flow. Eight radial vanes of extended length, located on diametral planes, are more effective in both the cylindrical and conical duct arrangements.

The addition of a central flat disk in the transverse entry plane is required in the conical diffuser case [4.41]. This device promotes stable vortex flows in the apexes of the triangular passages; the main stream is then diverted toward the diffuser wall. The diameter ratio of the disk is a function of diffuser angle, area ratio, and inlet flow features; as the outlet diameter is increased, the central jet flow is eventually replaced by an annular jet located in the vicinity of the outer wall. Maximum pressure recovery [4.42] occurs when this annular jet is just beginning to form. This corresponds with the closing of the "tripped" vortices a short distance upstream from the diffuser termination. Experimental determination of disk size, keeping the preceding flow features in mind, is recommended as insufficient design data are available. The ease with which disk diameter can be changed makes this a very practical approach to the optimization exercise.

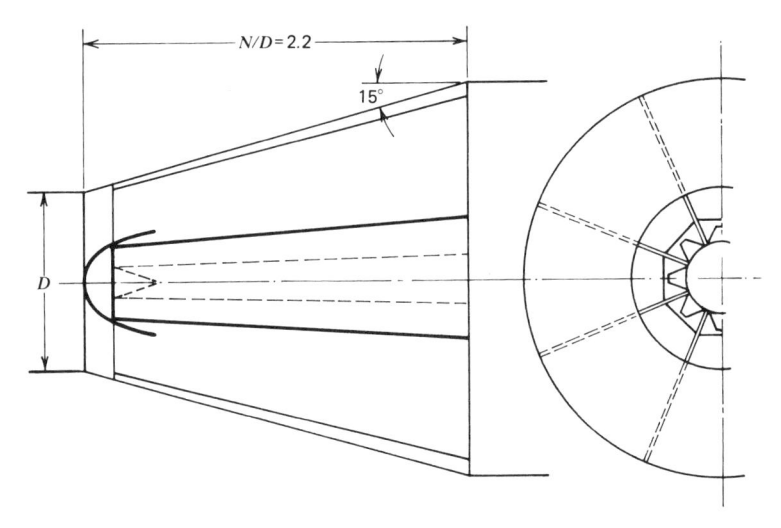

Figure 4.35 Diffuser with radial vanes, $2\theta = 30°$.

The effectiveness for an area ratio of 15, and divergence angles of 25, 38, and 50° [4.42], reaches a maximum when the vane leading edges are displaced downstream of the diffuser inlet by a distance of from 5 to 6% of diffuser length. The effectiveness (η_D) for each of these angles is then approximately 0.73, 0.61, and 0.42 respectively; these values apply to thin inlet boundary layer conditions and full-length vanes. The pressure recovery is measured at a downstream distance of $D_t/4$, where D_t is the tailpipe diameter. A feature of this device is the rapid return to stable boundary layer conditions.

A reduction in vane length, to two-thirds of diffuser length, has been successfully used in the application described in [4.41]; no performance details are available. Obviously there is still considerable scope for gaining additional design knowledge about this useful and fundamentally sound device. Any device that improves the outlet flow uniformity will improve system loss prediction with respect to downstream duct components.

Unpublished test work by the author has successfully demonstrated a useful development of the foregoing principle. The unit illustrated in Fig. 4.35 features a nose fairing on the "trigger" disk and tapered "fingers" extending from the disk into the eight segmental flow passages. In addition, a flat-sided centerbody has been introduced.

The purpose of this development is to replace the vorticity about a transverse axis by one in the stream direction. The fingers act as vortex generators of the counterrotating type, the vortices being attached to the inner wall. The rounded nose dome ensures low inlet losses and hence the maximum vortex strength. The fingers also act as deflectors, directing the flow toward the outer wall.

4.4 DIFFUSERS

In selecting the nose dome diameter and the finger geometry, the major factor to be considered is the scale of the desired vorticity. The latter must be large enough to maintain good turbulent mixing along the centerbody walls. Large fingers require a substantial nose dome diameter, which in turn increases the actual area ratio of the diffuser. A compromise usually has to be made on an experimental basis.

The wide-angle unit illustrated in Fig. 4.35, which was developed in connection with a gas turbine silencer assembly, provided a low loss device with good outflow properties, being free of low-frequency "woofing."

4.4.6 Resistance Control of Flow

By insertion of a wire screen or rod grid across a diffuser cross-section, the pressure drop associated with the resistance tends to counter the static pressure rise due to diffusion; separation is therefore avoided. However, when pressure recovery is a prime requirement, this device is counterproductive. Hence its main field of application is with respect to wind tunnels and similar facilities where flow uniformity rather than pressure recovery is a prime concern. Maintenance problems include the need for regular cleaning and replacement following inadvertent physical damage.

The transverse location of the first screen should approximately coincide with the onset of separation in the no-screen case. The need for additional screens will depend on the divergence angle and on the resistance of individual screens; when carefully designed according to [4.43] and [4.44] a reasonable economy with respect to added resistance can be attained. Best performance and effectiveness is achieved with screens possessing a K of approximately 2. Design recommendations on screen arrangements are provided in [4.44].

4.4.7 Boundary Layer Control Devices

The use of porous surfaces, or wall slots, through which the surface layer is removed, is a very effective method of flow control, as illustrated in Fig. 4.36. Alternatively, the boundary layer can be reenergized by a high-velocity two-dimensional air jet issuing from a tangential surface slot, with the so-called Coanda effect. However, because of the need for an auxiliary power source, and regular maintenance, these control methods are not generally favored by industry, being reserved for use in special circumstances.

The augmentation of boundary layer mixing by artificial means is a well-developed form of flow control. The devices are called "passive," as opposed to the "active" ones just outlined. The use of regular, and irregular, surface discontinuities and roughnesses constitutes a relatively crude control method; details are available in [4.45] and [4.46].

102 DUCT COMPONENT DESIGN AND LOSSES

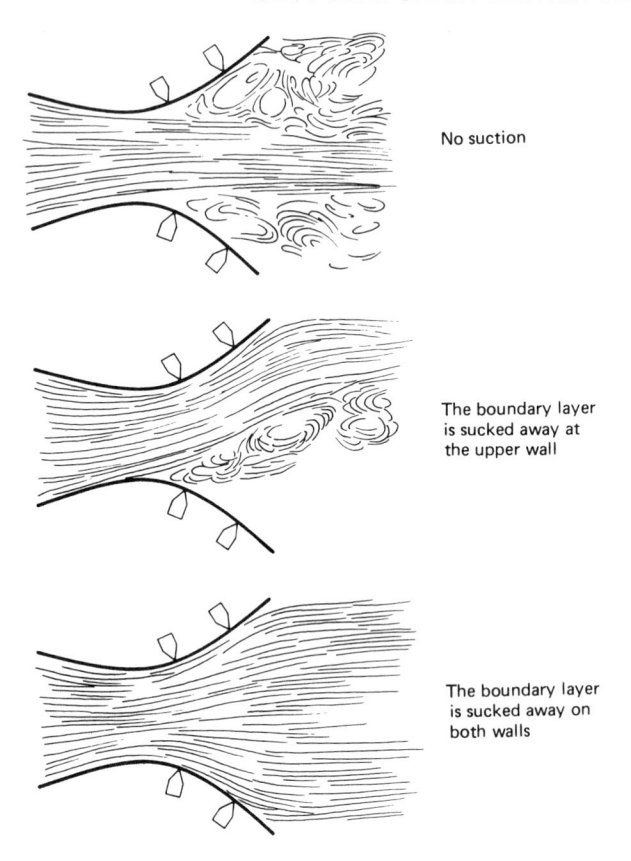

No suction

The boundary layer is sucked away at the upper wall

The boundary layer is sucked away on both walls

Figure 4.36 Diffuser flow patterns, with suction. (Sketch of photograph by Tietjens.)

Experimental studies of simple devices that combine flow deflection, due to blockage, with a measure of fluid mixing are described in [4.47]. The most promising of these consists of radially disposed circular rods attached to a central hub for the circular diffuser case [4.48]. The increasing percentage blockage toward the axis is responsible for an outward stream deflection, similar to that produced by the "trigger" plate of [4.42]. The vortices, or eddies, shed from the rod tips are believed to stimulate the momentum transfer mechanism.

An independent experimental study [4.49], aimed at obtaining a working knowledge of the device was conducted for identical values of area ratio and divergence angle, namely, 4 and 22°, respectively. The work of [4.47] was extended to obtain a measure of the effect of inlet boundary layer thickness on rod geometry and diffuser performance. With a thin inlet boundary layer and for the rod arrangement illustrated in Fig. 4.37, a diffuser effectiveness of

4.4 DIFFUSERS

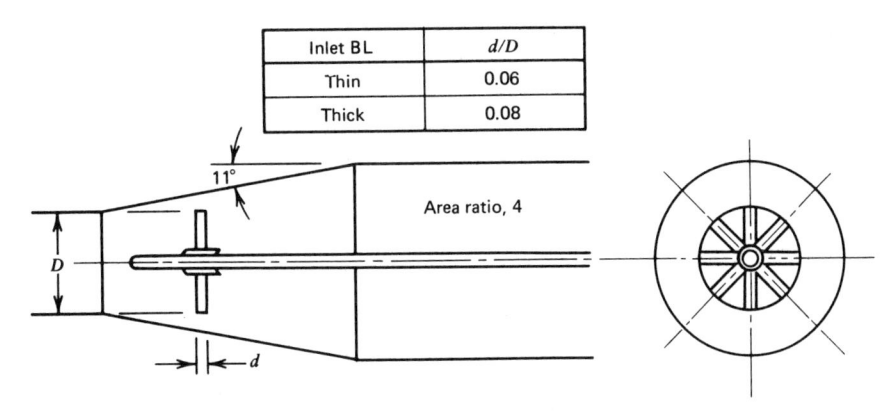

Inlet BL	d/D
Thin	0.06
Thick	0.08

Figure 4.37 Diffuser with flow control rods [4.49].

0.84 is obtained, which is slightly lower than that recorded in [4.48]; however, a greater blockage ratio, located further downstream, is required in providing this peak performance. With an increasing inlet boundary layer thickness, larger-diameter rods moved forward are necessary to retain a reasonable degree of effectiveness; the modified arrangement of [4.49], which yielded a value of 0.69, need not necessarily represent the optimum as insufficient arrangements were tested. The increased blockage can be related to the greater flow deflection required; the same argument would apply to larger diffuser angles.

The slower-moving inner core at diffuser outlet accelerates quickly to return the velocity profile to normal; pressure recovery is complete within two tailpipe diameters, in both test cases [4.49].

Tests [4.47] on a 30° divergence angle conical diffuser, of area ratio 6, have demonstrated the effectiveness of the device for thin inlet boundary layers.

Asymmetric inlet boundary layer conditions can be expected to require an uneven blockage arrangement for optimum performance. In view of the preceding limited findings an experimental approach to any given set of conditions is strongly recommended on an ad hoc basis. Diffusers of any cross-sectional shape will respond to blockage devices that achieve the preceding general objectives. The optimum longitudinal location of the "star" is not critical, being controlled by the balance between increased resistance and reduced effectiveness for forward and aft movements respectively. The star position for maximum flow stability is slightly upstream of that for best pressure recovery.

The application of the preceding flow-deflecting and mixing device to annular diffusers is reported in [4.50]. In this instance the study was restricted to expanding-centerbody, exhaust fan diffusers. The outward flow-deflecting action of the centerbody removes some of the potential gains from the blockage device. However, with an outer wall angle of 20°, eight radially orientated rods increased the pressure recovery coefficient by approximately 0.1, with improved flow stability.

The latter flow feature is also present for a wall angle of 15°, for which the peak η_D is 0.82.

Multiple stars can be used in a similar manner to resistance screens in controlling flow separation, with reduced loss penalties [4.48]. An added advantage is the avoidance of a major cleaning problem.

The author is aware of a number of successful applications of the star device in respect to remedial measures on diffusers that initially failed to possess satisfactory outflow properties.

Vortex generators provide a more organized and efficient form of boundary layer mixing than that provided by large-surface roughness. The initial use for the device was in controlling diffuser flow separation in a wind tunnel operated by United Aircraft [4.51]. The device was extensively studied and reported in [4.52] in which a preference for corotating rather than the original counterrotating vortices of [4.51] is expressed; various other systems used in vortex production are also discussed.

Counterrotating vortices scrub the low-energy inner boundary layer air into a common region adjacent to the surface. As the vortices travel downstream the size of this low-energy region expands until it underlies the vortices, forcing them off the surface. Therefore the device has limited downstream persistence and effectiveness against prolonged adverse pressure gradients. On the other hand, a corotating vortex scrubs the surface air toward the higher-energy side of the adjacent vortex (Fig. 4.38); the subsequent mixing process feeds dynamic energy into the surface layer, thus delaying separation. The vortices tend to stay near the surface until dissipated by viscosity. When the surface possesses longitudinal convexity, dynamic forces will progressively reduce vortex strength [4.23].

A major feature of turbulent boundary layer mixing is the existence of randomly appearing longitudinal vorticity on a scale matching the boundary layer thickness. As the initial vortices are dissipated, larger ones are appearing that match the boundary layer growth. Vortex generation is believed to augment these natural processes. In designing an appropriate array of generators the preceding fluid mechanism should be kept in mind. Design difficulties with respect to diffusers are at once apparent. For thin inlet boundary layers, the scale of introduced vorticity should be of the same order as the boundary layer thickness in the region of impending flow separation. However, the problems of dealing with thick inlet layers are self-evident when attempting to apply the foregoing rule. A fewer number of longer vanes, presumably of greater aspect ratio, will introduce a general swirl into the airstream, in addition to the individual vortices arising from the vane tips. The scale of this latter vorticity, because of the practical necessity of having to function within the boundary layer at the impending flow separation location, is smaller than the optimum for two-dimensional flow. In other words,

4.4 DIFFUSERS

(a) Corotating vortices   (b) Contrarotating vortices

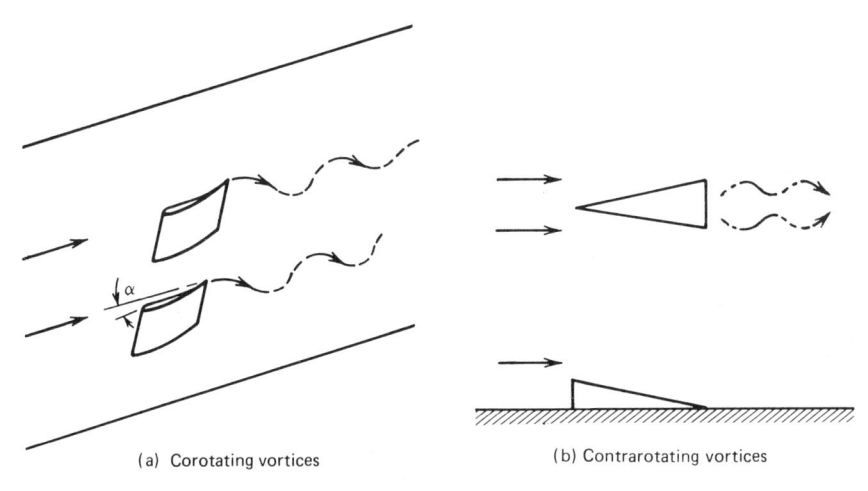

Figure 4.38 Vortex-generating devices.

the device is poorly matched to wide-angled diffusers with thick inlet boundary layers.

An experimental study [4.53] of the device on conical diffusers with divergence angles of 8, 12, 16, 20, and 30°, all of area ratio 4, showed the retention of good effectiveness and the maintenance of unseparated flow conditions up to a divergence angle of 16°, for thin inlet boundary layers. A threefold inlet boundary layer momentum thickness increase, however, resulted in a C_p loss in excess of 0.2 for the 16° diffuser fitted with vortex generators.

The preceding two methods of flow control on an effectiveness (η_D) basis, at the same area ratio (4), are compared in Table 4.3.

The author has successfully used wedges of the type illustrated in Fig. 4.38b to solve a flow separation problem in a short wide-angle annular diffuser developed for a specific practical purpose. The wedges were attached to an expanding centerbody surface creating a restriction in duct area which

Table 4.3 Diffuser Performance for Various Control Devices

Divergence Angle	Control Device	Effectiveness (η_D)	Reference
16°	VG	0.85 (0.61)[a]	[4.53]
20°	VG	0.70 (0.48)	[4.53]
22°	STAR	0.86	[4.48]
22°	STAR	0.84 (0.69)	[4.49]

[a] Effectiveness values in parentheses are for the thickened inlet boundary layer cases.

resulted in an outward flow deflection. The vortices control the inner wall flow.

4.4.8 Effect of Swirl

The introduction of a general rotational motion, of an arbitrary vortex nature [Eq. (2.32)], into the flow upstream of a conical diffuser inlet can improve the performance, as verified by many workers. A recent contribution [4.54] reports the effect of "solid body" rotation ($a = 0$, Eq. 2.32) for divergence angles and maximum swirl angles up to 30° and 15°, respectively (area ratio = 4). Fluid rotation of this type will be accompanied by an outward radial flow due to an imbalance with the radial pressure force distribution. This outflow increases the axial velocity in the vicinity of the wall (mass conservation) and must be a beneficial feature. The benefits persist up to both maximum wall and swirl angles but these are not large. The losses associated with the device for introducing the prescribed swirl, and the subsequent swirl energy loss, are difficult to establish with accuracy. The probable adverse effect of swirl on downstream duct component losses must also be taken into design consideration. Many theories have been put forward with respect to the flow mechanisms responsible for the beneficial influence of swirl but the question still awaits resolution. This type of orderly swirl should not be confused with the random large-scale turbulence resulting from flow separations.

The generally accepted rule is that swirl angles up to 10° can be beneficial. In practice, swirl is normally identified with fan and cyclone separation equipment and is often in excess of the preceding figure. Axial flow fans that possess no stator vanes, or inadequately designed ones, usually have a residual swirl in excess of 10°. Depending on the rotor design the swirl can be of either the free or arbitrary vortex type, but seldom of the "solid body" variety studied in [4.54]. Local or complete flow separation (fan stall) from the rotor blades will result in very large degrees of swirl.

In low-pressure exhaust fans some design advantage can be taken of the preceding swirl benefits. A greater equivalent angle for the diffuser, based on swirl considerations, can ensure a good exhaust fan installation efficiency with a short and cost-effective diffuser. The swirl angle will reduce toward the diffuser outlet, because of angular momentum conservation. The design swirl for this class of low-pressure fan, as employed in large, induced-draft cooling towers, will possibly possess a spanwise swirl variation from 5 to 20°. When a fan on a vertical axis is mounted on top of a bevel gear or electric motor system, axially aligned and faired support struts will provide minimum inlet flow disturbances. Downstream support systems can create flow problems. The foregoing is illustrated in Fig. 4.39.

Hence good practice will restrict the use of swirl, as a design aid, to suitably supported rotor-only low-pressure exhaust fans.

The swirl downstream of a cyclone separator will in normal circum-

4.5 CORNERS 107

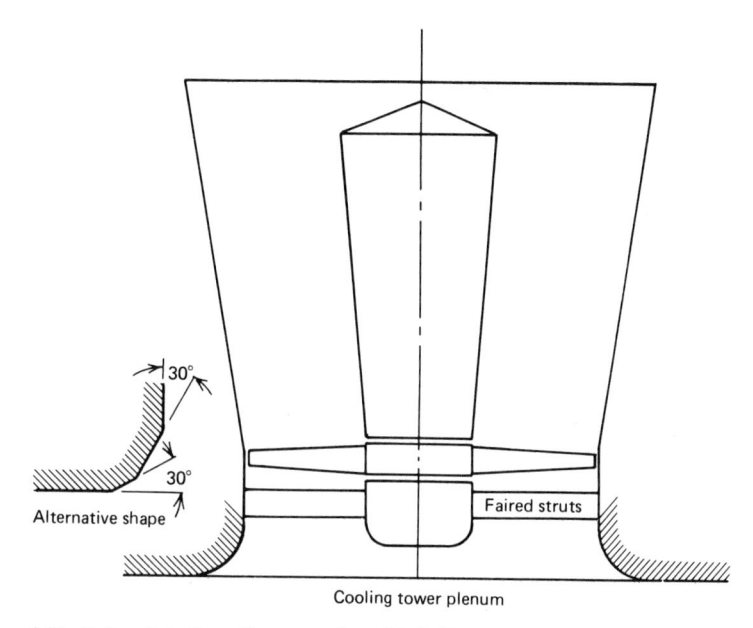

Figure 4.39 Induced-draft cooling tower fan with 50% pressure recovery diffuser ($C_p \approx 0.5$).

stances be greatly in excess of 10° and hence a downstream diffuser will probably produce a penalty rather than a benefit. However, a swirl-removing unit consisting of a centerbody, to reduce the swirl angles, fitted with pressure recovery stator vanes deserves design consideration. Axial velocity increases and angular momentum conservation considerations together result in swirl angle reductions. The vanes could be designed to provide a small residual swirl if this were required to improve downstream diffuser performance.

4.5 CORNERS

The variables defining the common bend are shown in Fig. 4.40. Practical difficulties in constructing units with circular cross-section have led to the widespread acceptance of the "lobster back" variant (Fig. 4.41). For small turning angles the mitered joint often provides a satisfactory solution, but as the angle increases the losses rise rapidly.

The losses associated with the foregoing types continue to be felt for considerable distances downstream of the bend termination. At two outlet duct diameters from the corner discharge the transverse static pressure gradient has virtually disappeared but the secondary flows (Section 2.7) continue to disturb the straight pipe flow for a distance of up to 30 pipe diameters. At this point the slope of the static pressure versus distance curve, for fully developed pipe flow, is reestablished. The additional duct loss that

108 DUCT COMPONENT DESIGN AND LOSSES

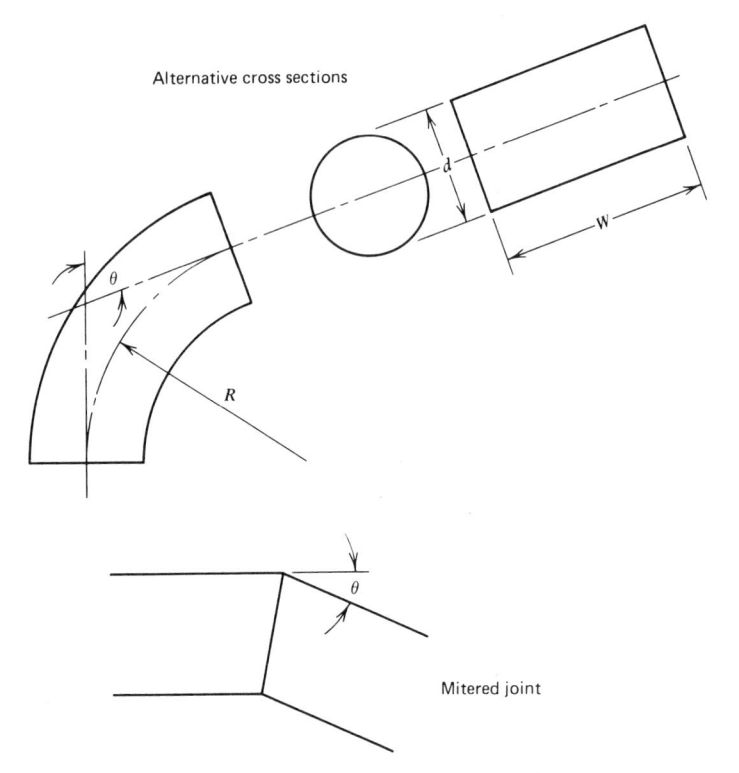

Figure 4.40 Definition of corner variables.

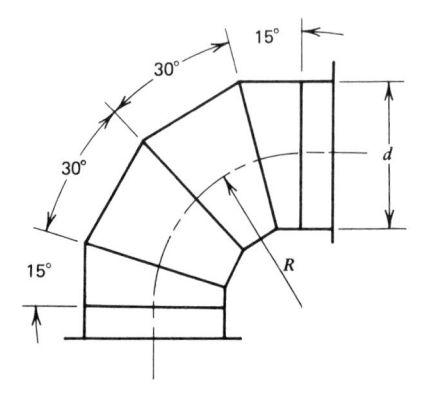

Figure 4.41 Four-piece elbow.

4.5 CORNERS

accrues from the corner outlet flow is termed the *outlet tangent loss*. When substantial flow separation is present in the bend, the secondary flows dissipate much earlier, reducing the duct length in which the outlet tangent losses are present.

The dual objectives of low loss and compactness are achieved when turning vanes are fitted. The greater number of smaller-scale secondary flows (Fig. 2.11) produces accelerated mixing that favors the rapid completion of the loss process.

Dividing the corner into separate passages by the use of splitters augments the radius ratio R/d, and hence units of greater compactness can be designed without incurring unacceptable losses.

Corners of rectilinear cross section are frequently of either a contracting or expanding variety. Provided the inner and outer walls are suitably shaped, the former presents few design problems. However, the latter requires special attention and, for large expansion ratios, vanes that combine flow splitting and turning are recommended. The flow that issues as a series of jets interspersed by separated flow regions undergoes a mixing process, resulting in a more uniform situation. The addition of screens will improve the downstream flow quality, provided blockage due to particulate matter is avoided.

4.5.1 Unvaned Circular Arc Corners

The "potential" or nonviscous flow around a constant area bend is illustrated in Fig. 4.42. Flow decelerations are present on both the inside and outside boundaries and these cause flow separations in a real viscous fluid (Fig. 4.43). Increasing the inside radius decreases the deceleration rate and eliminates separation; the same applies to the outer boundary (Fig. 4.43). The expanded bend will possess the lesser loss. When the R/d of the normal bend is sufficiently large, a minimum loss is attained. The outlet tangent loss will, however, constitute an appreciable percentage of the gross loss, because of secondary flow influences. The strength of the latter will be related to inlet flow properties. The boundary layer associated vorticity can minimize the extent of separation from the inner boundary on which, because of convex curvature, the production of turbulence is reduced.

In view of the preceding variables it is not surprising that large variations in published loss data exist. A satisfactory degree of correlation, however, was obtained in [4.55] when the gross loss was recorded with respect to interference-free inlet flows and smooth surfaces. The losses for square and circular ducts of the same R/d were also shown to possess similar magnitudes.

Taking the general case of a compact corner with flow separation on the inner wall (Fig. 4.44), corner loss is normally established by measuring static

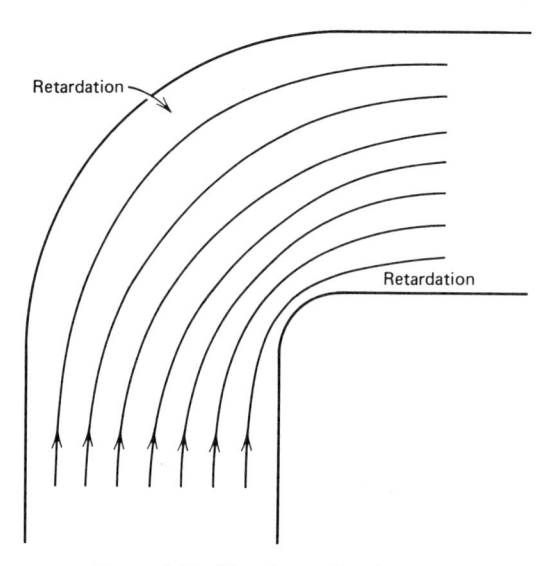

Figure 4.42 Nonviscous flow in corner.

pressures at stations A and B, in ducts of equal area. The downstream static pressure is then corrected back to the outlet flange by subtracting the pressure drop in an identical length (L/d) of interference-free straight ducting, often measured downstream of the region influenced by corner flow. For ducts of unequal area, the total pressure loss of the corner is obtained by taking the mean dynamic pressures in each duct into account.

When the corner discharges directly or through a short length of duct into a large volumetric space, the lack of a constant static pressure at the discharge plane makes the choice of ambient static pressure p_o at discharge an obvious reference one. However, when separation is present, e.g., at station B' (Fig. 4.44), the actual dynamic pressure exceeds the value calculated from outlet duct area and hence the resulting measurement does not represent the total pressure loss for the bend. Nevertheless, this differential can be converted readily to a combined corner/discharge loss by subtracting the final outlet total pressure (p_o) from the inlet value, namely,

$$\Delta H_c = p_A + \tfrac{1}{2}\rho V_A^2 - p_o \qquad (4.25)$$

or

$$K_c = 1 + \frac{(p_A - p_o)}{\tfrac{1}{2}\rho V_A^2} \qquad (4.26)$$

which is a common requirement for loss determination. This relationship is independent of inlet and outlet area equality.

4.5 CORNERS

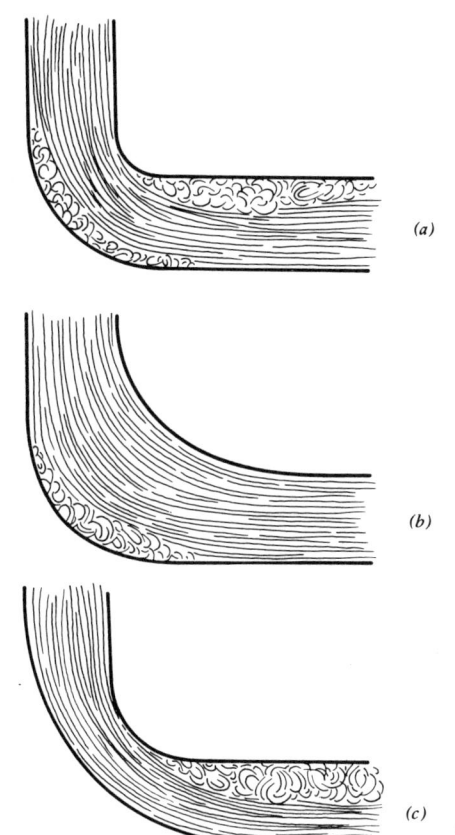

Figure 4.43 Effect of varying corner shape in rectangular ducting. (a) Normal bend. (b) Expanded bend. (c) Contracted bend. (Sketch of photograph by Nippert).

Corner losses determined in the preceding manner may possess a potential loss component due to the excess dynamic pressure resulting from flow separation; this reduces the effective flow area. This potential loss becomes an actual one for a discharge corner, but when downstream ducting is added a measure of static pressure regain occurs.

The general expression for corner loss can be expressed as

$$K_c = K_G C_{(L/d)} C_{R_d} C_{\text{rough}} C_{AR} \tag{4.27}$$

where K_G is the gross corner loss and the subsequent coefficients represent corrections for duct length, R_d, surface roughness, and rectangular duct aspect ratio, respectively.

The loss data presented here are from [4.9]. The results for the circular duct of constant cross-sectional area are given in Fig. 4.45 for an R_d of 1×10^6; the losses for square ducts are for all practical purposes the same. These

112　　　　　　　　　　　　　　　　　　DUCT COMPONENT DESIGN AND LOSSES

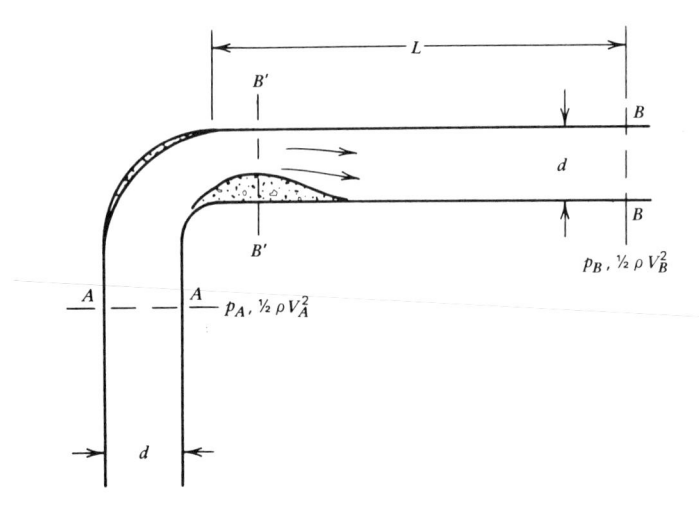

Figure 4.44 Corner loss determination.

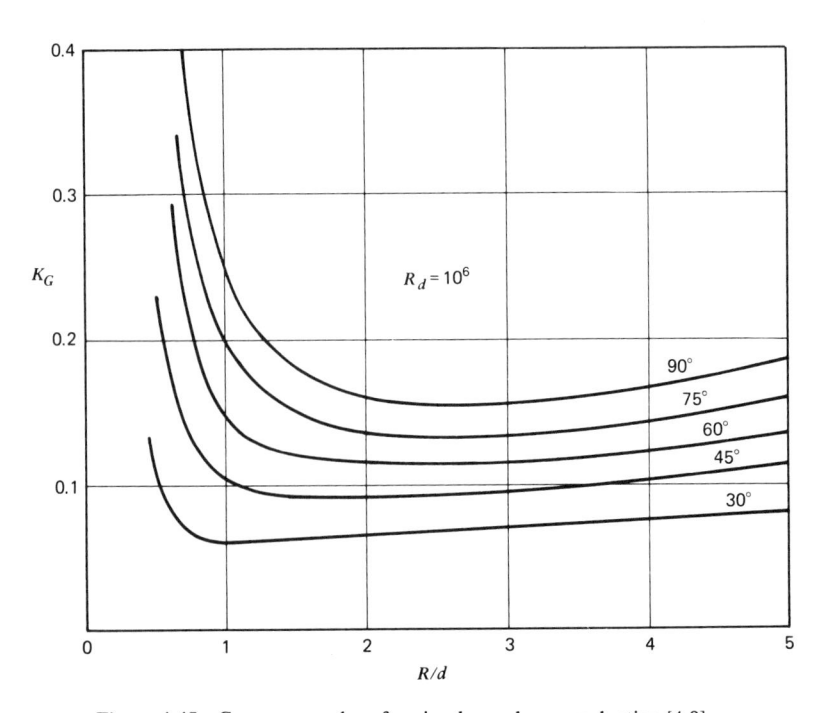

Figure 4.45 Gross corner loss for circular and square ducting [4.9].

4.5 CORNERS

gross losses, which include the interference loss in the downstream ducting, that is, the outlet tangent loss, can be corrected back to any given length of outlet duct by means of Fig. 4.46 and Eq. (4.27). For small K_G values the curve for 0.10 can be used without significant error. The correction required for high loss bends, with short outlet ducts, reflects the potential dynamic loss component discussed previously.

Corrections for R_d and surface roughness are also necessary. The curve presented in Fig. 4.47 is recommended by [4.56] for R/d greater than 1.5 whereas [4.9] gives 2.0 as the lower R/d limit. Below these limits and at the higher Reynolds numbers loss is assumed independent of R_d, thus suggesting a flow situation similar to that pertaining for fully rough pipe flow conditions. According to [4.9] the value of R_d at which this independence commences is lessened progressively with reducing R/d; this is obviously due to the growth and dominance of flow separation.

Since the Reynolds number is based on duct diameter and not inlet boundary layer thickness, the R_d correction will show the greatest consistency for fully developed inlet flows. Thin inlet boundary layers give rise to increased losses for a 90° bend of $R/d = 2$ when $L/d = 0$. The lesser turbulent mixing associated with reduced secondary flow strength leads to increased flow separation. The foregoing loss data are typical for relatively thick inlet boundary layers.

The R_d correction curve of both [4.9] and [4.56] follows the relationship $K_c \propto R_d^{-1/6}$, which indicates a reduced sensitivity to that for pipe flow where the index is either $-1/4$ or $-1/5$; the former applies below an R_d of 10^5 [Eqs. (3.58) (3.59)]. The growth of flow separation as R/d is reduced will entail a progressive trend toward R_d independence, for example, an index of $-(1/x)$ for an intermediate R/d value. This is different from the recommendation of

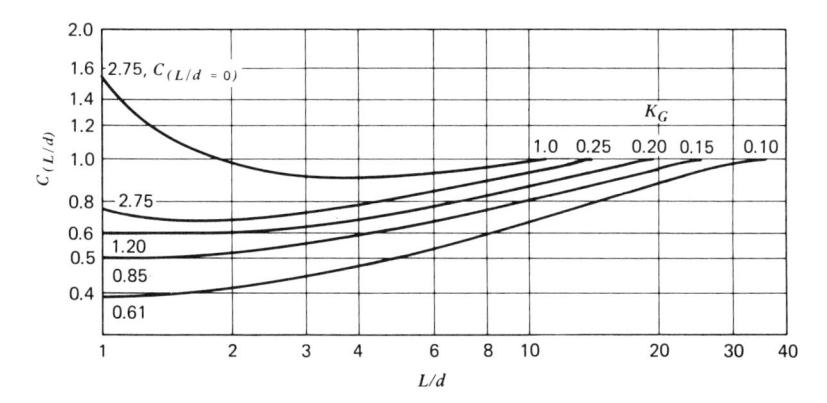

Figure 4.46 Outlet tangent loss correction factor for circular and square ducting. (Reproduced with permission from D. S. Miller, *Internal Flow Systems*, BHRA Fluid Engineering, Publishers, Bedford, U.K., from whom the book is obtainable.)

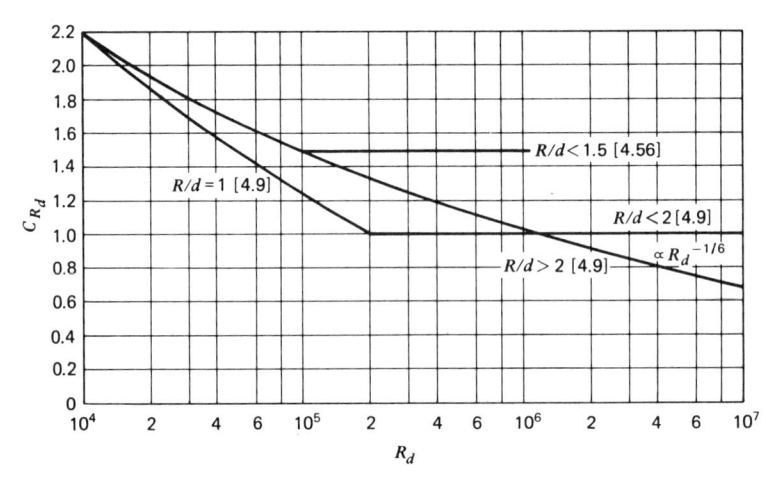

Figure 4.47 Reynolds number correction factor for corner loss determination. (Reproduced with permission from D. S. Miller, *Internal Flow Systems,* BHRA Fluid Engineering, Publishers, Bedford, U.K., from whom the book is obtainable.)

[4.9], which indicates a discontinuity in C_{R_d} for $R/d \approx 2$ at high R_d. Taking the argument further the curves for the lower R/d values will approach the unity line smoothly and at a small angle; the curve for each R/d value will have a different point of contact. The preceding uncertainties can be avoided by the design selection of the optimum R/d for the given turning angle.

The correction for surface sand roughness (excluding manufacturing imperfections) recommended in [4.9] is the ratio between rough and smooth friction factors as given by Fig. 3.8, for the bend inlet R_d. Since the R_d dependence is maintained, this correction procedure is preferable to that advocated in [4.56], where the correction factor for given roughness ratios is a function of R/d only.

An aspect ratio correction for rectangular ducts is suggested in [4.55] for loss coefficients; the curve presented in Fig. 4.48 is extracted from [4.56]. Superimposed are mean C_{AR} values for two aspect ratios when the data of [4.9] for the optimum R/d cases displayed on Fig. 4.49 are employed; for lesser R/d values the preceding relationship is far from reliable.

The procedure outlined above for outlet duct length correction, using Fig. 4.46, remains unchanged in the rectangular corner case, with few exceptions. One of these relates to the case of increasing aspect ratio where the outlet tangent loss is reduced because of declining spheres of secondary flow influence. However, acceptance of the recommendations of [4.9] will result in loss prediction discontinuities of doubtful accuracy.

The contribution of the corner flow to extra losses in the downstream length of straight ducting can be considered an interference loss. For an installation consisting of a bend followed by a short length of straight ducting joined to a diffuser, the reduced corner loss must be weighed against the

4.5 CORNERS

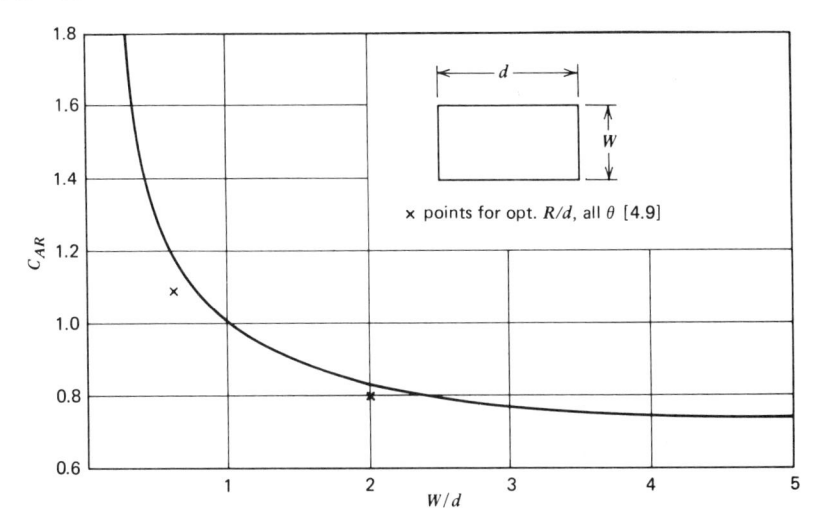

Figure 4.48 Aspect ratio correction factor for rectangular duct corners. (Adapted with permission of Pergamon Press from E. Ower and R. C. Pankhurst [4.16].)

corner flow influence on diffuser losses. In other words, we have substituted one interference with another of greater uncertainty.

From the information presented previously, the design, and loss predictions, of near-optimum constant area bends of the circular arc type can be undertaken with confidence. When special circumstances dictate corners of unusual and nonoptimum design, a special test program may be required to attain loss estimates of acceptable accuracy. Asymmetric and/or swirling inlet flows will modify the foregoing loss estimates; for obvious reasons no published data are available and hence test work is essential when prediction accuracy is required.

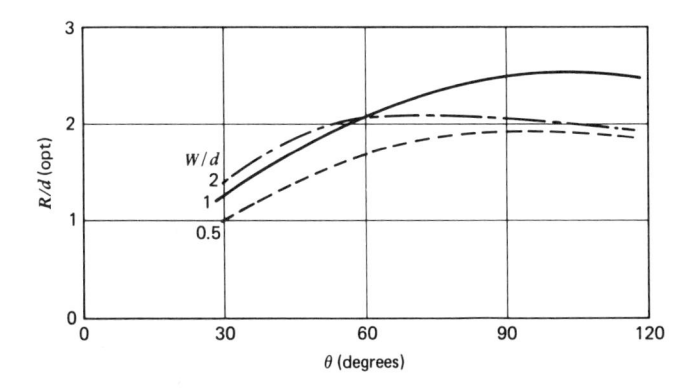

Figure 4.49 Optimum radius ratios for rectilinear ducts at various turning angles [4.9].

116   DUCT COMPONENT DESIGN AND LOSSES

4.5.2 Unvaned Corners with Specified Velocity Distributions

Rectilinear corners of the type illustrated in Fig. 4.43b can be designed with zero static pressure gradients on the inside wall [4.57]. Separation from this surface is thereby eliminated with reduced-loss benefits. However, because of skin friction and secondary flows the resultant loss is not substantially below that relating to corners of optimum R/d design.

4.5.3 Mitered Elbows

With the exception of small turning angles, relatively gross losses occur with this type of elbow, particularly for $L/d = 0$. Test data for such are usually for ducts of circular cross section, the type gaining most from manufacturing simplicity. The loss information presented in Fig. 4.50 is from [4.9], where its use is also considered approximately valid for ducts of rectangular cross section. However, data published in [4.58] indicate higher losses for 90° square duct elbows; despite a poor aerodynamic performance this bend type possesses some beneficial acoustic properties in reducing noise propagation.

Corrections for L/d are made with the aid of Fig. 4.46, where the explanations forwarded in Section 4.5.1 apply. However, a correction for surface roughness does not appear to be applicable to this type of elbow, because of the presence of separated rather than boundary layer flow conditions.

Loss independence of Reynolds number for the range, 0.5×10^5 to 5×10^5, is reported in [4.59], while [4.9] suggests corrections should be applied below 2×10^5. However, in view of arguments presented in Section 4.5.1, an R_d correction factor appears inappropriate.

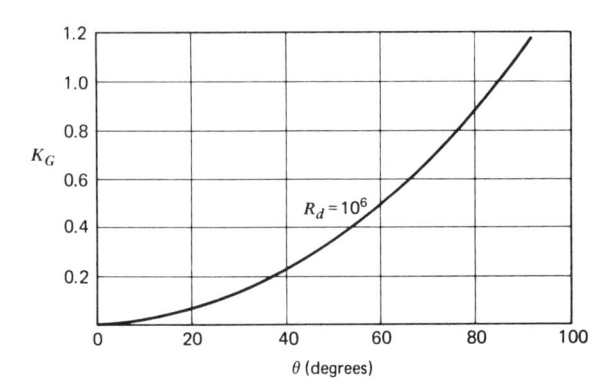

Figure 4.50 Gross loss coefficient versus turning angle for mitered circular and rectilinear ducts [4.9].

4.5.4 Lobster-Back Corners

When compactness is not a vital factor, a carefully manufactured multimitered bend provides a relatively low-loss design solution. In addition to data from [4.9], the results of [4.59] for $L/d = 3$ have been corrected for R_d and L/d in accordance with Fig. 4.46; these are presented in Fig. 4.51 for the 90° case. The corresponding curve for a smooth circular arc bend is superimposed for comparison.

The tests of [4.59] indicate R_d indices of $-(1/4)$, $-(1/25)$, and zero for R/d values of 3, 1, and 0.75, respectively. Since the tests were conducted over the R_d range, 0.46×10^5 to 5.2×10^5, the influence of the lower R_d is obviously present in the index for $R/d = 3$ [see Eq. (3.59)].

The optimum R/d for 90° turning is understandably slightly higher when compared to a circular arc unit.

The recommended surface roughness corrections of Section 4.5.1 are applicable in the present instance.

This design solution is exclusively related to ducts of circular cross section. The centerline dimension of the element between adjacent miters is given for a 90° corner by

$$a = \frac{2R}{\cot \frac{90°}{2n}} \quad (4.28)$$

where n is the number of mitered joints.

4.5.5 Bends with Splitter Vanes

This solution is usually reserved for compact, right-angle bends of rectilinear cross section. Two vanes provide an adequate low-loss solution for R/d values as low as 0.75; the recommended placement of [4.56] is given in Fig. 4.52.

In view of the lack of outlet tangent loss data it is recommended that (1) the loss coefficient for $R/d = 0.75$ and above be equated to K_G for circular arc corners of optimum R/d, (2) K_c be equated to K_G, (3) normal R_d and roughness corrections apply, and (4) the aspect ratio correction be unity.

The division of the bend into three passages has increased the number of secondary flows by a factor of 3, improved the individual radius ratios, and resulted in enlarged aspect ratios. The downstream tangent losses will be minimal as a consequence of the reduced strength of the individual secondary flows.

There are a number of gains and some penalties with respect to losses on a minimal-sized corner of this type. Experimental data suggest that these tend to balance one another. Hence for all practical purposes the previously suggested design approach will provide satisfaction.

118   DUCT COMPONENT DESIGN AND LOSSES

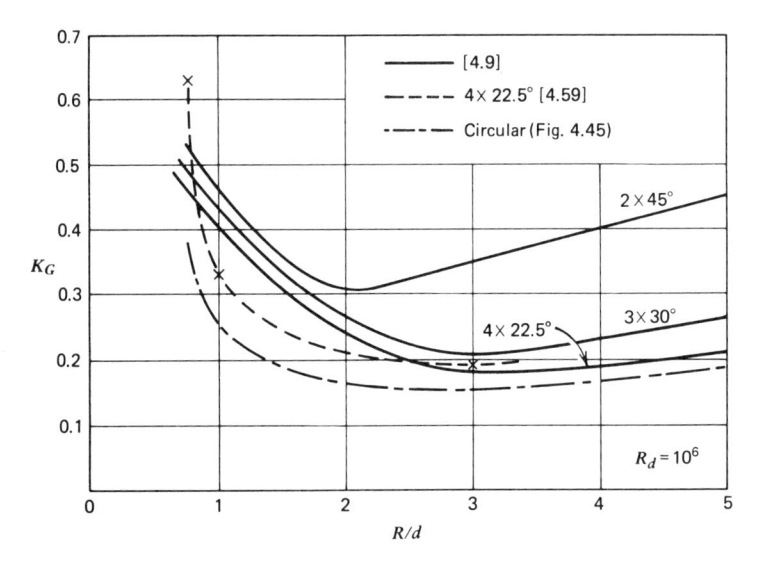

Figure 4.51 Loss in "lobster-back" right-angled corners. (Reproduced with permission from D. S. Miller, *Internal Flow Systems*, BHRA Fluid Engineering, Publishers, Bedford, U.K., from whom the book is obtainable.)

When asymmetric and swirling inlet flows are unavoidably present, the splitter arrangement of Fig. 4.52 may not provide the best solution. In many instances design acceptability may require experimentation and/or actual practical experience. Division of the flow ahead of the corner will necessitate a significant length of vane inlet tangent.

Many industrial installations feature a right-angled corner around which the flow is suddenly expanded into a working chamber. Without vanes the flow is strongest along the chamber floor and reversed along the ceiling. A greatly improved flow distribution in the chamber is attained when splitting/turning vanes are installed, as illustrated in Fig. 4.53. The limited width of the inlet airway will restrict the number of vanes as these must extend into the throat to achieve the best split of flow.

Flow separation will occur on the inside of each separate passage, with the extent varying with the geometric diffusion ratio and vane trailing edge angle. For instance, it may be preferable to seek a lesser turning angle than 90° at the vane exit, since the prospect of attaining a complete angular deflection of the airflow at this station is impossible for large area ratios. The object is to position the limited number of vanes in order to achieve early turbulent mixing within the chamber. The placement of grid-type resistance downstream of the vanes will hasten the process when a certain standard of flow quality is desired; the best results are obtained with grid nondimensional resistance coefficients of 1.5 to 2.

4.5 CORNERS

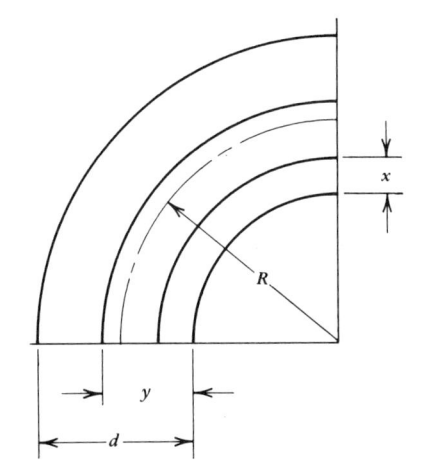

R/d	0.75	1.5
x/d	0.18	0.24
y/d	0.5	0.6

$K_c \approx 0.15$ at $R_d = 10^6$
(for above R/d range)

Figure 4.52 Recommended splitter plate arrangement. (Adapted with permission of ESDU International Ltd. from [4.56].)

The nondimensional resistance of individual flow paths will vary, and therefore flow deflection ahead of the vane leading edges may occur. Hence in obtaining the flow split that produces the desired distribution within the chamber, a measure of experimentation is advisable when determining the location of the vane leading edges.

4.5.6 Vaned Corners

This type of right-angled corner possesses the compactness of a mitered corner but with greatly reduced losses. The savings in power and space are considerable for large-scale industrial air-moving plants. Experimental studies have indicated that area increases of up to 20% can be obtained without incurring a significant increase in loss coefficient.

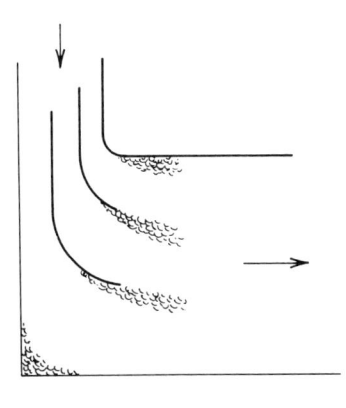

Figure 4.53 Vanes for flow redistribution after expansion.

120   DUCT COMPONENT DESIGN AND LOSSES

Vanes in corners of 45° or less are seldom warranted unless some special flow feature is required.

The constant area passage geometry (Fig. 4.54) developed by Collar [4.60] possesses the lowest published value of loss coefficient. The actual shape is centered on a s/c of 0.58, where s is $0.083d$, c is $0.143d$, and d is the distance between inner and outer walls (see Fig. 4.40). A K_c quantity of 0.05 is recorded at a Reynolds number of 1.8×10^5 based on vane chord, for thin inlet boundary layers, but unfortunately, fully developed flow conditions were not investigated.

A more common type of vaned corner employs cambered plates of constant thickness. The leading edge must be suitably aligned to the local flow to avoid vane nose separation. The angle of the trailing edge tangent to the duct axis is important in achieving the correct degree of air turning.

On occasion, the preceding camber line is used as a "backbone" for an airfoil thickness form. The leading-edge inclination is less critical, and the resulting thinner boundary layers make for a more consistent turning angle.

The convex inner surface of a corner is the most critical, for a variety of reasons. The first studies of vaned corners tended to concentrate on a uniform gap/chord spacing and hence the results have a bias toward the inner passage solution. However, it has been demonstrated that an arithmetical progression in gap/chord ratio results in a lower corner loss, particularly for round ducts.

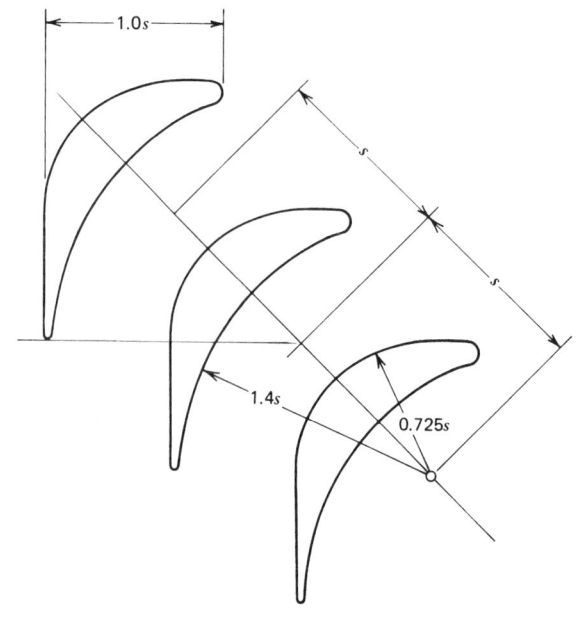

Figure 4.54 Constant area turning vanes for right-angled corner. (Taken from A. R. Collar [4.60] and reproduced with the permission of the Controller of Her Majesty's Stationery Office.)

4.5 CORNERS

The progression is represented by

$$\frac{\Delta}{d} = \frac{2\left(\frac{\sqrt{2}}{N} - \frac{s_1}{d}\right)}{N - 1} \qquad (4.29)$$

where Δ = constant distance component of the progression
d = distance between inner and outer walls of the inlet duct
s_1 = initial spacing
N = number of flow passages

Hence individual vane spacing will be given by

$$\frac{s_n}{d} = \frac{s_1}{d} + (n - 1)\frac{\Delta}{d} \qquad (4.30)$$

where n is the first, second, and nth vane passage counting from the inner wall.

Experiment has shown the gap/chord ratio to be of prime design importance. By assuming either the number of vanes or the turning-vane radius, the design for the constant thickness type can proceed.

A common design recommendation is to make the radius approximately equal to $d/4$. For a 90° sector the c/d ratio is $\sqrt{2}/4$, or 0.354. In [4.61] this ratio is 0.33. However, when 10% c straight extensions in the leading- and trailing-edge regions are present, the ratio has a value of 0.41.

When the vanes are evenly spaced, the number of passages for a given s/c will be

$$N = \frac{\sqrt{2}d}{s} \approx \frac{3.45}{s/c} \qquad (4.31)$$

Since N must be a whole number some adjustment to chord will usually be required in achieving a suitable s/c.

The use of an arithmetical progression involves the choice of an initial gap and the passage number. A successful combination for a rectilinear duct [4.16] is $N = 7$, $s_1 = 0.135d$, and $\Delta = 0.022d$; the corresponding figures for the circular duct [4.61] are 8, $0.103d$, and $0.021d$. In the first case, the initial and final passage s/c values are 0.33 and 0.65, with a mean value of 0.49, while quantities of 0.31, 0.75, and 0.53 apply to the latter.

Comprehensive studies of 90° corners with square cross section and possessing constant-thickness vanes of equal spacing are reported in [4.62] and [4.63]. The main findings of both investigations are presented in Fig. 4.55.

These indicate an optimum s/c range, 0.35 to 0.40. The s_1/c values for both the preceding arithmetical progression arrangements lie in the low-loss regimes when an identical value of 0.33 is substituted for c/d.

The best results in each case are obtained with similar vane geometries

122 DUCT COMPONENT DESIGN AND LOSSES

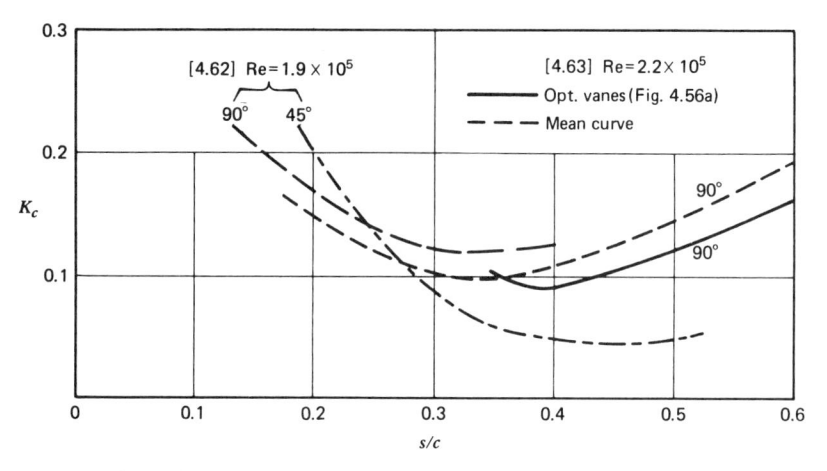

Figure 4.55 Corner losses, constant thickness vanes. (Left side, data taken from C. Salter [4.62] and reproduced with the permission of the Controller of Her Majesty's Stationery Office.)

(Fig. 4.56). The only modification necessary for expanding corners of up to 20% is an increase in the angle between tangents, by a degree or two [4.63]; expansion was not studied in [4.62]. The loss coefficients remain relatively unchanged for expansions up to 20%.

The 45° corner utilizes vanes with tangents aligned in the direction of the duct axes [4.62]; 10% extensions similar to those for the 90° corner are used.

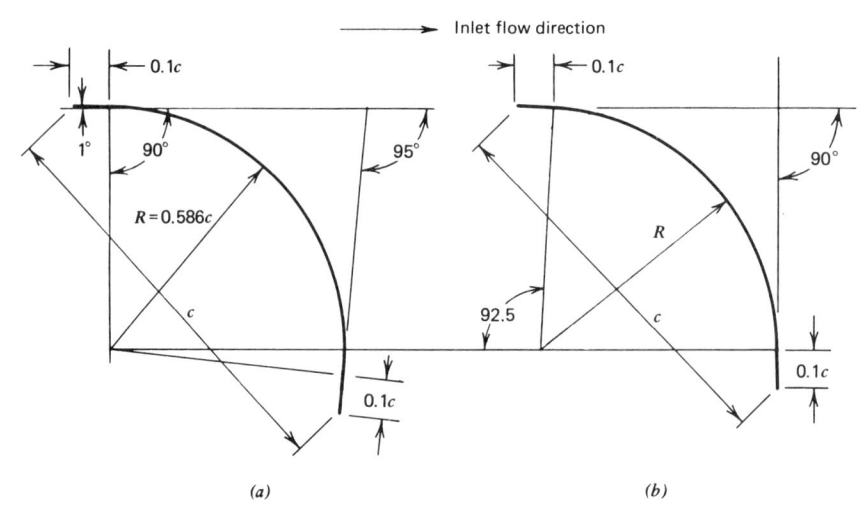

Figure 4.56 Constant thickness turning vanes for right-angled corners. (a) Recommendation of [4.63]. (b) Recommendation of [4.62]. (Sketch (b) taken from C. Salter [4.62] and reproduced with the permission of the Controller of Her Majesty's Stationery Office.)

4.5 CORNERS

In view of the lack of experimental data for expanding corners with an arithmetical progression vane geometry, a degree of design caution is advised. A lesser constant distance reducing the rate of gap increase, with provision for an extra vane(s), deserves consideration.

The use of spoilers upstream of the corner [4.63] as a boundary layer thickening measure failed to produce inner wall flow separation, although some weakening of the flow was experienced for a highly distorted inflow. The use of an extra vane with s/c of 0.2 in a grid of $s/c = 0.4$ vanes is suggested for the inner passage when poor quality inlet flow is present [4.63]. Since losses rise rapidly with decreasing s/c (Fig. 4.55) because of blockage factors, too close a spacing could prove counterproductive. A half- to one-chord trailing-edge extension to the inner vane, where $s/c \approx 0.3$, will reduce the extent of any local flow separations from the duct wall; this modification is suggested as a practical alternative.

The constant-thickness vane is far more sensitive to local leading-edge incidence than the Collar type. A range of geometric incidences from $-6°$ to $6°$ were studied in [4.63], with the finding that negative geometric angles gave significantly higher losses. As a consequence, $1°$ is recommended in practice. Since the flow is normally deflected inward ahead of the vanes, any small discrepancies should be in the positive angle sense.

The deviation angle, because of boundary layer growth, is approximately $5°$, necessitating a correction to the vane camber angle (Fig. 4.56). Small changes in vane shape have negligible influence on the loss coefficient but can influence the flow turning performance significantly [4.63].

Limited published data on Reynolds number effects are presented in Fig. 4.57, where chord replaces duct diameter as the characteristic length. Since the chord is normally about one-third of duct diameter, a simple conversion to R_d for comparative purposes can be made. The loss relationship for the [4.63] results is given approximately as being proportional to $R_d^{-1/3}$, and a similar scale effect is also present in [4.60]. In contrast, varying the air speed

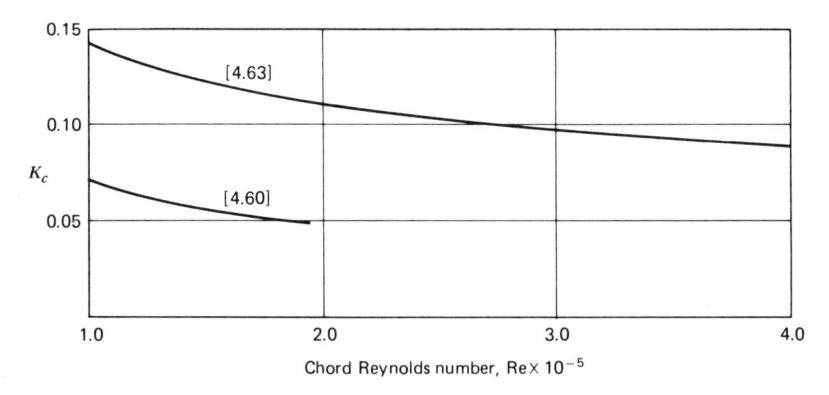

Figure 4.57 Reynolds number influence on vaned corner loss.

in [4.62] failed to produce any change in K_c; an $R_d^{-1/4}$ relationship is suggested in [4.64]. The preceding data apply to rectilinear ducts.

The loss coefficient for the circular duct corner [4.61] detailed previously was 0.16 at an R_d of 6.4×10^5. The unit is described as being of commercial rather than research standard manufacture.

Vanes with an airfoil-type thickness form will possess a larger s/c optimum value and slightly less camber than the foregoing type; losses are also lower [4.65].

Sufficient data are available for the successful design of a normal vaned corner. Although some doubt remains regarding the loss coefficient value for any given combination of vane type and R_d, the error will not normally be sufficiently large to vary the total system loss estimation by a significant amount.

Corrections for outlet tangent loss are negligible. Reliable corrections for roughness are not available, but adjustments based on experience should be applied.

A common geometry for corners in circular ducts is illustrated in Fig. 4.58. An elliptical duct segment is placed between the two mitered ducts and the vanes are positioned in this insert. The flat outer circumference has little effect on corner losses. The relative ease of manufacture is an important factor.

4.5.7 Interaction Between Adjacent Unvaned Corners

When a short length of parallel duct connects two 90° bends, correction has to be made to the gross loss coefficient (K_G) of a single bend. The procedure recommended in [4.9] is to apply a correction factor that is a function of L_s/d and R/d, where L_s is the length of intermediate duct. With the exception of mitered joints, the combined corner loss is always less than the addition of

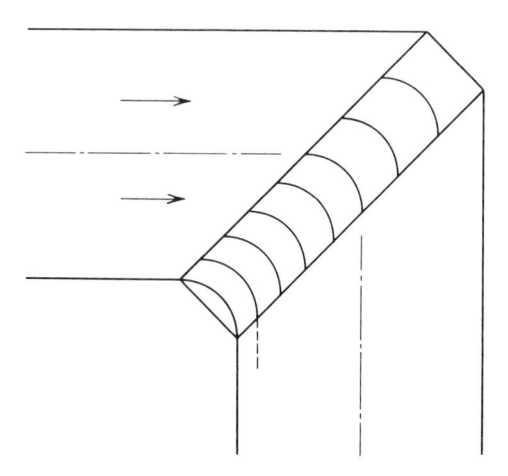

Figure 4.58 Suggested vaned corner arrangement for circular cross-section duct.

4.5 CORNERS

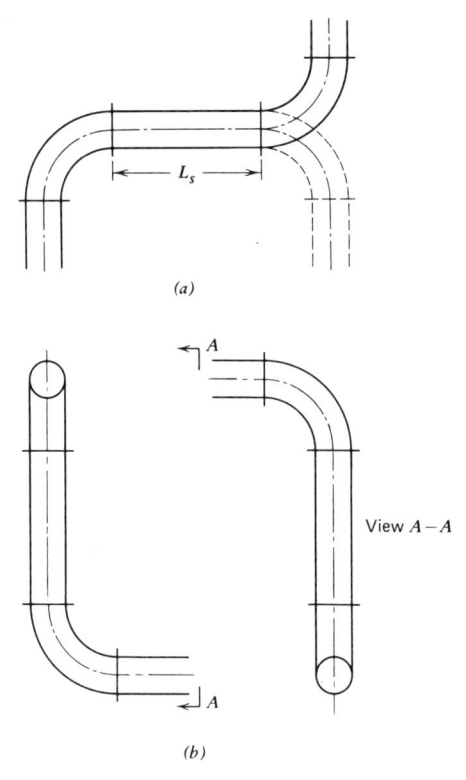

Figure 4.59 Three types of compound corners. (a) Gooseneck and U corners. (b) Offset corners.

the gross loss for the two bends when the spacer duct is less than 30 diameters long. For a L_s/d value of 4, the correction factor is approximately 0.75 in all three cases (Fig. 4.59). As the spacer duct length decreases, the factor remains constant for the offset case and reaches mean values of 0.9 and 0.65 for the gooseneck and U corners, respectively. Beyond a L_s/d of 4, the correction factors steadily approach unity at $L_s/d = 30$. The favorable loss characteristic of a 180° bend is related to the diminution of secondary flows that occurs after 80° of turning [4.9].

The combined corner loss, including the outlet tangent loss of the second bend, is given by

$$K_{2c} = \text{correction factor} \times (K_{G_1} + K_{G_2}) \qquad (4.32)$$

4.5.8 Interaction Between Corners and Diffusers

Considerable data on varying geometries and arrangements of these two components are presented in [4.9]. The combined loss for a diffuser–corner sequence is obviously much less than that for the reverse arrangement. Spacer lengths of at least $1d$, and preferably $1.5d$, are required in the latter instance for acceptable performance. Increased spacer lengths should be considered for rectilinear ducts.

The spacer length between a diffuser and downstream bend will depend on the diffuser geometry and outlet flow quality. A spacer length of $4d$ is generally satisfactory from a loss point of view, but $1d$ is markedly better than a direct connection.

Provided the spacer ducts have lengths in excess of the preceding recommendations, the addition of component losses for the diffuser and corner, as separate items, should prove reliable in most cases. However, where doubt exists, the data of [4.9] will provide more detailed information.

Experimental studies of 90° vaned corners with diffusing area ratios of 1.45 and 2.75 are reported in [4.66] and [4.67], respectively. In the first instance the loss coefficient rose from 0.11 to 0.24 as the boundary layer thickness was increased from 2 to 70% of half duct width. Constant-thickness cambered plates at an s/c of 0.5 were employed.

The higher area ratio corner [4.67] possesses features similar to those illustrated in Fig. 4.53. A wire screen with a resistance coefficient of from 1.4 to 3.0, placed at outlet, produced a relatively uniform velocity distribution in the discharge plane. The related overall corner/diffusion loss coefficient was not measured.

4.5.9 Curved Wall Diffusers

Design guidance on diffusion within a bend is presented in [4.68] and [4.9] for cases where space is restricted and the lowest possible loss for this imposed condition is sought. This development applies to rectangular ducts. Expansion about a circular arc centerline of greater radius than that associated with constant-area bends leads to losses in excess of those for carefully designed inner and outer walls. The achievable pressure recovery falls short of the value for a straight diffuser of equal area ratio. Hence this variant falls into a special ducting class. Because of its greater design complexity, detailed information is not included here.

4.6 INTERSECTING DUCTS

Two or more airstreams that intersect are a common feature in engineering processes, air conditioning systems, and mine ventilation. In view of the infinite number of possibilities in these situations, the emphasis here will be on good junction design rather than on flow resistance aspects. Where possible, the branch leg intersecting with a straight duct should be inclined at an angle of 45° or less. This arrangement prevents gross turbulent interference with the main airstream and minimizes the deflection angle for the branch flow.

4.7 INTERNAL BODIES

Converging airflows with substantially different velocities will possess an energetic dividing shear layer through which momentum is transferred to the stream of lesser velocity. The latter flow will therefore possess a negative loss, but the adjacent stream will be sharply penalized. The design adjustment of duct areas to match desired flows is therefore of great importance.

Depending on the required airflow ratios, a decrease or increase of main duct area may be advantageous downstream of the junction. This can be simply achieved with rectangular ducts, but a conical element is needed for the circular variety. For example, in a manifold-type situation, by exercising velocity pressure and hence static pressure control in the manifold, the required flow through each leg can be attained without the need for resistance regulation.

Right-angle intersections are sometimes unavoidable, an example of which is present in an exhaust airway/upcast shaft junction of underground mines. The local resistance can be substantially reduced by an appropriate radius or cutoff [4.9, 4.69, 4.70]. The 45° cutoff face should be wide enough to permit flow reattachment to the beveled surface following separation from its upstream edge.

In circular ducts this cutoff, or radius, can be accomplished by maintaining the cutoff contour along axial (to branch) rather than radial (to trunk) planes. In return for this small amount of additional expense, substantial power savings will accrue over the lifespan of a ventilation shaft.

Loss data for a number of representative duct intersections and flow conditions are presented in [4.9]. The model tests of [4.69] and [4.70] established complementary resistance data for the right-angled junction of rectilinear branches with a circular trunk duct. Additional loss data are presented in [4.13].

4.7 INTERNAL BODIES

The presence of fairings, struts, air-straightening honeycombs, and flow-smoothing grids adds significantly to system losses. The resistance of the former two is calculated from drag coefficient data and in the latter cases by a loss coefficient, K_x.

Drag coefficients can be translated into equivalent total pressure loss coefficients from the expression

$$K_x = \frac{D}{\frac{1}{2}\rho U_A^2 A_A} \tag{4.33}$$

where D is the drag force, A_A is the duct cross-sectional area, and U_A is the mean velocity. This relation applies in instances where the downstream duct length is sufficient to allow turbulent mixing to be substantially complete.

For the simple case of an obstruction of constant cross-sectional shape placed so that the longitudinal axis is at right angles to the airstream, the drag is obtained from the drag coefficient by the relation

$$D = C_D \tfrac{1}{2}\rho U_A^2 l d \tag{4.34}$$

where d is the diameter of a rod or the chord of an airfoil and l is the length of the obstruction. Hence,

$$K_x = \frac{C_D l d}{A_A} \tag{4.35}$$

The appropriate coefficient for an axisymmetric body is

$$K_x = \frac{C_D A}{A_A} \tag{4.36}$$

where A is the maximum cross-sectional area of the body. Information on the drag of various types of body is available in [4.71] for two-dimensional and finite aspect ratio cases. Most of the data relate to isolated bodies in an infinite stream and hence blockage factors that take into account the velocity increase that eventuates must be applied.

An additional problem arises in relation to high drag obstructions located just upstream of fan blading. The loss of total pressure is then localized and hence averaging it over the gross duct area, as assumed in the preceding equations, is inappropriate. The effect on fan behavior in such circumstances is discussed briefly in Section 23.3.

Bodies that involve substantial duct blockage factors, such as fan belt drive covers, should be faired in order to reduce their resistance to a practical minimum. Application of this principle to haulage mineshafts is discussed in [4.72] on a model test basis in relation to multiple I-beam support sections. Experimentally determined shaft resistance and C_D data for a number of actual arrangements illustrate the advantages of faired supports.

Figure 4.60 Flow around a cylinder.

4.7 INTERNAL BODIES

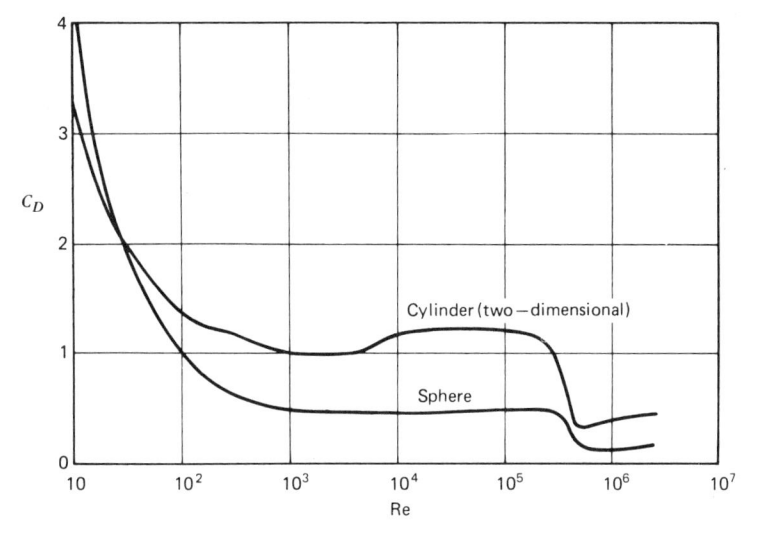

Figure 4.61 Drag coefficients for cylinder and sphere.

Bluff bodies, such as cylinders, possess an alternating vortex shedding of significant magnitude (Fig. 4.60), which can impose an unsteady load on fan blading when the axial separation length is small. This will increase fan noise and may initiate blade flutter.

The drag coefficient of a sharp-edged plate normal to the airstream is essentially independent of Re ($U_A d/\nu$) above 10^3 [4.71], having a value of 2.0 in the two-dimensional case. Aspect ratio effects are presented in [4.71], indicating that C_D remains relatively constant at 1.18 in the range 1 to 5.

Circular cylinder and sphere drag coefficients are given in Fig. 4.61 as a function of Re. In the Re range 10^4 to 10^5, the aspect ratio effect on a percentage basis is similar to that for flat plates (Fig. 4.62).

Roughness effects on cylinder drag can be obtained from [4.71] and [4.73] for sand- and pyramid-type projections. The latter data are uncorrected for tunnel blockage and must be divided by 1.2 to reduce the smooth surface coefficients to their customary values.

The application of blockage factor corrections to the preceding two-dimensional data must be on the basis of experience or actual experiment. However, with good duct design practice the blockage factor of support struts will be close to unity and hence may be ignored in estimating the duct system resistance.

The resistance of wire screens can be obtained from the Wieghardt expression, reproduced in [4.44].

$$K_x = \frac{6.5(1 - \beta)}{\beta^{5/3}} \left(\frac{Ud}{\nu} \right)^{1/3} \tag{4.37}$$

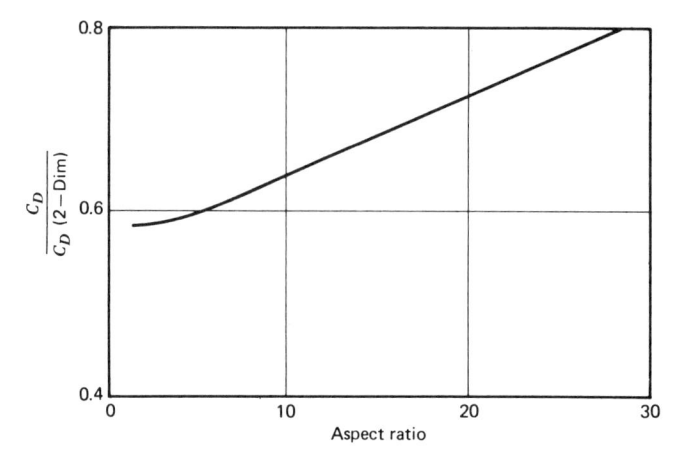

Figure 4.62 Influence of aspect ratio on drag for cylinders and plates normal to stream, Re range 10^4 to 10^5.

where β is the open area ratio given by

$$\beta = (1 - nd)^2 \tag{4.38}$$

and n is the number of cylinders of diameter d, in unit length. The preceding expression applies within the Reynolds number range of 60 to 600. The Collar relationship, namely,

$$K_x = \frac{C(1 - \beta)}{\beta^2} \tag{4.39}$$

is applicable for greater Reynolds numbers where C varies between 9 and 10.

In general, honeycombs have a loss coefficient less than 0.5 provided the swirl angle is less than 10° and the Reynolds number based on cell size, Re_h, is greater than 5×10^3. The loss coefficient according to [4.74] is given approximately for $Re_h > 6 \times 10^3$ by

$$K_x = \frac{(l/d)(Re_h \beta / Re_h^2)^{-0.83}}{\beta Re_h} \tag{4.40}$$

where l/d is the length to characteristic cell dimension ratio. Cell shape is apparently of minor relevance. For hexagonal cells the distance across the flats is taken as h.

4.8 CYCLONE SEPARATORS

Although the cyclone separator is simple to manufacture, the flow patterns within it for maximum separation efficiency attainment are often quite complex. As a result, most research work follows empirical rather than purely theoretical lines. Analytical methods for determining pressure drop and separation efficiency are outlined in [4.75] for one geometric class. These data are used in the presentation of design procedures for optimized units covering a range of operational requirements.

High-efficiency cyclones possess a reduced diameter and greater length than the high throughput variety. This augments the centrifugal force acting on the particles and increases the number of turns completed by the air prior to exhaust.

The presence of large-scale turbulent eddies, as distinct from the more orderly swirling flow, will tend to transport the smaller particles into the exhaust flow core, in a reingestion process. A development reported in [4.76] inhibits such turbulence by means of a mechanically rotated insert within the main chamber. As a consequence, higher separation efficiencies are obtainable. Good reviews of design theories and practice are contained in the preceding two references, along with extensive bibliographies.

Experimental values of pressure drop listed in [4.75] show a wide variation of from 3.4 to 11.4 times the inlet velocity pressure and show considerable scatter from the estimated values. Hence actual quantities determined experimentally should be used in system calculations. Swirl removal in the outlet duct, by an efficient diffusion process, will produce the greatest percentage improvement for cyclones in the lower resistance range. A stator row attached to a central streamline body, ahead of a normal diffuser, will ensure good pressure recovery (Section 4.4.8).

4.9 EJECTORS

When a high-energy air supply is available, this resource can be used to induce a secondary flow by virtue of the turbulent shear stresses existing around the boundary of the emerging jet. The efficiency of the energy trans-

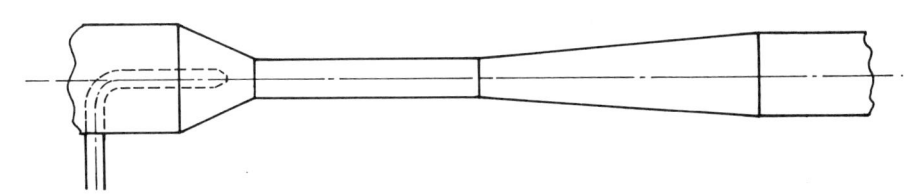

Figure 4.63 Typical ejector.

fer mechanism can be expressed as the gain in secondary flow energy to the energy lost by the primary fluid; the average efficiency is in the neighborhood of 0.30 for well-designed units.

A typical layout is illustrated in Fig. 4.63. The primary air nozzle is located a short distance upstream of the mixing chamber which is followed by a diffuser. Since the process is a dynamic one, a diffuser is required to conserve momentum by converting dynamic pressure into static pressure, thus avoiding excessive skin friction losses.

The mixing chamber may be either of a constant-pressure or of a constant-area type, where the former features a small contraction. According to [4.77] improved performance is obtained with constant-pressure mixing for supersonic jet flows, whereas [4.78] recommends the latter for normal subsonic primary streams. Most ejector devices possess a constant-area chamber.

For turbulent mixing to be complete before diffusion, a chamber dimension of from 4 to 10 chamber diameters is required [4.78]; the optimum lies between these two limits and is approximately 6 to 7 for most design cases [4.79].

Important parameters in the design of ejectors are the ratios of area, air density, pressure, and velocity with respect to the primary and secondary flows and the back pressure at the diffuser outlet. By adjusting these variables a wide range of volume flow/pressure characteristics can be attained. For example, a relatively large area ratio combined with a dense, high-velocity primary flow will result in a secondary mass flow that is a multiple of the primary value, provided the system back pressure is low. Design guidance is available in the literature quoted.

The majority of designs are in the subsonic flow range where the primary air is supplied by a high-pressure rise fan or a displacement-type blower. An important area of application is for gas flows that are inflammable, corrosive, dusty, or that convey granular materials and hence are unsuited to transit through a rotating air-moving device.

Injecting the air from an annular slot around the secondary air intake duct increases the circumferential length of the shear layer. This driving system has been used in relation to high-speed wind tunnels. Design information on this general type is available in [4.80].

REFERENCES

4.1 J. H. Horlock, The calculation of duct flows and the techniques of duct testing, Symposium on Subsonic Fluid Flows: Losses in Complex Passages and Ducts, *Proc. Instn. Mech. Eng.*, **182**, 3D, 1–4, 1967–68.

4.2 J. L. Livesey and T. Hugh, Suitable mean values in one-dimensional gas dynamics, *J. Mech. Eng. Sci.*, **8**, 374, Dec. 1966.

4.3 J. W. Richman and R. S. Azad, Developing turbulent flow in smooth pipes, *Appl. Sci. Res*, **28**, 419–441, Dec. 1973.

REFERENCES

4.4 J. Weir et al., The effect of inlet disturbances on turbulent pipe flow, *J. Mech. Eng. Sci.*, **16**, (3), 211–213, 1974.

4.5 S. T. McComas, Hydrodynamic entrance lengths for ducts of arbitrary cross section, *ASME J. Basic Eng.*, **89**, 847–850, 1967.

4.6 K. Sridhar et al., Settling length for turbulent flow of air in smooth annuli with square-edged or bellmouth entrances, *ASME J. Appl. Mech.*, **37**, 25–28, 1970.

4.7 ESDU International Ltd., Friction losses for fully developed flow in straight pipes, *Eng. Sciences Data Item No. 66027* (issued September 1966).

4.8 J. A. Brighton and J. B. Jones, Fully developed turbulent flow in annuli, *ASME J. Basic Eng.*, **86**, 835–844, 1964.

4.9 D. S. Miller, *Internal Flow Systems*, Vol. 5, BHRA Fluid Eng. Series, Cranfield, England, 1978.

4.10 K. Rehme, Simple methods of predicting friction factors of turbulent flow in non-circular channels, *Int. J. Heat Mass Transfer*, **16**, 933–950, Pergamon, 1973.

4.11 O. C. Jones, An improvement in the calculation of turbulent friction in rectangular ducts, *ASME J. Fluids Eng.*, **98**, 173–181, 1976.

4.12 G. E. Chmielewski, Boundary layer considerations in the design of aerodynamic contractions, *J. Aircraft*, **11** (8), 435–438, August 1974.

4.13 ASHRAE *Handbook of Fundamentals*, Chap. 33, ASHRAE, Atlanta, 1981.

4.14 T. Bernard, Inlet boxes of axial fans, Proc. 3rd Conf. on Fluid Mechanics and Fluid Machinery, Akadémiai Kiadó, Budapest 1969, pp. 37–43.

4.15 N. Yamaguchi, Axial flow type forced-draft fans for utility boilers, ASME, Paper 74-CT-130, 1974.

4.16 E. Ower and R. C. Pankhurst, *The Measurement of Air Flow*, Pergamon, Oxford, 5th ed., 1977.

4.17 N. Ruglen, A special inlet for vertical axial flow fan installations on downcast shafts E 21 and K 73, Mount Isa Mines Ltd., Aust. Dept. of Supply, *ARL Aero Note 276*, 1966.

4.18 I. S. Pearsall, Calibration of three conical inlet nozzles, Gt. Britain National Eng. Lab Report 39, 1962.

4.19 R. A. Wallis and N. Ruglen, On the aerodynamics of hanger type engine test facilities, *J. Roy. Aero. Soc.*, **70** (662), 312–320, January 1966.

4.20 H. Shuang, Optimisation based on boundary concept for compressible flows, *ASME J. Eng. Power*, **97**, 195–205, 1975.

4.21 J. J. Carlson et al. Effects of wall shape on flow regimes and performance in straight, two-dimensional diffusers, *ASME J. Basic Eng.*, **89**, 151–160, 1967.

4.22 J. Laufer, *New Trends in Experimental Turbulence Research*, Annual Review of Fluid Mechanics, Vol. 7, 1975, pp. 307–326, Annual Reviews Inc. Palo Alto, Calif.

4.23 P. Bradshaw, Effects of streamwise curvature on turbulent flow, *NATO AGARDOgraph 169*, 1973.

4.24 R. W. Fox and S. J. Kline, Flow regime data and design methods for curved subsonic diffusers, *ASME J. Basic Eng.*, **84**, 303–312, Sept. 1962.

4.25 J. P. Johnston and C. A. Powars, Some effects of inlet blockage and aspect ratio on diffuser performance, *ASME J. Basic Eng.*, **91**, 551–553, 1969.

4.26 A. T. McDonald and R. W. Fox, An experimental investigation of incompressible flow in conical diffusers, *Int. J. Mech. Sci.*, **8**, 125–139, 1966.

4.27 B. A. Russell and R. A. Wallis, A note on annular diffuser design, axial flow fans, Div. of Mech. Eng., *CSIRO Internal Rep. 63*, 1969.

4.28 G. Sovran and E. D. Klomp, Experimentally determined optimum geometries for rectilinear diffusers with rectangular, conical, or annular cross-section, Fluid Mechanics of Internal Flow, Elsevier, New York, 1967, pp. 270–319.

4.29 R. A. Wallis, Annular diffusers of radius ratio 0.5 for axial flow fans, Div. of Mech. Eng., *CSIRO Tech. Rep. 4*, 1975.

4.30 I. C. Shepherd, Annular exhaust diffusers for axial flow fans, 5th Australasian Conference on Hydraulics and Fluid Mechanics, Christchurch (N.Z.), 1974.

4.31 V. Kmonicek and M. Hibs, Results of experimental and theoretical investigations of annular diffusers, Div. of Mech. Eng., CSIRO, Translation No. 2, May 1974 (original publ. date 1962).

4.32 G. N. Patterson, Modern diffuser design, *Aircraft Eng.*, **10**, 267–273, 1938.

4.33 S. Wolf and J. P. Johnston, Effects of non-uniform velocity profiles on flow regimes and performance in two-dimensional diffusers, *ASME J. Basic Eng.*, 462–474, Sept. 1969.

4.34 I. H. Johnston, The effect of inlet conditions on the flow in annular diffusers, Gt. Britain, Aero. Research Council, ARC, C.P. 178, 1953.

4.35 R. A. Tyler and R. G. Williamson, Diffuser performance with distorted inflow, *Symposium on Subsonic Fluid Flow Losses in Complex Passages and Ducts*, Instn. Mech. Eng., London, 1967.

4.36 R. K. Duggins, The performance of truncated conical diffusers, *J. Mech. Eng. Sci.*, **13**, 103–109, 1971.

4.37 C. R. Fishenden and S. J. Stevens, Performance of annular combustor-dump diffusers, J. Aircraft, **14**, 60–68, Jan. 1977.

4.38 N. Ruglen, Annular diffuser studies for an upcast fan, Mount Isa Mines Ltd. (unpublished paper, Div. of Mech. Eng., CSIRO, 1974).

4.39 O. L. Cochran and S. J. Kline, Use of short flat vanes for producing efficient wide-angle two-dimensional subsonic diffusers, NACA TN 4309, Sept. 1958.

4.40 O. G. Feil, Vane systems for very-wide-angle subsonic diffusers, *ASME J. Basic Eng.*, **86**, 759–764, 1964.

4.41 D. M. Rao and S. N. Seshadri, Application of radial-splitters for improved wide-angle diffuser performance in a blowdown tunnel, *J. Aircraft*, **13**, (7), 538–540, July 1976.

4.42 D. M. Rao, A method of flow stabilization with high pressure recovery in short conical diffusers, *J. Roy. Aero. Soc.*, **75**, 330–339, May 1971.

4.43 G. G. Schubaeur and W. G. Spangenburg, Effect of screens in wide angle diffusers, *NACA Tech. Note 1610*, 1948.

4.44 R. D. Mehta, The aerodynamic design of blower tunnels with wide-angle diffusers, *Prog. Aerospace Sci.*, **18**, Pergamon, 59–120, 1977.

4.45 J. Persh and B. M. Bailey, Effect of surface roughness over the downstream region of a 23° conical diffuser, *NACA Tech. Note 3066*, 1954.

4.46 F. D. Stull and H. R. Velkoff, Flow regimes in two-dimensional ribbed diffusers, *ASME J. Fluids Eng.*, **97**, 87–96, 1975.

4.47 V. Kmonicek, Improvement of diffuser operation by simple means, CSIRO Div. of Mech. Eng., Translation No. 3, August 1974 (original publ. date, 1956).

4.48 V. Kmonicek, Flow through a conical diffuser with flow-stabilizing star-shaped insert, CSIRO Div. of Mech. Eng., Translation No. 4, June 1975 (original publ. date, 1956).

4.49 M. C. Welsh, Flow control in wide-angled conical diffusers, *ASME J. Fluids Eng.*, **98**, 728–735, December 1976.

4.50 M. C. Welsh et al., Straight walled annular exhaust diffusers of 0.5 boss ratio, axial flow fans, Inst. of Engrs., Aust., Thermodynamics Conference, Hobart, December 1976, Nat. Conf. Publ. No. 76/12.

4.51 U.S. Patent 2,558,816, Fluid mixing device, July 1951.

REFERENCES

4.52 H. H. Pearcey, Shock induced separation and its prevention by design and boundary layer control, *Boundary Layer and Flow Control*, Vol. 2, Pergamon, 1969, pp. 1190–1344.

4.53 Y. Senos and M. Nishi, Improvement of the performance of conical diffusers by vortex generators, *ASME J. Fluids Eng.*, **96**, 4–10, 1974.

4.54 R. S. Neve and N. E. A. Wirasinghe, Changes in conical diffuser performance by swirl addition, *Aero. Quart.*, **29**, 131–143, 1978.

4.55 A. J. Ward Smith, *Pressure losses in ducted flows*, Butterworths, London, 1971.

4.56 ESDU International Ltd., Pressure losses for incompressible flow in single bends, *Eng. Sciences Data Sheet Item 67040* (issued November 1967).

4.57 I. Harper, Tests on elbows of a special design, *J. Aero. Sci.*, **13**, 587, 1946.

4.58 G. N. Patterson, Corner losses in duct, *Aircraft Eng.*, **9**, 205, 1937.

4.59 M. S. Clarke et al., The effect of system connections on fan performance, *ASHRAE Trans.*, **84**, (2), 227–263, 1978.

4.60 A. R. Collar, Some experiments with cascades of aerofoils, Gt. Britain, Aero. Research Council, *ARC R & M 1768*, 1936.

4.61 N. A. Dimmock, Cascade corners for ducts of circular cross-section, *Brit. Chem. Eng.*, **2**, 302–307, 1957.

4.62 C. Salter, Experiments on thin turning vanes, Gt. Britain, Aero. Research Council, *ARC R & M 2469*, 1952.

4.63 J. Zwaaneveld, Onderzoek aan plaatschoepen voor bochten met 90° ombuiging bij 0, 10 en 20% expansie, National Aerospace Lab., *NLR Rep. A 1118*, 1950.

4.64 P. Bradshaw and R. C. Pankhurst, The design of low speed wind tunnels, *Prog. Aero. Sci.*, **5**, Pergamon, Oxford, 1964, pp. 1–64.

4.65 K. G. Winter, Comparative tests on thick and thin turning vanes in the R.A.E. 4ft × 3ft wind tunnel, Gt. Britain, Aero. Research Council, *ARC R & M 2589*, 1947.

4.66 D. Friedman and W. R. Westphal, Experimental investigations of a 90° cascade diffusing bend with an area ratio of 1.45:1 and with several inlet boundary layers, *NACA Tech. Note 2668*, 1952.

4.67 D. C. MacPhail, Experiments on turning vanes at an expansion, Gt. Britain, Aero. Research Council, *ARC R & M 1876*, 1939.

4.68 C. J. Sagi and S. P. Johnston, The design and performance of two-dimensional curved diffusers, *ASME J. Basic Eng.*, **89**, 715–731, 1967.

4.69 P. H. Wilson, Aerodynamic losses in downcast shaft/airway intersections, Div. of Mech. Eng., *CSIRO Tech. Rep. 13*, 1977.

4.70 P. H. Wilson, Aerodynamic losses in upcast shaft/airway intersections, Div. Mech. Eng., *CSIRO Tech. Rep. 14*, 1977.

4.71 S. Hoerner, *Fluid-Dynamic Drag*, Hoerner, Midland Park, N.J., 1965.

4.72 V. A. L. Chasteau and J. F. Kemp, Further results of scale model measurements of mine shaft resistance to air flow, by the CSIR, *J. Mine Ventilation Soc. S. Africa*, **15**, June, 1962.

4.73 E. Achenbach and E. Heinecke, On vortex shedding from smooth and rough cylinders in the range of Reynolds numbers 6×10^3 to 5×10^6, *J. Fluid Mech.*, **109**, 239–251, 1981.

4.74 J. Whitaker, P. G. Bean, and E. Hay, Measurement of losses across multi-cell flow straighteners, Gt. Britain, National Eng. Lab., Report 461, 1975.

4.75 D. Leith and D. Mehta, Cyclone performance and design, Atmospheric Environment, Pergamon Press, Vol. 7, 1973, pp. 527–549.

4.76 R. Razgaitis and D. A. Guenther, Separation efficiency of a cyclone separator with a turbulence-suppressing rotating-insert, *ASME J. Eng. Power*, **103**, 566–571, July 1981.

4.77 J. H. Keenan, E. P. Neumann, and F. Lustwerk, An investigation of ejector design by analysis and experiment, *ASME J. Appl. Mech.*, **17**, 299–309, 1950.

4.78 J. A. C. Kentfield and R. W. Barnes, The prediction of the optimum performance of ejectors, Instn. Mech. Eng., London, **186**, 671–681, 1972.

4.79 R. Jorgensen (Ed.), *Fan Engineering*, Buffalo Forge, Co., Buffalo, N.Y., USA 6th ed., 1961, pp. 543–547.

4.80 T. W. Van Der Lingen, A jet pump design theory, *ASME J. Basic Eng.*, **82**, 947–960, 1960.

CHAPTER 5

Duct System Design and Development

The design of optimum duct assemblies and the attainment of high accuracy in loss estimation undoubtedly call for the maximum degree of experimentally acquired aerodynamic skills. Since the benefits are substantial, it is important to outline a rational procedure that will ensure the closest approach to these goals.

Industry obviously possesses satisfactory manual and computer-centered design programs based on available published data, with [5.1] providing a representative example. However, because of different test procedures and flow conditions, the data from various sources frequently exhibit inconsistency for common duct components. When complex flows are involved, the available data are often totally inadequate.

The design approach of [5.2] represents a significant advance, since it possesses numerous illustrations of the flow features encountered in common duct units. Further, the loss data are graded with regard to probable accuracy. Class 1 data are considered reliable, provided good inlet and outlet flow conditions are assured in the application. The associated duct components are considered to be well researched, making further studies unprofitable.

When the results from different sources vary by an amount greater than experimental accuracy, or when corrections to Class 1 data are required, the ensuing data are given a Class 2 rating. For less reliable source data, or where corrections of uncertain accuracy are applied to loss coefficients in Classes 1 and 2, the grading is Class 3.

The high cost of conducting comprehensive test programs is a valid assertion made in [5.2]. However, when the objective is limited and related to a specific combination of duct components, a carefully planned test exercise need not be excessively expensive.

Restricted opportunity for experimental determination of system losses, dictated either by lack of test equipment or by management policies, is currently a matter for concern. Poorly conceived, managed, and conducted test programs have a poor cost effectiveness, and this no doubt has had its effect on the rate of present progress.

5.1 JOB SPECIFICATION

With few exceptions, current airflow equipment specifications contain a minority of aerodynamic data, largely because of the paucity of specialist advice. The customer consequently relies on the judgment and experience of the selected design and construction organizations to deliver equipment of satisfactory aerodynamic performance. This practice often results in installations possessing airflow deficiencies.

When preparing a specification for airflow equipment, it is best to seek collaboration with an aerodynamicist who can give advice on the various design options and on the possible need for supporting experimental work in developing the selected equipment items. The latter task either can be carried out by the contracting organization or can be the subject of a separate arrangement. The "feedback" benefit from such studies has a powerful influence on subsequent designs, reducing their development costs. The selected ducting arrangement should possess suitable locations for trial and operational test instrumentation.

The maintenance of aerodynamic advice to the customer is most desirable throughout the design, construction, installation, and test phases.

A specification that includes the preceding provisions will ensure close technical control over the contract, thus eliminating the need for contingency expenditure in rectifying any deficiency.

5.2 SYSTEM DESIGN

The design data of the preceding chapter relate generally to single duct components, each possessing a specified quality of inlet flow. This information is completely adequate for design and loss estimation purposes, provided the flow is of a low swirl, steady-state type with negligible duct component flow interference. The system designer makes a choice of preferred components, arranges them in the preferred or required sequence, and then sizes the units in accordance with aerodynamic, economic, and/or practical considerations.

When interference between duct component flows is unavoidable, the designer must introduce correction factors into the loss calculations. These should be based on relevant available data or on experimental findings obtained for a similar ducting arrangement. Guesswork is not recommended.

On estimating the K factors for each duct component, the system resistance can be calculated from Eqs. (4.9) or (4.10). In a parallel path system,

5.3 TEST-SUPPORTED DESIGN DEVELOPMENT

the resistance of each leg for the design throughflow should be estimated, thus establishing the most resistant leg. Measures to reduce losses in this leg by cross-sectional area increases, more efficient components, a rearrangement of component sequence, and greater attention to static pressure regain per the medium of improved flow diffusion should be carefully studied.

Flow control in the less-resistant branches of many industrial-type installations is obtained with readily adjusted damper vanes. Outlet registers play a similar role in air conditioning systems. However, in the latter instance the noise generated may be excessive. Increased resistance because of a smaller duct size can provide partial or complete flow control, depending on the loss increment required. A soft elastic membrane device, operated by air stagnation pressure, represents a current development. The membrane expands with increasing stagnation pressure, restricting the air flow passage.

The foregoing analysis comes under revision when the throughflow for the most resistant leg is a small percentage of the whole. In these circumstances, small in-line booster fans should be considered in one or more of the branch ducts. This allows attention to be focused on the main trunkline leg of the system in relation to fixing the pressure duty of the primary fan. Ventilation systems of underground mines represent an important instance where implementation of these broad principles results in substantial reductions in power costs.

When the information for insertion into Eqs. (4.9) and (4.10) is unreliable or nonexistent, test programs should be seriously considered. The objective of maintaining simple, predictable flow situations may be nullified in many instances by practical considerations. Complex flow, in which a mixture of flow types (Table 2.3) is present, can seldom be mathematically modeled or analytically assessed [5.3].

5.3 TEST-SUPPORTED DESIGN DEVELOPMENT

The value of model, full-scale, or pilot plant testing is freely acknowledged by all. However, the restricted availability of test facilities, time delays, and added costs often result in tenderers gambling in a tight, competitive market. Research organizations such as AMCA, ASHRAE, BHRA, and HVRA in the United States and the United Kingdom provide design information on some of the more commonly experienced complex flow situations. It is believed that most consulting engineers involved in airflow equipment design neither possess nor have access to adequate test facilities and personnel.

A greater appreciation by the customer of the cost effectiveness of obtaining sound advice from an experienced aerodynamicist at the specification stage is essential to significant progress in the test development of efficient equipment. The author can, from personal experience, readily attest to the truth of this statement.

Complex flows usually contain large-scale turbulence resulting either from the entrainment effect of jet flows or from boundary layer separation. Hence with few exceptions the flow patterns are not markedly affected by Reynolds number, since viscosity is of minor importance. In these circumstances small-scale model tests will provide both visual and quantitative loss data of acceptable accuracy, when correctly interpreted.

It is now known that the initial characteristics of a separating shear layer can substantially modify the subsequent flow downstream of sharp-edged bluff bodies. For example, an increase in free stream turbulence will accelerate the mixing rate, facilitating an earlier reattachment location. A small rounding of the separation initiating corners can also produce flow changes. Hence close attention should be given to turbulence and dimensional similarity.

When large Re differences between model and full-scale tests exist, resistance reductions of the order of 10% may be experienced in the latter case. Subsequent on-site tests will provide valuable "feedback" data on the accuracy of the model-based predictions.

For a nonrecirculating duct system the model throughflow is readily provided by any fan with the required duty characteristics. Since the data must be nondimensionalized, an accurate volume flow rate measurement is essential. The conical inlet device of Fig. 21.4 is suitable for a "blower" setup, whereas some form of orifice plate arrangement can be considered for an exhaust fan. When insufficient length is available to meet the requirements of an orifice plate code, the device must be calibrated for the proposed in situ arrangement. The termination of a short length of ducting attached to the fan outlet flange provides a convenient location for the device. When the model is replaced by the test assembly of Fig. 21.3, orifice calibration is established.

The modeling of recirculating systems can raise problems with respect to attaining and measuring the desired airflows. In many instances the flow patterns and associated resistance of various component assemblies may be studied separately in an open circuit test rig and the complete characteristic deduced from these observations. The requirement for modeling the inlet and outlet flow features as closely as possible must be satisfied. By changing the grouping of the individual components in the test program, the subsequent analytical assessment of the results will either confirm the validity of the tests or indicate the need for a more refined test model and/or procedure.

When carrying out model tests, it is important to relate flow features to physical measurements. Confusion and a loss of thought direction often accompany multiple geometric changes between test observations. The test program should promote a step-by-step advance toward a more complete understanding of the complex flow problem.

In duct model studies a fan can be likened to an electric battery; it introduces a source of energy. However, the fan may also be responsible for swirling and nonuniform flow features. An external flow-inducing source is permissible in models of systems fitted with in-line axial flow fans. When the

5.4 ON-SITE DUCTING TESTS

full-scale fan unit is fitted with stators no blading is required in the flow annulus. However, swirl vanes are required to simulate outlet conditions for the rotor-only case. The lack of a static pressure rise at this location is normally unimportant; flow features are not affected. Flow nonuniformity can be modeled by graded resistance.

System resistance should always be presented in terms of total pressure. Static pressure tappings are a quick and convenient test procedure, but they should be placed in cross-sectional planes that are free of flow separations. A low mean local velocity is a further desirable requirement, since flow non-uniformities will not noticeably influence the estimate of mean local total pressure. The direct measurement of mean total pressure can usually, but not always, be avoided. Yaw-insensitive total-pressure-measuring instruments are discussed in [5.4].

The use of vane anemometers as prime velocity-measuring instruments is not recommended. Nevertheless they fulfill an extremely useful purpose in the determination of flow relativity. Velometers also fall into this general usage category. When an accurate measuring device is impractical, or unavailable, a measure of greater uncertainty based on experience must be assumed with respect to flow measurements obtained by these instruments. A velometer calibrated immediately prior to testing is preferred.

The foregoing discussion has been centered on general issues. The infinite variety of flow combinations in complex situations makes a more detailed presentation of test procedures impractical, certainly in the context of this publication. In addition to the references of Chapter 4, the published literature is very wide-ranging, covering model test techniques and studies of many complex flow problems on a variety of airflow situations, including mine ventilation.

5.4 ON-SITE DUCTING TESTS

Acceptance or proving tests for ducts may be synonymous with the fan trials, since the fan total pressure rise must equal the duct system losses. However, when attention is focused at the specification stage on the performance of various elements, provision can readily be made for the desired studies. Instrument and viewing ports can be provided in the most suitable locations. Accessibility to these sites must also be planned and provided for.

Current practice in regard to on-site duct and fan testing is generally below a reasonable and reliable standard. This is highlighted by the AMCA assessment of the problem, as discussed in Chapter 21. Experienced aerodynamic advice at the job specification and design stages will provide satisfactory solutions. The required design changes are usually quite simple and logical, involving little if any additional cost. Equipment changes are difficult and costly to implement once installation is complete.

The location of suitable measurement planes and the associated instrumentation often require the attention of widely experienced aerodynamicists. An ability to visualize, in advance, the probability and nature of complex flows that could be present is absolutely essential.

5.5 TROUBLESHOOTING

This area of aerodynamic activity often involves a large measure of imaginative thinking. Unstable complex flows are present in many of the installations requiring investigation.

The best success rate accompanies a conceptual breakdown of the complex flow into its basic ingredients. Experimental studies can be designed to confirm or modify the concepts formulated. Progress is signified by the attainment of a more suitable rearrangement of flow features, with instrumentation playing a supporting rather than dominating role. The eradication of a severe engine test cell instability provides a published example of this approach [5.5]. The introduction of a frame that effectively separated the cell inlet flow from the recirculating propeller flow eradicated the cause of flow instability.

However, in many cases remedial measures lead to expensive partial or complete equipment replacements. It is important for an experienced engineer to report on the basic aerodynamic issues involved before any final decision is taken. For example, local and overseas test cell engineers could provide no answers to the problem of [5.5]; one even recommended the rebuilding of the facility. The remedial measures proved to be relatively simple and cost effective. In other instances the desired performance can be achieved only by new or heavily modified equipment.

Every case must be treated on its merits, within the experience of the aerodynamic adviser.

When the design test procedures outlined previously are followed, there is seldom any need for remedial studies and work.

REFERENCES

5.1 *ASHRAE Handbook of Fundamentals,* Chap. 33, ASHRAE, Atlanta, 1981.

5.2 D. S. Miller, *Internal Flow Systems,* Vol. 5, BHRA Fluid Eng. Series, Cranfield, England, 1978.

5.3 J. H. Horlock, The calculation of duct flows and the technique of duct testing, *Symposium on Subsonic Fluid Flows: Losses in Complex Passages and Ducts, Proc. Instn. Mech. Eng.* **182** (3D) 1–4, 1967–68.

5.4 E. Ower and R. C. Pankhurst, *The Measurement of Air Flow,* Pergamon, Oxford, 4th ed., 1969.

5.5 R. A. Wallis and N. Ruglen, On the aerodynamics of hangar type engine test facilities, *J. Roy. Aero. Soc.,* **70** (662) 312–320, 1966.

CHAPTER 6
Airfoil Data for Blade Design

Efficient fan design requires a thorough understanding of the many variables associated with airfoil shape and aerodynamic performance. The comprehensive study of [6.1] outlines the important features sought in relation to fan blading. Quantitative characteristic data are presented here for a number of airfoils selected in accordance with [6.1] findings.

6.1 DEFINITIONS

A two-dimensional lifting surface with a lift to profile drag ratio of 10 or more can be considered an airfoil. A flat plate qualifies under this definition.

Lift is a function of airfoil incidence, defined as the angle between the approaching airstream and some reference line. The angle between the mean resultant velocity vector relative to the blade and the chord line is the incidence angle for fan blades. Published two-dimensional isolated airfoil data can then be used directly in the design process.

The line joining the two camber line extremities is the normally accepted chord line with the exception of flat undersurface airfoils where incidence is measured with respect to the lower surface.

When the flow field around a blade is modified by adjacent blading, the airfoils are considered to be in cascade. The British and American axial flow compressor design methods employ a local leading-edge incidence and angle of attack, respectively. The inlet velocity vector relative to the blade constitutes a common feature but the leading-edge camber line tangent and the blade chord line are the alternative reference lines. Neither method uses isolated airfoil data.

6.2 TYPES OF AIRFOIL SECTION

An orthodox airfoil may conveniently be considered as a mean line on which is superimposed some thickness form such as a streamlined shape. With a flat or cambered plate, the thickness remains constant along the chord except near the leading and trailing edges, where suitable local shaping is often carried out.

When the mean line is straight, the airfoil is a symmetrical one; with the exception of a certain type of stator blade and reversible-flow rotors, such airfoils are rarely encountered in fan design. By introducing camber into the mean line, greater working loads can be obtained from the blades.

The camber line employed may be of a completely arbitrary shape but circular or parabolic arcs are often used. In the latter case, the curvature decreases as the trailing edge is approached.

The streamlined form with which the camber line is "clothed" is a function of

1. maximum thickness to chord ratio.
2. chordwise position of maximum thickness.
3. leading-edge radius.

These have an influence on maximum lift, drag, and critical Mach number, but do not significantly affect general lift properties, which are determined by the camber line.

The contour over the forward part of the airfoil is usually elliptical, whereas aft of the point of maximum thickness the curvatures are small with this portion of the airfoil resembling a wedge. Small leading-edge radii produce large local flow accelerations when the airfoil is at workable incidences and the subsequent retardations downstream of the leading edge can lead to boundary layer separation from the airfoil nose. Most airfoil sections suitable for fan blades possess a moderately large leading-edge radius but constant thickness airfoils such as cambered plates are, of course, an exception. The efficient working range of fans employing cambered plate blades is limited by local flow separations of the type just discussed.

There are in existence numerous families of airfoils that have been developed with some specific aim in mind, mainly in connection with aircraft applications. Since, however, the blading requirements of low-pressure-rise fans are markedly similar, it is possible, in design, to standardize on a very limited number of types; manufacturing expediency may play an important role in the actual choice.

The design of high-pressure-rise equipment, on the other hand, calls for a more refined approach to the development of suitable airfoil sections. A comprehensive study in [6.1] confirms the general applicability of British axial flow compressor airfoils, with an important leading-edge modification.

6.3 LIFT

An airfoil at zero incidence constitutes a flow obstruction and hence local speeding up of the air takes place. With increasing incidence the velocity at any point on the top surface is progressively increased, whereas, in general, the reverse is true on the lower surface. These velocity changes give, according to Bernoulli's equation, decreasing and increasing static pressures on the top and lower surfaces, respectively. The difference in pressure between the two surfaces governs the lifting force acting on the airfoil.

When the pressure forces are integrated over the entire airfoil, one force normal to the chord line and one parallel to it are obtained. Resolution of these forces in directions normal and parallel to the oncoming stream gives the lift and drag forces, respectively.

The nondimensional lift coefficient C_L has been previously defined in Section 2.4.3 as

$$C_L = \frac{L}{\frac{1}{2}\rho U_0^2 A}$$

where C_L is a function of airfoil incidence and Reynolds number.

There is a limit to the amount of lift that can be obtained from increased airfoil incidence. The angle at which maximum lift is obtained is known as the stalling incidence.

6.4 PITCHING MOMENT

The locality in which the resultant lift force acts is generally between the 25% and 50% chord positions, depending largely on airfoil camber and fan blade solidity. This force produces a turning moment that is usually presented as a moment about the 25% chord line, namely,

$$C_{M_{c/4}} = \frac{PM}{\frac{1}{2}\rho U_0^2 A c}$$

In fan design the twisting loads so applied are normally small in comparison with the blade-bending loads, and the tensile loads due to centrifugal force. Hence, with the exception of flexible and variable-pitch blading, pitching moment data can be ignored.

6.5 DRAG

The drag associated with a two-dimensional airfoil is known as profile drag and a three-dimensional wing possesses, in addition, induced or secondary drag.

6.5.1 Profile Drag

This drag arises from skin friction and pressure forces. The air flowing over the top surface of an airfoil is initially accelerated but then suffers retardation as the trailing edge is approached. Owing to the presence of a rapidly growing boundary layer, the rate of diffusion is reduced through the displacement effect of the layer (Section 2.3.3). As a result, the pressure forces acting on the airfoil surfaces downstream of the point of maximum thickness are less than they would be if the boundary layer were absent. When resolved along the chord line, the integrated pressure forces upstream of maximum thickness are greater than those downstream and hence there is a resultant down-chord component of force. This force gives rise to the pressure drag, toward which the top surface makes the larger contribution.

With increasing incidence, the adverse pressure gradient on the top surface becomes more severe and the boundary layer thickens; this increases the pressure drag. When boundary layer separation eventually takes place, the static pressure over the rear of the airfoil is sharply reduced and the pressure drag rises steeply. This phenomenon of flow separation is directly responsible for the stall, that is, lift limitation, of an airfoil (Fig. 6.1).

The skin friction and pressure drags can also be considered as forces arising from the tangential and normal stresses, respectively.

6.5.2 Secondary Drag

On an airfoil of finite span (e.g., an aircraft wing) the air tends to flow from the high-pressure region on the lower surface around the extreme tip to the low-pressure region on the top surface. This phenomenon introduces flows with spanwise components and these combine to form the well-known tip vortex that is present when lift is being obtained from a finite wing. The momentum of this vortex flow passes downstream and is finally dissipated as heat. Vortices are also being shed from the trailing edge along the entire span since conditions at the tip influence the whole flow field. The momentum loss due to this cause is known as the induced drag.

The preceding vortex flow produces an effective change in the direction of the incident airstream and hence the lift force is inclined backward at a small angle, thus giving a component in the direction of the drag force. In this way the momentum losses associated with the secondary flow are experienced by the wing as a pressure force.

Ducted fan installations possess a variety of secondary flows, as intro-

6.6 GENERAL DEVELOPMENT OF BLADING SECTIONS

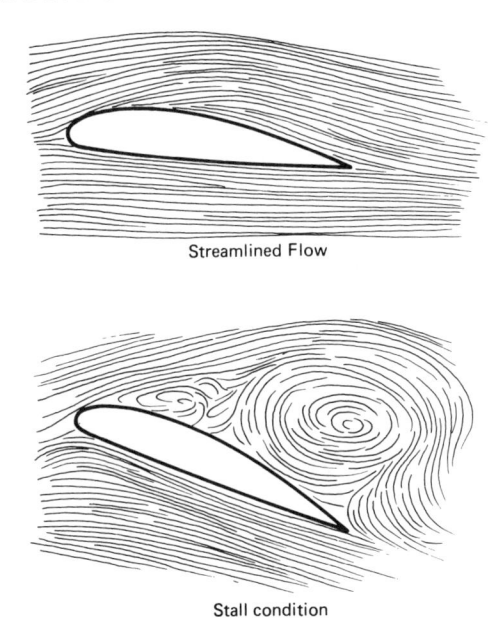

Streamlined Flow

Stall condition

Figure 6.1 Airfoil flow patterns.

duced in Sections 2.7 and 7.5. In each instance the secondary flows are associated with three-dimensional conditions at the blade extremities and constitute momentum losses, reducing fan efficiency.

6.6 GENERAL DEVELOPMENT OF BLADING SECTIONS

Many different types of airfoil section can be used in the design of fan blading but some are less efficient than others. Hence, for the designer seeking to develop blading of a preferred form to meet some specified duty requirement, the following guidance is provided.

A blade acts as a deflecting device and hence experiences a resultant force that has lift and drag components. Hence for any given set of local blade circumstances there is a unique relation between lift and air deflection as both express a change in the air momentum condition. Since this relationship is dependent on the absolute flow angles, which in turn influence the optimum blade camber line, its nature must be explored. For present purposes the drag force component can be ignored.

6.6.1 The Blade Lift/Flow Deflection Relation

A section of a cascade possessing an infinite number of airfoils is presented in Fig. 6.2. The velocity component V_a normal to the cascade plane will

148　　　　　　　　　　　　　　　　　　　AIRFOIL DATA FOR BLADE DESIGN

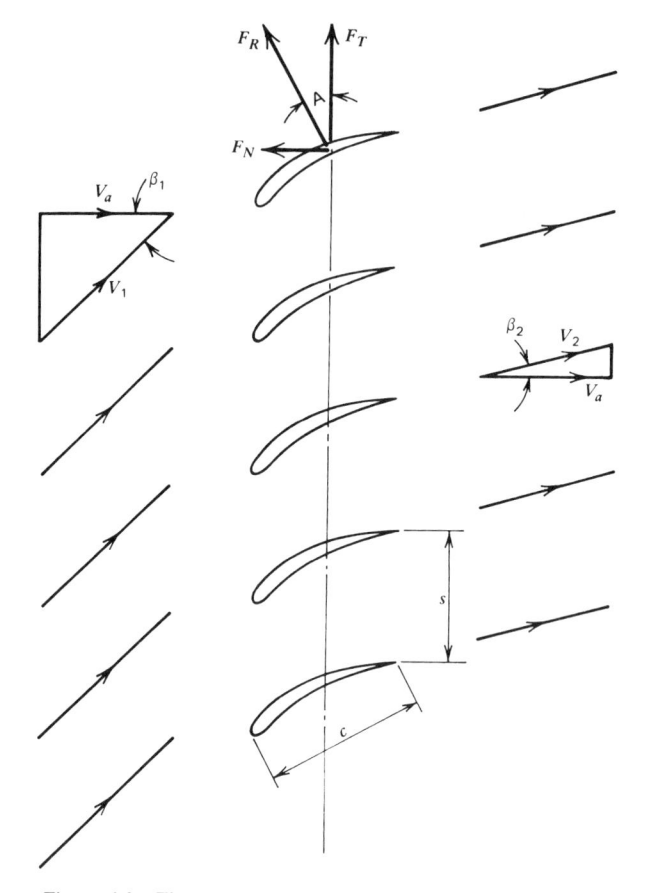

Figure 6.2　Flow through two-dimensional cascade of airfoils.

remain constant for incompressible and two-dimensional flow, in accordance with the continuity equation. The air deflection, however, reduces the component parallel to the plane; this reduction involves a rate of change of momentum equal to the force applied by the blades in this plane. The force F_T per blade is given by

$$F_T = \rho s V_a (V_a \tan \beta_1 - V_a \tan \beta_2) \tag{6.1}$$

The preceding assumption of zero drag implies no loss in total pressure; this permits a direct application of the Bernoulli relation. The normal force F_N is therefore given by

$$F_N = s(p_2 - p_1)$$

$$= s(\tfrac{1}{2}\rho V_1^2 - \tfrac{1}{2}\rho V_2^2) \tag{6.2}$$

6.6 GENERAL DEVELOPMENT OF BLADING SECTIONS

It can be shown that the angle by which the resultant force F_R is inclined to the cascade plane is equal to β_m, the angle the vector mean of V_1 and V_2, namely, V_m, makes with the normal to the cascade (Fig. 6.3). This angle is then given by

$$\tan A = F_N/F_T = \tan \beta_m$$

$$= \frac{1}{2}\left(\frac{\sec^2\beta_1 - \sec^2\beta_2}{\tan \beta_1 - \tan \beta_2}\right)$$

$$= \tfrac{1}{2}(\tan \beta_1 + \tan \beta_2) \qquad (6.3)$$

when $V_a \sec \beta_1$ and $V_a \sec \beta_2$ are substituted for V_1 and V_2, respectively.

Since the drag force is assumed to be zero, the lift and resultant forces can be equated. The former is given by

$$L = \tfrac{1}{2}\rho c C_L V_m^2 \qquad (6.4)$$

and hence

$$F_T = \tfrac{1}{2}\rho c C_L V_m^2 \cos \beta_m$$

Equating this expression to Eq. (6.1),

$$C_L = 2\left(\frac{s}{c}\right) \cos \beta_m (\tan \beta_1 - \tan \beta_2) \qquad (6.5)$$

An equivalent equation, which includes the drag coefficient, is developed in Chapter 8 from related reasoning.

The lift coefficient can therefore be expressed as a function of blade solidity c/s, flow deflection angle $(\beta_1 - \beta_2)$, and the outlet flow angle β_2.

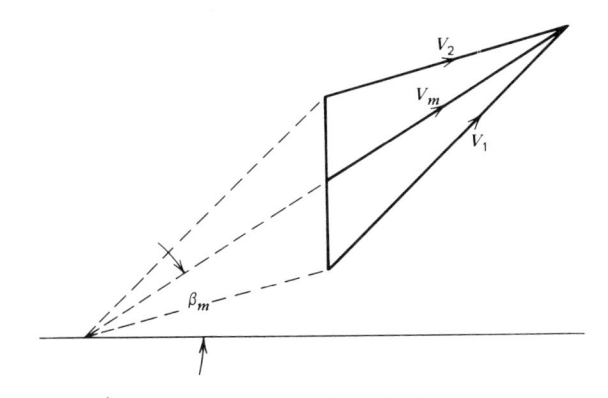

Figure 6.3 Mean relative velocity vector.

6.6.2 Relationship Between Blade Camber and Flow Deflection

From the foregoing it can be correctly postulated that the required blade camber must bear a definite relation to the design degree of air deflection. This is illustrated in Fig. 6.4, where the camber angle θ is shown as

$$\theta = (\beta_1 - \beta_2) + (\delta - i) \tag{6.6}$$

The angle the inlet flow makes with the leading-edge camber line tangent, known as the local incidence, has been shown to possess an optimum value, depending on cascade properties [6.1]; however, this value does not vary greatly from zero, the so-called shock-free inlet condition.

The deviation angle δ, being dependent on airfoil boundary layer growth, is a relatively difficult variable to evaluate with precision. However, since the selection of airfoil camber does not require a high degree of exactitude, approximate estimates are acceptable.

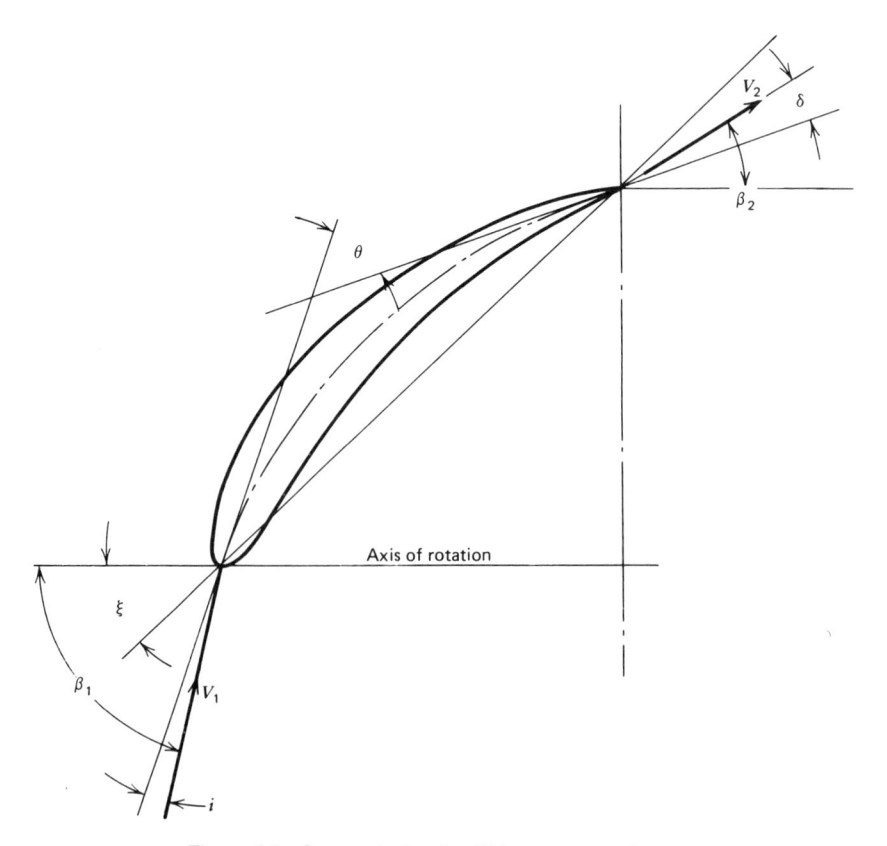

Figure 6.4 Geometric details of blade or vane element.

6.7 FAN BLADING

The geometric relationships for the camber line are

$$\text{Radius of curvature} = \frac{c}{2 \sin \theta/2} \tag{6.7}$$

$$\frac{b}{c} = \tfrac{1}{2} \tan \theta/4 \tag{6.8}$$

or

$$\frac{b}{c} \approx 0.00221\theta, \quad \text{in the range } \theta = 0 \text{ to } 50° \tag{6.9}$$

In fan design, the fan radius is used to nondimensionalize Eq. (6.7), giving

$$\frac{R_{\text{cur}}}{R} = \frac{c}{R} \frac{1}{2 \sin \theta/2} \tag{6.10}$$

6.6.3 Local Airfoil Nose Camber

The addition of a small amount of nose camber, or droop, improves the maximum, or stalling, lift coefficient and delays the drag rise to higher lifts, when compared with two-dimensional airfoils of a relatively low drag type [6.1]. The amount of nose droop remains a function of both blading and flow considerations [6.2].

6.6.4 Clothing the Camber Line

The various streamlined bodies specified here, when bent around the camber line spine of a blade section, have no significant influence on the design lifting or deflection properties of the blade element. However, the thickness/chord ratio aspect does result in slight profile drag changes with this ratio for the normal design thickness range.

6.7 FAN BLADING

A stage of rotor or stator blades is fundamentally a circular rather than a linear cascade of airfoils. Although the flow deflection will usually vary along the blade, flow conditions at a given radial station can be considered as being "two-dimensional," that is, independent of adjoining stations. This design feature, which holds for "free vortex" flows, stems from the theoretical work of Glauert [6.3], and has been validated in design practice.

The first axial flow machines designed were of a small pressure rise variety and employed low-solidity blading. As a consequence the blading design was successfully based on two-dimensional, isolated airfoil lift and drag data. Increased pressure duties, however, resulted in larger blade solidities with a progression toward mutual flow interference between adjacent blades. An interference factor C_L/C_{L_i} was introduced on the basis of theoretical studies. However, the use of this factor failed to predict accurately the cascade effect on blade performance because of the inviscid nature of the investigations. In addition, the evaluation of the maximum, or stalling, lift coefficient as a function of solidity remained outside the scope of the foregoing studies. The designers of axial flow compressors, therefore, were obliged to procure deflection, lift, and drag data from two-dimensional cascade wind tunnel tests. When combined with design and operating experience, this procedure became the accepted one.

Fans often experience multiplane interference toward the blade root, demanding a "blend" design solution. The method recommended here makes use of interference factors derived from experimental data for "tailored" airfoils with circular arc camber lines [6.4].

Any airfoil for which two-dimensional lift and drag data are available can be used in fan design. However, design accuracy is improved by the use of carefully selected sections for which experimentally established performance information is available. The latter permits extensive use of graphically presented design data in terms of the flow and swirl coefficients.

6.8 BLADE DATA

Four general categories of airfoil are herein recommended for use, namely,

1. A high-performance type of circular arc blading, being particularly suitable for high-pressure rise units.
2. A high-performance type of flat undersurface blading for low-pressure rise units.
3. Elliptical sections for flow-reversing design requirements.
4. A less efficient and cheaper variety of blading featuring cambered constant thickness plates.

6.8.1 F-Series Airfoils

This series meets the need for a unified approach to fan blade design, irrespective of blade solidity requirements. The selection followed a wide-ranging study of all relevant published experimental data, such as that presented in [6.5] and [6.6].

The section is the well-known circular-arc C4 airfoil modified to incorporate nose droop of varying amounts (Fig. 6.5). The coordinates of a 10% thick, 10° camber angle and 1% nose droop airfoil are listed in Appendix D.

6.8 BLADE DATA

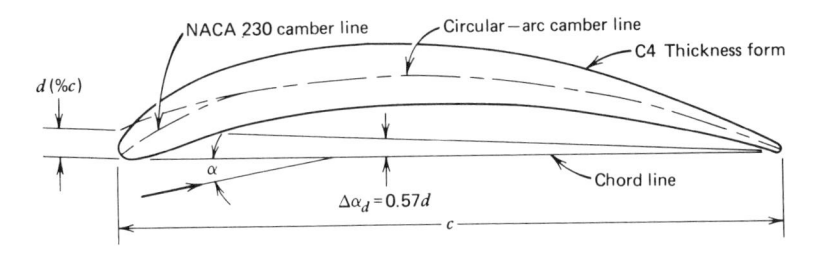

Figure 6.5 F-series airfoil geometry.

Linear relationships permit the coordinates of other sections to be established, namely,

$$Y_U = Y'_U + k_U(\theta - 10) + l_U(t - 10) + m_U(d - 1.0) \qquad (6.11)$$

$$Y_L = Y'_L + k_L(\theta - 10) + l_L(t - 10) + m_L(d - 1.0) \qquad (6.12)$$

where θ is in degrees and the thickness t and nose droop d are in percentage of chord. These equations are accurate for the normal rotor blade design limits of $\theta = 10$ to $36°$, $t = 7$ to 13%, and $d = 0$ to 3%.

Complete freedom of choice with regard to blade camber, blade thickness, and extra nose camber is therefore available. This linearity carries over into airfoil lift characteristics, for varying camber angles. Within the thickness and nose droop limits specified, the lift coefficient versus incidence curve remains unaltered, except in the vicinity of stall. The resulting design lift coefficient data are presented in Fig. 6.6 [6.2]; the no-lift versus camber angle relation is given in Fig. 6.7 [6.1]. The lift curve slope remains constant at 5.7 per radian of incidence within the design range ($C_L < 1.0$); it is relatively independent of Reynolds number for values above 3×10^5.

This unique presentation of lift and incidence data is based on the experimentally established feature of nonsubstantive changes in lift curve slope with first, blade camber, and second, thickness/chord ratio [6.1], for the normal design range of these variables. However, the incidences of Figs. 6.6 and 6.7 have to be corrected for nose droop since the former are with respect to the chord line of the basic circular-arc airfoil (Fig. 6.5). The corrections are

$$\alpha = \alpha_{0d} - \Delta\alpha_d = \alpha_{0d} - 0.57d \qquad (6.13)$$

and

$$\alpha_N = \alpha_{N,0d} - 0.57d \qquad (6.14)$$

154   AIRFOIL DATA FOR BLADE DESIGN

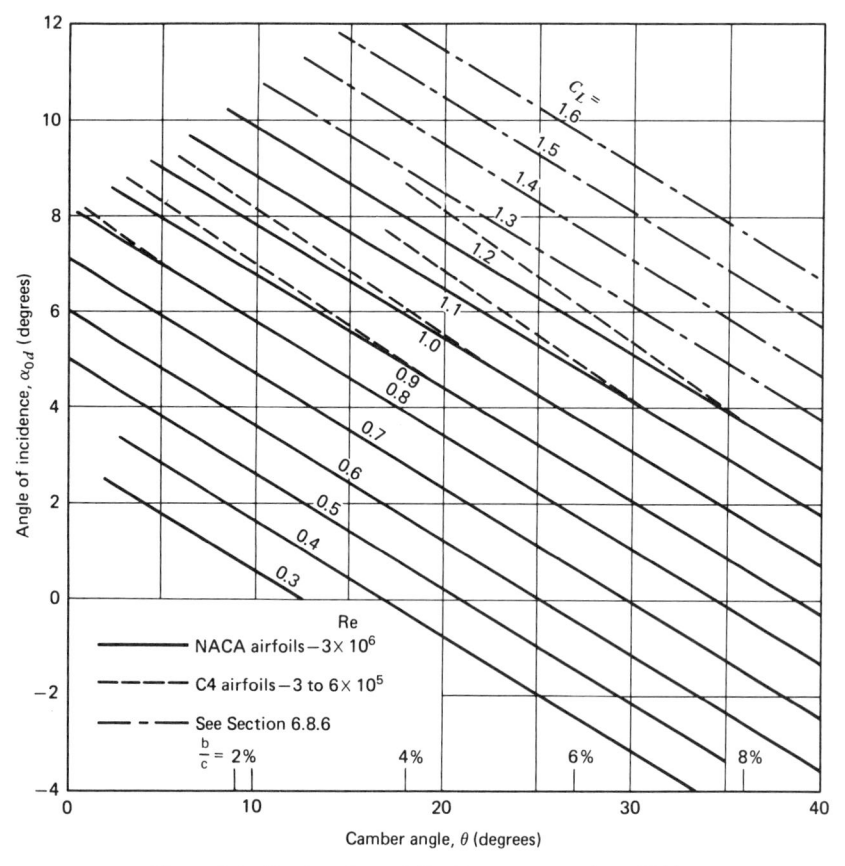

Figure 6.6 Cambered aerofoils, C_L versus α_{od}, data, [6.2].

where the subscript $0d$ signifies the zero nose droop case, d is in percentage chord, and N denotes the no-lift condition.

The predominance of the trailing-edge angle in determining lift is responsible for the preceding simplifying development (Kutta-Joukowski condition).

Since this airfoil series has not been subjected to actual wind tunnel tests, the $C_{L,\max}$ data are necessarily of an indirect type. However, keeping the whole fan design exercise in proper perspective, this limitation is not considered serious. A number of practical factors such as surface irregularities, roughness, dust erosion, spanwise lift loading, secondary flows, multiplane interference, Reynolds number, stream turbulence, and so on, have a more vital influence on the attainable $C_{L,\max}$ at fan stall. To provide a measure of guidance to the designer, Figs. 6.8 and 6.9 are presented; these illustrate the effects of camber, and thickness/chord ratio, for closely comparable smooth-

6.8 BLADE DATA

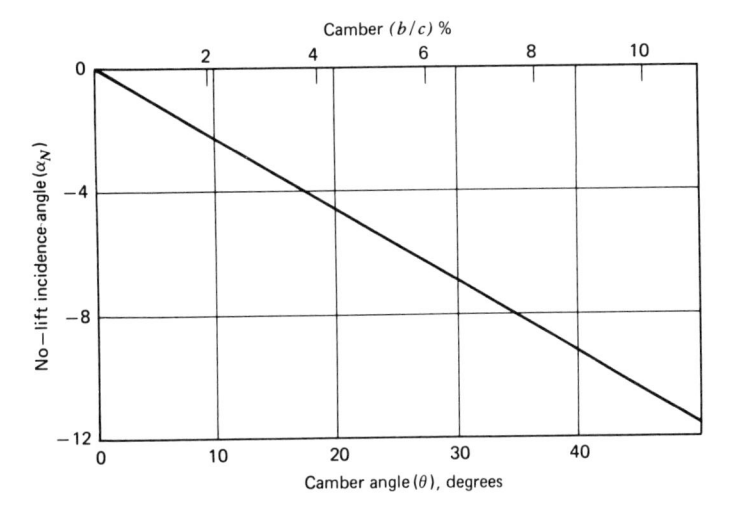

Figure 6.7 No-lift angle for circular-arc airfoils (zero LE droop).

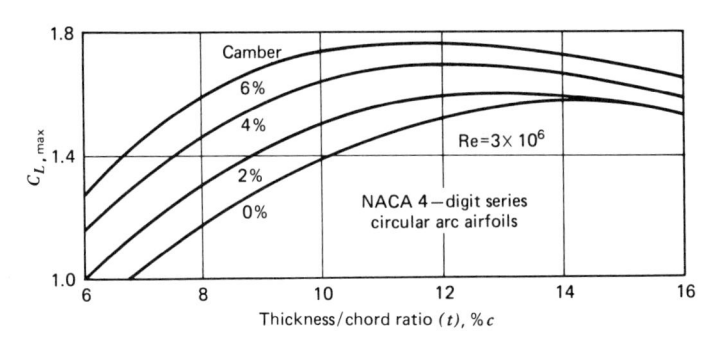

Figure 6.8 $C_{L,\max}$ versus thickness and camber ratios (smooth surfaces).

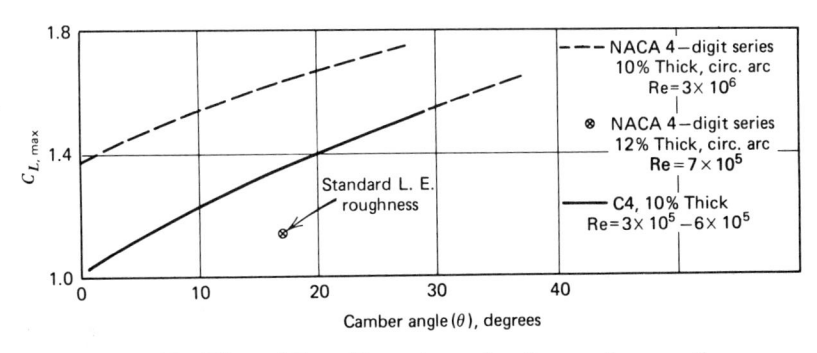

Figure 6.9 Effects of Reynolds number and surface roughness on $C_{L,\max}$.

AIRFOIL DATA FOR BLADE DESIGN

surfaced sections. The F-series airfoils, which have nose droop, and a slightly larger leading-edge radius than the NACA airfoils, will possess improved $C_{L,\max}$ properties when employed as fan blades.

Drag coefficient data adequate for design use are available in Figs. 6.10 and 6.11 [6.2]. The use of minimum drag data in design is supported by experiment in [6.7] where, for similarly developed airfoils, the minimum-loss condition was closely related to optimum leading-edge incidence. These data apply to smooth-surfaced airfoils only; allowances based on experience should be made for manufacturing imperfections and operational circumstances.

6.8.2 NACA 65-Series Airfoils

Although the F-series airfoils are favored here, worldwide research and development of the NACA 65-series sections necessitate their present inclu-

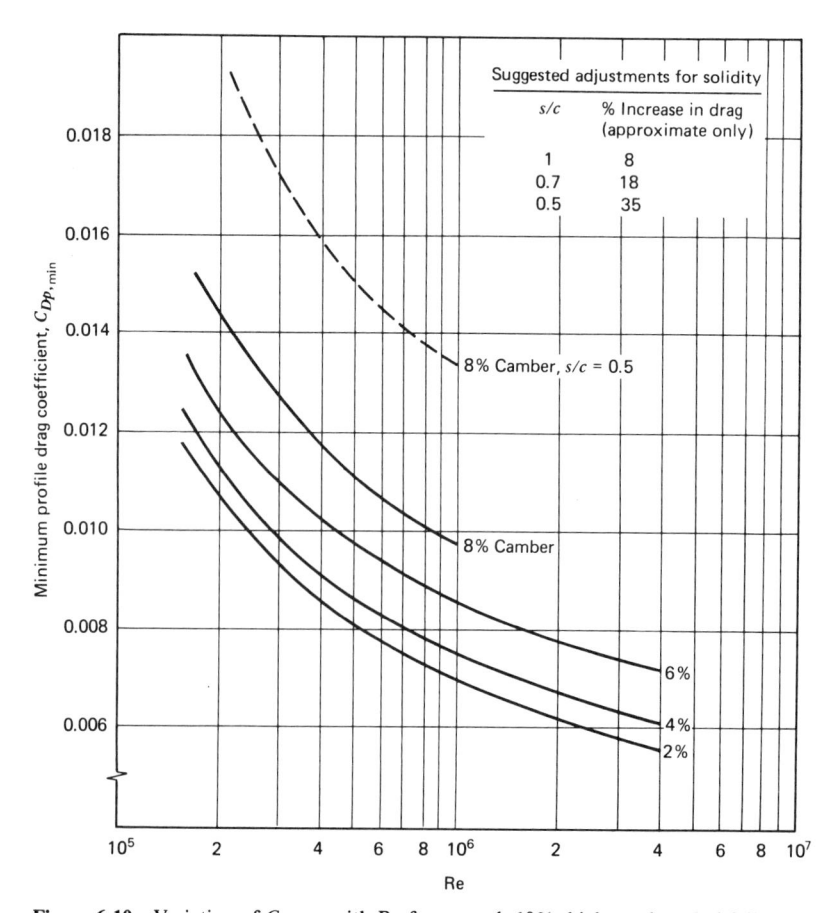

Figure 6.10 Variation of $C_{Dp,\min}$ with Re for smooth 10% thick cambered airfoils [6.2].

6.8 BLADE DATA

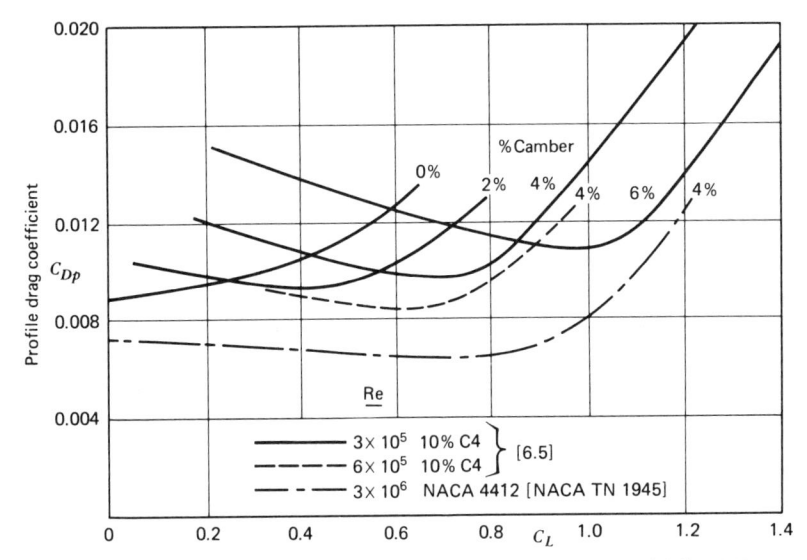

Figure 6.11 C_{D_p} versus C_L, smooth 10 to 12% thick cambered airfoils [6.2].

sion. Since these airfoils were initially selected for high Mach number operation in compressors, their maximum thickness occurs at 40% chord, whereas a 30% value applies to the low-speed airfoils discussed previously.

The published data on these airfoils as isolated lifting surfaces are confined to cambers of 1.1 and 2.2% chord [6.8]; these values are less than the normal fan blading ones. However, this problem can be tackled by adopting similar procedures to those previously described.

Because of reflex curvature toward the trailing edge and the associated airfoil thinning, these sections have been modified to provide a more practical aft thickness distribution, particularly for small chord blades [6.9].

The modified basic thickness form for a 10% thick section is presented in Appendix D. For other thickness ratios, the coordinates are adjusted on a linear basis. The coordinates of the base camber line are listed in Appendix D for a theoretical isolated airfoil lift coefficient C_{L_0} of unity. A constant chordwise pressure coefficient was assumed in calculating this mean line, which possesses a camber/chord ratio of 0.05515.

Camber is always specified in terms of C_{L_0}. For example, when the design requires a camber represented by $C_{L_0} = 0.6$, the camber line coordinates are obtained by multiplying the basic y_c quantities by 0.6. The section coordinates as a percentage of chord are then calculated from the relationships given in Appendix D.

In fan applications the aft profile is often modified still further, to attain a more rugged and practical section, particularly from a casting viewpoint.

The camber line is symmetrical about the mid-chord, featuring increasing curvature toward the leading and trailing edges. This departure from constant curvature bestows small amounts of additional leading-edge droop and discharge angle when compared with airfoils of circular-arc camber. The resulting "flap" effect produces a slight lift curve shift for an identical camber/chord ratio. Hence the equivalent camber is greater than the actual.

The curve due to Horlock [6.10] acknowledges this feature whereas that given in [6.7] ignores the contribution (Fig. 6.12). When the isolated airfoil data for the 65-210 and 65-410 sections at Reynolds numbers from 3 to 9 × 10^6 are extracted from [6.8], a more pronounced shift is evident. However, the modifications to the aft region of the airfoil are expected to reduce the "flap" effect and hence the relationship of [6.10] may be reasonably accurate for production blades operating at normal Reynolds numbers. The slight leading-edge droop tends to compensate for the smaller leading-edge radius, when comparison is made on maximum-pressure duty issues with C4 airfoils.

Significant improvements to the 65-series airfoils for fan use would follow the introduction of a composite camber line similar to the one adopted in relation to the F-series development.

The recommended fan design procedure once the C_{L_0} distribution is selected is to establish the equivalent camber angles from Fig. 6.12 and then to proceed along identical lines to those advocated for the F-series airfoils, using common lift and drag data, including those for multiplane interference.

The design procedure developed at NEL [6.11] evolved out of an analytical study of the experimental cascade airfoil data of [6.9]. It establishes the angle of attack $(\beta_1 - \xi)$ and C_{L_0} as functions of the inlet angle β_1. Linear

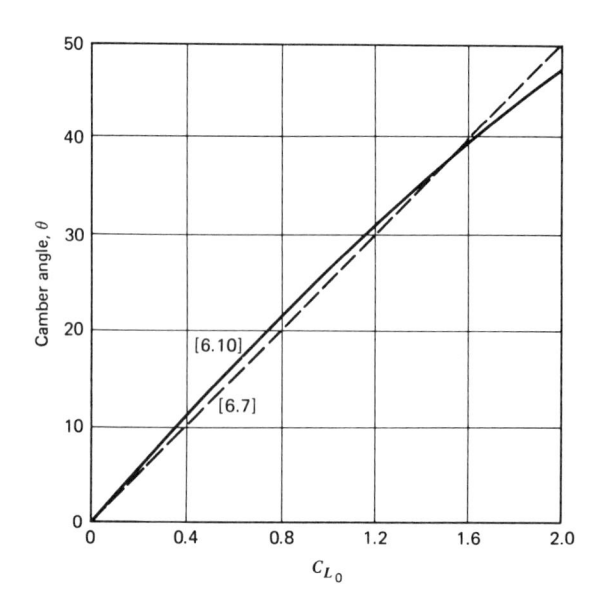

Figure 6.12 Relationship between C_{L_0} and θ, NACA 65-series compressor airfoils (Reproduced with permission of J. H. Horlock [6.10]).

6.8 BLADE DATA

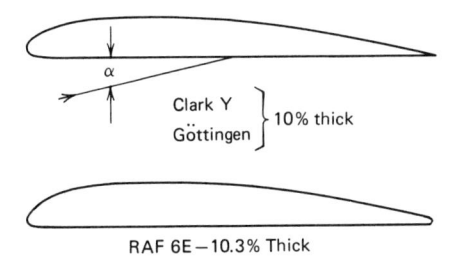

Figure 6.13 Flat undersurface airfoils.

extrapolation to zero solidity is included in the design method development. It can be shown that this latter expedient results in angles of attack that are too small, failing to generate the design C_L at low solidities. However, tests on a relatively high-solidity fan demonstrated the reliability of the design method at this end of the solidity range [6.12].

In view of the preceding limitation, a design method based on equivalent camber lines is the preferred one for general design purposes.

6.8.3 Airfoils with Flat Undersurfaces

Airfoils in this category are the Clark Y, Göttingen, and RAF 6 profiles (Fig. 6.13). All possess maximum thickness at 30% chord and varying this quantity without altering the flat undersurface is equivalent to changing the camber. For 10% sections the Clark Y and Göttingen airfoils possess additional leading-edge droop of approximately 1% c, while a 3% c value is more relevant to the RAF 6 case.

Since the sections aft of the maximum thickness location are very similar for all three types, the lift curves will be virtually identical for any selected thickness. Hence the no-lift angles and the lift can be plotted as a function of maximum thickness to chord ratio, for all practical purposes, when the lower surface constitutes the reference chord line. As before, the lift curve slope is

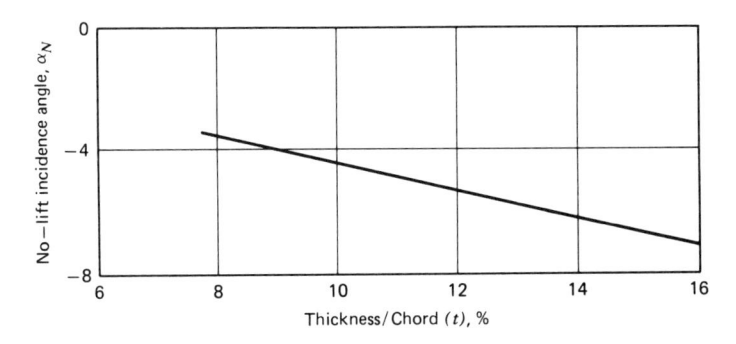

Figure 6.14 No-lift angle for flat undersurface airfoils.

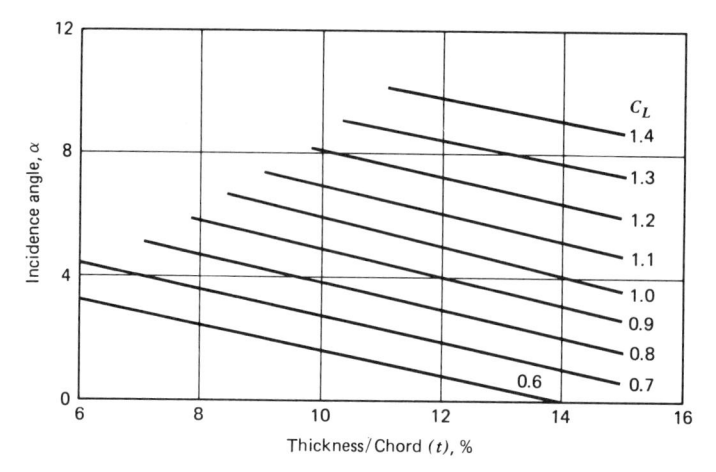

Figure 6.15 Lift versus incidence for flat undersurface airfoils.

constant at 5.7 per radian of incidence, in the design C_L range. The essential data for these presentations (Figs. 6.14, 6.15) are extracted from [6.13] to [6.16].

The drag coefficient data presented in Fig. 6.16 disclose differences related to leading-edge shape. The RAF 6E airfoil with its greater nose droop suffers a drag penalty at low C_L but surpasses the Clark Y at high C_L, for

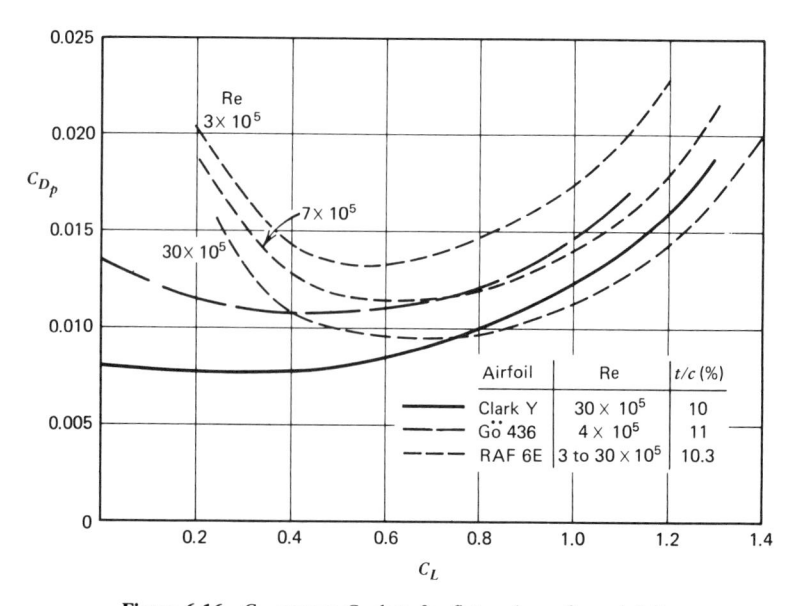

Figure 6.16 C_{D_p} versus C_L data for flat undersurface airfoils.

6.8 BLADE DATA

identical Reynolds numbers. The Göttingen and Clark Y sections possess similar characteristics when Re effects are taken into account. For blades requiring a design C_L of about 0.8 to 0.9, there is little significant drag difference between all three sections.

A small delay in the stall onset is indicated for the RAF 6E section.

An approximate estimate of the drag coefficient increment that accompanies the blade blockage effect is suggested in Fig. 6.10. However, much larger increases must be expected for thick airfoils of high solidity and large stagger angle.

The available $C_{L,\max}$ data on this type of airfoil possess inconsistencies. However, since 10% thick sections have approximately 3% camber, and roughly parallel the performance of 3% cambered circular arc sections, the use of Fig. 6.8 is recommended. The resulting estimates are in broad agreement with selected experimental data.

These sections are particularly suited to small air-turning angle situations where multiplane interference is nonexistent.

6.8.4 Elliptical Airfoils

Ellipses possess the same airfoil characteristics for reversed directions of airflow approach. They are therefore used in design situations where an identical duty condition is required for reversible fan assemblies, such as utilized in a number of preheat furnaces and drying kilns. Flat plates, which represent one minor to major axis ratio limiting case, are often used when the nondimensional pressure rise, and blade forces, are both small.

The amount of experimental data available for increasing thickness to chord ratio are limited to ratios of 0.167 and 0.5 [6.17, 6.18]. Lift and profile drag information for the former ratio is reproduced in Figs. 6.17 to 6.19 and shows a strong Reynolds number dependency, particularly in the range 0.3×10^6 to 2×10^6. In this Re bandwidth greater variations at 10° of incidence were present for the 0.5 ratio case. Hence it can be argued that for lesser ratios than 0.167 these large differences in lift and drag with Re will progressively diminish.

Flow characteristics in the region of the trailing edge determine lift. In the low Re case a large transverse vortex, due to boundary layer separation on the suction surface, is obviously entraining air in a trailing-edge downwash situation; the aft surface curvature assists this latter flow feature. With increasing Re the separation line moves aft, reducing the scale of vorticity and eliminating the induced downwash effect.

The lesser lift curve slope at the higher Reynolds numbers, in relation to the customary airfoil value, is due to unequal lengths of attached flows on upper and lower surfaces.

When the trailing-edge radius to chord ratio and aft surface curvature are both reduced, the preceding Re influences must merge with the usual airfoil ones. The lift curve slopes for the higher Re cases are given in terms of

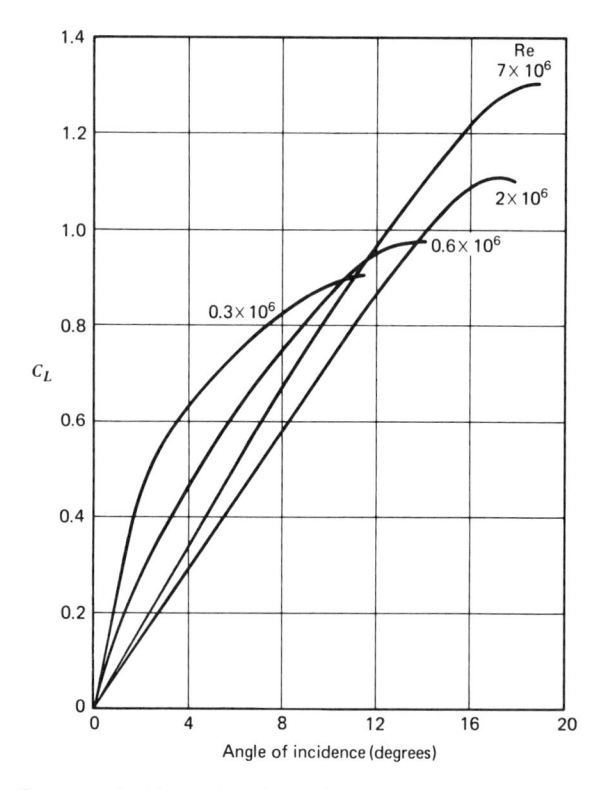

Figure 6.17 C_L versus incidence data for various Re, 16.7%c thick ellipse. (Taken from D. H. Williams and A. F. Brown [6.17] and reproduced with the permission of the Controller of Her Majesty's Stationery Office.)

thickness to chord ratio (Fig. 6.20). Included is the value for Re = 2×10^6, the lowest Re at which a linear lift curve exists for a t/c of 0.167. Stream turbulence will increase the effective Re. In view of the preceding flow prediction difficulties, a conservative and flexible design approach should be adopted.

The profile drag coefficient at design lift coefficients will, of course, be greater than the $C_{Dp,\text{min}}$ of Fig. 6.19, for $t/c = 0.167$. The design estimates of drag for this and lesser ratios must be made on the basis of design judgment and experience.

6.8.5 Cambered Plate Airfoils

This type of blading has many obvious advantages. When correctly shaped and constructed (Fig. 6.21), the resulting operational characteristics and peak efficiency can be good, provided high-quality inlet flow is assured in the

6.8 BLADE DATA

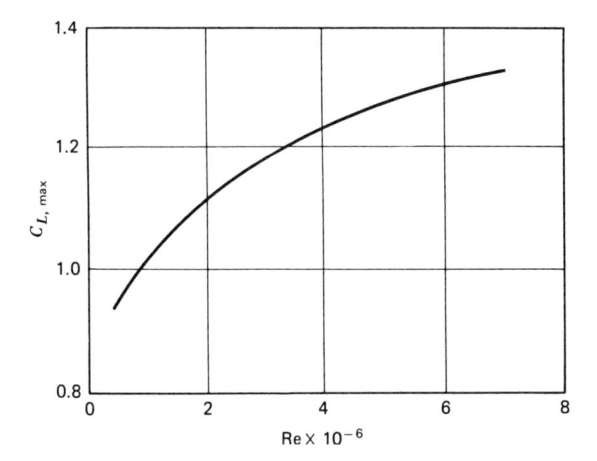

Figure 6.18 Maximum C_L versus Re, 16.7%c thick ellipse. (Taken from D. H. Williams and A. F. Brown [6.17] and reproduced with the permission of the Controller of Her Majesty's Stationery Office.)

duct/fan installation. Unfortunately, many existing fans fail to meet these basic requirements and consequently are noisy, inefficient, and inadequate.

Cambered plate fans have a reduced band of efficient operation compared with fans possessing sections of faired thickness form. When the local incidence i exceeds $\pm 2°$, flow disturbances at the leading edge cause local nose flow separation of varying chordwise extent, depending on local and overall section geometries. With increasing camber the preceding disturbances increase the upper-surface boundary layer thickness for positive local incidence and increase the severity of the local lower-surface separation for negative values. The former progressively reduces the magnitude of the lift curve slope while the latter is accompanied by a rapid falloff in lift with

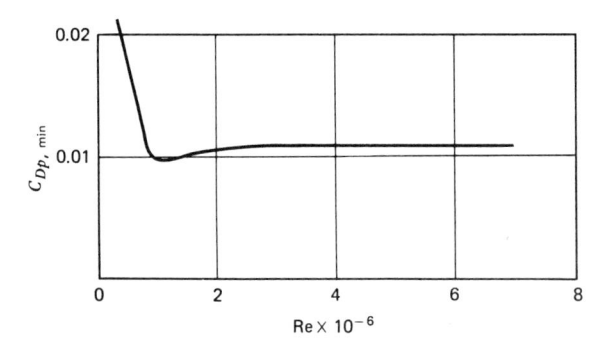

Figure 6.19 Minimum C_{D_p} versus Re, 16.7%c thick ellipse. (Taken from D. H. Williams and A. F. Brown [6.17] and reproduced with the permission of the Controller of Her Majesty's Stationery Office.)

164 AIRFOIL DATA FOR BLADE DESIGN

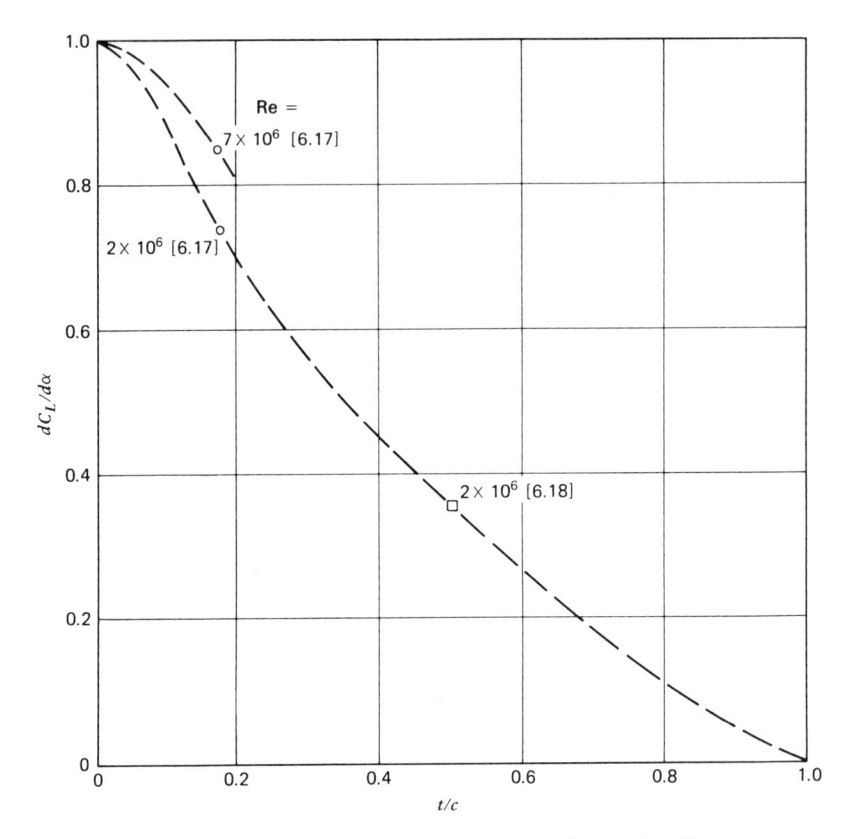

Figure 6.20 Lift curve slope versus thickness/chord ratio, ellipse.

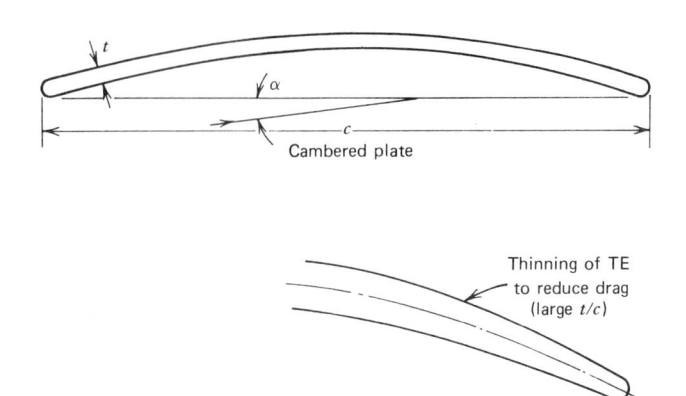

Figure 6.21 Cambered plate blade details.

6.8 BLADE DATA

decreasing incidence. The reduced lift curve slope is associated with an increased angle of deviation, due to the thickened boundary layer and separation moving forward with incidence from the trailing edge. The lift versus incidence data are presented in Figs. 6.22 and 6.23 [6.19].

However, despite the foregoing characteristic features, the experimental data can be rationalized for design and analysis, with a good degree of accuracy and confidence (Figs. 6.24 to 6.26). The straight line lift versus camber relations of Fig. 6.24 were established with less than 0.25° deviation from the experimentally determined values of incidence. Partial design guid-

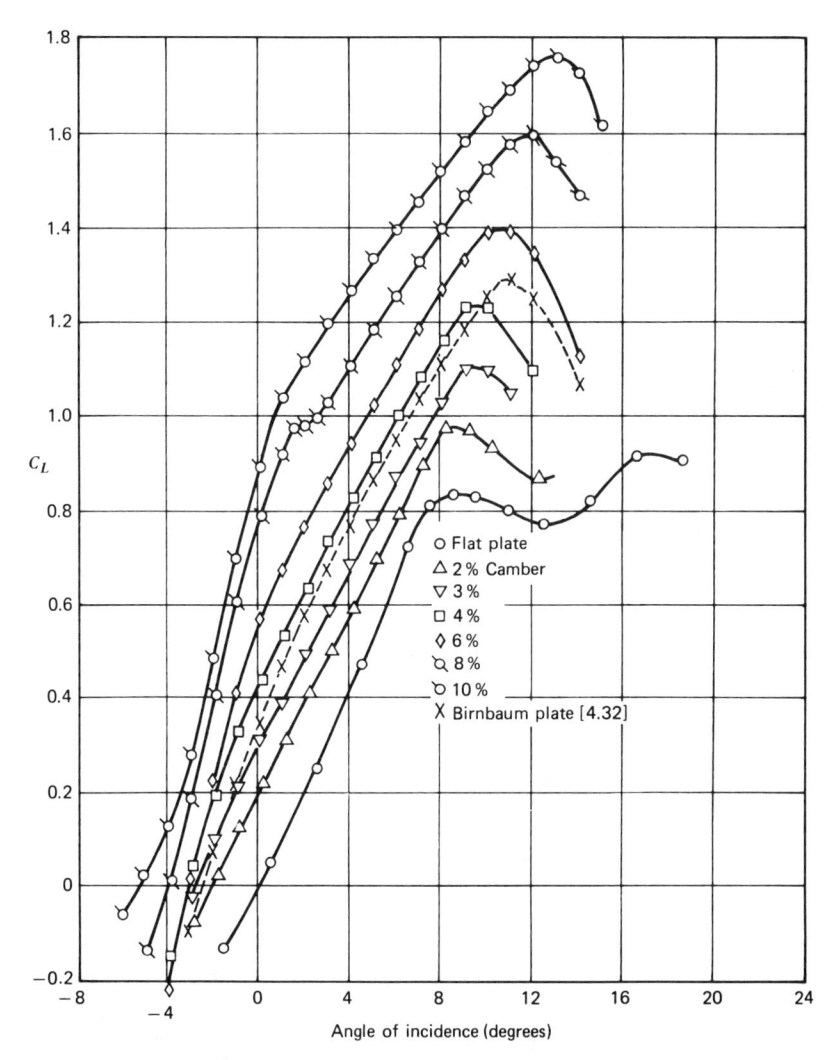

Figure 6.22 Cambered plate lift data, Re = 3×10^5 ($t/c = 0.02$).

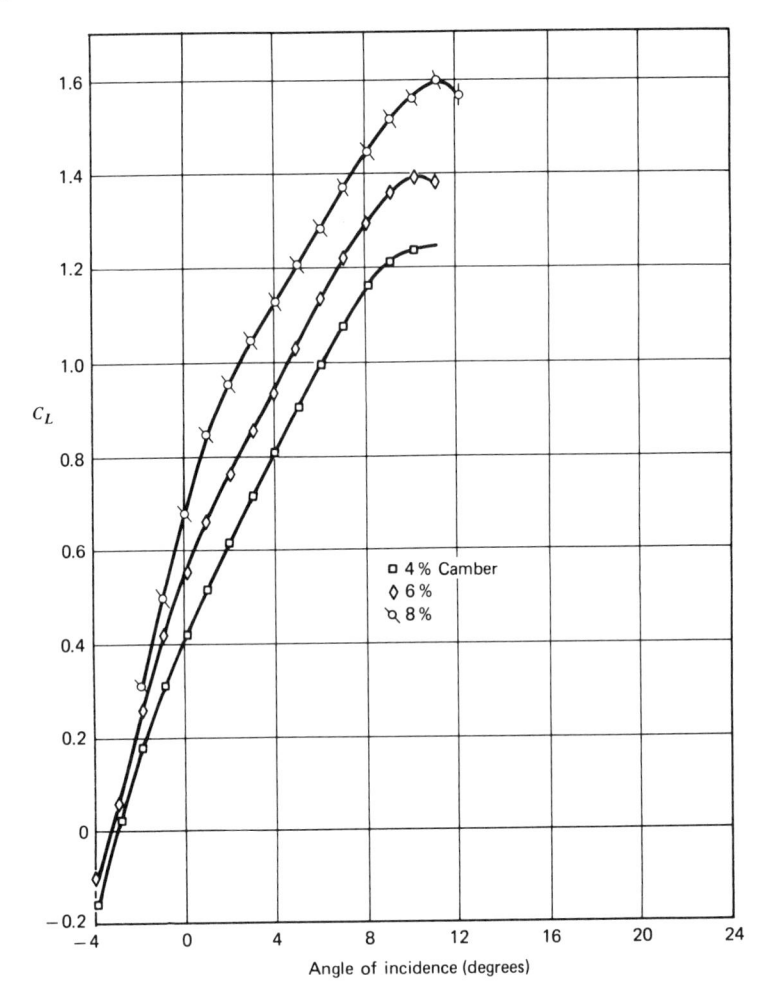

Figure 6.23 Cambered plate lift data, Re = 6×10^5 ($t/c = 0.02$).

ance is provided by Fig. 6.27, which graphs the lift coefficient versus blade camber, for maximum C_L/C_{D_P}.

Profile drag data are presented in Fig. 6.28 for 2% thick cambered plate sections at Re of approximately 3×10^5. Data for Re of 6×10^5, contained in [6.19], are not shown as these are now considered to be slightly excessive. The minimum profile drag coefficient at Re = 3×10^5 is approximately 0.017 for all cambers up to 8% with a design value of 0.02 at maximum C_L/C_{D_P}. These two values are subject to changes in t/c but the amount is unknown. The main importance of this design aspect, however, is in power estimation where a degree of conservatism is appropriate.

6.8 BLADE DATA

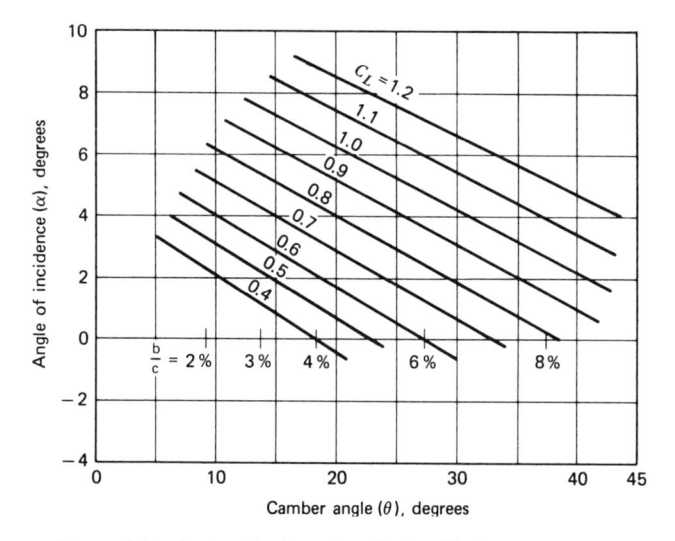

Figure 6.24 Rationalized cambered plate lift data ($t/c = 0.02$).

These sections are not normally suitable for fans absorbing large powers or where noise must be kept to an absolute minimum.

6.8.6 Interference Factor

In the case of high blade solidity, the two-dimensional lift and drag coefficient data will be modified by the flow around adjacent sections, the so-

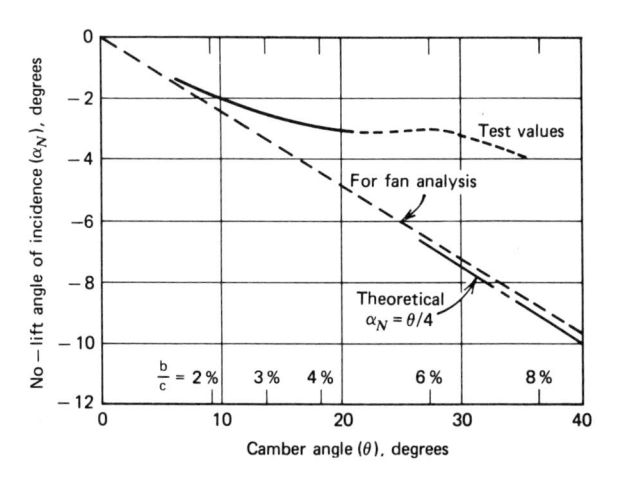

Figure 6.25 No-lift angle versus camber, for fan analysis ($t/c = 0.02$).

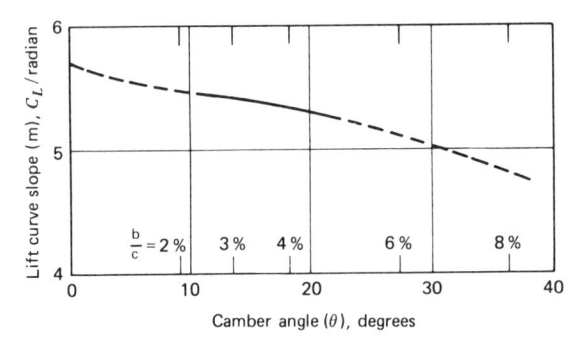

Figure 6.26 Lift curve slope versus camber, for fan analysis ($t/c = 0.02$).

called multiplane interference effect. The large air-turning angles associated with such blading necessitate the use of substantial blade cambers, particularly toward the blade root.

When attention is focused on optimum fan and blading situations, acceptable quantitative design answers emerge, as evidenced by the studies of [6.4] and [6.2] for faired airfoils possessing a circular-arc camber line. A generalized solution to the problem covering parabolic camber line airfoils is not available.

The lift curve of a carefully designed blade section is adequately characterized by the lift curve slope $dC_L/d\alpha$ and by the no-lift angle; both are modified by increasing solidity. However, when the enclosed design techniques are employed, the multiplane interference effect can be expressed as a single factor, namely, C_L/C_{L_i}, where C_{L_i} is the lift of an isolated section.

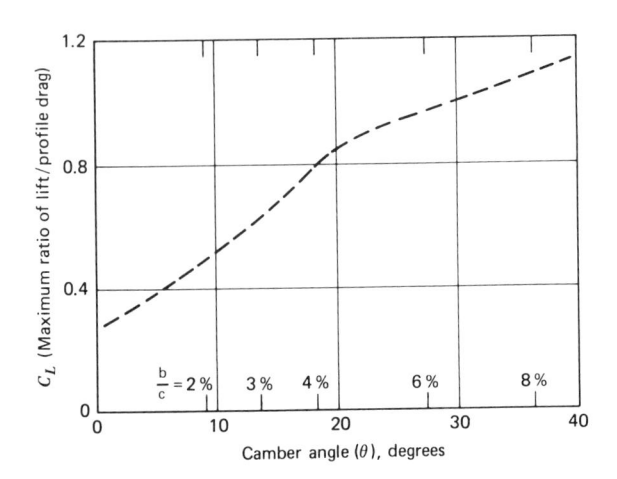

Figure 6.27 C_L for peak C_L/C_{D_p} versus camber ($t/c = 0.02$).

6.8 BLADE DATA

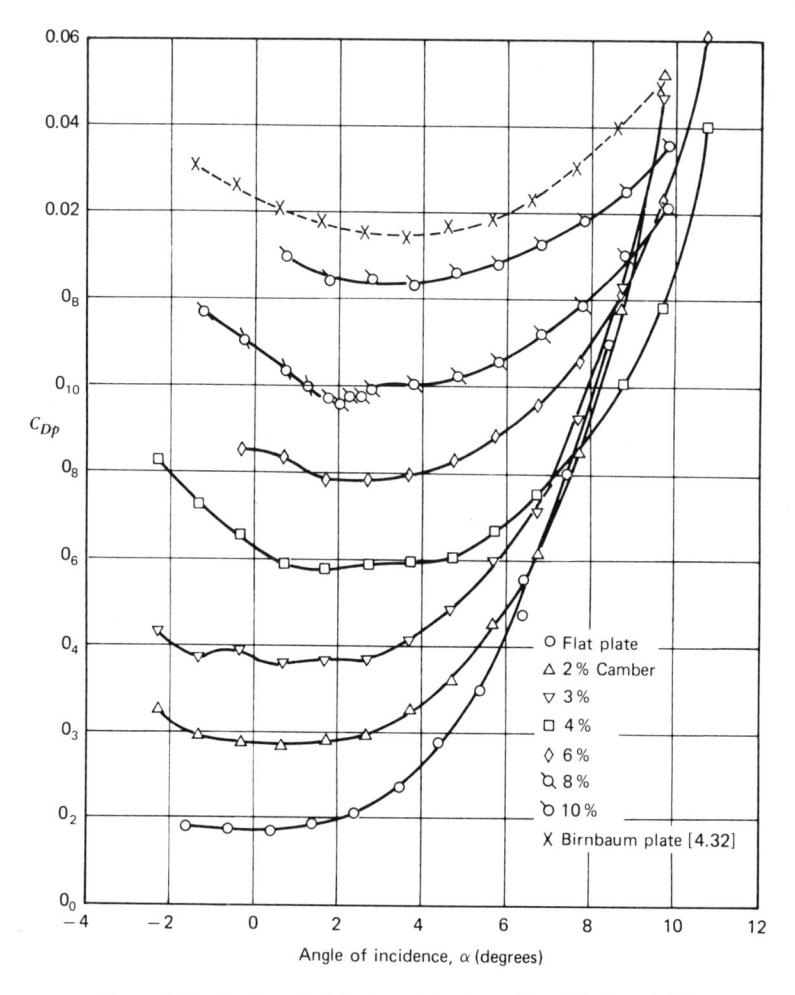

Figure 6.28 Cambered plate drag data, Re = 3×10^5 ($t/c = 0.02$).

In developing a reliable interference factor for design use, the nominal design lift coefficient expression due to Howell, namely,

$$C_L^* = 2 \left[\frac{\cos \beta_1}{\cos \beta_2} \right]^{2.75} \tag{6.15}$$

was adopted [6.4]. With increasing air-turning angle ($\beta_1 - \beta_2$) and blade solidity σ, the maximum lift of the section decreases, forcing the adoption of a lower design C_L. The nominal condition is defined, on the basis of British cascade airfoil experiments, as that associated with an air deflection 80%

of the maximum, where the stall is assumed to occur when C_{D_P} equals twice the minimum value. Working from the U.S. cascade data [6.9], Myles et al. [6.11] arrive at similar design C_L values centered on maximum lift to total drag ratios. The data obtained by the use of cascade design methods were modified at high blade stagger angles, on the basis of theoretical studies, since the former underestimates the deviation angle δ to a marked extent in such instances. An additional consideration involved the acceptance of unity as the upper limit of C_L/C_{L_i}. The latter condition is founded on practical experience since there are no documented examples of factors exceeding unity. The tendency of theoretical methods to predict lift ratios in excess of 1 for high stagger angles also appears linked to deviation angle discrepancies between theory and practice.

Assuming a smooth transition from cascade to isolated airfoil data, and working within the preceding framework, a series of simple design curves (Fig. 6.29) has been derived. Accuracy is expected to be good with inadvertent blade setting angle errors likely to exert a greater influence in practice. The method involves the calculation of the hypothetical C_{L_i} value and subsequently its related incidence, assuming a continuation of the linear lift curve; this ignores the onset of stall for the isolated airfoil concerned. The chain dotted lines in Fig. 6.6 represent the C_{L_i} values for the proposed lift curve extensions.

The blade loading factor, $(c/s)C_L$ [Eq. (6.5)] at a particular radius will allow C_L to be calculated once c/s is selected. In choosing the latter a slight divergence from values derived from Eq. (6.15) is recommended, as dis-

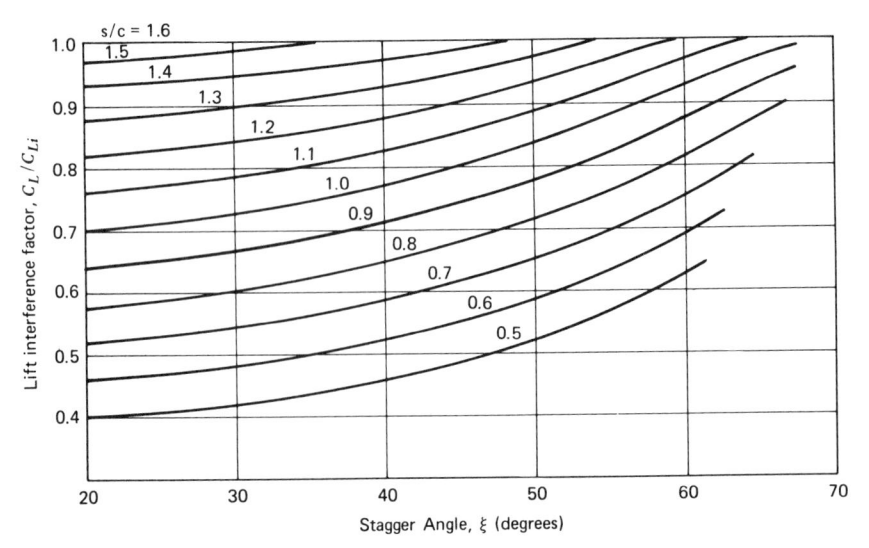

Figure 6.29 Multiplane interference factor for airfoils with near-optimum solidity and camber [6.2].

6.9 EFFECT OF REYNOLDS NUMBER ON AIRFOIL DATA

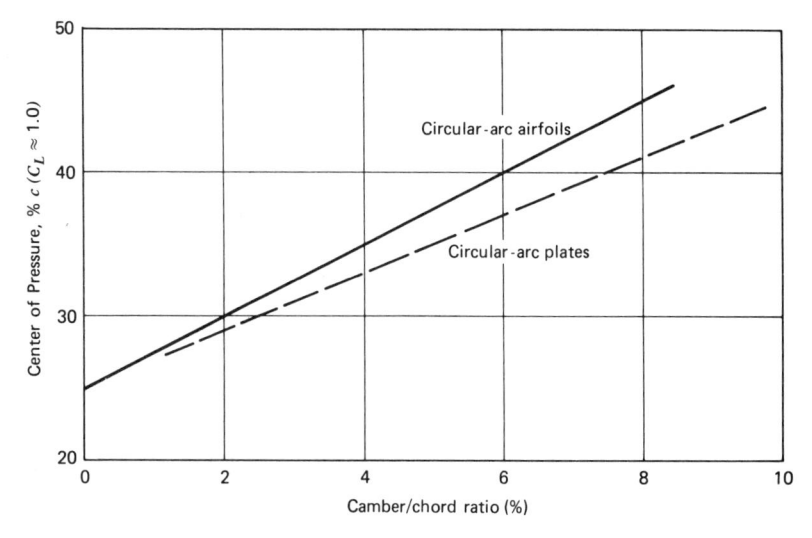

Figure 6.30 Center of pressure as function of camber.

cussed in Section 9.3.3; the degree of tolerance in the foregoing development is more than adequate to cover these variations.

6.8.7 Center of Lift

The displacement of the center of pressure from the blade section centroid constitutes a moment arm that is of some importance in either flexible or variable pitch blade operation. The lift center is a function of airfoil camber and lift coefficient, for isolated airfoils.

When the lift coefficient approximates unity, the rearward movement with airfoil camber is indicated in Fig. 6.30. This relationship also holds for the flat undersurface airfoils, which for $10\%c$ thickness have an equivalent camber/chord ratio of approximately 3%.

The center of pressure for ellipses at incidence lies slightly ahead of the $0.25c$ position [6.17].

When the pressure field around an airfoil is modified by an adjacent blade, a change in lift center location might be expected. The author is unaware of any published data on this matter.

6.9 EFFECT OF REYNOLDS NUMBER ON AIRFOIL DATA

The fundamental significance of Reynolds number has already been outlined in Section 2.4. As mentioned there, airfoil force data are usually presented as functions of the chord Reynolds number.

172 AIRFOIL DATA FOR BLADE DESIGN

The major influence on lift coefficient usually occurs at the higher incidences where the stall is imminent. The minor variations at lesser incidences can be explained in terms of the boundary layer growth. The thinner upper surface boundary layer that exists for larger Reynolds numbers increases the effective airfoil camber slightly due to a lesser boundary layer displacement effect. Provided the airfoil is matched to the flow deflection requirement, the lift curve will not be significantly displaced, with the lift curve slope remaining virtually unchanged.

The profile drag coefficient will respond more positively to changes in boundary layer thickness. However, the main interest in the resulting effects will lie in operational performance rather than in design procedure changes. This is further emphasized by problems associated with estimating the effect of unknown amounts of surface roughness and "free stream" turbulence on drag. The magnitude of changes in estimated fan performance will be kept to a minimum when the airfoil recommendations put forward here are adhered to.

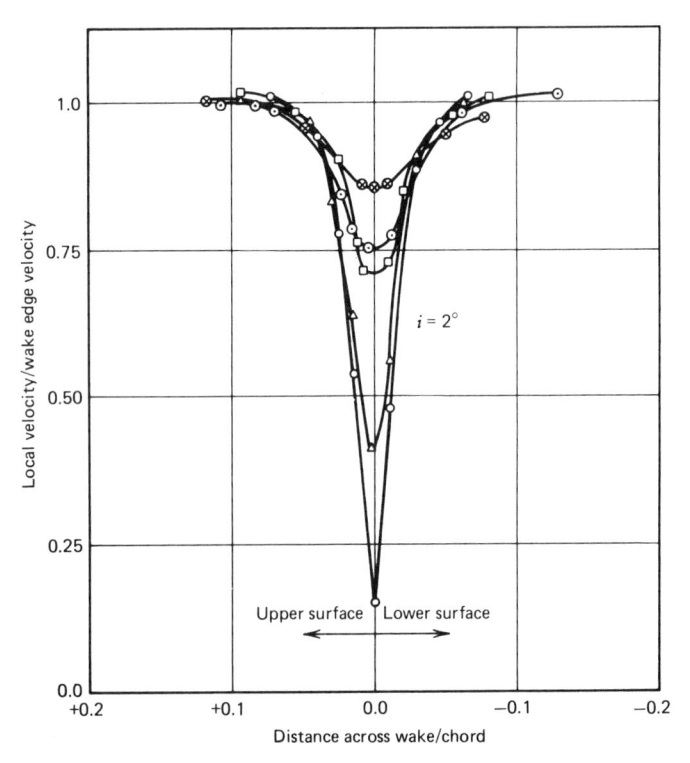

Figure 6.31 Wake profile at various downstream locations. (Reproduced with permission of Cambridge University Press from R. Raj and B. Lakshminarayana [6.20].)

6.10 WAKES

In the case of fans working at very low Reynolds number, large-scale laminar separations can greatly modify the force data; a greater degree of additional leading-edge camber will tend to alleviate the problem but actual operating experience must provide the overall characteristics of the unit. Lift curve slope will also become a variable.

The force data presented in this chapter are adequate for designs using the recommended airfoils. Further guidance is provided in Section 10.2.2.

6.10 WAKES

As the flow leaves the trailing edge of an airfoil at working incidences, the boundary layers from the two surfaces of the airfoil join to form a wake or region of low velocity (Fig. 8.1). Initially the gradient of velocity across the wake is very large, but as the flow progresses downstream this gradient is

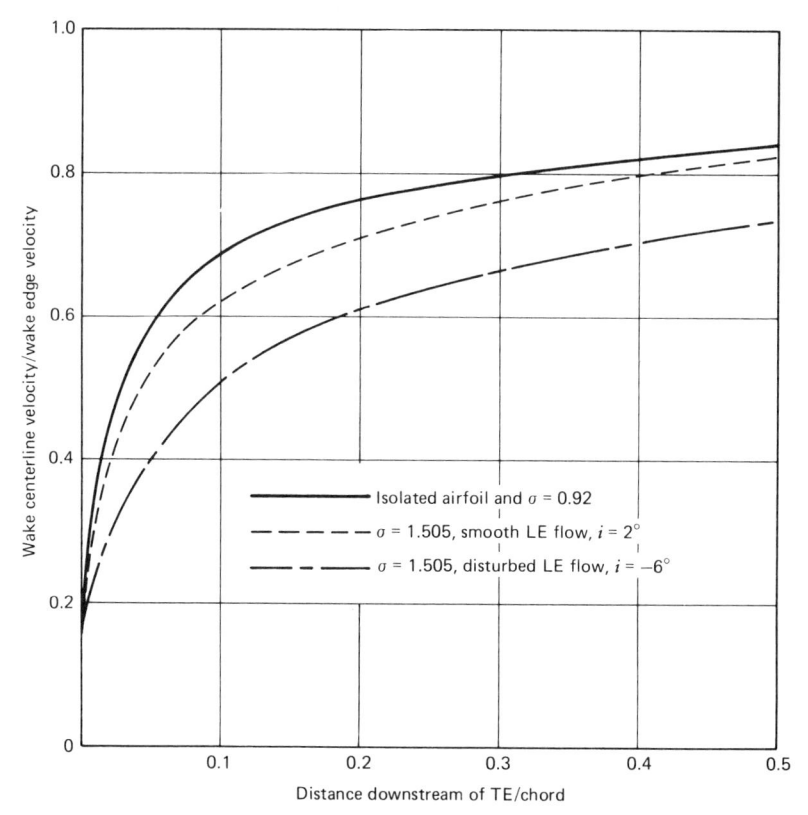

Figure 6.32 Maximum velocity defect with downstream distance. (Reproduced with permission of Cambridge University Press from R. Raj and B. Lakshminarayana [6.20].)

rapidly reduced under the influence of turbulent mixing. This process involves a progressive increase in the width of the wake until at an infinite distance downstream a uniform velocity field is again established and hence the wake ceases to exist. (Incidentally, the profile drag of an airfoil can be determined by measuring the momentum deficiency in the wake a little distance downstream of the trailing edge.)

Raj and Lakshminarayana [6.20] have experimentally established the wake flow characteristics for a cascade of airfoils (Figs. 6.31 and 6.32). The rate at which the wake centerline velocity recovers is influenced by both blade solidity and leading-edge flow conditions, with the latter influencing both wake width and shape. As the ratio, width of wake flow/width of free stream, increases because of the preceding influences, the Bernoulli equation gives rise to a small but significant downstream adverse pressure gradient. This gradient is believed to be responsible for the velocity recovery retardation apparent in Fig. 6.32.

When turbulent boundary layer separation is present on the suction surface of the airfoil, the wake is greatly widened; the stall is characterized by large, unsteady eddying motions (see Fig. 6.1) similar to those commonly observed downstream of a pier in a river. This phenomenon is, of course, indicative of a large force to which the airfoil is subjected, mainly in the form of pressure drag.

Provided the designer follows the design recommendations contained here, a detailed knowledge of the blade wake is not required. However, ignoring the qualitative understanding of wake phenomena can result in unsatisfactory fan performance, blade vibration and subsequent failure, and grossly excessive fan noise.

6.11 REVERSED AIRFOIL LIFT

When highly loaded fans such as mine ventilation units have to provide a reversed airflow for possible emergency fire-fighting purposes, the trailing edge becomes the leading edge when rotational direction is changed. The ensuing performance cannot be estimated since the fan will be stalled. This is broadly indicated in Fig. 6.33 [6.21, 6.22] where lift curves for airfoils with 0 and 2% camber are presented; the $C_{L,\max}$ values are relatively small. The main fire fighting requirement is for a positive pressure and restricted airflow and hence fan stalling is of little consequence from a performance viewpoint.

The increase in lift for the lesser Re is due to a similar phenomenon to that discussed for the elliptical sections, namely, a change in the effective location of the rear stagnation point. Symmetrical airfoils therefore lack stability at low Re.

However, fan blading will normally possess camber; predictions indicate a greater stability with these sections. Nevertheless, the lift will decrease progressively as the camber is increased, resulting in small $C_{L,\max}$ and high

REFERENCES

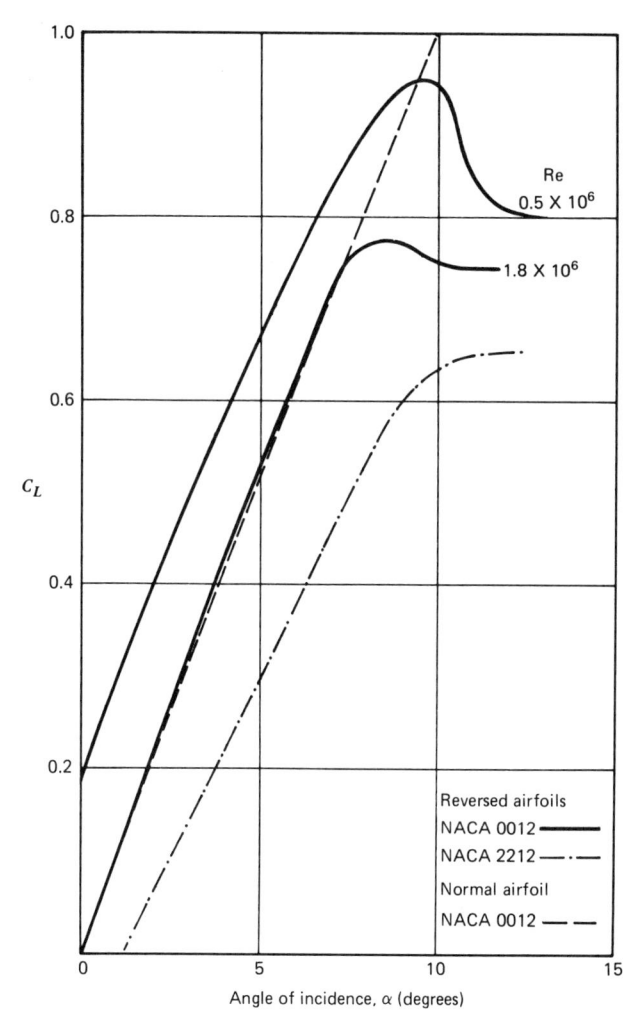

Figure 6.33 Reversed airfoil lift data.

drag. As a consequence the fan must be able to withstand high levels of unsteady load and rotor vibration.

REFERENCES

6.1 R. A. Wallis, The development of blade sections for axial flow fans, *Mech. Chem. Eng. Trans. I. E. Aust.*, **MC8** (2), 111–116, 121, Nov. 1972.

6.2 R. A. Wallis, The F-series airfoils for fan blade sections, *Mech. Eng. Trans. I. E. Aust.*, **ME2** (1), 12–20, 1977.

- 6.3 H. Glauert, *The Elements of Aerofoil and Airscrew Theory*, Cambridge University Press, London, 1926, p. 208.
- 6.4 R. A. Wallis, A rationalized approach to blade element design, axial flow fans, *Proc. Third A'asian Conf. Hydraulics and Fluid Mechanics*, Sydney, 1968, pp. 23–29.
- 6.5 N. Ruglen, Low speed wind tunnel tests on a series of C4 section aerofoils, Aust. Dept. of Supply, Aero. Research Labs, *ARL Aero Note 275*, 1966.
- 6.6 E. N. Jacobs, K. E. Ward, and R. M. Pinkerton, The characteristics of 78 related airfoil sections from tests in the variable density wind tunnel, *NACA Report 460*, 1933.
- 6.7 S. Lieblein, Experimental flow in two-dimensional cascades, in I. A. Johnsen and R. O. Bullock, (Eds.), *Aerodynamic Design of Axial Flow Compressors*, National Aeronautics and Space Administration, Rep. NASA SP-36, 1965, Chap. VI.
- 6.8 I. H. Abbott and A. E. von Doenhoff, *Theory of Airfoil Sections*, Dover, New York, 1959.
- 6.9 J. C. Emery, L. J. Herrig, J. R. Ervin, and A. E. Felix, Systematic two-dimensional cascade tests of NACA 65-Series compressor blades at low speeds, *NACA Report 1368*, 1958.
- 6.10 J. H. Horlock, *Axial Flow Compressors*, Butterworth, London, 1958.
- 6.11 D. J. Myles, R. W. Bain, and G. H. L. Buxton, The design of axial flow fans by computer, Part 2, *Gt. Britain National. Eng. Lab. Report 181*, 1965.
- 6.12 J. E. Hesselgreaves and M. R. Jones, The performance of an axial flow fan designed by computer, *Gt. Britain National Eng. Lab. Report 447*, 1970.
- 6.13 R. M. Pinkerton and H. Greenberg, Aerodynamic characteristics of a large number of airfoil tests in the variable density wind tunnel, *NACA Report 628*, 1938.
- 6.14 G. N. Patterson, Ducted fans: design for high efficiency, Australian Council for Aeronautics, ACA Report 7, 1944.
- 6.15 F. W. Riegels, *Aerofoil Sections*, Butterworth, London, 1961.
- 6.16 E. N. Jacobs and I. H. Abbott, Airfoil section data obtained in the NACA variable-density tunnel as affected by support interference and other corrections, *NACA Report 669*, 1939.
- 6.17 D. H. Williams, and A. F. Brown, Experiments on an elliptic cylinder in the Compressed Air Tunnel, *Gt. Britain Aero. Research Council, ARC R & M 1817*, 1937.
- 6.18 E. C. Polhamus, E. W. Geller, and K. J. Grunwald, Pressure and force characteristics of non-circular cylinders as affected by Reynolds number with a method for determining the potential flow about arbitrary shapes, *NASA Tech. Rep. R 46*, 1959.
- 6.19 R. A. Wallis, Wind tunnel tests on a series of circular arc plate aerofoils, Aust. Dept. of Supply, Aero. Research Labs, *ARL Aero Note 74*, 1946.
- 6.20 R. Raj and B. Lakshminarayana, Characteristics of the wake behind a cascade of aerofoils, *J. Fluid Mech.*, **61**, (4), 707–730, 1973.
- 6.21 C. C. Critzos, H. H. Heyson, and R. W. Boswinkle, Aerodynamic characteristics of NACA 0012 sections at angles of attack from 0° to 180°, *NACA Tech Note TN 3361*, 1955.
- 6.22 A. Naumann, Pressure distribution on wings in reversed flow, *NACA Tech. Memo. TM 1011*, 1942.

CHAPTER 7
Introduction to Fan Design Methods

Fan design consists in designing two-dimensional blade sections at various radii; a moderate number of stations usually suffices, as additional information can be obtained by interpolation.

Although the basic equations and assumptions are often identical, published design methods can vary in detail. The design method presented here is one that has been evolved by the author from the original work by Patterson [7.1]. The most important feature claimed for it is the *expression of most design parameters in terms of two basic velocity ratios*. This, of course, permits a complete graphical representation of the design and performance equations. These ratios have clear physical significance, which adds to their value as major design parameters.

7.1 MAJOR DESIGN PARAMETERS

It can be shown that the ratio between the swirl velocity and the axial velocity component, denoted by ϵ, at a given radius is an important variable. As stated previously, the swirl imparted, or removed, by the rotor is a measure of the rotor torque. Assuming for a given flow and pressure rise that the efficiency remains constant, it is apparent that the amount of swirl for a given power input can be changed by altering the design speed of the rotor. This introduces a second parameter, the flow coefficient λ, which is the ratio between the axial component and the rotational rotor speed at a given radius.

Rotor design is almost exclusively a function of the preceding coefficients of swirl and flow; stator design is largely a function of the former.

The quantity common to the preceding two coefficients is the axial velocity component, and for consistency the axial velocity pressure has been used in reducing static pressure and total pressure changes to a nondimensional form. In many existing design methods the blade tip velocity is chosen as the reference velocity; it is felt, however, that the axial velocity is the more suitable quantity as it eliminates one variable and is more meaningful from the fluid dynamic viewpoint.

7.2 SPECIFIC SPEED

The idea of specific speed was first evolved in connection with water turbines as a means of classifying the wide range of hydrodynamic machines, from Pelton wheels, using small flows under heads of hundreds of meters, to reaction turbines, using very large quantities under heads of a few meters.

The original conception was the unit machine, geometrically similar in all respects to the given machine, but of such a size that under similar operating conditions, it would develop unit power on unit head. The speed of this unit machine was called the specific speed.

The phrase "similar operating conditions" means that the velocity vector diagrams for the flow through the rotors of the two machines are similar, assuming that the efficiency of the unit machine is identical with that of the given machine.

As a result of accumulated test experience, the turbine manufacturer was able to plot parameters relating to important design data against the specific speed; these data could then be used to assist in new designs. One of the main advantages was the information provided concerning the most suitable types of turbine for specific duties.

The success obtained with turbines led to the use of specific speed in pump and fan design. The parameter, for fixed density, is usually expressed as

$$n_s = n Q^{1/2} \Delta H^{-3/4}$$

where n is fan speed, Q is rate of flow, and ΔH is head rise.

In the early days it proved a very useful guide as to whether an axial flow or centrifugal fan should be used for a given task. Specific speed also provided a basis on which empirical methods of design were evolved. With the development of more precise methods, and the evolution of the axial flow machine as one capable of doing work normally considered suitable only for a centrifugal fan, the use of specific speed as a parameter has lapsed to some extent.

The designer can now select rotor and boss diameters, and rotational speed, from the combination of a series of simple graphs and computations [7.2], as outlined in Chapter 15; the designs that possess peak efficiency are clearly identified. This approach is preferable to a specific speed one, being based on actual detailed design parameters and features.

The growing importance of energy conservation will accelerate the trend toward equipment designed for peak efficiency operation.

7.3 SPANWISE LOAD AND VELOCITY DISTRIBUTIONS

Free vortex flow is achieved when both the total pressure rise and axial velocity component remain constant along the blade span. These conditions are never present in practice due mainly to the annulus wall boundary layers. A review of experimental studies of units designed with spanwise axial velocity variations is available in Chapter 23. From these tests it can be deduced that, for fans particularly, the preceding spanwise constancies provide the most satisfactory design starting point.

The design flow and swirl coefficients are consequently inversely proportional to the blade element radius ratio.

A steady, swirl-free, and axisymmetric fan entry flow is a necessary prerequisite for the implementation of the preceding design approach. The flow should be axial and unseparated on the annulus walls at entry to a blade row.

Practical considerations sometimes dictate a reducing total pressure rise toward the blade root. This design decrease is in keeping with the lesser local blade work capability for fans of small boss ratio. Specific design procedures are presented in Appendixes A and B.

In multistage fans a progressive peaking of the velocity distribution will occur as the number of stages is increased. This will decrease the work output based on a uniform velocity distribution. The introduction of a "work-done" factor, as discussed in Chapter 23, is recommended for design use.

7.4 THE BLADE ELEMENT

In the foregoing we have considered, in broad outline, overall flow changes without discussing the ability or otherwise of the fan blade to produce these changes.

A blade element at a given radius can be defined as an airfoil section of vanishingly small span. It is assumed in design that each such element can be considered as a two-dimensional airfoil, that is, one completely independent of conditions at any other radius. In addition, it can be assumed that flow conditions are steady over the whole area swept by the rotor blades. This is equivalent to assuming an infinite number of small chord blades. In practice, the preceding assumption has proved satisfactory even when the number of blades is small. It applies equally to rotor and stator blades.

There is an upper limit to the amount of total pressure rotor blades are capable of introducing into the flow. The total pressure rise, which is closely related to the lift on the blades, is dependent on four main factors, namely, the dynamic pressure due to the relative velocity, the blade chord, the blade camber, and the angle of incidence the velocity vector makes with the blade. An increase in each of these will, within limits, increase the attainable total pressure rise.

For a given mass flow, the relative velocity may be increased by increasing either the axial or the rotational component. The former is achieved by increasing the boss diameter and/or by reducing the fan diameter. The subsequent diffusion that is often necessary downstream of the fan can, however, reduce the overall efficiency substantially. The second expedient, that of increasing the rotor speed, also has its limitations as the loss in efficiency at small flow coefficients, that is, large relative rotational speeds, can be appreciable. In addition, compressibility trouble may arise if the relative velocity at the tip exceeds 160 m/s. Noise is augmented with increasing tip speed.

Instead of blade chord, it is more convenient to consider the ratio of blade chord to circumferential gap, namely, blade solidity. Increasing solidity will eventually result in a marked reduction in lift for a given incidence, as discussed in Chapter 6. Furthermore, there is a limit to the loading factor $C_L\sigma$, which in turn is a function of stagger angle. Adherence to the design recommendations contained here will ensure that the appropriate design limits are not exceeded.

Blade camber in excess of the desirable value for high efficiency operation will increase the attainable total pressure rise by a useful amount provided the loading factor is less than the limit value. However, this expedient should be restricted to special cases for which the normal design alternatives cannot be seriously considered. Excessive camber should be avoided in general use because of reduced efficiency, increased noise, and greater susceptibility to dust erosion.

Increasing blade incidence to obtain higher design pressure rise capacity will, of course, reduce the interval between the design and maximum duty points. In addition to the associated drag increases, the operational dangers are considerable as inadvertent entry into stall conditions may occur.

Fan stall is a function of many factors, such as loading factor, inlet flow quality, blading inaccuracies, surface roughness, and blade extremity flow conditions. As a result, owing to the operational unknowns, this phenomenon cannot normally be predicted with accuracy during the design phase.

Greater blade incidence as a means of increasing the design duty should therefore be sparingly used and then only by an experienced designer for special fan/duct installation purposes. The manufacturing tolerances have to be correspondingly tightened.

Consideration must also be given to the blade element of the stator. The swirl produced by the rotor in the case of the rotor-straightener unit is removed by the stators. As the process is one of diffusion (i.e., an increasing static pressure in the direction of flow), the maximum design angle through which the air can be deflected is approximately 45°. Design deflections of 60° or more are possible with the prerotator type of stator as in this case the air is accelerated. The greater the prerotation, however, the larger the static pressure rise across the rotor; rotor stalling may therefore limit the amount of preswirl that can be used. Increased rotor blade drag due to the diffusing

7.5 FAN EFFICIENCY

passage effect at high stagger angles also tends to limit the maximum pre-rotator design swirl angle.

7.5 FAN EFFICIENCY

The attainment of high fan unit efficiency demands careful attention to design detail. Unlike the alternative radial flow fan equipment, which produces a substantial proportion of its pressure rise through the medium of centrifugal forces, the axial flow unit relies entirely on an interchange between static and velocity pressures. For best efficiency it is important to ensure that diffusion is efficiently carried out within the blading and that the downstream diffuser reduces the velocity to the required value with a minimum of loss. In addition, the downstream flow must be free of significant swirl.

A high-efficiency axial flow unit therefore possesses optimized rotor and stator blades, nose and tail fairings, and an effective downstream diffuser; contrarotating units do not require stators. (In contrast, the well-designed radial flow fan achieves high efficiency without stators or fairing. This result can be attributed to lower blade speeds and throughflow velocities, and to the diffusing volute, which recovers a proportion of both the tangential and radial velocity pressures. However, these features substantially increase the physical size of the unit, for a given duty condition.)

In seeking the highest possible efficiency for compact axial flow units the various sources of loss must be itemized and studied. As discussed earlier, total pressure loss results from shear, separated, and secondary flows. The first two, when taken in relation to blading, are termed *profile drag,* which is a function of blade shape, surface roughness, air turbulence, and Reynolds number.

Secondary flow drag is dependent on wall boundary layer properties, blade surface condition, air-turning angle, blade solidity, aspect ratio, tip clearance, and Reynolds number. In view of the complexity of the estimation exercise [7.3] simple ad hoc expressions derived for a particular set of circumstances are in general design use.

Howell [7.4] and Carter [7.5] achieved limited success by assuming a crude analogy with the induced drag of a finite aircraft wing. In presenting the measured cascade aerofoil drag component as a function of C_L^2, a reasonable degree of consistency was achieved. However, this expression was developed in relation to a set of variables common to axial flow compressor geometries and may be subject to variation when relevant fan data become available.

Assuming that the complete expressions reviewed by Horlock and Perkins [7.3] can eventually be evaluated for compressors, a simplified ad hoc presentation of the solutions will be required for design use. The present data on fans are extremely limited and hence the design rules can best be described as intelligent guesses.

The annulus drag component introduced by Howell and Carter has now been replaced by a revised form of secondary drag component that includes tip clearance drag; this combination provides a sounder basis for development.

There exists an optimum tip clearance for which the secondary drag is minimized. Although the secondary losses are given for this condition, efficiency adjustments for other tip clearances can be applied on the basis of experimental studies.

For well-designed fan units the estimation of pressure losses due to nose and tail fairings, downstream diffuser, and residual or planned swirl present a minimum of difficulty. However, for fans exposed to other than a steady, nonswirling inlet flow, the related matters of fan performance and efficiency are greatly compromised.

The techniques for calculating losses are outlined in the relevant chapters. This information is used, first, in obtaining an estimate of the theoretical total pressure rise and, second, in calculating fan power. The former constitutes the starting point of detailed blade element design.

7.6 DESIGN CASES

Basically, the design theory presented is for a rotor with both preswirl and afterswirl. The preswirl may be supplied either by the first of a pair of contrarotating rotors or by a set of stator blades, and the afterswirl may be removed by a contrarotating rotor or by stators, or may be permitted to pass downstream unchanged. Provided attention is paid to the design convention given in Section 8.3, the direction of the design swirl relative to the rotor direction can be adjusted as desired. When appropriate, the preswirl or afterswirl may be equated to zero; for example, the preswirl is zero for a rotor-straightener unit.

The design theory is developed for the case of free vortex flow through the rotor and stators. Peak efficiency configurations, for given fan duty requirements, are given special consideration.

A simple method of fan analysis is included as an essential aid in the design of adjustable- or variable-pitch fan installations.

Approximate design methods are presented in Appendixes A to D in relation to equipment for which the free vortex flow condition does not apply.

7.7 PROCEDURES

The required total pressure rise and volume flow capacity can be converted into suitable nondimensional coefficients when the fan and boss diameters and the rotational speed are known or assumed. If a tentative estimate of efficiency is made, the swirl coefficients at selected spanwise stations can be provisionally established. Using the swirl and flow coefficients so obtained, fairly accurate estimates of efficiency and theoretical total pressure rise can be achieved; design values of swirl can then be determined at various radii for the appropriate values of flow coefficient.

Most design variables can be estimated as functions of the flow and swirl coefficients. Hence extensive use has been made of graphical representations in which the coordinates are the unknown and the flow coefficient, together with lines of constant swirl coefficient.

In graphing the computed data, the following arbitrary limits are set:

Flow coefficient. An upper limit of 1.4.

Swirl coefficient. Upper limits of 1.4 and 1.1 for prerotators and straighteners, respectively; fan blade loading and blockage limits have an overriding influence on the usable design quantities.

In the initial design steps the loss in efficiency for each component part of the fan unit is assessed separately. This procedure has the advantage that the designer, from a knowledge of the magnitude of the various losses, possesses valuable information that can be used with advantage in making the final choice of fan unit type and dimensions to ensure efficient and satisfactory operation. The time required to carry out these separate estimations, with the aid of graphs, is negligible. It is considered that this procedure is therefore preferable to one that attempts to obtain a direct estimate of total efficiency by the use of one or two equations in which the losses are combined.

Recommended design procedures are fully described and outlined in Chapter 18, and in Appendixes A to D.

REFERENCES

7.1 G. N. Patterson, Ducted fans: design for high efficiency, Australian Council for Aeronautics, ACA Report 7, 1944.

7.2 R. A. Wallis, Optimisation of axial flow fan design, *Mech. Chem. Eng. Trans. I. E. Aust.,* **4** (1), 31–37, May 1968.

7.3 J. H. Horlock and H. J. Perkins, Annulus wall boundary layers in turbomachinery, *AGARD-AG-185, NATO,* 1974.

7.4 A. R. Howell, Fluid dynamics of axial compressors, *Proc. Instn. Mech. Eng.,* London, **153**, 441–452, 1945.

7.5 A. D. S. Carter, Three-dimensional flow theories for axial compressors and turbines, *Proc. Instn. Mech. Eng.,* London, **159**, 255–268, 1948.

CHAPTER 8

Rotor: Momentum Considerations, Free Vortex Flow

Commencing with the general momentum theorem, as applied to airfoil cascades, the total pressure relationships at each stage of a fan unit are given, the increase in total pressure is equated to the change in flow momentum through the rotor, the velocity vector diagrams are considered, and finally the change in momentum is related to the lift and drag forces exerted by the blades.

The design equations are extracted and presented in Section 8.7.

8.1 GENERAL MOMENTUM THEOREM

The boundaries of the "control box" are assumed to be the inlet and outlet stations to a stationary blade row, and two flow streamlines a gap width s apart (Fig. 8.1). The axial force Z, acting on an element of blade dr in length for constant inlet velocity and pressure, can be shown as

$$Z = \left[\rho s \, dr \, V_{a_1}^2 - \int_0^s \rho \, dr \, V_{a_2}^2 \, dy \right] + \left[p_1 s \, dr - \int_0^s p_2 \, dr \, dy \right] \quad (8.1)$$

when assuming zero momentum flow across the streamlines forming two of the box boundaries [8.1]. The tangential force Y is then given by

$$Y = \rho s \, dr \, V_{a_1} V_{\theta_1} - \int_0^s \rho \, dr \, V_{a_2} V_{\theta_2} \, dy + E \quad (8.2)$$

where E is the shear stress term due to the wake flow, as defined in [8.1]. For symmetrical wakes the contributions from both sides of the wake, each

8.1 GENERAL MOMENTUM THEOREM

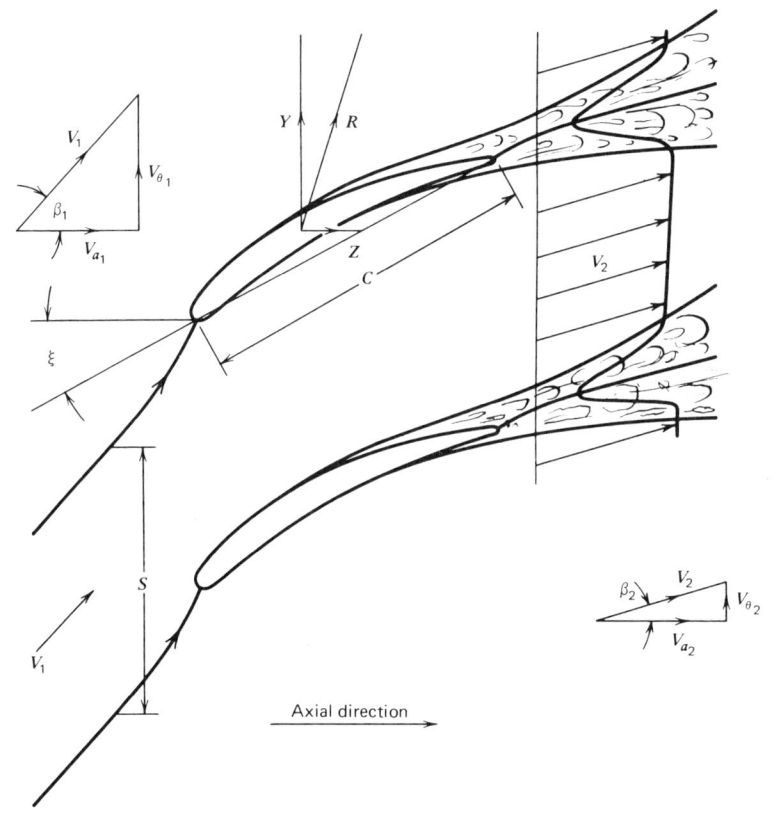

Figure 8.1 Schematic illustration of cascade flow.

being equal and opposite in sign, cancel out. This condition is approximately met by blading designed with camber matching the required flow deflection and as a consequence the term E may be neglected. However, an increasing degree of error will accompany large differences in upper and lower surface contributions such as accompany a thick upper surface shear flow resulting from gross camber, large surface excrescences, and flow separation.

Since the wake flow detail is normally unknown, Eqs. (8.1) and (8.2) cannot be solved without further simplification. This is achieved by writing

$$Z = (p_2 - p_1)s\, dr \tag{8.3}$$

and

$$Y = s\rho \bar{V}_a(V_{\theta_1} - V_{\theta_2})\, dr \tag{8.4}$$

on the assumptions that the axial velocity component $\bar{V}_a$ is a constant mean quantity throughout the flow field, p_1 and p_2 are constant within their planes,

186 ROTOR: MOMENTUM CONSIDERATIONS, FREE VORTEX FLOW

and the wake flow terms can be represented by a mean total pressure loss $\bar{\omega}$, namely

$$(p_2 - p_1) + \bar{\omega} = \Delta p_{th} \tag{8.5}$$

where Δp_{th} is the theoretical static pressure rise. The quantity $\bar{\omega}$, when correctly determined from experimental data, can be converted to a blade drag coefficient, in which form it normally appears in the blade design equations.

Expressions (8.3) and (8.4), along with other relationships, are now developed in an alternative manner.

8.2 DESIGN ASSUMPTIONS

In free vortex flow there is, of course, no radial component of velocity. Two sufficient conditions for achieving this type of flow involve design assumptions making both the theoretical total pressure rise Δh_{th}, and the axial velocity component V_a constant with radius. The absence of radial flow will ensure constancy of V_a throughout the blading annulus.

8.3 PRESSURE RELATIONS AND VELOCITY VECTORS

Conditions in an elementary annulus of width dr and constant radius r are given in Fig. 8.2 at various stations in the fan unit. Air entering the prerotators axially is deflected tangentially in a direction opposite to the rotation of the fan. It is assumed that the air leaves the fan rotor with a swirl component in the direction of rotation, and hence the straighteners are left with the task of deflecting it back into the axial direction.

It is convenient to take the preswirl and afterswirl as both positive although strictly they are of opposite sense.

The Bernoulli relationships at the four stations (Fig. 8.2) assuming spanwise constant total pressure at each cross-sectional plane are

$$H_0 = p_0 + \tfrac{1}{2}\rho V_a^2 \tag{8.6}$$

$$H_1 = p_1 + \tfrac{1}{2}\rho V_a^2 + \tfrac{1}{2}\rho V_{\theta_p}^2 \tag{8.7}$$

$$H_2 = p_2 + \tfrac{1}{2}\rho V_a^2 + \tfrac{1}{2}\rho V_{\theta_s}^2 \tag{8.8}$$

$$H_3 = p_3 + \tfrac{1}{2}\rho V_a^2 \tag{8.9}$$

where H, p, V_a, and V_θ are the total pressure, static pressure, axial velocity component, and swirl velocity component, respectively. The overall change

8.3 PRESSURE RELATIONS AND VELOCITY VECTORS

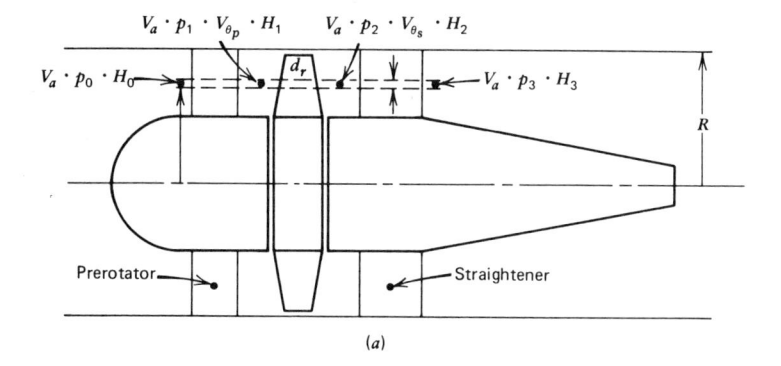

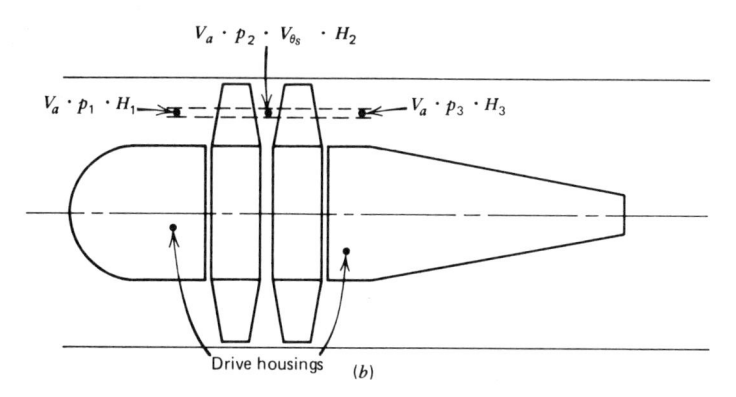

Figure 8.2 Blading arrangements. (a) General blading case. (b) Contrarotating case.

in total pressure in the annulus can be written

$$H_3 - H_0 = \Delta H_{th} - \Delta h_R - \Delta h_P - \Delta h_S \qquad (8.10)$$

where ΔH_{th} is the theoretical mean total pressure rise, the other terms denoting the losses in the rotor, prerotators, and straighteners, respectively, which have a spanwise variation.

The following nondimensional equation is obtained by dividing Eq. (8.10) by $\frac{1}{2}\rho \bar{V}_a^2$, where $\bar{V}_a$ is the mean axial velocity through the fan:

$$\frac{H_3 - H_0}{\frac{1}{2}\rho \bar{V}_a^2} = K_{th} - k_R - k_P - k_S \qquad (8.11)$$

where $k = \Delta h / \frac{1}{2}\rho \bar{V}_a^2$, for example, $k_R = \Delta h_R / \frac{1}{2}\rho \bar{V}_a^2$

From Eqs. (8.6) and (8.9) it follows that $(H_3 - H_0)/\frac{1}{2}\rho \bar{V}_a^2$ is also the nondimensional static pressure rise for the unit. Defining the swirl coefficient

188 ROTOR: MOMENTUM CONSIDERATIONS, FREE VORTEX FLOW

as $\epsilon = V_\theta/\bar{V}_a$, the static pressure rise across the fan rotor at radius r is then given by

$$p_2 - p_1 = \Delta p = \tfrac{1}{2}\rho \bar{V}_a^2 (K_{th} - k_R + \epsilon_p^2 - \epsilon_s^2) \tag{8.12}$$

The Bernoulli relationships having been discussed, it is now appropriate to consider the work done by the rotor. The output of work from the rotor in the elementary annulus is

$$(H_2 - H_1)\, 2\pi r\, dr\, V_a \tag{8.13}$$

and the input is $\Omega\, dT$, where Ω is the rotational speed of the rotor in radians per second and dT is the element of torque. From the rate of change of angular momentum,

$$dT = \rho V_a 2\pi r\, dr\, (V_{\theta_s} + V_{\theta_p}) r \tag{8.14}$$

when the sign convention adopted at the beginning of this subsection is observed.

Replacing $(H_2 - H_1)$ by the theoretical total pressure rise in Eq. (8.13) and equating the new relation to $\Omega\, dT$,

$$(H_2 - H_1 + \Delta h_R) 2\pi r\, dr\, V_a = \rho V_a 2\pi r\, dr\, (V_{\theta_s} + V_{\theta_p})\Omega r \tag{8.15}$$

and therefore

$$\Delta H_{th} = \rho \Omega r (V_{\theta_s} + V_{\theta_p}) \tag{8.16}$$

Nondimensionally,

$$K_{th} = \frac{2}{\lambda} (\epsilon_s + \epsilon_p) \tag{8.17}$$

where the flow coefficient is defined by $\lambda = \bar{V}_a/\Omega r$.

The resultant velocity vector V_m, which determined the lift on the blade element, is shown in Fig. 8.3 together with the velocity components at inlet and outlet from the blade element.

The tangential velocity component of the air relative to the blade is given by

$$\Omega r - \tfrac{1}{2}(V_{\theta_s} - V_{\theta_p})$$

where the second term is the mean swirl between the rotor inlet and outlet. The angle β_m that the resultant velocity V_m makes with the plane of rotation is obtained from

$$\tan \beta_m = \frac{\Omega r - \tfrac{1}{2}(V_{\theta_s} - V_{\theta_p})}{\bar{V}_a} \tag{8.18}$$

8.4 BLADE ELEMENT FORCES

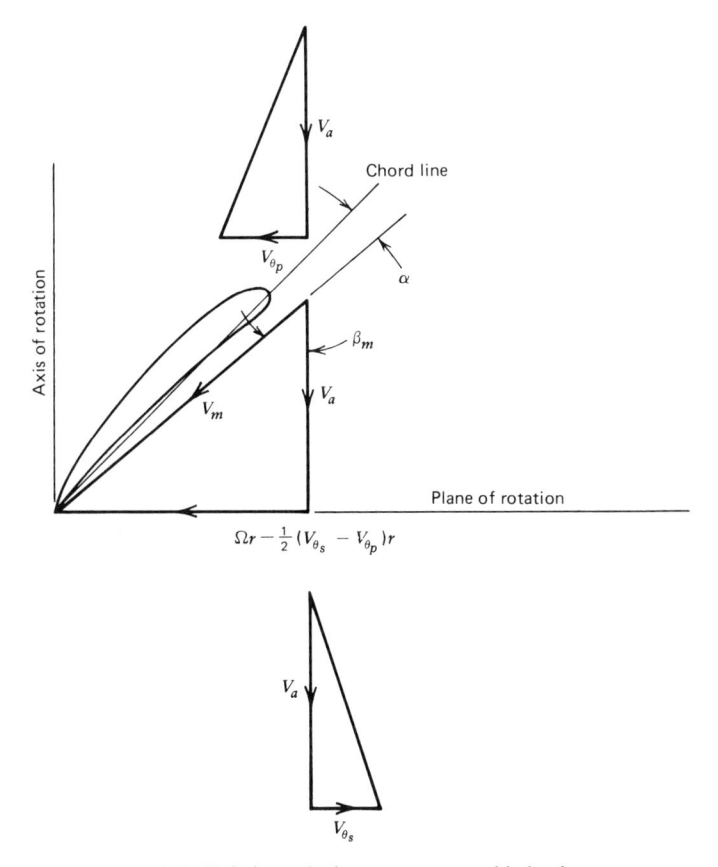

Figure 8.3 Relative velocity vectors, rotor blade element.

or

$$\tan \beta_m = \frac{1 - \frac{1}{2}(\epsilon_s - \epsilon_p)\lambda}{\lambda} \tag{8.19}$$

Finally, it can be shown from Eq. (8.16) that $(V_{\theta_s} + V_{\theta_p})$ is inversely proportional to the radius r when ΔH_{th} is constant along the blade. It is usual to make one of these swirls zero or inversely proportional to the radius; in both cases the flow will satisfy the condition for free vortex flow.

8.4 BLADE ELEMENT FORCES

The general momentum theory is now applied to a blade possessing relative motion, as illustrated in Fig. 8.4.

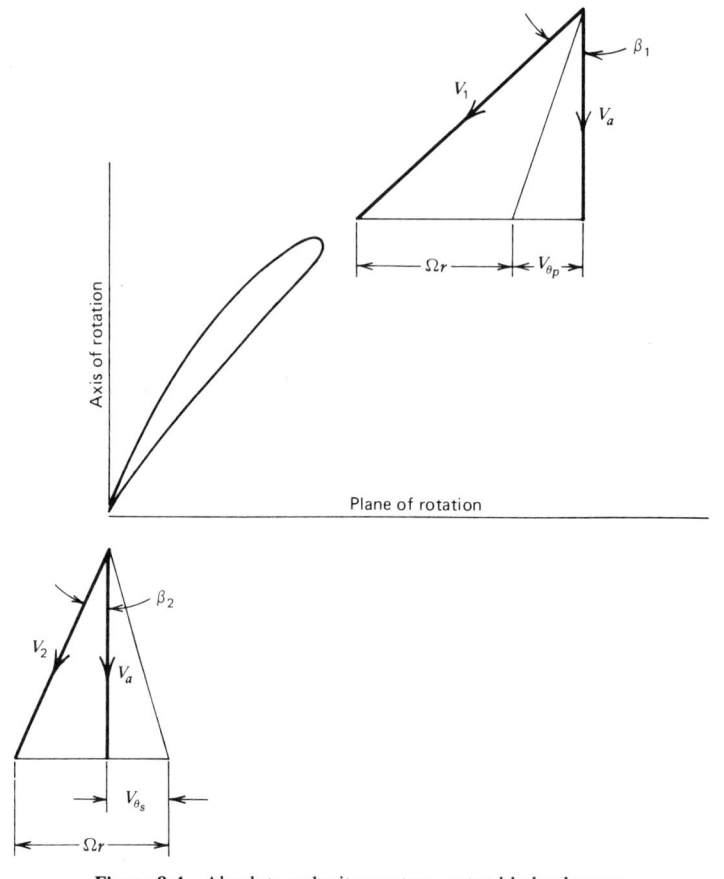

Figure 8.4 Absolute velocity vectors, rotor blade element.

The change in the tangential velocity component from blade inlet to outlet is

$$(\Omega r + V_{\theta p}) - (\Omega r - V_{\theta s}) = V_{\theta p} + V_{\theta s}$$

In addition, the theoretical static pressure rise Δp_{th} is equal to the change in dynamic pressure, namely,

$$\Delta p_{th} = \tfrac{1}{2}\rho V_1^2 - \tfrac{1}{2}\rho V_2^2 \qquad (8.20)$$

a relationship that can alternatively be obtained by combining Eqs. (8.12) and (8.16) and by using the vector diagrams of Fig. 8.4.

From the force vector diagram (Fig. 8.5) it follows that the axial force Z is given by

8.4 BLADE ELEMENT FORCES

$$Z = \Delta p \, s \, dr \tag{8.21}$$

and the tangential force Y by

$$Y = s\rho \overline{V}_a (V_{\theta_p} + V_{\theta_s}) \, dr$$

$$= s\rho \overline{V}_a^2 (\epsilon_s + \epsilon_p) \, dr \tag{8.22}$$

From Fig. 8.5, the drag of a blade element is

$$D = Y \sin \beta_m - Z \cos \beta_m \tag{8.23}$$

and substituting for Y and Z,

$$D = s\rho \overline{V}_a^2 (\epsilon_s + \epsilon_p) \, dr \sin \beta_m - s \, \Delta p \, dr \cos \beta_m$$

Substituting for Δp from Eq. (8.12), writing V_a/V_m as $\cos \beta_m$ (Fig. 8.3) and dividing by $\frac{1}{2}\rho V_m^2 c \, dr$ in order to obtain a drag coefficient,

$$C_D = \frac{s}{c} \cos^2 \beta_m [2(\epsilon_s + \epsilon_p) \sin \beta_m - (K_{\text{th}} - \epsilon_s^2 + \epsilon_p^2) \cos \beta_m] + \frac{s}{c} k_R \cos^3 \beta_m$$

Eliminating K_{th} by Eq. (8.17) and using Eq. (8.19)

$$C_D = 2\frac{s}{c} \cos^2 \beta_m (\epsilon_s + \epsilon_p) \left[\sin \beta_m - \frac{\cos \beta_m}{\cot \beta_m} \right] + \frac{s}{c} k_R \cos^3 \beta_m$$

and hence

$$C_D = \frac{s}{c} k_R \cos^3 \beta_m \tag{8.24}$$

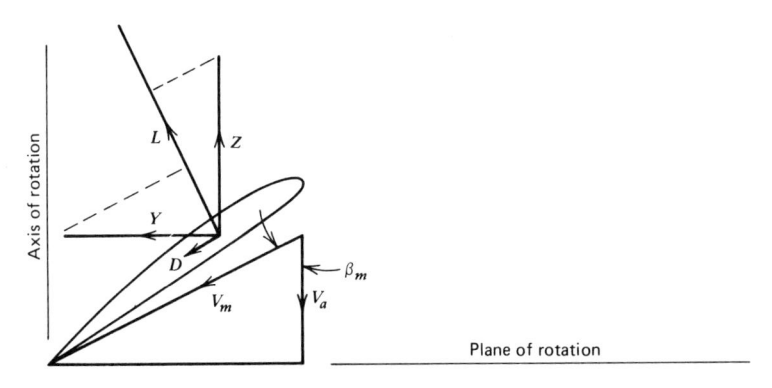

Figure 8.5 Rotor blade-element force vectors.

Similarly, the lift is

$$L = Y \cos \beta_m + Z \sin \beta_m \tag{8.25}$$

$$= s\rho \overline{V}_a^2 (\epsilon_s + \epsilon_p) \, dr \cos \beta_m + s \, \Delta p \, dr \sin \beta_m$$

and hence

$$C_L = \frac{s}{c} \cos^2 \beta_m \, [2(\epsilon_s + \epsilon_p) \cos \beta_m + (K_{\text{th}} - \epsilon_s^2 + \epsilon_p^2) \sin \beta_m]$$

$$- \frac{s}{c} k_R \cos^2 \beta_m \sin \beta_m$$

Making substitutions similar to the previous ones and using Eq. (8.24)

$$C_L = 2 \frac{s}{c} (\epsilon_s + \epsilon_p) \cos \beta_m - C_D \tan \beta_m \tag{8.26}$$

or

$$C_L \sigma = 2(\epsilon_s + \epsilon_p) \cos \beta_m - \sigma C_D \tan \beta_m \tag{8.27}$$

where $\sigma = c/s = $ solidity.

8.5 ABSOLUTE INLET AND OUTLET ANGLES

When developing preferred blading sections of the F-series or similar type, it is desirable to have the blade inlet and outlet angles in terms of λ and ϵ. From Fig. 8.4 it follows that

$$\tan \beta_1 = \frac{1 + \epsilon_p \lambda}{\lambda} \tag{8.28}$$

$$\tan \beta_2 = \frac{1 - \epsilon_s \lambda}{\lambda} \tag{8.29}$$

$$\tan \beta_1 - \tan \beta_2 = \epsilon_s + \epsilon_p \tag{8.30}$$

and when combined with Eq. (8.19)

$$\tan \beta_m = \tfrac{1}{2}(\tan \beta_1 + \tan \beta_2) \tag{8.31}$$

In determining the sign of the angles, the usual convention that $(\beta_1 - \beta_2)$ is always positive is accepted. It can be shown that β_1 is always positive for

8.7 DESIGN EQUATIONS

flow-retarding blading and β_2 always negative for the accelerating case. For the type of rotor design considered here, the inlet angle is always positive and, provided the product of ϵ_s and λ is less than unity, β_2 will also be positive.

8.6 THRUST AND TORQUE GRADIENTS

The element of thrust developed by the blade element due to the flow in the annulus is

$$dTh = \Delta p \, 2\pi r \, dr$$

Defining a thrust coefficient, Th_c as

$$Th_c = \frac{Th}{\frac{1}{2}\rho \overline{V}_a^2 \pi R^2} \tag{8.32}$$

where R is the tip radius. Hence,

$$dTh_c = \frac{\Delta p}{\frac{1}{2}\rho \overline{V}_a^2} 2x \, dx \tag{8.33}$$

where $r/R = x$.

Substituting from Eq. (8.12)

$$\frac{dTh_c}{dx} = 2x(K_{th} - k_R + \epsilon_p^2 - \epsilon_s^2) \tag{8.34}$$

The element of torque is given by Eq. (8.14), and when a torque coefficient is defined as

$$T_c = \frac{T}{\frac{1}{2}\rho \overline{V}_a^2 \pi R^3} \tag{8.35}$$

it follows that

$$\frac{dT_c}{dx} = 4x^2(\epsilon_s + \epsilon_p) \tag{8.36}$$

8.7 DESIGN EQUATIONS

In the foregoing, a number of simple design equations have been derived from momentum considerations. These are

194 ROTOR: MOMENTUM CONSIDERATIONS, FREE VORTEX FLOW

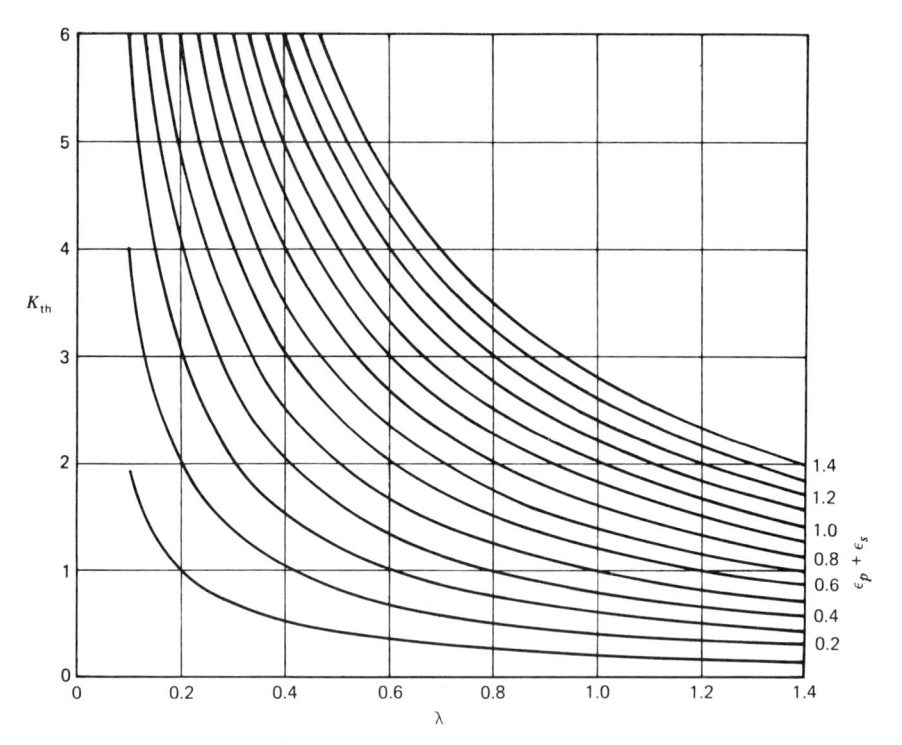

Figure 8.6 K_{th} as function of λ and ϵ.

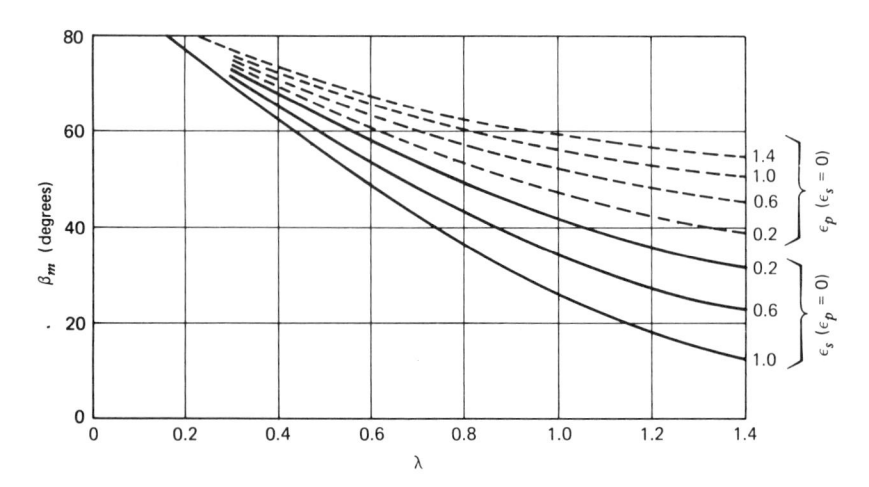

Figure 8.7 Relative velocity angle as function of λ and ϵ.

8.7 DESIGN EQUATIONS 195

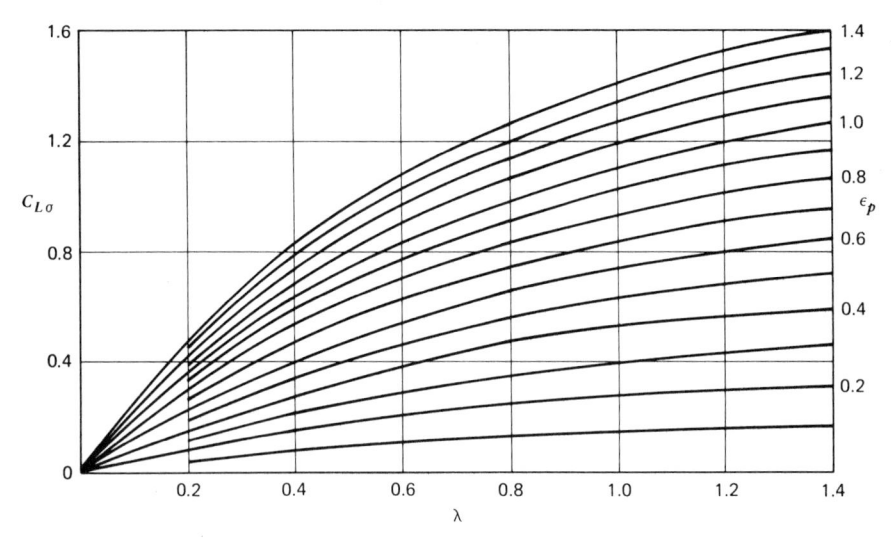

Figure 8.8 Loading factor as function of λ and ϵ_p, $\epsilon_s = 0$.

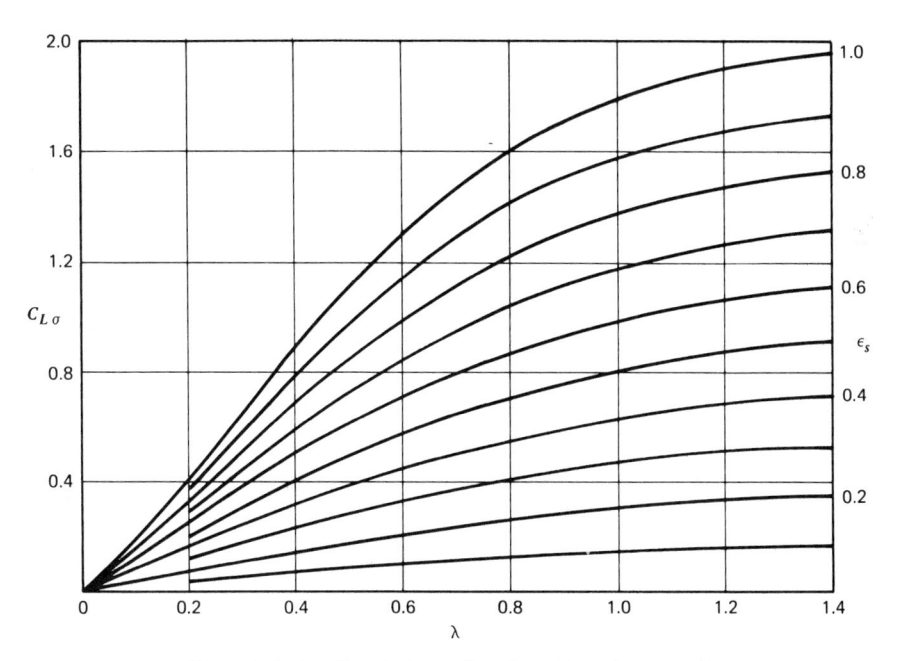

Figure 8.9 Loading factor as function of λ and ϵ_s, $\epsilon_p = 0$.

$$K_{th} = \frac{2}{\lambda}(\epsilon_s + \epsilon_p) \qquad (8.17)$$

$$\tan \beta_m = \frac{1 - \frac{1}{2}(\epsilon_s - \epsilon_p)\lambda}{\lambda} \qquad (8.19)$$

$$C_L \sigma = 2(\epsilon_s + \epsilon_p) \cos \beta_m - \sigma C_D \tan \beta_m \qquad (8.27)$$

The last term is usually small, being ignored in most design methods. Hence

$$C_L \sigma = 2(\epsilon_s + \epsilon_p) \cos \beta_m \qquad (8.37)$$

which is identical to Eq. (6.5) after substitution from Eq. (8.30). When either ϵ_s or ϵ_p is zero, these equations can be presented in simple graphical form (Figs. 8.6 to 8.9).

Blade loss is calculated from

$$C_D = \frac{s}{c} k_R \cos^3 \beta_m \qquad (8.24)$$

REFERENCE

8.1 M. H. Vavra, *Aero Thermodynamics and Fluid Flow in Turbomachinery*, Chap. 5, Wiley, New York, 1960.

CHAPTER 9

Rotor Blade Design

A selection of the best available information relating to optimum blading design and blade loading limits is presented. Application of these data will lead to a consistently better and more uniform class of efficient fan.

9.1 BLADE CAMBER

9.1.1 Air-Turning Angle

Spanwise variations in camber are virtually mandatory for design cases where substantial differences in air deflection exist along the blade. However, constant camber blading can provide a satisfactory alternative when all spanwise air-turning angles are relatively small.

9.1.2 Local Incidence

Blade camber from Eq. (6.6) has a dependence on i and δ. Since airfoil boundary layer growth is influenced by leading-edge flow conditions, these variables are interrelated.

Carter in [9.1] presented i_{opt} (defined as the angle for which the lift/profile drag ratio is a maximum) as a function of blade camber covering a general range of design situations. The inadequacy of this relation was later admitted by Carter [9.2], who undertook further studies. This revision was prompted by the realization that local incidence was a very important parameter, particularly for compressor stall. By relating these latter recommendations to nominal design conditions [9.3], as expressed by Eq. (6.15), Fig. 9.1 results.

Alternative definitions of optimum local incidence, namely, minimum drag coefficient [9.4], and maximum lift/drag ratio where the secondary drag is included [9.5], result in the incidence recommendations of Fig. 9.2, for nominal design conditions; these data apply to the NACA 65-series blower sections.

198   ROTOR BLADE DESIGN

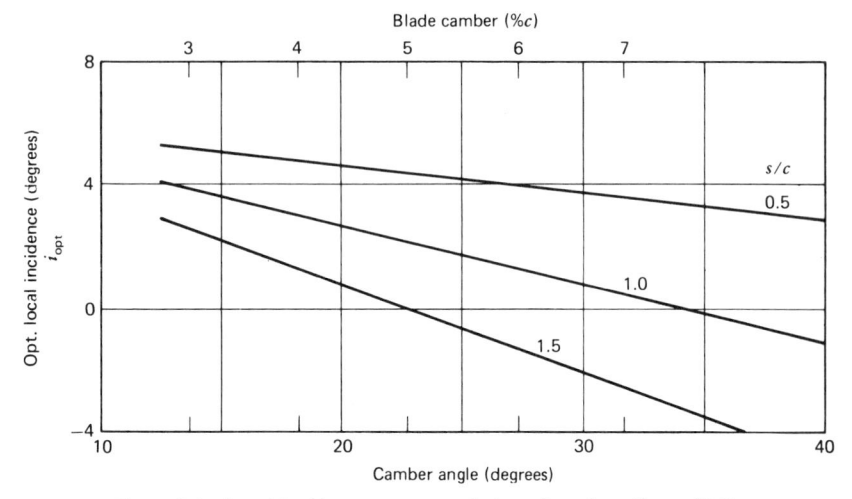

Figure 9.1 Local incidence recommendations, based on Carter [9.2].

Keeping in mind, first, the different criteria for i_{opt} and, second, Carter's remark [9.2] that his figures are slightly on the high side for design use, because of stall proximity, the agreement can be considered satisfactory. The use of $i = 0$ in preliminary design studies is therefore justified.

The actual flow incidences may differ slightly from the preceding geometric values. As indicated in [9.6], increasing blade solidity and thickness both create flow blockage, and hence an increase in geometric local incidence is

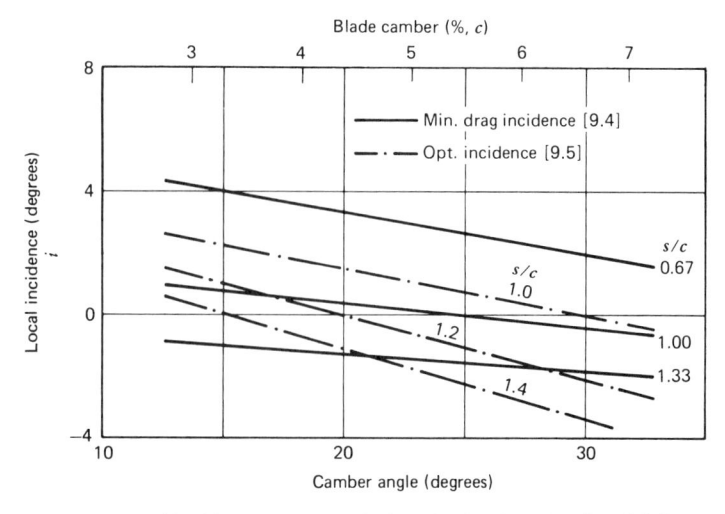

Figure 9.2 Local incidence recommendations for NACA 65-series airfoils [6.1].

9.1 BLADE CAMBER

required. The reverse is true for increasing camber and a related decrease in stagger angle. These trends are reflected in Figs. 9.1 and 9.2

Since a large proportion of fan design is concerned with the isolated airfoil characteristics, attention can be focused on the lower left-hand side of these graphs; a small negative angle is affirmed. The related performance improvement that accompanies small additional nose camber, or droop, on isolated sections is not at variance with this finding, and hence continuity between cascade and isolated airfoil data is maintained.

Physical reasoning [9.6] suggests that with increasing solidity the improvements due to nose droop are reduced. However, on the other hand, small additional nose camber is not expected to affect high solidity blading performance adversely.

9.1.3 Deviation Angle

The deviation angle is dependent on both viscous and nonviscous phenomena, making its precise determination a matter of some difficulty and complexity. In most design instances, an accurate estimate is not required as the cascade influence is treated by the "interference factor" technique. Isolated airfoil data record the loss of lift due to boundary layer growth as a reduced slope in the lift versus incidence curve. Since increasing solidity changes both the lift curve slope and no-lift angle of a section, it is apparent that a useful interference factor must incorporate both these variables. The data presented in Fig. 6.29, on the basis of actual test information, must therefore apply chiefly to the airfoil types concerned (e.g., the C4, F-series, and closely related sections) at normal design lift coefficients. Small departures from these conditions will not result in substantial design errors.

The estimates for deviation angle in the preceding development are based on the Carter relation, namely,

$$\delta = m\theta \sqrt{\frac{s}{c}} \qquad (9.1)$$

where m is a function of blade stagger angle. However, the function presented by Carter [9.1] was of necessity modified and extended to higher stagger angles to give δ data that blended with those for isolated airfoils [9.7] (Fig. 9.3).

Since δ is approximately 10° for F-series sections [9.8], a preliminary design estimate of camber angle can be obtained from

$$\theta = 10 + (\beta_1 - \beta_2) \qquad (9.2)$$

The final choice of camber will depend on the designer's assessment of matters relating to blade stiffness, blade erosion, fan efficiency and power, and design selection of C_L. Within reason, increasing camber results in a

ROTOR BLADE DESIGN

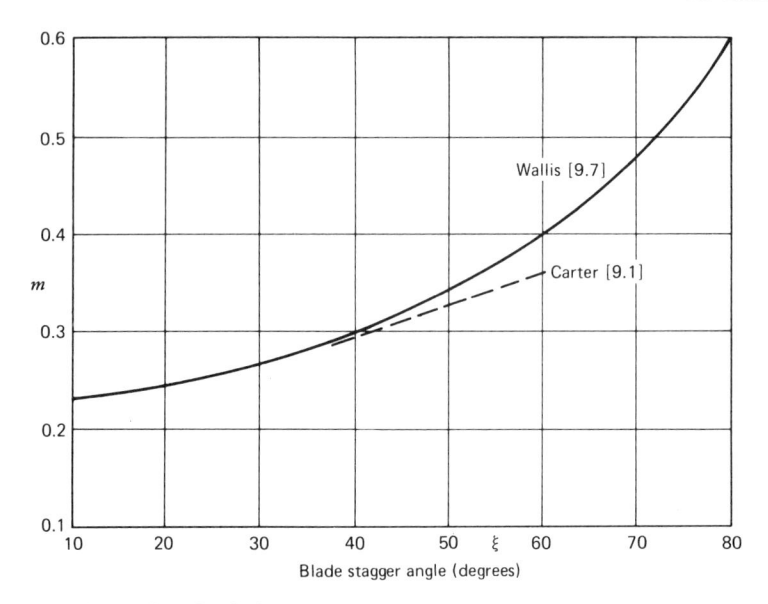

Figure 9.3 Deviation angle coefficient versus stagger angle [6.4].

higher design C_L and $C_{L,\max}$, provided the loading factor limit is not exceeded. A minimum camber-angle of from 14 to 15° is suggested that incorporates a small negative local incidence angle when $(\beta_1 - \beta_2)$ is minimal.

Cambered plate blades will require a greater θ value than conventional airfoils. A decreased angle of local incidence, to control leading-edge separation, and an increased deviation angle are the relevant parameters. Since the blading is not normally of a high-solidity type, isolated airfoil data on lift/drag ratio versus lift (Fig. 6.27) provide accurate guidance on the appropriate camber. For small turning angles, 18° of camber is the recommended minimal value. Blade root values will normally be larger, as the air-turning angle is greatest in this locality. A requirement for adequate blade stiffness can also be met with modest camber increases.

A guide to plate camber can be obtained from the relation

$$\theta = 15 + (\beta_1 - \beta_2) \tag{9.3}$$

9.2 ADDITIONAL LEADING-EDGE CAMBER

The recommendation concerning nose droop (see Section 6.6.3) is based on the facts presented in [9.6], and on the favorable maximum pressure duty characteristic of fans designed with 1% c droop (e.g., see [9.8]). The design

9.3 BLADE LOADING LIMITS IN DESIGN

choice is usually noncritical and hence the following broad guidelines will suffice:

One percent: applicable to large fans of high Re possessing smooth blade surfaces or operating under dust erosion conditions. Also appropriate to high-solidity blading cases at intermediate to high Re.

Two percent: for use when Re is in excess of approximately 2×10^5, the blade surfaces are smooth, and the airstream is free of erosive dust.

Three percent: desirable for cases of Re less than 2×10^5 and for rough blade surfaces resulting from a solids buildup or manufacturing imperfections, at all Re.

Intermediate percentage droops may be used, at the discretion of the designer. Nose droop is particularly effective at low Reynolds numbers.

9.3 BLADE LOADING LIMITS IN DESIGN

The design C_L for blading of high solidity and hence large load factor is selected on the basis of stall properties rather than on the lift/drag ratios obtained from isolated airfoil tests. The simple Keller rules [9.9] that the product of C_L and σ, and the value of σ, should not exceed unity and 1.1, respectively, only apply over a limited design range.

Variables affecting the design choice of C_L are air-turning angle, blade solidity and aspect ratio, tip clearance, stagger angle, and inlet flow conditions, where the latter often involve skewed wall boundary layers. The secondary flows resulting from the preceding combination of parameters exert a major influence over the stall onset. As a result, the designer is forced to be conservative. A survey of recommended C_L values in fan design indicates a range from 0.6 to 1.0 [9.10]. However, a more definitive selection method is obviously desirable.

9.3.1 Low-Solidity Blading

The maximum ratio of lift/profile drag increases with camber angle, along with the C_L at which the former is reached; for the isolated C4 airfoil of 8% camber, this C_L value is 1.4 [9.11]. However, when secondary drag is included in the lift/drag ratio, the figure is reduced (Fig. 9.4) to unity or less. The use of 0.018 as the constant in the secondary drag relation given in Fig. 9.4 will be pessimistic for high aspect ratio, low-solidity blading operating at moderate to high Re.

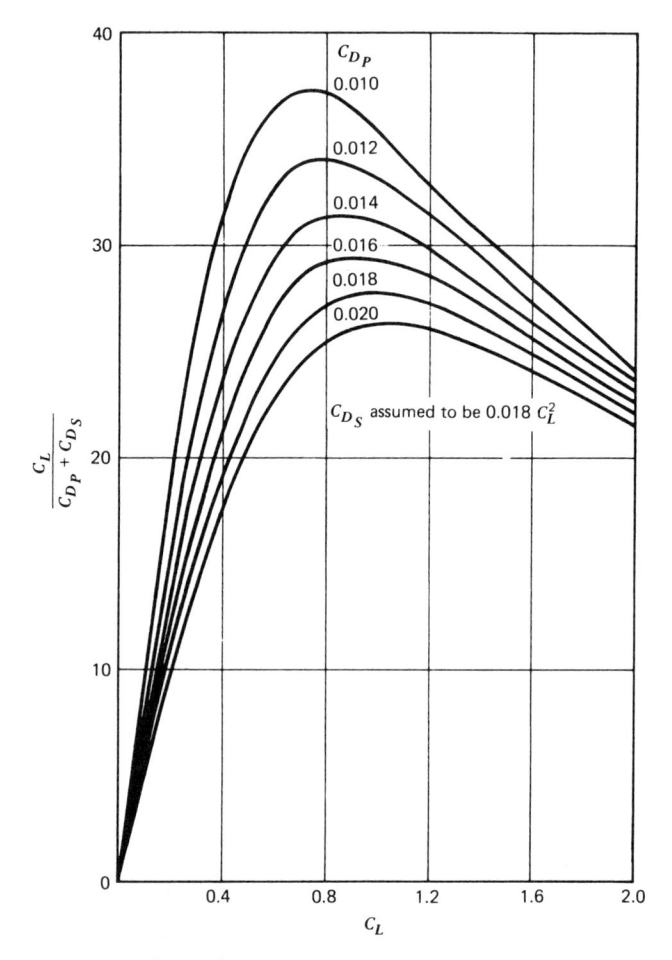

Figure 9.4 Influence of lift coefficient on lift/drag ratio.

However, the development of a section along previously discussed lines, or the selection of the flat undersurface variety, is recommended particularly as high degrees of camber are not required in achieving a design C_L approaching unity. Lower powers and noise levels, a wider band of high-efficiency operation, and a safe stall limit are advantages of this approach. In addition, small duct design errors, or fluctuations in operational duct resistance, are of lesser importance when the C_L is conservatively selected.

Reduced design lift coefficients must be adopted in cases of blade surface roughness and imperfections. Poor-quality inlet flow containing appreciable swirl, velocity fluctuations, and nonuniformities can be particularly damaging. The low Reynolds number case is one for special attention, as outlined in Section 6.9.

9.3 BLADE LOADING LIMITS IN DESIGN

9.3.2 Loading Factor Limits

Because of the importance of the $(C_L \sigma)$ design limit relationship for high-solidity blading, a summary of various approaches to the problem will be outlined.

Boundary layer separation, which foreshadows the onset of blade stalling, is related to the product of boundary layer momentum thickness and the prevailing adverse pressure gradient. Starting with the momentum equation [Eq. (3.5)], Lieblein [9.4] developed a diffusion factor D. Assuming that all the best blade sections possessed an approximately linear convex-surface velocity distribution at the desired design incidence, and applying a number of plausible assumptions, he arrived at

$$D = \left(1 - \frac{\cos \beta_1}{\cos \beta_2}\right) + \frac{\cos \beta_1}{2\sigma} (\tan \beta_1 - \tan \beta_2) \tag{9.4}$$

When D is plotted against momentum thickness to chord ratio for all available test data, a relatively unique curve is obtained; for D values in excess of 0.6, the slope of the line progressively increases, suggesting that $D \approx 0.6$ represents the design limit.

Carter [9.12], starting from an identical velocity distribution assumption, proceeded to argue that the lift coefficient, based on blade outlet velocity, is a constant. Hence

$$\frac{2 \cos \beta_2}{\sigma \cos \beta_m} (\tan \beta_1 - \tan \beta_2) = 1.35 \tag{9.5}$$

for circular-arc C4 blading of zero blockage. Introducing the blockage factor,

$$B = 1 - \frac{2}{\sigma} \left(\frac{t}{c}\right)_{max} \sec \beta_m$$

where for 10% thick blades $(t/c)_{max} \sec \beta_m$ can be taken as equal to 1/12. Therefore

$$B = 1 - (1/6)\sigma$$

Hence, for 10% thick blades

$$\frac{\cos^2 \beta_2}{\sigma \cos \beta_m} (\tan \beta_1 - \tan \beta_2) = 0.675 B. \tag{9.6}$$

The blockage correction is based on very limited experimental evidence.

Following an unsuccessful theoretical attack on the problem, Howell discovered that experimental results could be closely approximated by the empirical relation

$$C_L^* = 2\left(1 - \frac{\Delta p_{th}}{\tfrac{1}{2}\rho V_1^2}\right)^{1.375} \qquad (9.7)$$

which converts to the relationship Eq. (6.15), namely,

$$C_L^* = 2\left(\frac{\cos \beta_1}{\cos \beta_2}\right)^{2.75} \qquad (6.15)$$

when Eq. (8.20) and Fig. 6.4 are used.

A comparison with Carter's expression can be made when Eq. (9.5) is written as

$$C_L^* = 1.35\left(\frac{\cos \beta_m}{\cos \beta_2}\right)^2 \qquad (9.8)$$

where the asterisk denotes nominal design conditions.

European practice [9.13] makes use of an expression developed by Zweifel. With arguments centered on Δp_{th} and surface velocity distributions, the following equation was derived:

$$\frac{\cos^2 \beta_2}{\sigma \cos \xi}(\tan \beta_1 - \tan \beta_2) = 0.4 \qquad (9.9)$$

Myles et al. [9.5], starting with the experimental data of [9.14] for the 65-series blower sections, developed a tabulated set of data relating to design values of C_L. These quantities, which represent peak lift/total drag conditions, were graphically obtained by the fairing of experimental data, and subsequent interpolation and extrapolation procedures.

Because of section similarity and the extensive use of experimental data in developing the preceding relations, all are in reasonable agreement for the Reynolds number range covered in [9.14], namely, 2×10^5 to 4×10^5.

When translated to fan design needs the preceding information will in many instances provide conservative answers. Blade sections possessing nose droop in accord with the recommendations of Section 9.2 will acquire drag rise delay benefits akin to those associated with higher Reynolds number. The related design lift gains for sections of low camber are expected to lessen as blade camber is increased.

A more basic approach to optimum leading edge contour determination is outlined in [9.15] for axial flow compressors. At a given duty a chordwise pressure distribution that potentially results in the least boundary layer growth is specified.

However, the promise of a higher design C_L can be offset by lack of adequate manufacturing care resulting in excessive tip clearance, rough

9.3 BLADE LOADING LIMITS IN DESIGN

blade surfaces, and/or inaccurate blade angles. Fans operating in dusty duty conditions will also stall earlier. Allowances based on experience must be made for equipment in these categories.

9.3.3 Graphical Development of Load Limits Recommendations

The development of the foregoing information for ready design use is now outlined. After substitution, Eqs. (9.8) and (6.15) can be presented as in Figs. 9.5 and 9.6. For the normal design range, namely, $\beta_2 > 20°$, there is little difference between the two sets of curves. However, a similar presentation in use by a British engine firm in 1956 (Fig. 9.7) presents no data for high-solidity blading at large flow outlet angles. Presumably, this is related to blade blockage difficulties, as discussed later. Other useful information derived from Eq. (6.15) is displayed in Figs. 9.8 and 9.9.

When the data for $\epsilon_p = 0$ are presented in terms of λ and ϵ (Figs. 9.10 and 9.11), the design usefulness is increased. These curves result from combined

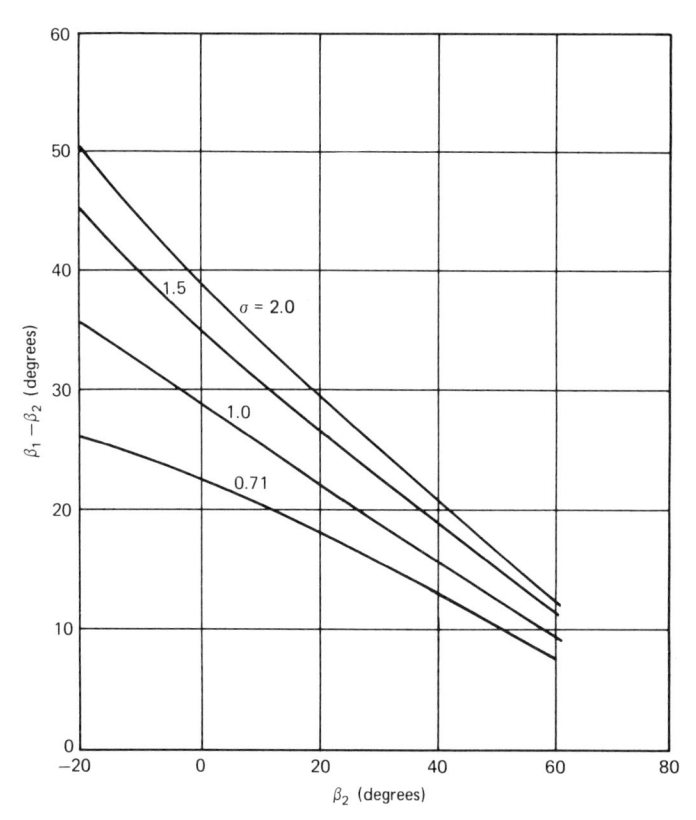

Figure 9.5 Blade loading relations of Carter.

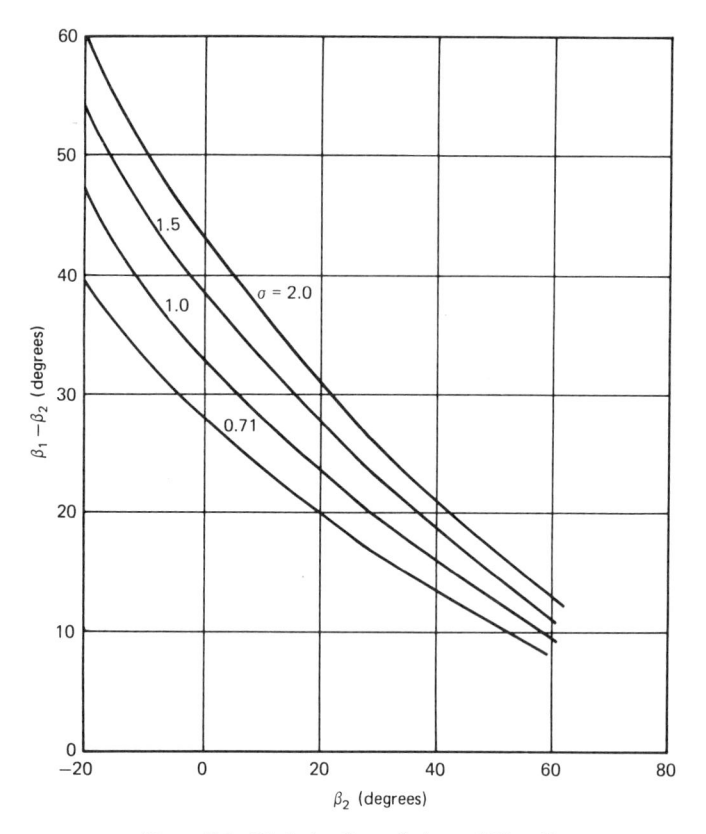

Figure 9.6 Blade loading relations of Howell.

computational and graphical procedures using the momentum relationships of Chapter 8, particularly Eq. (8.37). The curve for $\epsilon_s = 1.0$ approximately represents the blading design limit. The two sets of curves are in reasonable agreement at low λ but differ at high λ because of the influence of including secondary drag in the NEL design criterion [9.5].

The ratio between the reaction and impulse forces decreases as λ increases, resulting in higher design lifts for flow coefficients above approximately 0.8. For such design cases the straighteners are responsible for an increasing proportion of the static pressure rise through the blading unit. The rotor blades, which are highly cambered, induce large air-turning angles for a reducing static pressure rise. The resulting tangential velocity component is recovered as static pressure in the stators.

For ϵ and λ values approaching unity and 0.8, respectively, the presentation of Figs. 9.10 and 9.11 display a measure of marked sensitivity. This is due to the relatively large values of loading factors ($C_L \sigma = 1.6$). However,

9.3 BLADE LOADING LIMITS IN DESIGN

the ratio of reaction to impulse contributions is approaching 50%, easing the adverse pressure gradient on the blade surface. At $\lambda = 1.4$ the blade element is virtually of an impulse type, with a negligible static pressure rise.

Design experience and optimum fan and blade design studies have indicated the desirability of easing these solidity limit restrictions. The modified relationships of Fig. 9.12 are the result of basic reasoning, test experience, and interference factor considerations. However, it should be emphasized that these recommendations are only applicable to carefully developed blading of an optimum type, working at above critical Re, and subject to good-quality inlet airflow and clean air.

A limited trade-off between solidity and lift can be undertaken. Because of

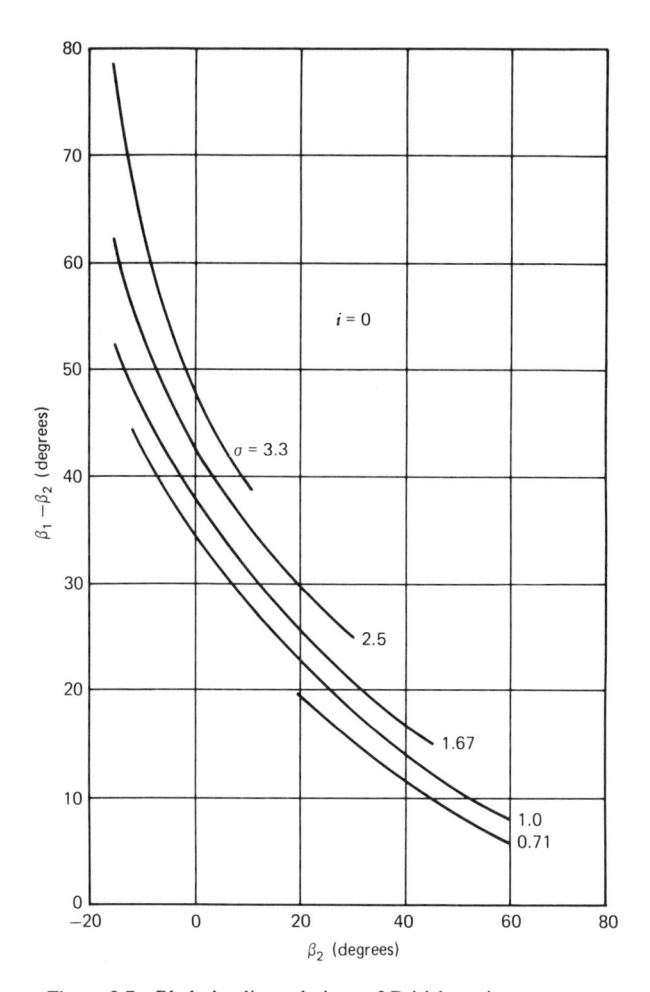

Figure 9.7 Blade loading relations of British engine company.

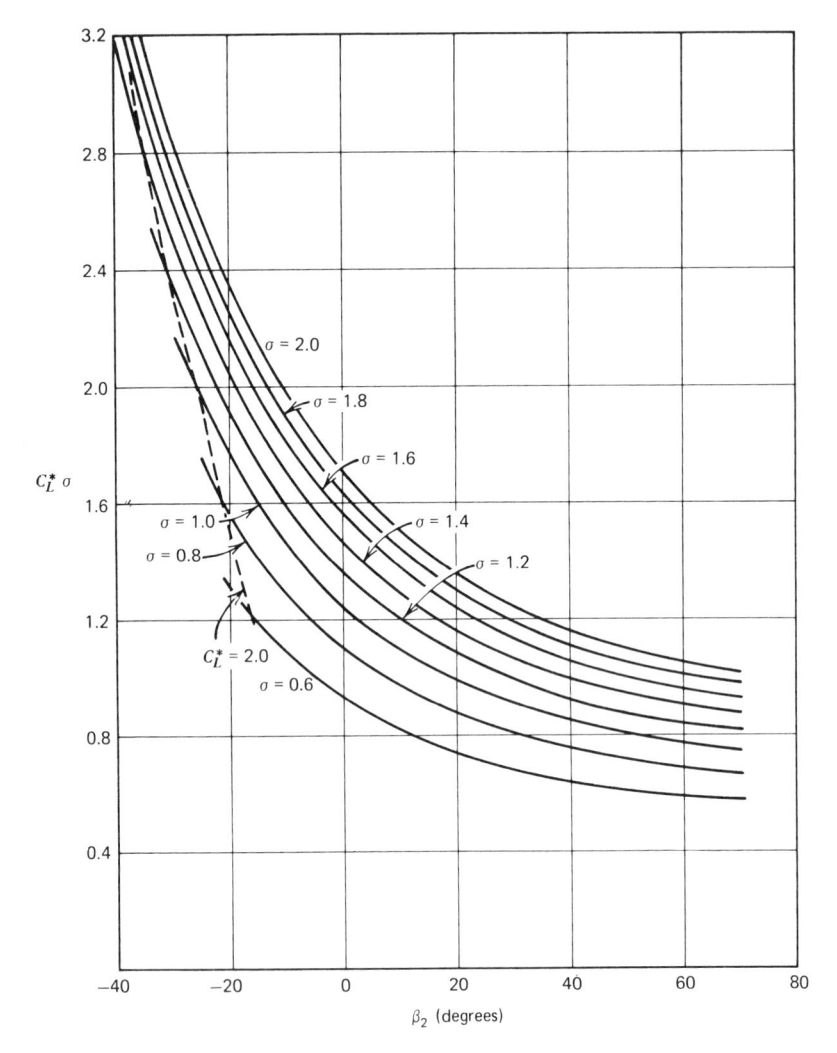

Figure 9.8 Howell relations for loading factor $C_L\sigma$.

multiplane interference effects (see Fig. 6.29), increasing the solidity up to a value of 2 gives restricted gains in the load factor $C_L\sigma$. Hence for fans possessing a high loading, a nonlinear increase in blade chord toward the root must be expected for free vortex design conditions. This can lead to impracticable blade planforms.

Blade root chord reductions can be undertaken when full consideration is given to the extremely local nature of the design difficulty, to the boundary layer and secondary flow features, and to the inconsequential result of a

9.3 BLADE LOADING LIMITS IN DESIGN

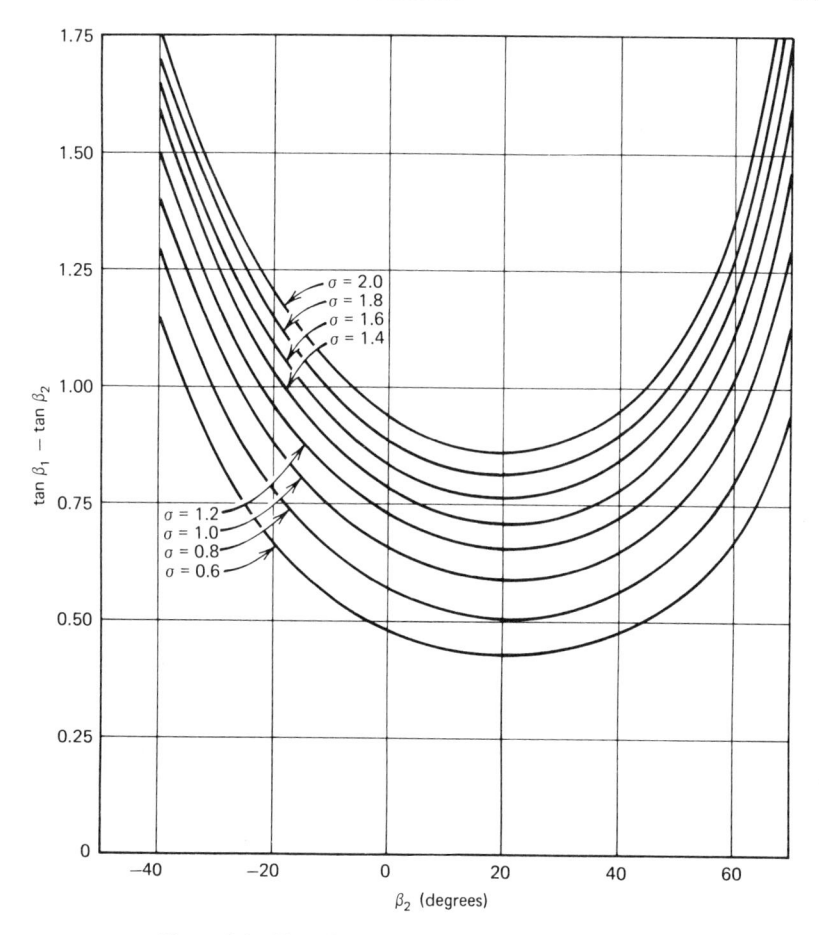

Figure 9.9 Howell relations for swirl, $\tan \beta_1 - \tan \beta_2$.

slight "wash-out" in blade incidence. Small local increases in both general and nose camber will tend to improve $C_{L,\max}$ capability and hence delay the local flow separation that could result from chord reduction; this would also reduce secondary drag (Section 10.3).

At a spanwise station just outside the complex root flow region, which is present for highly loaded fans, any tradeoff between σ and C_L should be well considered if reductions in solidity are desired; these should be minimal when using the data of Fig. 9.12.

Prerotator-rotor unit designs, because of an increased relative blade velocity, result in a substantially lower loading factor $(C_L \sigma)$ for a given duty condition (ϵ, λ). In addition, the rotor blades are solely concerned with static pressure recovery having no impulse component. The design recommendations based on the data of both Howell and Myles for this fan type are

ROTOR BLADE DESIGN

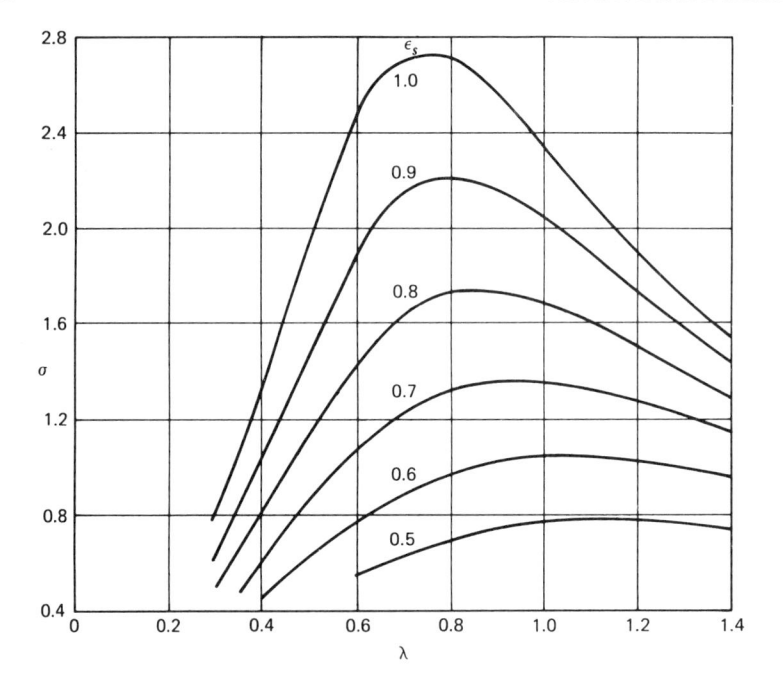

Figure 9.10 Nominal solidity from Howell relation, $\epsilon_p = 0$.

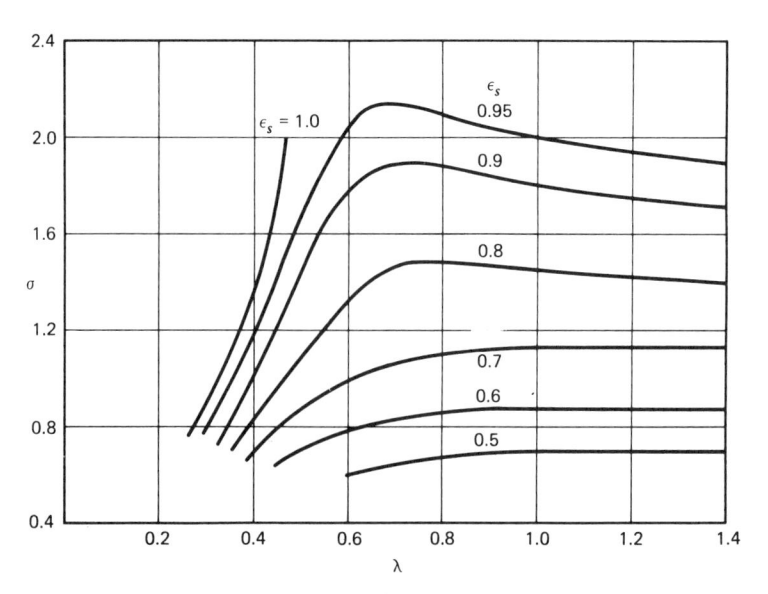

Figure 9.11 Optimum solidity from NEL development, $\epsilon_p = 0$.

9.3 BLADE LOADING LIMITS IN DESIGN

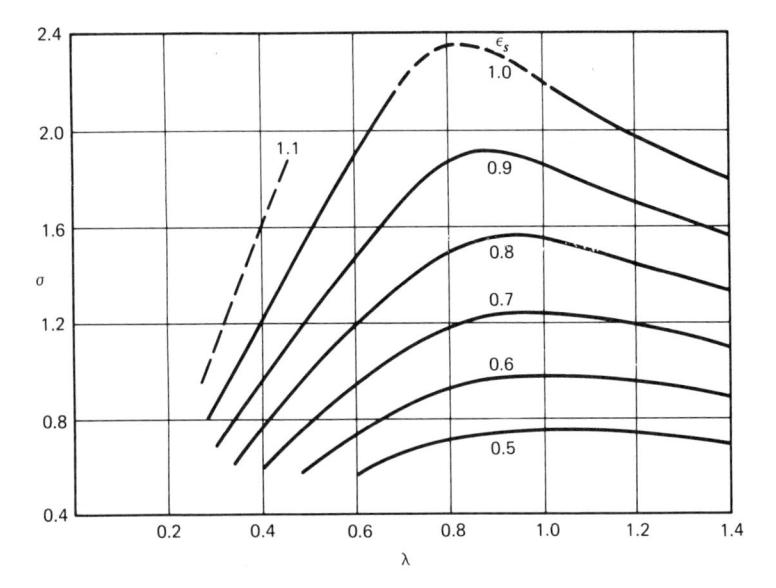

Figure 9.12 Recommended nominal solidity, $\epsilon_p = 0$.

presented in Figs. 9.13 and 9.14, respectively. On the basis of practical experience, the author favors the use of the former. The C_L^* values related to these curves are lower than those for a rotor-straightener unit due to a relatively larger static pressure rise [Eq. (9.7)].

The majority of prerotator-rotor designs is for duties where $K_{th} > 4$. This restricts the design λ range to the lower values, and therefore, high stagger angles. In many instances the blade blockage becomes a new and vital

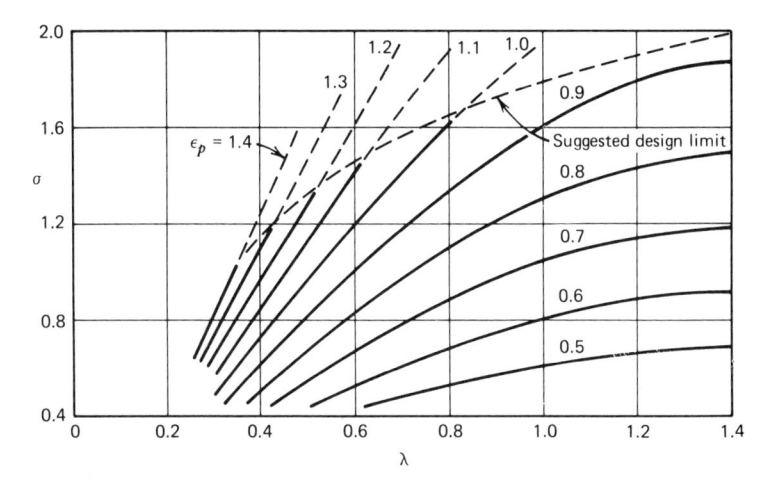

Figure 9.13 Nominal solidity from Howell relation, $\epsilon_s = 0$.

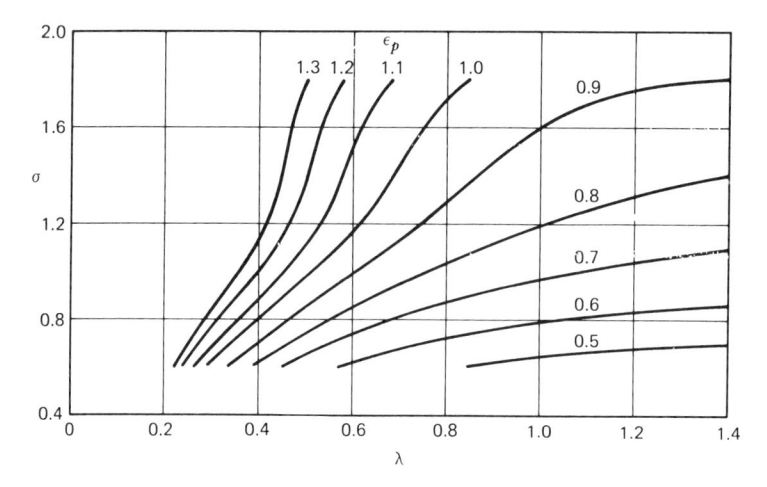

Figure 9.14 Optimum solidity from NEL development, $\epsilon_s = 0$.

parameter, introducing its own design restrictions, particularly on permissible swirl.

The design of prerotator-rotor-straightener units of high root solidity requires a return to the general Howell equation and Figs. 9.6, 9.8, and 9.9. The design approach will be governed by the relative degrees of preswirl and afterswirl, which will in turn determine the interpretation of the graphical data; the preceding and subsequent discussions should prove helpful in this regard.

9.3.4 Blade Blockage Considerations

The flow passages between high-solidity blading become constricted at large stagger angles and hence the diffusing passage flow assumes growing and vital importance. Increases in the drag coefficient and related deviation angle will accompany this phenomenon. Reducing the blade thickness ratio will help minimize the problem.

Experimental data relating to performance penalties are given in [9.10] for actual machines and in [9.16] for two-dimensional cascades. However, design guidance cannot be directly extracted from these papers. After careful consideration of all information, the following advice is offered:

1. *Rotor-straightener units.* The solidity recommendations of Fig. 9.12 will provide sufficient protection from this phenomenon, but it is suggested that the blade root thickness be 10% c or less for $\epsilon_s = 1$, the approximate design limit.

9.4 CASCADE BLADING DESIGN

2. *Prerotator-rotor units.* It can be inferred from Fig. 8.7 that blade stagger angle is substantially increased, at a given duty condition (λ, ϵ), for this alternative design arrangement. Consequently, loss in performance will accompany designs that lie above the "safe" limit suggested by Fig. 9.13 for 10% c thick sections. This thickness ratio should be reduced for designs that exceed this recommendation; cambered plate blading is one suggestion.

9.4 CASCADE BLADING DESIGN

Multistage fans possess an optimum flow coefficient that is high when related to general industrial practice. The advantages of high Λ are discussed in Section 15.5.1.

The optimum blading for such a fan is characterized by small stagger angles and by substantial blade camber and solidity. When the boss ratio is large, these conditions exist over the complete span. Therefore the cascade design procedure must be employed. The simple basic technique that has been exceedingly successful in British aircraft engine development is herein adopted.

The blade loading factors and angular air deflections required to produce the desired pressure and flow are calculated in accord with Chapter 8.

The nominal solidity is obtained from Fig. 9.12 or Fig. 9.13. Combining Eqs. (6.6) and (9.1),

$$\theta = \frac{(\beta_1 - \beta_2) - i}{1 - m\sqrt{s/c}} \tag{9.10}$$

As a preliminary measure, values of zero and 0.26 will be assigned to i and m, respectively. This permits an approximate value of stagger angle to be calculated from

$$\xi = (\beta_1 - i) - \frac{\theta}{2} \tag{9.11}$$

With these preliminary values of θ and ξ, more accurate values of i and m can be obtained from Figs. 9.1 and 9.3. The final design values of θ and ξ can then be computed.

When nose droop is desired the specified correction to blade stagger angle [Eq. (6.13)] is applied.

Greater accuracy in the determination of blade section coordinates, for cambers exceeding the recommended limits suggested in Table E.1, can be obtained when the equations of Appendix E are programmed for computer use.

9.5 WALL STALL

A similar pressure rise coefficient has been used in predicting the onset of wall stall to that developed in relation to blade loading limitation. In the present case, the inlet boundary layer thickness constitutes an important additional variable.

Correlation of cascade wind tunnel test data [9.17] for inlet flows of zero swirl was obtained with the relation

$$\left(\frac{\delta^*}{c} + 0.0285\right) \frac{\Delta p}{\frac{1}{2}\rho V_1^2} \not> 0.0185 \tag{9.12}$$

where δ^* is the displacement thickness of the inlet boundary layer (Section 3.4). Replacing Δp by Δp_{th} introduces a slight degree of conservatism into the relationship.

Hence Eq. (9.12) becomes

$$\left(\frac{\delta^*}{c} + 0.0285\right)\left(1 - \frac{\cos^2 \beta_1}{\cos^2 \beta_2}\right) \not> 0.0185 \tag{9.13}$$

when Eq. (8.20) and Fig. 6.4 are used.

The test airfoils of [9.17] were cambered plates of three different chord lengths but constant camber angle (30°), and 65-series sections of constant chord possessing 0, 30, and 60° camber. Within the investigation range, end wall losses were relatively independent of s/c, θ, ξ, $\beta_1 - \xi$, and airfoil shape. (However, despite some scatter in the test results, cambered plates of equal chord length to the 65-series airfoils appear to possess the greater mean losses probably because of blade local leading-edge separations at certain duty conditions.)

For negligible values of δ^*/c, the pressure rise at which end wall losses rise rapidly [Eq. (9.13)] is directly related to a $\cos \beta_1/\cos \beta_2$ ratio of 0.60; this value increases with δ^*/c. In [9.18] the wall stall is considered imminent when this ratio drops below 0.72, suggesting a corresponding value of 0.01 for δ^*/c.

Highly loaded fans of an optimized free vortex type will feature a large root solidity. Hence the related blade chord length will be far greater than the inlet displacement thickness for a fan unit possessing an adequate nose fairing or spinner.

The wall boundary layer thickness at the blade tip is generally in excess of the root value but the pressure rise, characterized by $\cos \beta_1/\cos \beta_2$, is always less. The difference is, however, reduced as the design value of Λ is increased (Chapter 18).

The tests of [9.17] were confined to much larger values of δ^*/c than those normally associated with the usual critical region for fans, namely, the boss. Relevance of the findings to free vortex fans can be gauged by considering

9.6 BLADE THICKNESS RATIO

the design and test data in relation to the 0.914 m diameter research fan described in [9.8] and [9.19]. Root section details of the F-series airfoil blade were $c = 0.217$ m, $\sigma = 1.82$, $\theta = 36°$, $\xi = 38°$, and $t/c = 0.08$. A hemispherical nose spinner of 0.457-m diameter produced a thin skewed boundary layer at fan inlet. The δ^*/c ratio was therefore minimal.

The design value of $\cos \beta_1/\cos \beta_2$ at the blade root is 0.64. For this condition stroboscopic observations of attached tufts reveal an orderly secondary vortex flow positioned on the fan boss. As the fan stall is approached, the orderly secondary flow suddenly becomes irregular, heralding the onset of wall stall. The resulting disturbed flow that is dispersed over the inner blade surface grows in extent with increasing pressure load until the complete suction surface experiences separated flow. (The rate at which the separation spreads is a function of the spanwise C_L distribution.)

Since the permissible $\cos \beta_1/\cos \beta_2$ is 0.60 for zero inlet boundary layer thickness, the relevance of Eq. (9.13) to the preceding fan type appears reasonable. Extended upstream centerbodies will thicken the inlet boundary layer, which in some cases could result in wall stall at the duty condition. This event will hasten fan stall but not necessarily precipitate the phenomenon.

The relationship between the wall stall and blade loading data for thin inlet boundary layers is emphasized in Chapter 18, where the blade root value of $\cos \beta_1/\cos \beta_2$ for three different highly loaded fan designs is approximately the same (0.65 to 0.67). Hence for fans of this type with nose fairings or spinners of appropriate aerodynamic length, wall stall will not occur at the design duty. This explains the lack of importance that has been ascribed to wall stall in the past. It should not be ignored, however, in instances where separated or thick boundary layers exist at blading inlet.

For fans possessing a high design value of Λ the possibility of a blade stall progressing from the tip region should be kept in mind (Section 18.4).

Small chord blades are undesirable from wall stall, Reynolds number, stiffness, and manufacturing considerations.

When the flow is of a marked arbitrary vortex type, the preceding conclusions will not necessarily apply. A stalling pattern observed in relation to such equipment is described in Appendix C.

9.6 BLADE THICKNESS RATIO

The most commonly used thickness to chord ratio is 0.10. This thickness approximates the optimum for circular-arc sections. However, reductions down to 0.07 have to be implemented when blade blockage problems are present or when weight reductions are necessary.

Structural and vibrational factors are usually responsible for thickness increases. Variable-pitch mechanisms require a circular root shaft of adequate strength and/or rigidity to avoid blade flutter phenomena. In such instances local root blade thicknesses of up to 13% can be considered without significant aerodynamic penalty, provided the stagger angles are not large enough to create blade blockage problems.

REFERENCES

9.1 A. D. S. Carter, The low speed performance of related aerofoils in cascades, Gt. Britain Aero. Research Council, *ARC CP 29*, 1950.

9.2 A. D. S. Carter, The calculation of optimum incidence for aerofoils, Gt. Britain Aero. Research Council, *ARC CP 646*, 1961.

9.3 R. A. Wallis, The development of blade sections for axial flow fans, *Mech. Chem. Eng. Trans., I. E. Aust.*, **MC8**(2), 111–116, 121, 1972.

9.4 S. Lieblein, Experimental flow in two-dimensional cascades, in I. A. Johnsen and R. O. Bullock (Eds.), *Aerodynamic Design of Axial Flow Compressors*, National Aeronautics and Space Administration, Report NASA SP-36, 1965,Chap VI.

9.5 D. J. Myles, R. W. Bain, and G. H. L. Buxton, The design of axial flow fans by computer, Part 2, Gt. Britain National Eng. Lab., *NEL Report 181*, 1965.

9.6 R. A. Wallis, The F-series aerofoils for fan blade sections, *Mech. Eng. Trans., I. E. Aust.*, **ME2**(1), 12–20, 1977.

9.7 R. A. Wallis, A rationalized approach to blade element design, axial flow fans, *Proc. Third Australasian Conf. Hydraulics and Fluid Mechanics*, Sydney, pp. 23–29, 1968.

9.8 J. H. Perry, A study of the flow characteristics of an axial flow fan, *I. E. Aust. Nat. Conf. Publ. No. 74/7*, Melbourne, 1974.

9.9 C. Keller, *The Theory and Performance of Axial Flow Fans*, McGraw-Hill, New York, 1937.

9.10 R. C. Turner, Notes on ducted fan design, Gt. Britain Aero. Research Council, *ARC CP 895*, 1964.

9.11 N. Ruglen, Low speed wind tunnel tests on a series of C4 section aerofoils. Aust. Dept. of Supply, Aero. Research Labs., *ARL Aero Note 275*, 1966.

9.12 A. D. S. Carter, The axial compressor, in H. Roxbee-Cox, (Ed.), *Gas Turbine Principles and Practice*, Newnes, London 1955.

9.13 B. Eck, *Fans*, translated by R. S. Azad and D. R. Scott, Pergamon, Oxford, 1973.

9.14 J. C. Emery, L. J. Herrig, J. R. Erwin, and A. E. Felix, Systematic two-dimensional cascade tests of NACA 65-series compressor blades at low speeds, *NACA Report 1368*, 1958.

9.15 G. J. Walker, A family of surface velocity distributions for axial compressor blading and their theoretical performance, *ASME J. Eng. Power*, **98**:229–241, April 1976.

9.16 M. Matsuki, K. Takahara, H. Nishiwaki, and M. Morita, Cascade tests of high stagger compressor blades, *National Aero. Lab.* TR-10, 1961 (in Japanese).

9.17 W. T. Hanley, A correlation of end wall losses in plane compressor cascades, *ASME J. Eng. Power* **90**:251–257, July 1968.

9.18 J. H. Horlock, J. F. Levis, P. M. E. Percival, and B. Lakshminarayana, Wall stall in compressor cascades, *ASME J. Basic Eng.*, pp. 637–648, 1968.

9.19 R. A. Wallis, Design procedure for optimal axial flow fans, *I. E. Aust. Nat. Conf. Publ. No. 76/12*, Hobart, 1976.

CHAPTER 10

Rotor Losses

The numerous research papers dealing with secondary losses in turbomachines all emphasize the many interrelated variables present and hence the difficulty in achieving problem resolution. However, experimental and theoretical work under the direction of Horlock at Liverpool and Cambridge Universities has attempted separate studies into various facets of the problem, [10.1]. A review of knowledge in this subject is available in [10.2].

Design practice requires the complexities of the problem to be crystallized into empirical-type relations where the loss estimates are expressed in terms of known major design parameters.

Test data relate strictly to the set of geometric and flow parameters studied. Hence the results of compressor studies must be applied with caution to the wider range of fan design possibilities. However, with little relevant information on secondary flow losses in fan equipment, the designer must rely on compressor studies in serving his needs.

In the case of "optimized" fan equipment (Chapter 15), the blade root solidities and stagger angles tend to approximate those of compressors. Hence in this instance the design loss recommendations are applicable.

The current procedure of combining losses resulting from passage secondary flow, large boundary layer skew at blade entry, end wall friction, and tip clearance flow, under the secondary flow heading [10.2] is adopted here. These losses are expressed in terms of midspan design conditions, for convenience. However, this does not imply that the losses are evenly distributed along the blade.

In contrast, the estimation of the profile drag loss component presents fewer problems. As before the component is computed from midspan data.

10.1 MOMENTUM CONSIDERATIONS

Because of the nonuniformity of losses along the blade, the actual total pressure rise across the rotor will vary with spanwise station, resulting in the blade element characteristic curves of pressure rise versus flow rate differing from each other. However, experience has shown that, for equal mean flow conditions, the procedures adopted in design result in close agreement between the estimated and actual mean total pressure rise values, provided tip clearances remain small.

The drag coefficient for a blade element is related to the loss in total pressure ($\bar{\omega}$) by the relation of Eq. (8.24), namely,

$$C_D = (s/c) k_R \cos^3 \beta_m$$

where $k_R = \bar{\omega}/\tfrac{1}{2}\rho \bar{V}_a^2$. Multiplying by C_L/K_{th} and writing γ for C_L/C_D,

$$\gamma \frac{k_R}{K_{th}} = \frac{C_L \sigma}{K_{th} \cos^3 \beta_m}$$

Substituting for $C_L \sigma$ [Eq. (8.27)] and K_{th} [Eq. (8.17)]

$$\gamma \frac{k_R}{K_{th}} = \frac{\lambda}{\cos^2 \beta_m} - \frac{\sigma C_D \tan \beta_m}{K_{th} \cos^3 \beta_m} \qquad (10.1)$$

Since the last term is normally much smaller than the preceding one, this expression for design purposes reduces to

$$\gamma \frac{k_R}{K_{th}} = \frac{\lambda}{\cos^2 \beta_m} \qquad (10.2)$$

Making the design assumption that the mean drag coefficient for the rotor is given by

$$\bar{C}_D = C_{D_P} + C_{D_S} \qquad (10.3)$$

where C_{D_P} is the profile drag coefficient at mid blade span, and the secondary drag coefficient C_{D_S} is suitably calculated, then the mean total pressure loss coefficient K_R for the rotor follows from Eq. (10.2), namely,

$$\frac{K_R}{K_{th}} = \left(\frac{C_D \lambda}{C_L \cos^2 \beta_m} \right)_{MS} \qquad (10.4)$$

This expression can alternatively be written

$$\frac{K_R}{K_{th}} = \frac{K_{R_P}}{K_{th}} + \frac{K_{R_S}}{K_{th}} = \left(\frac{C_{D_P}}{C_L} + \frac{C_{D_S}}{C_L} \right) \left(\frac{\lambda}{\cos^2 \beta_m} \right)_{MS} \qquad (10.5)$$

where MS denotes the midspan.

10.2 PROFILE DRAG

Since the angle β_m in this particular case is a function of the midspan values of λ and ϵ [Eq. (8.19)], it is possible to display rotor efficiency losses in terms of these variables and lift/drag ratio (Fig. 10.1), for both rotor-straightener and prerotator-rotor arrangements. For a given design condition, in terms of λ and ϵ, the former arrangement is the more efficient because of lower relative blade velocities.

The secondary drag coefficient is assumed to include a component for minimum tip clearance loss. Adjustments to K_R/K_{th} for excessive tip clearance, and to K_{th} for the accompanying loss in pressure at a given fan throughflow quantity, are suggested in Chapter 14.

10.2 PROFILE DRAG

The variables that influence this component of drag at the design incidence are profile shape, surface roughness, blade solidity, Reynolds number, and air turbulence. Since the design incidence is for all practical purposes synonymous with minimum profile drag coefficient (see Chapter 6 and [10.3]), for correctly cambered blades with "shock-free" entry flow conditions, the estimation problem is simplified for F-series or similar type blading. The drag

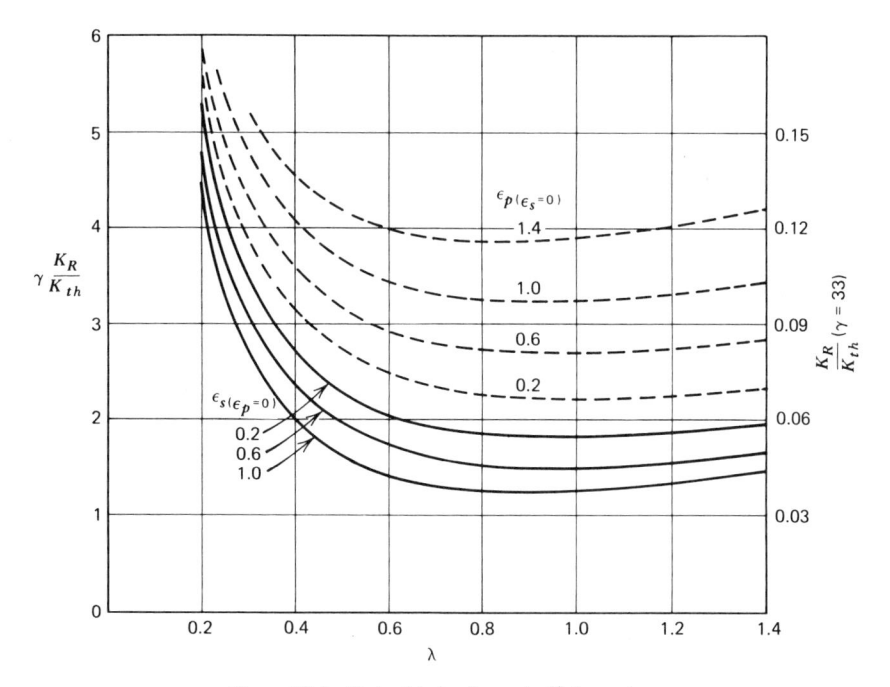

Figure 10.1 Rotor blade element efficiency loss.

data presented in Section 6.8.3 are also sufficient for the design of airfoils with flat undersurfaces.

The only variables not covered in Chapter 6 are surface roughness and air turbulence.

10.2.1 Surface Roughness

Loftin and Smith [10.4] measured increases in the minimum drag coefficient of 0.0055 and 0.004 for a NACA 4412 airfoil at Reynolds numbers of 7×10^5 and 6×10^6, respectively, when "standard" leading-edge roughness is applied. This roughness consists of 0.28 mm mean diameter carborundum grains spread over 8% chord surface lengths, from the leading edge, on upper and lower surfaces of a 610 mm chord airfoil. Since the roughness/boundary layer thickness ratio controls the drag increase (Chapter 3), the same-sized particles will have a much reduced influence when located toward the blade trailing edge. Experience is essential for the designer seeking to estimate drag increases because of surface irregularities and roughness, but some guidance can be obtained from Chapter 3.

Surface excrescences will accelerate the onset of boundary layer separation and hence precipitate the fan stall. Hence to provide a reasonable margin between the design and stall points in such circumstances, conservative design lift coefficients should be chosen; this is again a matter for experience.

10.2.2 Air Turbulence and Re_{crit}

Air turbulence has a marked influence on the critical Reynolds number, that is, the number above which the fan performance is appreciably improved. Downstream of a blade row there is always a substantial turbulence increase and a corresponding reduction in Re_{crit} and hence data obtained from multistage compressor tests, where the turbulence intensity rises and levels out at approximately 6% [10.5], are not generally applicable to fans.

In contrast, the air turbulence intensity ahead of fan blading is normally in the range 0.5 to 1.0%, for obstruction-free inlet flows.

For low inlet turbulence the Reynolds number below which performance begins to deteriorate is 2.5×10^5. This is supported by wind tunnel tests on isolated C4 airfoils [10.6] and by the rotor-only fan tests reported in [10.7]; this quantity refers to smooth, well-formed 10% thick airfoil sections.

Increasing the turbulence level to 3% reduces the preceding specified number to 1.6×10^5 [10.5], for the case of an 8% thick NACA 65-608 section; this reference also reports a higher Re_{crit} with increasing thickness to chord ratio. Leading-edge droop, by reducing the adverse static pressure gradients over the nose, reduces Re_{crit} by a small amount.

A commonly reported value of Re_{crit} is 2×10^5 with marked performance penalties below 1×10^5. In the case of many industrial installations where the fan inlet flow is disturbed, the turbulence level will exceed 1%; hence

10.3 SECONDARY DRAG

these latter values are approximately correct for the general run of "in-line" fans with 10% thick blading.

When selecting blade chords from the calculated design distribution of solidity, consideration should be given to Re_{crit} data in order to attain the best operational characteristics.

The flow features that are responsible for the abrupt change in drag values for streamlined profiles are absent in the case of thin, constant thickness plates. Except for a small incidence range, the flow is always separated from the leading edge. The Reynolds number has a minor influence on the reattachment point, producing slight but trivial changes to the lift and drag properties. Tests on a flat plate ($t/c = 0.02$) over the Re range, 1.5×10^5 to 4.0×10^5, show no significant change in these force properties [10.8]. An authority quoted in [10.8] also reports little change in flat plate characteristics over the range 3×10^5 to 1.7×10^6. Hence for all practical purposes, thin cambered plate force data will not be subject to variation with Re.

The minimum drag coefficient remains constant over the camber/chord range from 0% to 8% (Fig. 6.28), as distinct from faired airfoils (Fig. 6.10).

10.3 SECONDARY DRAG

The emphasis herein is on achieving acceptable estimation accuracy in relation to the general run of well-designed fans; a literature survey of this complex subject is not attempted.

The current practice of combining the secondary and annulus drags results in the Howell data [10.9] being presented as

$$C_{D_S} = aC_L^2 + 0.02 s/h$$
$$= aC_L^2 + 0.02 c/\sigma h \qquad (10.6)$$

where h is the blade height and a is a function of Re and varies between 0.019 at $Re = 1 \times 10^5$ and 0.015 at 5×10^5. The figure commonly quoted is 0.018, which matches the normal compressor design condition.

Vavra [10.10] has suggested an expression of the form

$$C_{D_S} = \frac{0.04\, C_L^2}{\dfrac{h}{c}} \qquad (10.7)$$

This equation omits the solidity parameter of Howell. Griepentrog [10.11] while acknowledging that solidity can influence the secondary flows reports contradictory experimental evidence. This would suggest a minor role for this variable.

The influence of air-turning angle ($\beta_1 - \beta_2$) on secondary flow and losses is briefly introduced in Section 2.7. Since C_L^2 is used in expressing the drag component, it is apparent that the relationship between flow angles and C_L [Eq. (6.5)] is a related factor. However, since an increase in stagger for a fixed value of ($\tan \beta_1 - \tan \beta_2$) increases C_L, and reduces ($\beta_1 - \beta_2$), a reason for frustration in this area of research can be seen. When the midspan stagger angle has a relatively limited range of values, for example, in compressor or fan designs of an optimum type the C_L^2 relationship assumes a greater relevance, provided the coefficient can be adjusted on the basis of experimental data. Phenomena other than flow turning are of course covered by the secondary drag expression.

The C_L^2 dependency has been partially verified in relation to large fans for primary mine ventilation [10.12]. The characteristic curves and efficiencies were estimated for various blade settings by the methods outlined in Chapter 20. Primarily because of the C_L^2 term, closures of the highest constant efficiency curves, ahead of the stall, were predicted with acceptable accuracy.

The data listed in Table 10.1 refer to the preceding prerotator-rotor-straightener unit and to a rotor-straightener fan designed and tested in the United Kingdom [10.13]. Both parallel-walled duct units possessed nose and tail fairings.

The preceding limited design and test data are insufficient to permit a more definitive approach to secondary drag calculations. However, the loss estimation procedure employed previously appears to possess a degree of conservatism. As a consequence no increase in the secondary drag estimate, to embrace the annulus drag component, is required for design purposes.

While acknowledging that small aspect ratios will have a significant influence on C_{D_S}, the data available do not permit its explicit inclusion in an analytical expression. For example, predictions based on Eq. (10.7) would have produced substantial errors in the preceding two fan cases. A lesser influence is predicted by Eq. (10.6), but the inclusion of blade solidity in the expression is open to question, particularly for fans.

An expression of the form

$$C_{D_S} = b C_L^2 \tag{10.8}$$

is therefore proposed where b is an undefined function of blade aspect ratio, Re, and stagger angle. The preceding experiences (Table 10.1) would suggest that b will approximate the a values proposed by Howell.

It is recommended that a value of 0.018 be allotted to b for general use reducing to 0.015 or less for fans possessing high operational Reynolds number. Provided the aspect ratio is greater than 1.5, these values should produce efficiency predictions within the normally desired limits of $\pm 2\%$.

When the aspect ratio nears 0.7, an additional loss of around 2% in fan efficiency is likely. The characteristic curve near stall will be more rounded because of the influence of the larger secondary flows.

Table 10.1

Fan Unit	Rotor Drag Assumptions C_{Dp}	a	Ann. Drag	Mid-span Values ξ	σ	Mean Re	Aspect Ratio	Tip Clear %c	Design (Unit) Est. η_T	Design (Unit) Test η_T	Design (Rotor) Est. η_R	Design (Rotor) Test η_R
Mine Fan $D = 6100$ mm $x_b = 0.5$	0.010	0.015	2%	66°	0.46	2.4×10^6	3.2	0.6	0.875 (0.895)	0.88^a	0.90 (0.92)	—
NEL Fan $D = 610$ mm $x_b = 0.716$	0.016	0.018	0.7%	31°	0.79	3.6×10^5	1.4	1.0	0.848^b (0.868)	0.863	0.92 (0.927)	0.924

[a] See Chapter 20 for additional data.
[b] 2% annulus loss distributed over rotor, stators, and tail fairing. ( ) Without annulus drag loss estimate.

Downstream of prerotators, the duct wall flows possess a larger tangential deflection than that associated with the mainstream. The resulting vorticity is of opposite sign to the rotor-induced secondary flow. The effect of this feature on fan efficiency is not known, but an increase in b is unlikely.

The preceding recommendations regarding b will provide a mean basis for estimating C_{D_S}. However, adjustments based on personal experience will be required in achieving accuracy over the whole range of design possibilities. The preceding empirical approach is considered the most practical from the design viewpoint.

In the case of constant thickness plate sections, increased secondary drag losses are present because of local leading-edge flow disturbances that induce a greater flow of low-energy air into the secondary vortices. However, the suggested quantities of [10.14], namely,

$$C_{D_S} = 0.025 \, C_L^2$$

$$C_{D_A} = 3\% \text{ loss in fan efficiency}$$

are now known to be excessive when used in relation to fans of improved design to those studied.

In suggesting amended values for b, consideration must be given to the increased boundary layer sensitivity of plate sections, with incidence changes, when comparison is made with the situation for optimum airfoil shapes. Hence the values

$$b = 0.025 \text{ to } 0.040$$

are tentatively proposed, with the former applying to well-designed units operating at Reynolds numbers in excess of 2×10^5 and with aspect ratios approximating 2. The designer must be guided by actual experience in his choice of the appropriate b value, keeping in mind likely trends for differing sets of design variables. For well-designed and installed units the resulting estimates of rotor efficiency should fall within the required limits of $\pm 2\%$.

Means of minimizing the secondary drag are listed in [10.11]. One suggestion with relevance to the recent F-series development is for additional leading-edge camber, in order to accommodate the higher-flow angles in the blade root boundary layer.

10.4 BLADE END CLEARANCES

Provided the tip clearance does not exceed 1% of blade height (i.e., span), no adjustments to design pressure duty or fan efficiency are necessary. However, this clearance cannot be maintained for all fan installations, because of

10.4 BLADE END CLEARANCES

thermal or mechanical reasons; hence in the interests of accuracy some corrections are needed.

The existence of an optimum tip clearance, for minimum fan losses, is widely accepted, but the actual value has yet to be uniquely expressed in terms of blading and flow parameters. According to [10.15] the optimum tip clearance/blade chord ratio is 0.04 for a stationary blade cascade. However, Horlock concedes the possibility of rotation reducing this value, and certainly the fan test data support this view.

The secondary vortex that exists on the convex surface at the blade tip, because of flow turning, has a rotational direction that involves a duct wall flow toward the blade tip. The tip leakage flow opposes this motion, restricting the secondary flow; hence the concept of an optimum tip clearance is well founded. The effect of the clearance passage shape on the actual optimum value is experimentally demonstrated in [10.16]. Prerotators would increase the beneficial effect of the duct wall flow.

The pressure rise load is not greatly reduced in the clearance region, provided the latter is small [10.15]; the load appears to be supported by the local vorticity.

Since the blade chord rather than the span will determine the eventual scale of the secondary vortex, research workers express their findings in terms of clearance/chord ratio [10.15]. However, the effects of extra tip clearance on efficiency and pressure duty differ substantially and hence are unlikely to respond to a common treatment method. Efficiency loss is due mainly to local increased swirl velocities, whereas pressure duty reductions are the product of changes over an increasing extent of the blade span as the clearance grows. For the larger clearance ratios these load changes can be considerable, bearing a small resemblance to spanwise loading on an aircraft wing.

In an attempt to correlate efficiency and pressure duty losses, Hesselgreaves [10.17] utilized a sensitivity factor that was derived from blade geometry and flow parameters. Loss data from a number of fan and pump test sources were plotted against the product of sensitivity factor and tip clearance/blade height ratio. However, the correlation leaves a lot to be desired.

The loss data from [10.18], [10.19], and [10.20] plotted in Figs. 10.2 and 10.3 appear to illustrate the difference between the efficiency and pressure duty loss phenomena. The former tends to increase linearly with clearance gap, whereas the latter possesses a varying loss rate. The limited information prevents the establishment of trends due to spanwise blade loading and aspect ratio and other possible variables. However, the data support the design recommendation that, provided the clearance ratio is less than 1%, no special design considerations are necessary. For the best performance, however, the preferred ratio would appear to vary between 0.5% and 1.0%.

With the exception of separate fan and duct mountings such as those for car cooling systems, tip clearance ratios can be restricted, thus reducing the adjustment quantities for loss. The limited data presented in Figs. 10.2 and

×----× AR 1.4 Kahane [10.19]
□----□ AR 2.8 Smith [10.20] four-stage compr., average η
○——○ AR 1.4 Ruden [10.18]

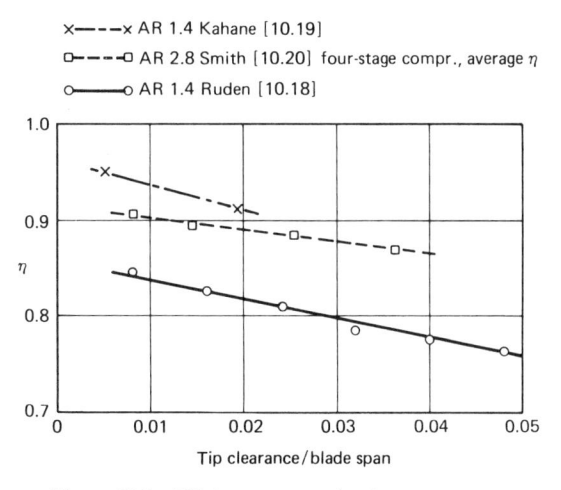

Figure 10.2 Efficiency versus tip clearance ratio.

10.3 can therefore be used with an added degree of confidence. For normal free vortex fan designs the suggested correction is

$$\text{Efficiency loss} = 2\left(\frac{\text{tip clearance}}{\text{blade span}} - 0.01\right) \tag{10.9}$$

The pressure duty loss adjustment can be read off the intermediate curve, due to Ruden [10.18], presented in Fig. 10.3. The errors resulting from omission of a probable aspect ratio effect are expected to be small, when using the preceding tip clearance to blade span ratio.

Blade root clearance has received little research effort. This aspect is extremely relevant to adjustable pitch fans, where in practice the root clearance must be adequate to deal with large angular blade changes and possible hub dirt accumulation. Some designers keep the clearances uniformly small by machining a spherical surface on the fan hub, but this measure will induce earlier wall stall and a subsequent deterioration in blade root flow conditions.

Root clearance, for fans that have a root-initiated stall, seldom produces an unfavorable influence. The air bleed tends to achieve a significant static pressure rise relief on the inner wall as well as providing a partial counter to the secondary flows, as discussed earlier. As before, no separate loss allowances are required for clearances of 0.5 to 1.0% of blade height; the former is preferred.

The author's only experience with larger root clearances was in connection with the aforementioned large mine fans [10.12]. The 850 mm root chord of a 3.2 aspect ratio blade was cut normal to the blade axis, to permit blade movements in excess of 90°. Model tests showed an efficiency loss of 2%

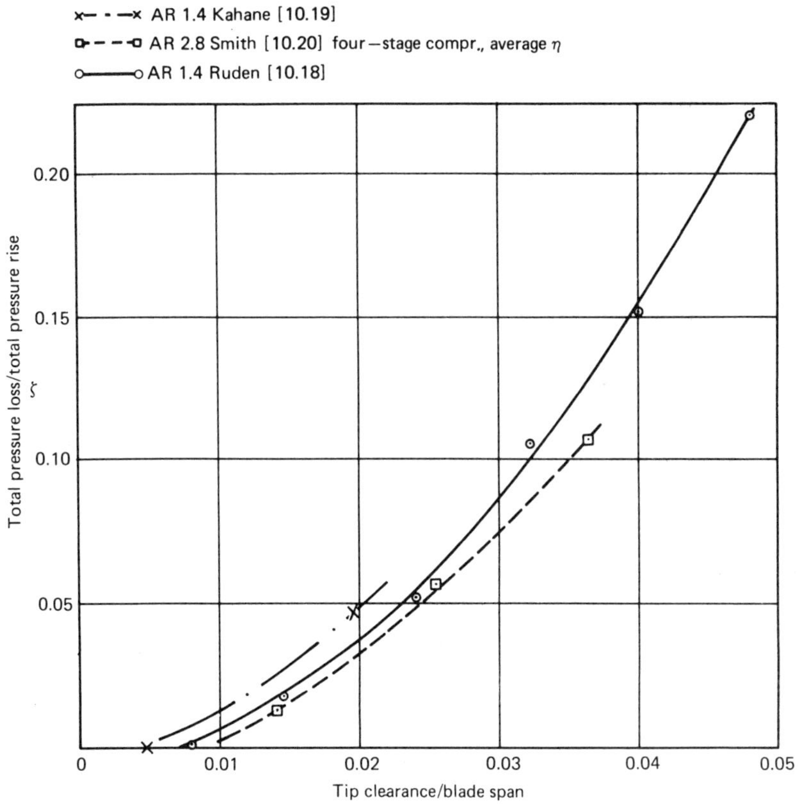

10.5 OTHER ROTOR LOSSES

Figure 10.3 Pressure loss factor versus tip clearance ratio.

with no change to either the characteristic curve or the fan stall point. The smaller annulus area for a given clearance and a lesser moment arm of the tangential force associated with the leakage flow, in comparison with the tip clearance case, both contributed to reduced performance loss.

In the absence of published data the designer must be guided by experience or, for the case of major equipment, by model tests.

The adjustments to the design computation of the theoretical total pressure rise coefficient are given in Chapter 14.

10.5 OTHER ROTOR LOSSES

Through boundary layer shear force action, a rotating fan disk will induce a rotary flow. Provided this flow remains unobstructed by stationary objects no adjustment is necessary to rotor power. However, in variable pitch change mechanisms that remain unfaired or unenclosed, substantial losses will occur when the induced flows are impeded by bearing support systems or other longitudinal structural members. Instead of a body of air rotating at close to rotor speed, with little power dissipation, the stationary members absorb considerable rotor power that is supplied by pressure as well as by tangential shear forces. An adequate power margin must be provided when these circumstances are present. These pressure decrements are commonly referred to as windage losses.

All weight lightening openings and mechanism apertures in the rotor disk must be sealed against throughflow. Centrifugal forces acting on a friction-induced rotating air mass can also set up an internal pumping system, with varying degrees of air interchange with the main stream.

The drag contribution of various configurations of nose fairing are discussed in Chapter 13 as efficiency and pressure losses.

Parallel continuity of the blading annulus surfaces should be maintained. Minimum practical clearance gaps between the rotor hub and adjacent fairing surfaces can be considered drag free.

REFERENCES

10.1 B. Lakshminarayana and J. H. Horlock, Leakage and secondary flows in compressor cascades, Gt. Britain Aero. Research Council, *ARC R&M 3483*, 1967.

10.2 J. H. Horlock and H. J. Perkins, Annulus wall boundary layers in turbomachinery. *AGARD-A6-185*, NATO, 1974.

10.3 S. Lieblein, Experimental flow in two-dimensional cascades, in I. A. Johnsen and R. O. Bullock (Eds.), *Aerodynamic Design of Axial Flow Compressors*, National Aeronautics and Space Administration, Report NASA SP-36, 1965, Chap VI.

10.4 L. K. Loftin and H. A. Smith, Aerodynamic characteristics of 15 NACA airfoil sections at seven Reynolds numbers from 0.7×10^6 to 9.0×10^6 *NACA Tech. Note 1945*, 1949.

10.5 H. Schlichting and A. Das, On the influence of turbulence level on the aerodynamic losses of axial turbomachines, in L. S. Dzung (Ed.), *Flow Research on Blading*, Elsevier, Amsterdam, 1970.

10.6 N. Ruglen, Low speed wind tunnel tests on a series of C4 section aerofoils, Aust. Dept. of Supply, Aero. Research Labs, *ARL Aero Note 275*, 1966.

10.7 W. M. Schulze, J. R. Erwin, and G. C. Ashby, NACA 65-Series compressor rotor performance with varying annulus-area ratio, solidity, blade angle, and Reynolds number, and comparison with cascade results, *NACA Tech. Note 4130*, 1952.

10.8 R. A. Wallis, Wind tunnel tests on a series of circular arc plate aerofoils, Aust. Dept. of Supply, Aero. Research Labs, *ARL Aero Note 74*, 1946.

10.9 A. R. Howell, The present basis of axial flow compressor design, Part II: Compressor theory and performance, *Gt. Britain RAE Report E3961*, 1942.

10.10 M. H. Vavra, *Aero-Thermodynamics and Flow in Turbo Machinery*, Wiley, New York, 1960.

10.11 H. Griepentrog, Secondary flow losses in axial compressors, *AGARD Lecture Series No. 39*, 1970.

10.12 K. E. Mathews et al., Development of the primary ventilation system at Mount Isa, *Aus. I. M. and M. Proc.* No. 222, p. 1–61, 1967.

REFERENCES

10.13 J. E. Hesselgreaves and M. R. Jones, The performance of an axial flow fan designed by computer, Gt. Britain, National Eng. Lab., *NEL Report 447,* 1970.

10.14 R. A. Wallis, Performance of sheet metal bladed fans, Aust. Dept. of Supply, Aero. Research Labs, *ARL Report A90,* 1954.

10.15 J. H. Horlock, Some recent research in turbo-machinery, *Instn. Mech. Eng.,* London, **182,** Part 1, Proc., 1967–68.

10.16 R. Hürlimann, discussion of paper by W. Traupel on *Ergebnisse von turbinenversuchen,* in L. S. Dzung (Ed.), *Flow Research on Blading,* Elsevier, Amsterdam, 1970.

10.17 J. E. Hesselgreaves, A correlation of tip-clearance/efficiency measurements on mixed-flow and axial-flow turbomachines. Gt. Britain National Eng. Lab., *NEL Report 423,* 1969.

10.18 P. Ruden, Investigation of single stage axial flow fans, *NACA Tech. Mem. No. 1062,* 1944.

10.19 A. Kahane, Investigation of axial-flow fan and compressor rotors designed for three-dimensional flow, *NACA Tech. Note No. 1052,* 1948.

10.20 L. H. Smith, Jr., Casing boundary layers in multistage axial flow compressors, in L. S. Dzung (Ed.), *Flow Research on Blading,* Elsevier, Amsterdam, 1970.

CHAPTER 11

Stators: Design Considerations

Stator design is a function of a single parameter, namely, the swirl coefficient. This permits the presentation of recommended vane geometry data in simple design graphs. The spanwise distributions of swirl are calculated in accordance with the free vortex flow relationship.

11.1 MOMENTUM CONSIDERATIONS

Momentum arguments of a similar nature to those employed in Section 8.4, when used in conjunction with Figs. 11.1 and 11.2, result in the following relationships:

$$\beta_{m_s} = \tan^{-1}\left(\frac{\epsilon_s}{2}\right) \tag{11.1}$$

$$\beta_{m_p} = \tan^{-1}\left(\frac{\epsilon_p}{2}\right) \tag{11.2}$$

Straighteners

$$C_L = 2\frac{S}{C}\epsilon_s \cos \beta_{m_s} - C_D \frac{\epsilon_s}{2} \tag{11.3}$$

$$C_D = \frac{S}{C} k_S \cos^3 \beta_{m_s} \tag{11.4}$$

Prerotators

$$C_L = 2\frac{S}{C}\epsilon_p \cos \beta_{m_p} - C_D \frac{\epsilon_p}{2} \tag{11.5}$$

11.2 STRAIGHTENER DESIGN

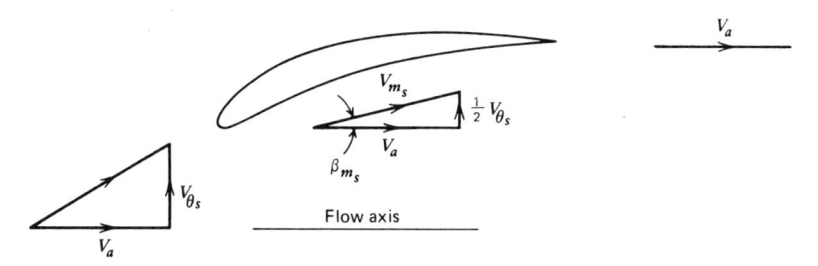

Figure 11.1 Velocity vectors for straightener vane-element.

$$C_D = \frac{s}{c} k_P \cos^3 \beta_{m_p} \tag{11.6}$$

Since the last term in Eqs. (11.3) and (11.5) is usually much smaller than the preceding one

$$C_L \sigma = 2\epsilon_s \cos \beta_{m_s} \tag{11.7}$$

$$C_L \sigma = 2\epsilon_p \cos \beta_{m_p} \tag{11.8}$$

It can be deduced from these expressions that flow deflection ($\beta_1 - \beta_2$) is a unique function of C_L, for equal solidities. However, in practice the permissible C_L for prerotators exceeds that for straighteners, at a given ϵ, because of the accelerating nature of the flow.

The loading factor, presented in Fig. 11.3 as a function of either ϵ_s or ϵ_p, can reach higher values than in the rotor blade design case, because of the zero and negative outlet flow angles for straighteners and prerotators, respectively (see Fig. 9.8).

11.2 STRAIGHTENER DESIGN

The important flow and geometric parameters controlling straightener design are defined in Fig. 11.4. In accordance with normal practice the design value

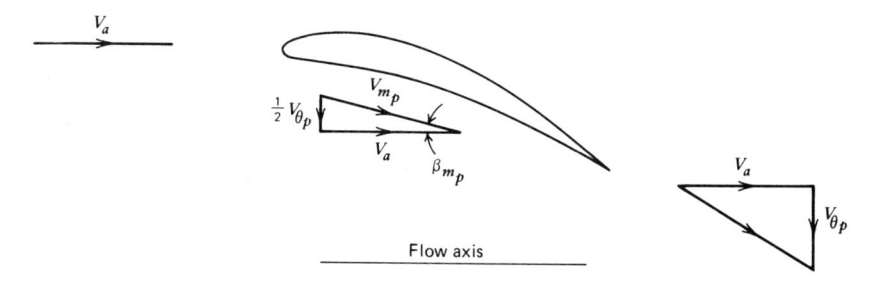

Figure 11.2 Velocity vectors for prerotator vane-element.

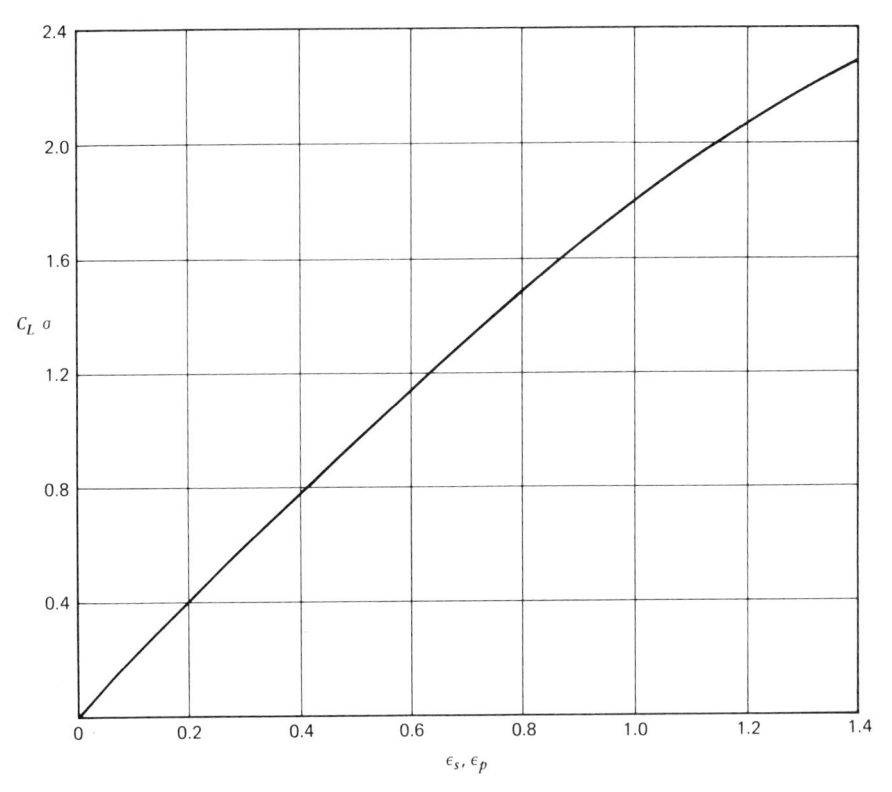

Figure 11.3 Loading factor as a function of ϵ_s or ϵ_p.

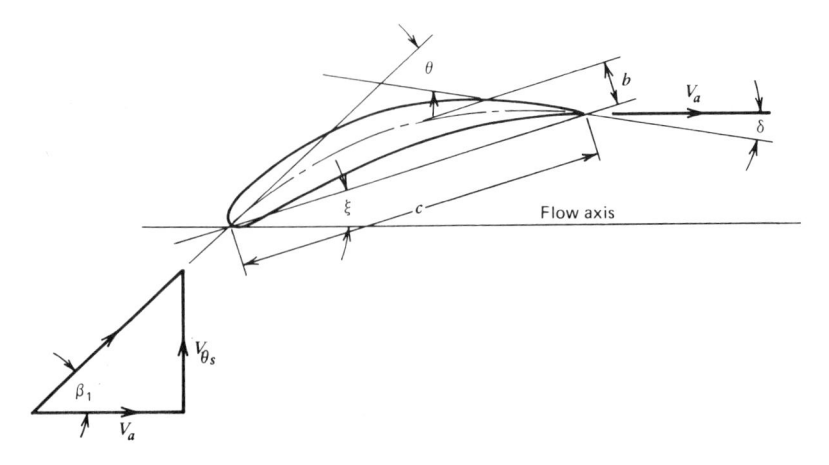

Figure 11.4 Straightener vane-element geometry.

11.2 STRAIGHTENER DESIGN

of local incidence i has been equated to zero. For the deviation angle δ to remain substantially constant, i must not exceed $\pm 5°$ [11.1]. Because of this characteristic straighteners can provide low swirl outlet conditions over a significant fan duty range.

From Fig. 11.4 and Eqs. (6.6) and (9.10) we have the following relationships:

$$\theta = \beta_1 + \delta \qquad (11.9)$$

$$\theta = \frac{\beta_1}{1 - 0.26\sqrt{s/c}} \qquad (11.10)$$

The radius of curvature and percentage camber are obtained from Eqs. (6.10) and (6.8), respectively.

The stagger angle, from Fig. 11.4, is

$$\xi = \beta_1 - \frac{\theta}{2} \qquad (11.11)$$

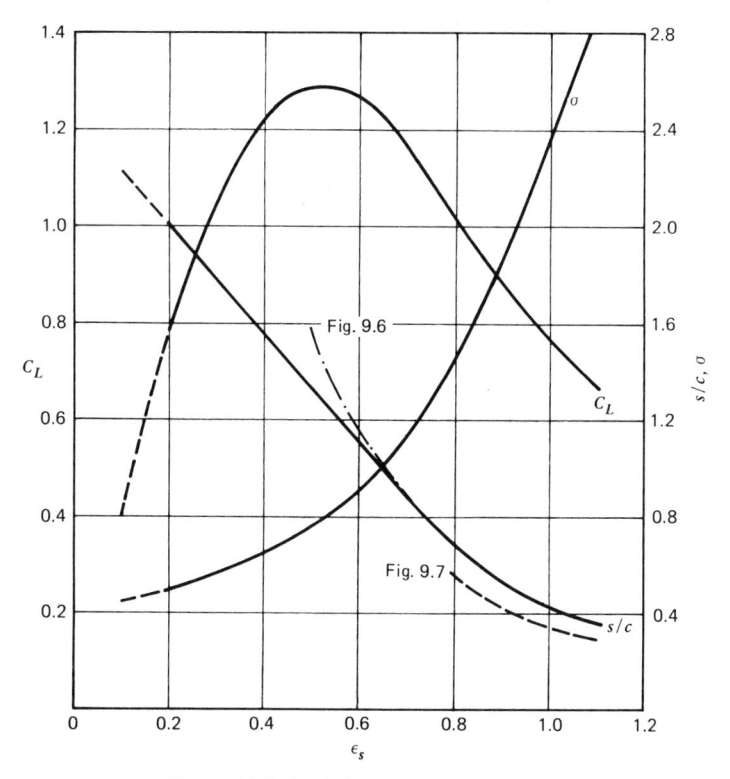

Figure 11.5 Straightener vane design data.

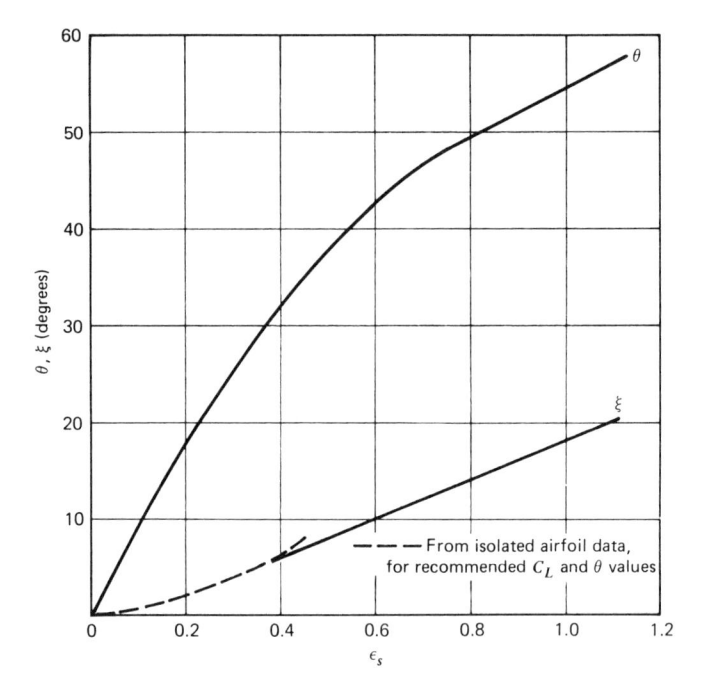

Figure 11.6 Straightener vane camber and stagger angles.

The design data presented in Figs. 11.5 and 11.6, in the ϵ_s range 0.7 to 1.1, follow from Fig. 9.6; between 0.2 and 0.7 a linear distribution of s/c has been arbitrarily adopted. This composite curve (Fig. 11.5) ensures a smooth blending of cascade and isolated airfoil data [11.2] at $\epsilon_s = 0.3$ (Fig. 11.6) and restricts the design C_L to 1.3 at $\epsilon_s = 0.5$.

The arbitrary s/c range for application of the Howell expression [Eq. (6.15)] is from 0.5 to 1.5. However, the $(\beta_1 - \beta_2)$ data of Fig. 9.7 possess a lower s/c limit of 0.3; the resulting design information for the straightener case ($\beta_2 = 0$) is given on Fig. 11.5. The proposed design relation between s/c and ϵ_s is therefore reliable up to ϵ_s of 1.1, as the slopes of the two curves are similar. For these limit conditions, however, a reduction in the thickness/chord ratio below 10% should be seriously considered, as a drag-saving measure.

The design limit on ϵ_s is normally determined by rotor considerations. For example, in the case of optimized fan arrangements blade loading limits may on occasion restrict ϵ_s to slightly less than unity. Reserve capacity is therefore available in the straightener design, if required.

Design is often dictated by a desire for a linear or zero chord variation along the span. In such instances care should be taken to ensure the design

11.3 PREROTATOR DESIGN

remains conservative. The geometric design properties then follow from the foregoing equations rather than from Figs. 11.5 and 11.6.

11.3 PREROTATOR DESIGN

Flow conditions and related vane geometry requirements are outlined in Fig. 11.7. The local incidence i is assumed to be zero, in accord with usual design practice. A variation of at least $\pm 5°$ in i can be tolerated without significant changes in induced flow deflection.

From Fig. 11.7 and the assumption that $i = 0$,

$$\theta = \beta_2 + \delta \tag{11.12}$$

For accelerating cascades and nominal design conditions

$$\delta = 0.19 \frac{s}{c} \theta \tag{11.13}$$

where 0.19 is the mean value of m as obtained from [11.3] for accelerating flow stators. Hence

$$\theta = \frac{\beta_2}{1 - 0.19 s/c} \tag{11.14}$$

The radius of curvature and camber/chord ratio are given by Eqs. (6.10) and (6.8), respectively.

From Fig. 11.7 the stagger angle is given by

$$\xi = \frac{\theta}{2} \tag{11.15}$$

The favored design relations are centered on the findings of Ainley and Mathieson [11.4]. The ensuing designs are slightly more conservative than those based on the data presented in Fig. 9.6. The s/c curve from $\epsilon_p = 0.5$ to 1.4 (Fig. 11.8) represents the flow condition that precedes an experimentally established rise in profile drag for a 10% thick, circular-arc airfoil section [11.4]. This results in a constant C_L value of 1.76.

The s/c curve for values of ϵ_p below 0.5 has been influenced by the desire to achieve reasonable C_L values at low ϵ_p, blending with isolated airfoil data. Because of the flow accelerating feature of this stator type, the upper surface boundary layer will be thinner and hence the incidence for a given C_L will be slightly smaller than for the flow retardation case. As a consequence, the blend will be better than that indicated on Fig. 11.9. The ensuing s/c distribu-

236   STATORS: DESIGN CONSIDERATIONS

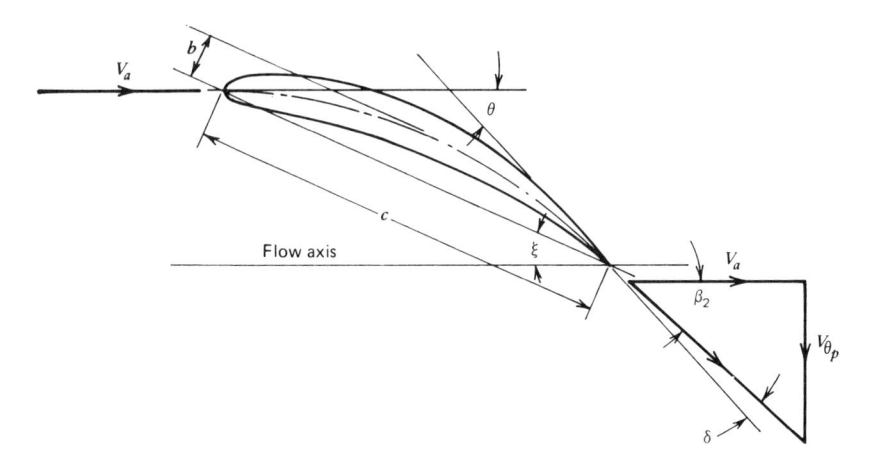

Figure 11.7 Prerotator vane-element geometry.

tion of Fig. 11.8 is therefore confirmed, providing a computational basis for the additional geometric information on Figs. 11.8 and 11.9.

The nozzle vane design recommendations of [11.4] are based on "throat" geometry. In the present instance, however, full reliance has been placed on the deviation angle expression of Eq. (11.13), in view of confidence expressed in [11.5] and favorable fan design experiences.

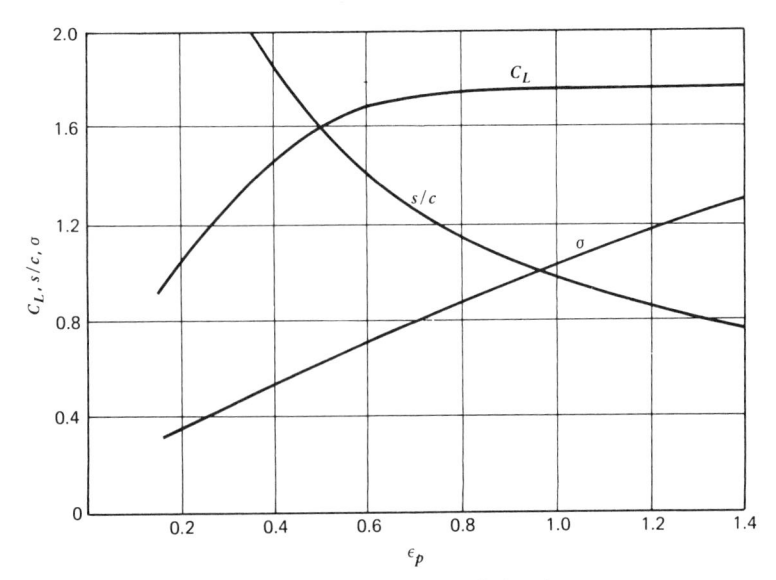

Figure 11.8 Prerotator vane design data.

11.4 SYMMETRICAL VANES PARALLEL TO FAN AXIS

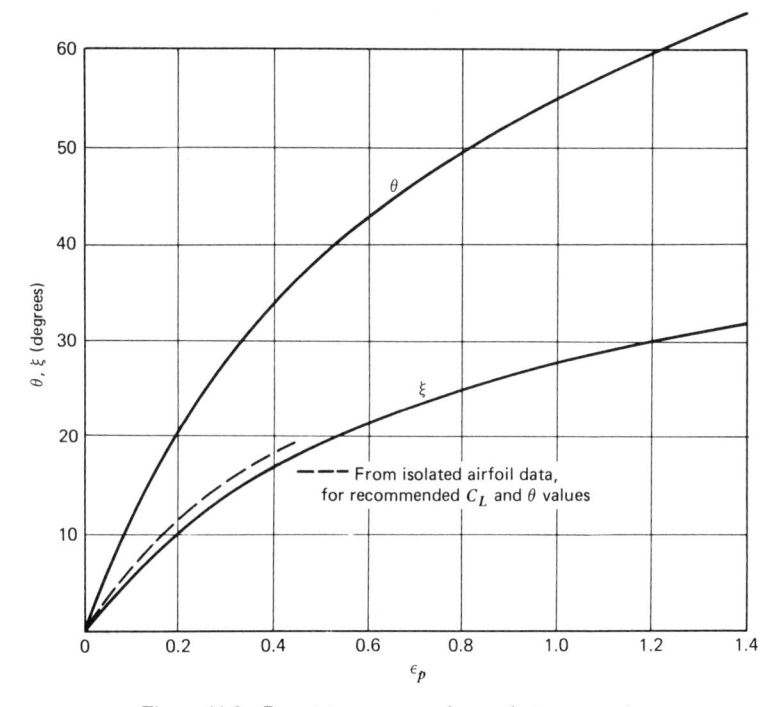

Figure 11.9 Prerotator vane camber and stagger angles.

The amount of prerotation required for any prerotator-rotor fan unit will be governed by rotor blade considerations, being always within vane capability limits.

The spanwise distribution of chord can be altered to suit practical requirements provided the design data are derived by calculation from the foregoing equations. The selected values of solidity should equal or exceed the values indicated in Fig. 11.8.

11.4 SYMMETRICAL VANES PARALLEL TO FAN AXIS

Inadequate aerodynamic consideration of the consequences of certain bearing support and/or rotor drive systems often leads to operational problems. For example, the electric drive motor is often mounted on a bench plate spanning the duct, incorporating one or more radial stiffening plates. This limited array of plates is assumed, incorrectly, to perform a flow-straightening function. Instead flow separation from each plate leading edge will lower fan efficiency and create downstream flow problems. This example is only one of many resistance-prone mechanical arrangements.

The configuration that provides the correct aerodynamic alternative to the preceding example possesses a motor or bearing enclosure of equal outside diameter to the fan boss; this is in turn supported by a number of aerodynamically designed radial vanes. These vanes can, in certain low swirl circumstances, consist of uncambered symmetrical airfoils [11.6].

The so-called NPL-type straighteners (Fig. 11.10) were proposed by Collar after an analysis of experimental cascade data. The results indicated that cambered airfoils of unit solidity, and with the "no-lift" line of the section parallel to the duct axis, would remove all swirls within the working range of the airfoil. He therefore proposed the use of symmetrical sections, such as the NACA 0012. This airfoil, according to [11.7], can be considered satisfactory for swirls within the ϵ_s range 0 to 0.4.

The preceding axial vane arrangement will be associated with a small residual swirl, because of the presence of a deviation angle of unknown magnitude. However, small swirls are usually of little practical consequence (Sections 13.5 and 14.2). The preceding minor design flaw will not be aggravated by the assumption of constant chord rather than constant solidity. This will result in a more constant spanwise distribution of C_L (Fig. 11.3). The assumption of unit solidity at the inner spanwise extremity of ϵ_s of 0.4 will require an approximate C_L of 0.8; for constant chord the C_L value will decrease toward the tip.

Symmetrical 12% thick airfoils of conventional shape will operate successfully at the previously nominated C_L. Retaining this design value of 0.8 for lesser ϵ_{s_b} values than 0.4, the required stator root solidity will follow from Fig. 11.3, thus enabling the constant chord dimension to be chosen.

Hollow straighteners of the preceding airfoil type provide ducts for motor cooling air and electrical cables. A longitudinally slotted motor enclosure and open-ended tail fairing provide an alternative method of motor cooling.

Symmetrical inlet guide vanes of adjustable pitch setting are often used for controlling fan duty. However, since these accelerate the flow, higher C_L and ϵ_p values are permissible than for straighteners of equal solidity. In the absence of relevant published data, the designer must be guided by experience.

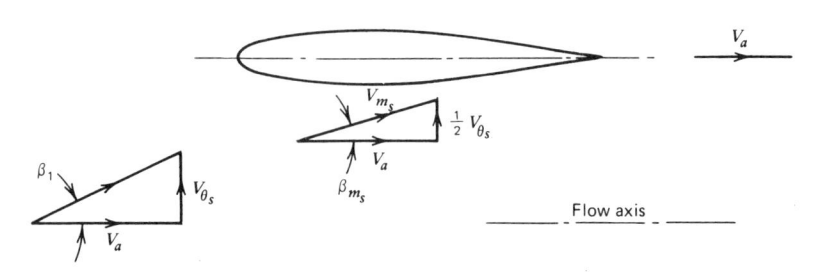

Figure 11.10 Velocity vectors for NPL-type straightener.

11.5 CONSTANT THICKNESS STATORS

Fixed symmetrical upstream vanes supporting the bearing and/or drive arrangement can be designed to perform an aerodynamic air-straightening function in cases of small inadvertent upstream swirl.

11.5 CONSTANT THICKNESS STATORS

The clothing of the camber line with a constant thickness plate section will give satisfactory performance at the design condition since the local incidence i is zero, thus minimizing the chances of leading-edge separation. The width of the satisfactory working range, however, will compare unfavorably with that for correctly designed streamlined sections. Resistance losses and noise levels will both be significantly increased for off-duty operation.

On many occasions manufacturing costs and other practical considerations may demand stators of this general type. When the design of the complete fan unit is along correct aerodynamic and acoustic lines, cambered plate blading can provide a relatively efficient and quiet industrial unit.

The choice of solidity should be more conservative than that for optimum-type sections, the degree depending on the expected duty range of the ensuing fan. Since the design method recommended for accelerating flow prerotators is already conservative, the need for increased solidity is less obvious.

The angle of deviation will exceed that associated with C4 type vanes. An inspection of Figs. 6.6 and 6.24 at $\theta = 40°$ and $C_L = 1$ indicates a difference in incidence angle α of approximately $1\frac{1}{2}°$. Hence the stagger angle should be decreased by this amount as an approximate δ correction.

Since it is difficult to roll curvature into the plate edges the sections can be provided with extra chord, as indicated in Fig. 11.11. In the case of large

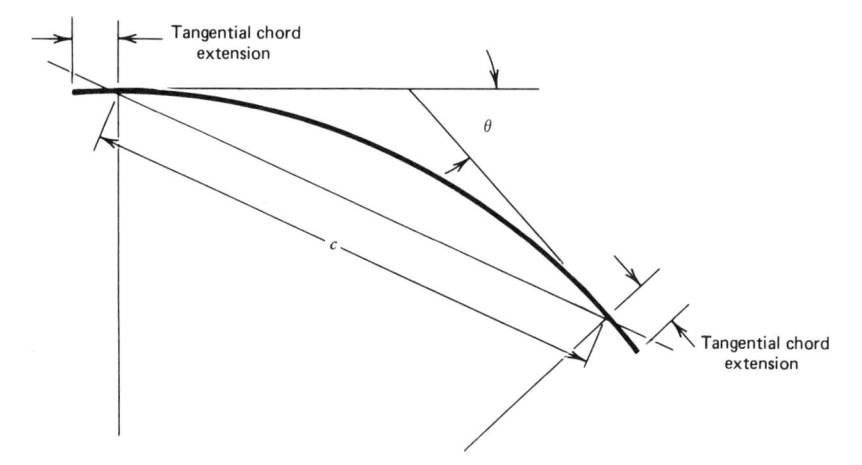

Figure 11.11 Cambered plate vane-element.

curvatures these chord extensions can provide the degree of conservatism desired as well as producing a small improvement in the flow-deflecting characteristics of the vane. A real practical difficulty is thus eliminated.

11.6 AXIAL GAP

When the axial distance separating blade rows is small, interference between the pressure and velocity fields surrounding the respective rows will be present. Howell [11.1] reports no change in load characteristics for clearances from 1/6 to 1 chord and a small increase in blade output for lesser values. In [11.8] the output continued to fall up to the experimental limit of one-half blade chord. These studies are related to compressors.

However, it is recommended in Chapter 16 that the clearance should be at least 50% chord in order to minimize the interference noise created; this figure is based mainly on blade wake considerations. Performance losses are not expected in the fan case, as opposed to the multistage compressor where radial flow redistributions will influence the output from successive rotor stages. Fan designers seeking increased work output, and unit compactness, have produced excessively noisy fans; on occasions these have suffered serious structural failure.

REFERENCES

11.1 A. R. Howell, Fluid dynamics of axial compressors, *Proc Inst. Mech. Eng.*, London, **153**, 441–452, 1945.

11.2 R. A. Wallis, A rationalized approach to blade element design, axial flow fans. *Proc. Third Australasian Conf. Hydraulics and Fluid Mechanics*, Sydney, pp. 23–29, 1968.

11.3 A. D. S. Carter, The low speed performance of related aerofoils in cascades, Gt. Britain Aero. Research Council, *ARC CP 29*, 1950.

11.4 D. G. Ainley and G. C. R. Mathieson, An examination of the flow and pressure losses in blade rows of axial flow turbines, Gt. Britain Aero. Research Council, *ARC R&M 2891*, 1955.

11.5 A. D. S. Carter, The calculation of optimum incidences for aerofoils, Gt. Britain Aero. Research Council, *ARC CP 646*, 1961.

11.6 A. R. Collar, Cascade theory and the design of fan straighteners, Gt. Britain Aero. Research Council, *ARC R&M 1885*, 1940.

11.7 G. N. Patterson, Ducted fans: design for high efficiency. Australian Council of Aeronautics Report ACA 7, 1944.

11.8 L. H. Smith, Jr., Casing boundary layers in multistage axial flow compressors, in L. S. Dzung, (Ed.), *Flow research on blading*, Elsevier, Amsterdam, 1970.

CHAPTER 12

Stator and Strut Support Losses

Stator blading is subject to smaller resultant air velocities than rotor blades and hence the losses are greatly reduced. As a consequence, average values of the losses can be assumed in design estimates. The ensuing losses are presented as a function of the midspan swirl coefficient. This simplification has been made possible by the complete design dependence of stators on ϵ_s and ϵ_p and by the adoption of Chapter 11 design recommendations.

A method for estimating the losses due to support struts or rods of an acceptable aerodynamic type is also presented.

12.1 MOMENTUM CONSIDERATIONS

Adopting a similar procedure to that outlined in Chapter 8, and developing the pertinent relationships for the loss coefficients in the general design case where both prerotators and straighteners are employed, it follows that for straighteners

$$\frac{k_S}{K_{\text{th}}} \left(\frac{C_L}{C_D} + \frac{\epsilon_s}{2} \right) = \frac{\lambda}{\cos^2 \beta_{m_s}} \frac{\epsilon_s}{\epsilon_s + \epsilon_p}$$

and since $C_L/C_D \gg \epsilon_{s/2}$

$$\frac{k_S}{K_{\text{th}}} \frac{C_L}{C_D} = \frac{\lambda}{\cos^2 \beta_{m_s}} \frac{\epsilon_s}{\epsilon_s + \epsilon_p} \qquad (12.1)$$

and similarly for prerotators,

$$\frac{k_P}{K_{\text{th}}} \frac{C_L}{C_D} = \frac{\lambda}{\cos^2 \beta_{m_p}} \frac{\epsilon_p}{\epsilon_p + \epsilon_s} \qquad (12.2)$$

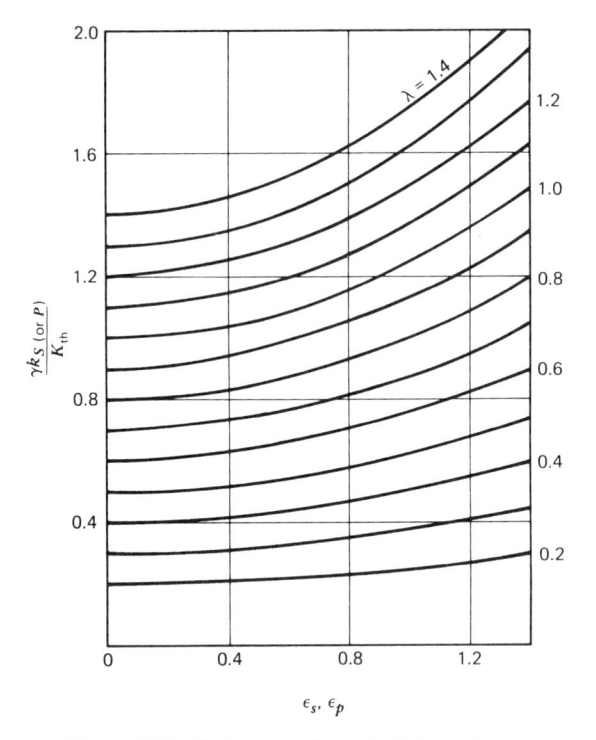

Figure 12.1 Stator vane-element efficiency loss.

Hence the fan efficiency loss is a linear function of the flow coefficient λ, which is a measure of axial velocity and fan duty conditions.

The product of the efficiency loss and lift/drag ratio, for either ϵ_p or ϵ_s equal to zero, is presented in Fig. 12.1.

The preceding general equations also apply to the NPL type of straightener, which may be installed in conjunction with prerotators, removing a small amount of either positive or negative swirl. The appropriate lift and drag coefficients are obtained from Fig. 11.3 and airfoil data, respectively.

When symmetrical sections are used as support struts, or to remove residual swirl, a modified approach to design loss estimates must be adopted, since C_L is nominally zero for design conditions.

Subject to the limitations and assumptions outlined in Section 4.7, then from Eq. (4.35) it follows that the local nondimensional total pressure loss k, for internal obstructions can be expressed for n vanes as

$$k = \frac{C_D n c \, dr}{2\pi r \, dr} \qquad (12.3)$$

12.2 DRAG COEFFICIENTS

Hence for symmetrical airfoils uniformly spaced around the circumference and at zero flow incidence

$$k = \sigma C_{D_{\alpha=0}} \tag{12.4}$$

or, in terms of fan efficiency loss

$$\frac{k}{K_{th}} = \frac{\sigma C_{D_{\alpha=0}}}{K_{th}} \tag{12.5}$$

Similarly, for uniformly spaced cylindrical rod supports of diameter d in an axial flow,

$$\frac{k}{K_{th}} = \frac{C_D}{K_{th}} \frac{d}{s} \tag{12.6}$$

The ratio d/s is normally quite small and, when this is the case, the associated total pressure loss can be ignored in design estimates of fan efficiency. When the resultant velocity is v, due to a swirl component,

$$\frac{k}{K_{th}} = \frac{C_D}{K_{th}} \frac{d}{s} \left(\frac{v}{V_a}\right)^2$$

The drag coefficient is obtained from Section 4.7.

12.2 DRAG COEFFICIENTS

The profile and secondary drag coefficients for prerotators differ significantly in magnitude from those applicable to straighteners but the component losses in fan efficiency are comparable.

1. Straighteners. The profile drag coefficient C_{Dp} is assumed constant at 0.016. This value corresponds to blading of moderate to high solidity at Re of 2×10^5 approximately. Reductions of 20 to 25% can be expected at Reynolds numbers of around 6×10^5. The adoption of the above value for C_{Dp} therefore ensures an increasing degree of conservatism, for advancing Re.

The secondary drag coefficient C_{Ds} is assumed to be given by

$$C_{Ds} = 0.018 \, C_L^2 \tag{12.7}$$

2. Prerotators. The profile drag coefficient is a function of air-turning angle, namely, β_2 (Fig. 12.2). The curve from $\beta_2 = 40$ to $65°$ is derived from the data analysis of [12.1], which, according to [12.2], provides a more

244   STATOR AND STRUT SUPPORT LOSSES

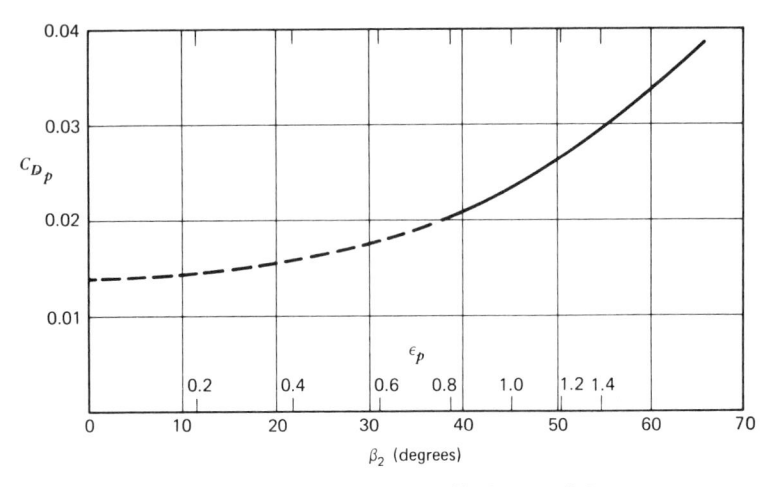

Figure 12.2 Prerotator vane profile drag coefficient.

accurate prediction of C_{D_P} than other published work. The curve has been arbitrarily extrapolated below 40° to correspond to expected drag values as the symmetrical airfoil case is approached. These drag data are for Re of 2×10^5.

The higher surface velocities associated with this accelerating type of stator are the probable reason for the profile drag increase with β_2. From data presented in [12.2] the decrease in C_{D_P} at Re = 6×10^5 appears to approximate that suggested earlier for straighteners.

The degree of uncertainty as regards secondary drag prediction for accelerating blading is greater than that for the decelerating type. According to [12.2], none of the expressions put forward cover the whole range of possible conditions satisfactorily. A later review of the subject is available in [12.3] in which correlation with the vane loading factor ($C_L \sigma$), inlet and outlet angles, and vane aspect ratio has been established. However, for present purposes Eq. (10.8) provides a more desirable expression.

Calculations based on the design recommendations of Chapter 11 together with the C_{D_S} data plotted in Fig. 9 of [12.3] result in a relatively constant value of b over the ϵ_p range 0.5 to 1.4, when representative values of the design aspect ratio are assumed to be 1.8 and 1.0, respectively, at these ϵ_p values. Hence it is suggested that the expression

$$C_{D_S} = 0.015 \, C_L^2 \tag{12.8}$$

will provide an adequate and conservative measure of the secondary drag coefficient for design estimation purposes.

12.3 EFFICIENCY LOSS ESTIMATES

The value of b, derived from the relationship established in [12.1], varies between 0.010 and 0.012 for the recommended design geometries; aspect ratio is not a specified variable.

12.3 EFFICIENCY LOSS ESTIMATES

Applying similar averaging techniques to those adopted in Chapter 10, the mean drag coefficient is given by

$$\bar{C}_D = (C_{D_P} + C_{D_S})_{\text{MS}} \quad (12.9)$$

and hence

$$\frac{K_S}{K_{\text{th}}} \frac{C_L}{C_D} = \left(\frac{\lambda}{\cos^2 \beta_{m_s}} \frac{\epsilon_s}{\epsilon_s + \epsilon_p} \right)_{\text{MS}} \quad (12.10)$$

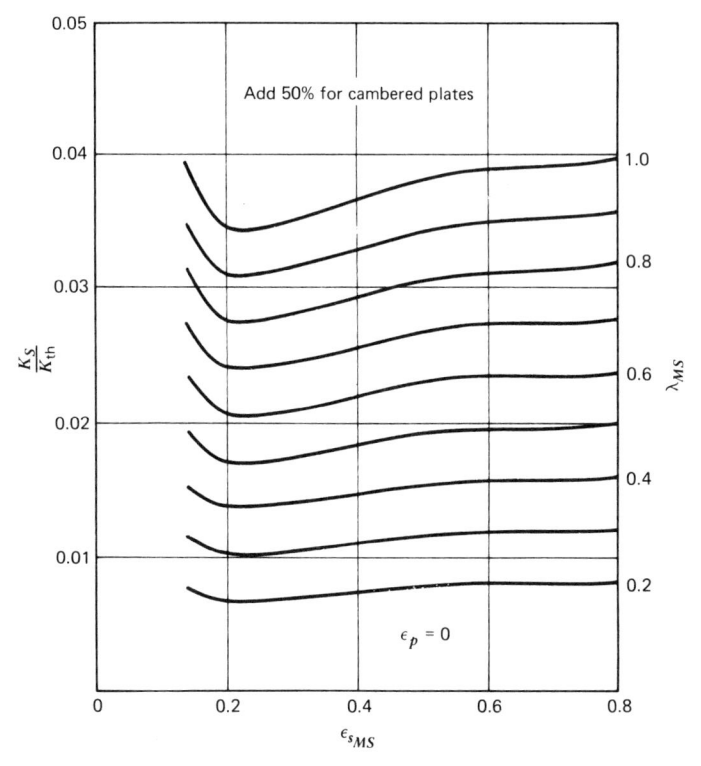

Figure 12.3 Straightener efficiency loss for specified designs.

STATOR AND STRUT SUPPORT LOSSES

$$\frac{K_P}{K_{th}} \frac{C_L}{C_D} = \left(\frac{\lambda}{\cos^2 \beta_{m_p}} \frac{\epsilon_p}{\epsilon_s + \epsilon_p} \right)_{MS} \quad (12.11)$$

where MS denotes the midspan.

For the normal design cases of either ϵ_p or ϵ_s equal to zero, and following the design guidance of the preceding chapter, the efficiency losses are as displayed on Figs. 12.3 and 12.4.

A feature of Fig. 12.3 is the relative constancy of efficiency loss with $\epsilon_{s_{MS}}$, for a given λ_{MS}. This is due in part to limited lift/drag ratio changes over the relevant C_L range (Figs. 11.5 and 9.4). The corresponding curves for prerotators (Fig. 12.4) show increased slopes. However, both types of stator possess efficiency losses of similar magnitude, being equal at $\epsilon_{MS} = 0.6$.

The choice of blade geometries that differ marginally from those recommended in Figs. 11.4 to 11.9 will not alter the above loss estimates by a significant amount.

When constant thickness cambered plate sections are used, the efficiency losses will be greater; it is suggested that the foregoing loss estimates be multiplied by a factor that could reach 1.5; for design conditions the actual

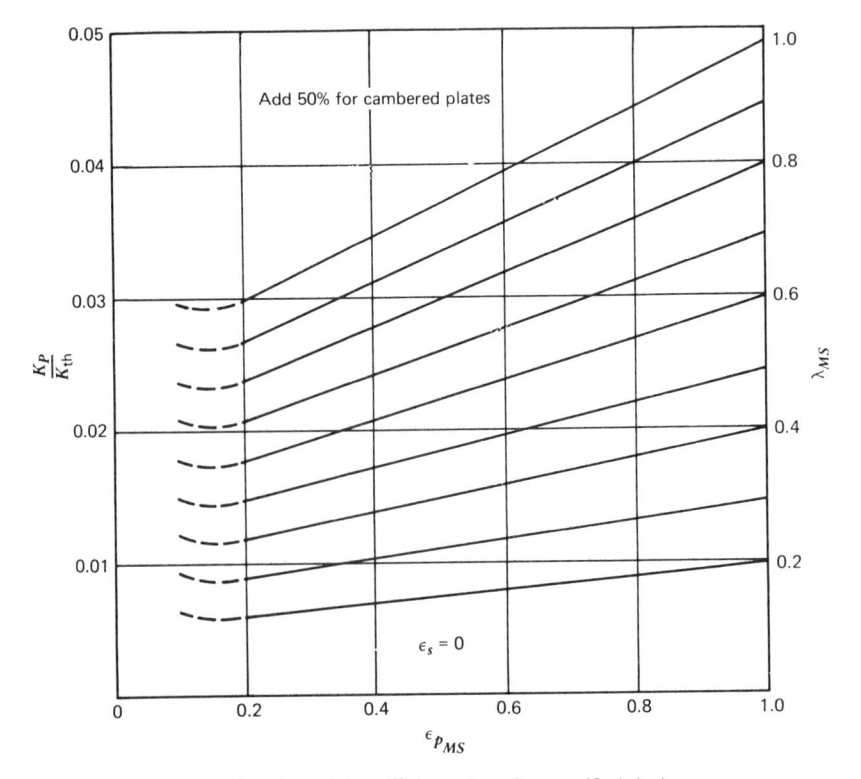

Figure 12.4 Prerotator efficiency loss for specified designs.

figure will be a matter for experience. Increased losses will be present for off-design straightener operation, because of leading-edge separation.

The minimum profile drag coefficient for symmetrical straighteners occurs at zero incidence. Guidance on the appropriate value for estimation purposes can be obtained from Fig. 6.10; when flow incidence is present, the drag increment can be assessed from Fig. 6.11, for the design C_L. The efficiency loss estimates then follow from Eqs. (12.9) and (12.10).

REFERENCES

12.1 D. G. Ainley and G. C. R. Mathieson, An examination of the flow and pressure losses in blade rows of axial flow turbines, Gt. Britain Aero. Research Council, *R&M 2891*, 1955.

12.2 J. H. Horlock, *Axial Flow Turbines*, Butterworths, London, 1966.

12.3 J. Dunham, A review of cascade data on secondary losses in turbines, *J. Mech. Eng. Sci.*, **12** (1), 48–59, 1970.

CHAPTER **13**

Ancillary Component Design

To ensure the blade inlet flow approximates the design assumption of a uniform axial velocity, careful attention must be given to inlet duct geometry and to the nose fairing. Because the flow is an accelerating one the design difficulties are normally of a minor nature.

Efficient flow diffusion aft of the blading outlet demands a closer attention to design detail. The actual operating requirements of various fan/duct systems may require a differing downstream duct assembly; several possibilities are outlined herein.

Since overall fan efficiency is dependent on resistance losses within these ancillary components, loss estimation techniques are given.

13.1 NOSE FAIRINGS

The ideal nose fairing shapes [13.1] are based on streamlined bodies of revolution (Fig. 13.1), where the portion forward of the maximum diameter provides the fairing contour. A fairing length/diameter ratio of 0.75 will ensure good flow along this inlet surface.

When the fairing is attached to the rotor, as a spinner, hemispheres are commonly employed. This shape experiences a degree of overacceleration followed by flow retardation in the downstream junction region. However, this appears to have little effect on fan test performance [13.2] as demonstrated in Fig. 13.2. Shorter fairings, or spinners, are subject to increasing loss, which for the no-fairing case reaches substantial proportions. The preceding test data relate to a particular blading and boss arrangement; hence the loss data of Fig. 13.2 cannot be applied in a general manner.

When space restrictions apply, the use of short spinners is preferable to none at all. However, allowance should be made in design for fan pressure

13.3 INLET PLENUMS

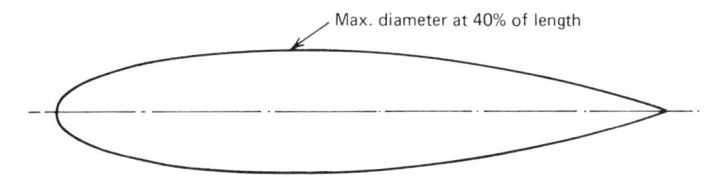

Figure 13.1 Streamline body of revolution.

and efficiency losses; the adjustments must be based on personal design experience.

For length to diameter ratios of 0.5, or greater, the design allowance is zero.

13.2 BELLMOUTH ENTRY

The ideal inlet duct shape for blower fans has a contour approximating that illustrated in Fig. 4.19. This avoids the overacceleration and subsequent retardation associated with semicircular shapes. However, provided a generous radius of between 0.25 and 0.3 times the duct diameter is adopted, the latter alternative ensures a satisfactory airflow with negligible resistance; losses become appreciable when the ratio is less than 0.15.

Failure to fit an appropriate inlet shape can result in substantial fan performance penalties, the magnitude of which will depend on blading proximity to the inlet. Losses of from 10 to 15% are reported in [13.2].

The conical arrangements of Section 4.3 will restrict fan efficiency losses.

13.3 INLET PLENUMS

Because of the low velocity pressure pertaining in a large plenum chamber, the total pressure is approximately equal to the static pressure. Provided the fan inlet ducting is suitably contoured in the manner just described, and the chamber air possesses zero angular momentum, fan performance will remain unaffected. The recommended inlet radius is 25% of duct diameter. A two- or three-element conical contraction also provides a satisfactory low loss solution.

However, the existence of angular momentum in the upstream air space, because of an off-center side feed to the chamber, can seriously compromise fan performance. Every effort should be made to ensure that the tangential velocity components are removed within the chamber or at entry to the fan annulus by flow-straightening guide vanes. The latter should be located with an axial gap of at least one vane chord between the vanes and the rotor blades.

250   ANCILLARY COMPONENT DESIGN

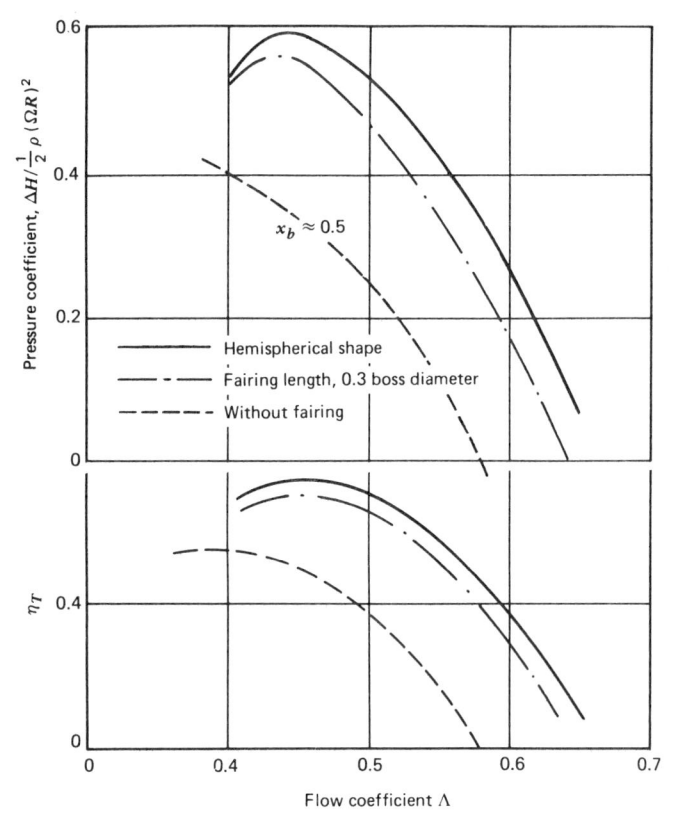

Figure 13.2 Nose fairing influence on fan performance (Reproduced with permission of Railway Gazette International from J. L. Koffman [13.2].)

Installations in conflict with the preceding self-evident rules have poor and unstable performance, are noisy, and can suffer noise-induced structural damage.

13.4 INLET BOXES

Data on inlet box design and losses are presented in Section 4.3. This component is a type of side-entry plenum of limited volume. Most installations of this type have a bearing upstream of the rotor with vanes carrying the load across the annulus to the outer casing and its support base. These vanes are aerodynamically developed for the removal of swirl resulting from flow asymmetry at box inlet and for the control of flow features associated with the central tunnel shroud. Flow separation from diffusing ducts feeding the

13.5 DOWNSTREAM LOSS ESTIMATION

inlet box will result in unsatisfactory box and fan performance. Upstream damper vanes or butterfly valves for resistance control of flow rate should be located well ahead of the box inlet. Stable flow at box inlet and outlet are vitally important if the losses are to remain small and the fan is to operate satisfactorily.

The losses are in excess of those for a centrally fed plenum chamber. However, designs developed along model testing lines usually possess an acceptable loss component.

13.5 DOWNSTREAM LOSS ESTIMATION

A fan unit devoid of downstream components represents the lowest capital cost solution. However, some measure of the performance and power penalties is required.

The downstream losses can be represented by the coefficients K_D and K_{DL}, expressed as

$$K_D = (1 - C_p) - \left(\frac{A_A}{A_B}\right)^2 \tag{13.1}$$

$$= (1 - \eta_D)\left[1 - \left(\frac{A_A}{A_B}\right)^2\right] \tag{13.2}$$

and

$$K_{DL} = K_D + K_K = 1 - C_p \tag{13.3}$$

$$= (1 - \eta_D)\left[1 - \left(\frac{A_A}{A_B}\right)^2\right] + \left(\frac{A_A}{A_B}\right)^2 \tag{13.4}$$

where

$$C_p = (p_2 - p_1)/\tfrac{1}{2}\rho \bar{V}_a^2 \tag{13.5}$$

$$\eta_D = C_p \bigg/ \left[1 - \left(\frac{A_A}{A_B}\right)^2\right] \tag{13.6}$$

or

$$\eta_D = C_p/C_{p_i} \tag{13.7}$$

The coefficient K_D expresses the diffusion losses in the normally accepted manner, whereas K_{DL} includes a kinetic energy loss term represented by K_K.

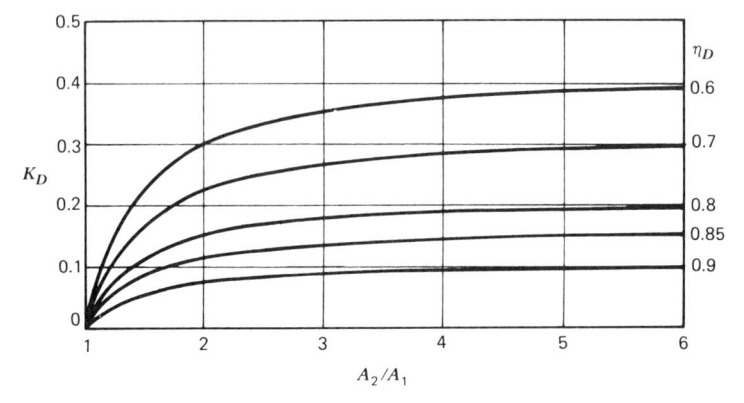

Figure 13.3 Diffuser loss coefficient as function of area ratio and diffuser efficiency.

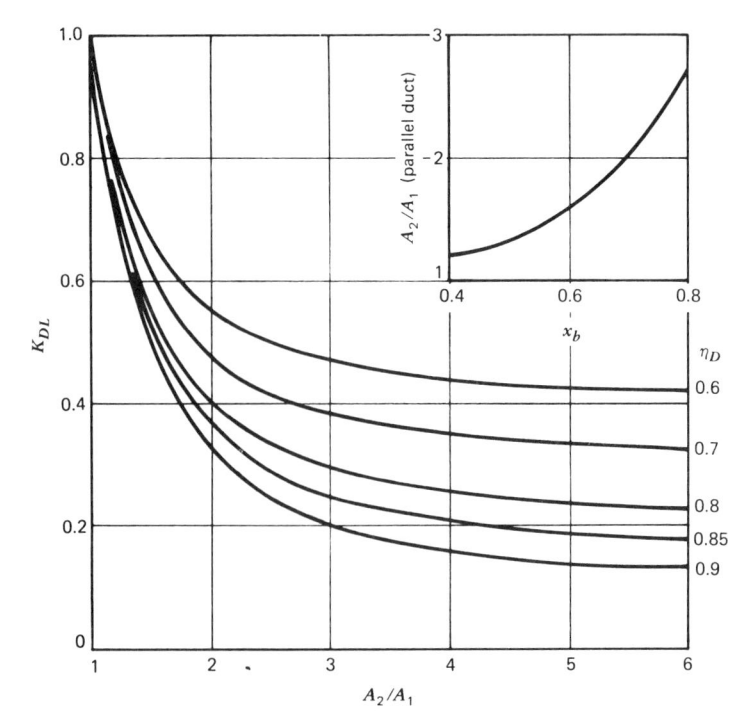

Figure 13.4 Downstream loss coefficient, exhaust units.

13.6 CYLINDRICAL OUTLET DUCT

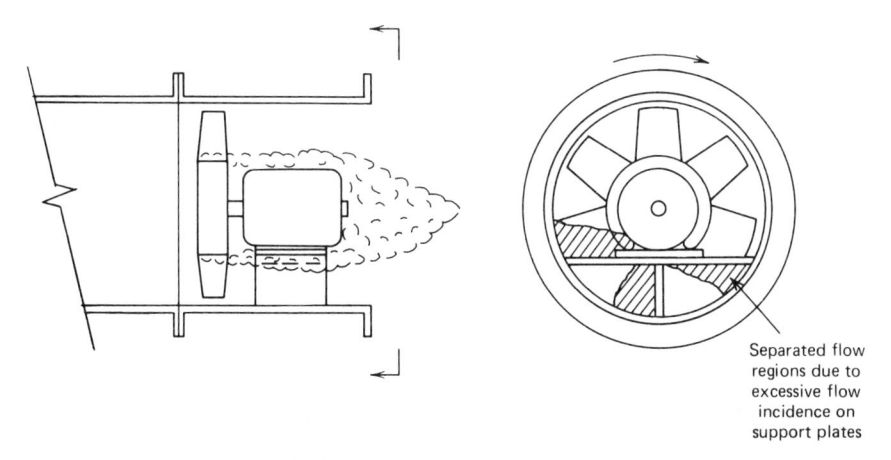

Figure 13.5 Separated flow in substandard assembly.

Equations (13.2) and (13.4) are illustrated in Figs. 13.3 and 13.4. Superimposed are the area ratios for various boss ratios when circular downstream ducting is planned.

The ratios K_D/K_{th} and K_{DL}/K_{th} represent the efficiency losses due to any downstream fan components.

Downstream losses will be at a minimum for a uniform velocity field aft of the fan blading, and for small residual swirl. Provided the fan is of a free vortex flow type and operating near the design point, these conditions are approximately achieved. Experiments have shown annular diffuser performance to be independent of blading components upstream when the above flow features are present [13.3], [13.4]. However, sheared inlet flow arising from off-design duty conditions will eventually lead to diffuser performance penalties [13.3].

Loss estimates for installations of the type illustrated in Fig. 13.5 are not included in this Chapter, because of flow unpredictability.

13.6 CYLINDRICAL OUTLET DUCT

In the absence of a tail fairing, the bluff termination creates a region of highly turbulent eddying flow. However, mixing processes reduce the pressure loss to less than the theoretical dynamic pressure difference between upstream and downstream stations (Fig. 4.30). Existing low-pressure-rise commercial units seldom possess tail fairings, but since the boss ratios are 0.4 or less the losses are not substantial. Nevertheless, in all cases, the tail fairing should only be omitted after a careful analytical study has established the optimum practical solution.

The preferred tail fairing consists of a conical surface that may be truncated at a plane 10 to 20% of cone length from the apex. The losses are virtually the same as those for streamlined bodies of varying curvature. The recommended minimum cone wall angles for boss ratios of 0.5 and 0.7 are 17° and 14°, respectively [13.3]; angles of 18° and 12° are suggested for the ratios 0.4 and 0.75.

The diffuser effectiveness will fall within the range 0.8 to 0.9, depending on inlet flow quality and Reynolds number. Omitting the tail fairing will reduce effectiveness to values suggested in Section 4.4.4.

13.7 CONICAL DIFFUSER WITH TAIL FAIRING

This downstream arrangement is common to both in-line and exhaust fan applications. The design guidance provided by [13.3] is reproduced in Figs. 13.6 and 13.7 and relates to a conical fairing extending the full length of the diffuser. However, for the 0.5 boss ratio case, experiment has demonstrated that shorter fairings, or truncation, have a minor influence on diffuser effectiveness.

By reference to Fig. 13.7 it will be noted that the recommended outer wall angle for a boss ratio of 0.5 is very similar to that for a conical diffuser with an area ratio of 2.85 [13.3]. This suggests that for boss ratios of 0.5 and less the outer and inner wall angles may be established independently of each other. Test experience would tend to support this tentative assumption in relation to the preceding area ratio.

Truncated tail fairings of reduced wall angle will represent a more practical approach for fans with boss ratios approaching 0.7, for the larger diffuser area ratios. The "equivalent included angle" data of Fig. 4.25, in conjunction with Fig. 13.6 can be used in establishing the diffuser wall angles.

Diffuser effectiveness for tail fairing units peaks at approximately 0.9, for design conditions and high Re. The effect of entry flow on performance is illustrated in Fig. 13.8 [13.5].

13.8 CONICAL DIFFUSER WITH DIVERGENT CENTERBODY

This arrangement has advantages for exhaust fans because of reduced sensitivity to inlet flow and to greater compactness. A minor disadvantage is a slight decrease in the maximum effectiveness when compared with converging centerbody diffusers.

An in-line system, where the fan discharges into a plenum chamber, is also well served by this diffuser type.

The empirical work of [13.6] indicates that all optimum annular and conical diffusers can be represented by a single straight line on an area ratio versus $N/\Delta R_1$ plot, where ΔR_1 is either the annulus width or conical diffuser

13.8 CONICAL DIFFUSER WITH DIVERGENT CENTERBODY

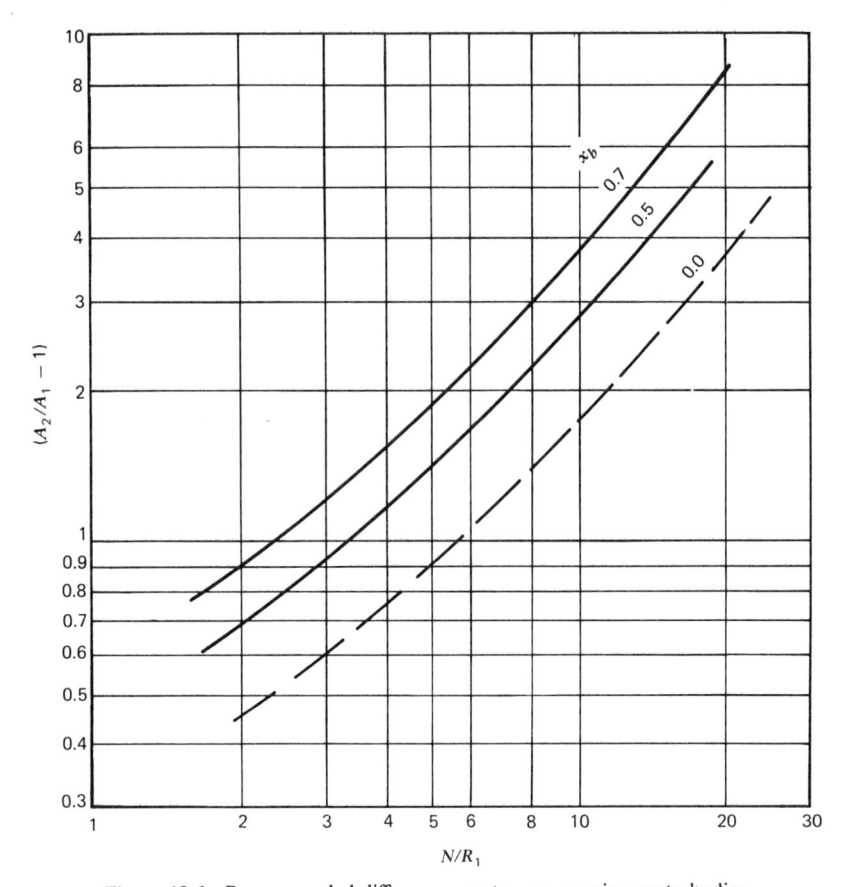

Figure 13.6 Recommended diffuser geometry, converging centerbodies.

radius, at inlet. Subsequent studies [13.7] have shown this rule to be in error for all but the annular diffuser class centered on the expression

$$\tan \theta_i = x_b \tan \theta_o \qquad (13.8)$$

where i and o denote inner and outer walls, respectively. The inner and outer cones possess a common projected apex. Acceptance of a unique curve of area ratio versus $N/\Delta R_1$ is herein adopted.

Using the plain conical diffuser data of [13.8] as the base, the curves of Fig. 13.9 have been prepared using the relation

$$\theta_i = x_b \theta_o \qquad (13.9)$$

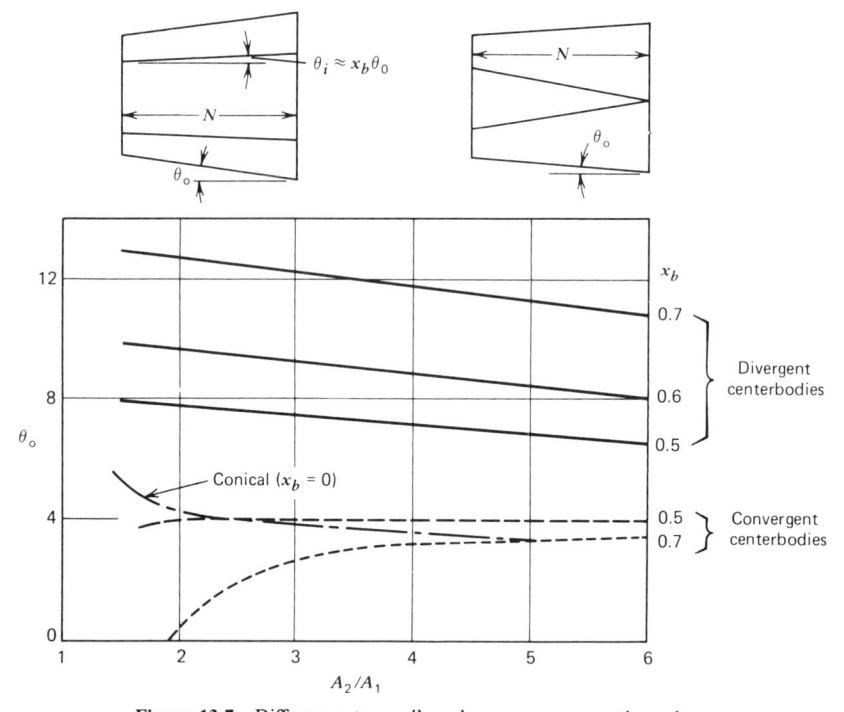

Figure 13.7 Diffuser outer wall angles versus area ratio and x_b.

At the lower area ratios, however, the curves have been straightened, making linear relationships between this ratio and outer wall angle possible (Fig. 13.7); this represents a conservative design approach.

Limited test data indicate that any diffuser designed in accordance with Fig. 13.9 will possess a peak effectiveness close to 0.9, provided the inlet flow is of good quality. Adjustments based on personal experience must be made when this latter condition is not met.

The data of Fig. 13.9 can be used when the desired wall angles differ by small amounts from the recommendations of Fig. 13.7.

Increased wall angles will result in a shorter diffuser for a fixed area ratio. Experimental data on diffusers with outer wall angles of 20° and 15°, for a boss ratio of 0.5, are presented in [13.9]. Although some reduction in peak effectiveness was expected, the loss was not excessive as a value of 0.82 was recorded for one test configuration.

13.9 NONCONICAL DIFFUSER WITH TAIL FAIRING

When the immediate downstream ducting requires a change in cross-sectional shape to suit a specific application, the "equivalent included angle"

13.10 DIFFUSER INLET SWIRL

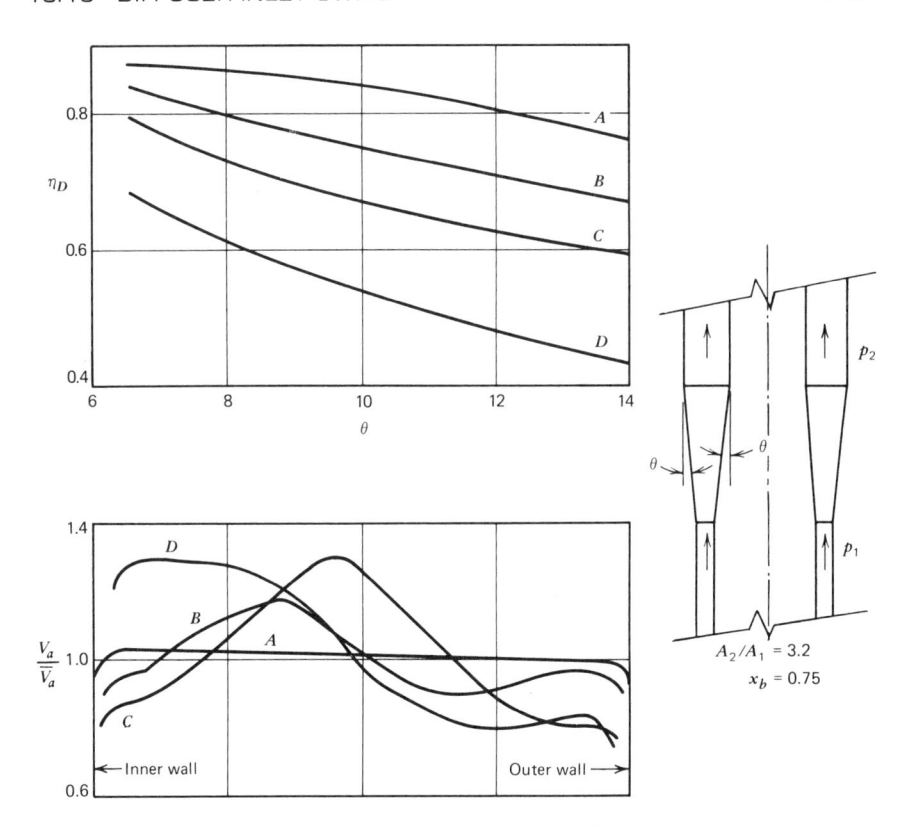

Figure 13.8 Effect of velocity distribution on diffuser efficiency. (Taken from I. H. Johnston [13.5] and reproduced with the permission of the Controller of Her Majesty's Stationery Office.)

procedure is recommended. The shape of the downstream termination can be either square or rectangular.

In general, the area ratios are not large and hence the diffusers are of limited length. A lesser value of equivalent angle than those displayed on Figs. 4.25 and 4.26 should be selected. The degree of conservatism exercised will be at the discretion of the designer. Asymmetric geometries and increasing aspect ratios of the diffuser termination need special consideration; model testing as a design aid is advised.

13.10 DIFFUSER INLET SWIRL

The "equivalent included angle" is increased when inlet swirls of approximately 10° are present. The energy contained in these tangential flow compo-

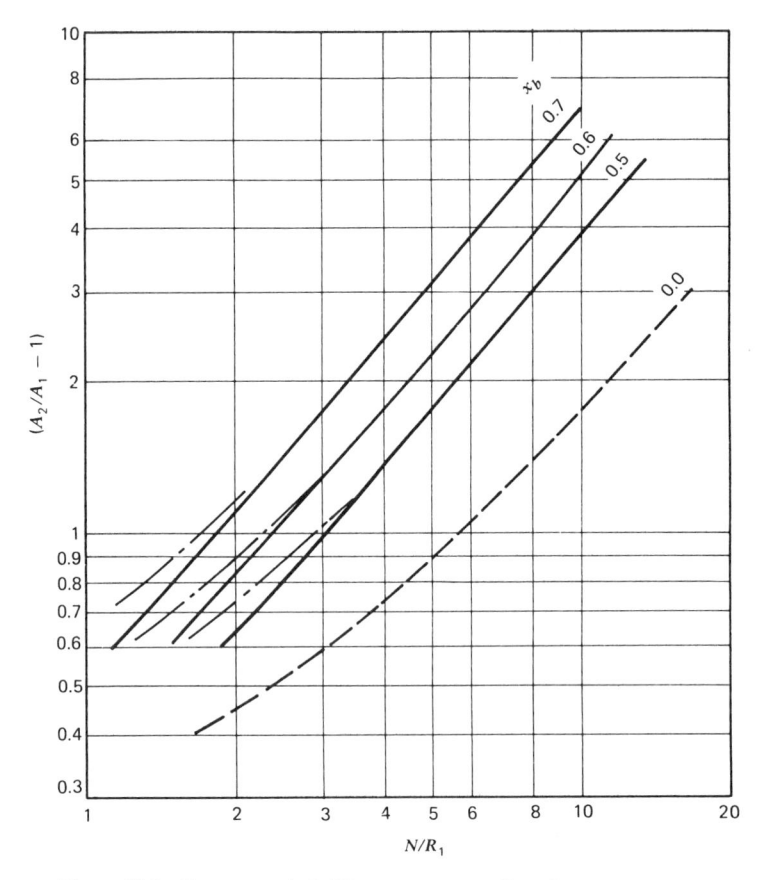

Figure 13.9 Recommended diffuser geometry, diverging centerbodies.

nents is very small and hence consideration might be given to stators that are designed with a residual swirl feature, for exhaust fans particularly.

Fans of low nondimensional pressure rise design seldom possess stators (e.g., induced draft cooling tower units). The discharge diffuser should be designed to take full advantage of the resultant swirl as a means of increasing the area ratio for a given length. Further data are available in Section 4.4.8 and in [13.10].

13.11 BOUNDARY LAYER CONTROL

It is recommended that the use of flow control devices be restricted to either remedial or special design cases. Vortex generators are a proven device for flow correction in large wind tunnel diffusers. Reference should be made to Section 4.4.7 for further data on this flow control subject.

REFERENCES

13.1 E. P. Warner, *Airplane Design—Performance*, 2nd ed., McGraw-Hill, New York, 1936, p. 349.

13.2 J. L. Koffman, Fans for traction applications, *Diesel Rly. Traction,* **5**:87–94, 1951.

13.3 R. A. Wallis, Annular diffusers of radius ratio 0.5 for axial flow fans, Div. Mech. Eng. CSIRO, *Tech. Rep. No. 4,* 1975.

13.4 S. D. Adenubi, Performance and flow regimes of annular diffusers with axial turbomachine discharge inlet conditions, *ASME J. Fluids Eng.* **98**:236–243, 1976.

13.5 I. H. Johnston, The effect of inlet conditions on the flow in annular diffusers, Gt. Britain Aero. Research Council, *ARC CP 178,* 1954.

13.6 G. Sovran and E. D. Klomp, Experimentally determined optimum geometries for rectilinear diffusers with rectangular, conical or annular cross-section, *Proc. Symposium Fluid Mech. Int. Flow,* Elsevier, New York, 1967.

13.7 I. C. Shepherd, Annular exhaust diffusers for axial flow fans, *Fifth Australasian Conf. Hydraulics and Fluid Mechanics,* Christchurch, N.Z., December 1974; and private communication, 1979.

13.8 A. T. McDonald and R. W. Fox, An experimental investigation of incompressible flow in conical diffusers, *Int. J. Mech. Sci.*, **8**:125–139, 1966.

13.9 M. C. Welsh, I. C. Shepherd, and C. J. Norwood, Straight walled annular exhaust diffusers of 0.5 boss ratio axial flow fans, *Inst. Eng. Aust. Thermodynamics Conf.*, Hobart, 1976, *Nat. Conf. Publ. No. 76/12.*

13.10 A. H. Elgammal and A. M. Elkersh, A method for predicting annular diffuser performance with swirling inlet flow, *J. Mech. Eng. Sci.*, **23**:107–112, 1981.

CHAPTER 14

Overall Efficiency, Torque, Thrust, and Power

14.1 SWIRL MOMENTUM

In addition to the component losses discussed in previous chapters, swirl energy is a loss in a fan system that is devoid of stators. The momentum loss at any spanwise station is given by $\frac{1}{2}\rho V_\theta^2$ and hence the blade element efficiency loss is given by

$$\frac{\frac{1}{2}\rho V_{\theta s}^2}{\Delta h_{\text{th}}} \quad \text{or} \quad \frac{k_{\text{swirl}}}{K_{\text{th}}} = \frac{\epsilon_s^2}{K_{\text{th}}}$$

When free vortex flow exists the swirl momentum of the midspan station is the mean for the fan. Hence

$$\frac{K_{\text{swirl}}}{K_{\text{th}}} = \left(\frac{\epsilon_s^2}{K_{\text{th}}}\right)_{\text{MS}} \tag{14.1}$$

14.2 OVERALL EFFICIENCY

A ready method of calculating fan system losses is required as a starting point for detailed design calculations. This enables K_{th} to be established from either

$$K_{\text{th}} = \frac{K_T}{\eta_T}, \text{ for in-line fans} \tag{14.2}$$

14.2 OVERALL EFFICIENCY

or

$$K_{th} = \frac{K_{IT}}{\eta_{IT}}, \text{ for exhaust fans} \tag{14.3}$$

when tip clearance is small.

The overall efficiency is expressed by the following relationships:
rotor unit:

$$\eta_T = \frac{K_{th} - K_R - K_{swirl} - K_D}{K_{th}} \tag{14.4}$$

rotor-straightener unit:

$$\eta_T = \frac{K_{th} - K_R - K_S - K_D}{K_{th}} \tag{14.5}$$

prerotator-rotor unit:

$$\eta_T = \frac{K_{th} - K_R - K_P - K_D}{K_{th}} \tag{14.6}$$

prerotator-rotor-straightener unit:

$$\eta_T = \frac{K_{th} - K_R - K_P - K_S - K_D}{K_{th}} \tag{14.7}$$

contrarotating rotor unit:

$$\eta_T = \frac{K_{th} - K_{R_1} - K_{R_2} - K_D}{K_{th}} \tag{14.8}$$

When the preceding units are used in the exhaust role, the efficiency based on inlet total pressure is given by

$$\eta_{IT} = \eta_T - \frac{K_K}{K_{th}} \tag{14.9}$$

where K_K is the discharge velocity pressure coefficient.

In the case of excessive tip clearance, the value of K_{th} obtained from Eq. (14.2) using corrected efficiency will fall short of that required to meet the K_T objective, when the design is based on a duct radius of R. Hence

$$K_{th} = K_T(1 + \zeta)/\eta_T \tag{14.10}$$

where ζ is the ratio of the tip clearance total pressure loss to required total pressure rise and η_T is the efficiency corrected for tip clearance excesses.

All the preceding loss coefficients can be estimated from data presented in Chapters 10, 12, and 13.

A study of the individual loss coefficients allows assessment of the relative merits of alternative designs as well as exposing areas of undue loss.

14.3 TORQUE

The torque acting on the rotor shaft can be expressed in terms of the swirl momentum added to the stream. For free vortex flow it follows from Eq. (8.36) that

$$T_c = \int_{x_b}^{1} 4(\epsilon_s + \epsilon_p) x^2 \, dx \qquad (14.11)$$

substituting for $(\epsilon_s + \epsilon_p)$ from Eq. (8.17)

$$T_c = K_{th} \Lambda (1 - x_b^2) \qquad (14.12)$$

The torque then follows from Eq. (8.35).

Alternatively, when the power is known, torque can be obtained from the relation

$$\text{Shaft power} = 2\pi TN \qquad (14.13)$$

where N is in rev/s.

14.4 THRUST

In fan design the main interest in the thrust produced by the rotor is in relation to the design of thrust bearings and supports. For these purposes an estimate based on the static pressure rise across the rotor and the swept area is usually adequate.

When more detailed information is required, the equations of Section 8.6 must be developed. From Eq. (8.33)

$$\text{Th}_c = \int_{x_b}^{1} \frac{\Delta p}{\frac{1}{2}\rho \bar{V}_a^2} 2x \, dx \qquad (14.14)$$

For free vortex flow and after substitution

14.5 POWER

$$\text{Th}_c = \int_{x_b}^{1} 2x \left(\eta_R - \frac{\epsilon_s^2}{K_{th}} + \frac{\epsilon_p^2}{K_{th}} \right) dx \qquad (14.15)$$

This expression can be integrated for the following three swirl conditions:

$\epsilon_s = \epsilon_p$

$$\text{Th}_c = K_{th}\eta_R(1 - x_b^2) \qquad (14.16)$$

$\epsilon_p = 0$

$$\text{Th}_c = K_{th} \int_{x_b}^{1} \left(2x\eta_R - K_{th} \frac{\Lambda^2}{2x} \right) dx$$

$$= K_{th}\eta_R(1 - x_b^2) + \tfrac{1}{2} K_{th}^2 \Lambda^2 (\ln x_b) \qquad (14.17)$$

$\epsilon_s = 0$

$$\text{Th}_c = K_{th}\eta_R(1 - x_b^2) - \tfrac{1}{2} K_{th}^2 \Lambda^2 (\ln x_b) \qquad (14.18)$$

The total thrust follows from Eq. (8.32).

14.5 POWER

The nondimensional power coefficient is

$$W_c = \frac{W}{\tfrac{1}{2}\rho \bar{V}_a^3 \pi R^2} \qquad (14.19)$$

This relationship is of value in relating the powers of similar units of varying size.

An alternate coefficient used extensively by industry is

$$\Phi = \frac{W}{\rho N^3 D^5} \qquad (14.20)$$

where $D = 2R$

Power is normally determined from the relation

$$\text{Shaft power} = \frac{\Delta H_T Q}{\eta_T} \text{ (watts)} \qquad (14.21)$$

where ΔH_T is fan total pressure duty in pascals and Q is volume flow rate in m³/s; this product constitutes the air power.

Alternatively, when the fan torque is known the power can be calculated from Eq. (14.13).

CHAPTER 15

Design Optimization

A combination of aerodynamic, mechanical, electrical, structural, acoustical, and operational factors is involved in any overall optimization exercise. The attainment of high fan efficiency is, however, a major objective and hence this aspect will receive prior attention.

The following free vortex flow development establishes a platform for peak efficiency determination, in terms of the flow and swirl coefficients. It is restricted to rotor/stator configurations with appropriate downstream duct components [15.1].

15.1 FAN COMPONENT LOSSES

Since η_T and η_{IT} differ only in respect to downstream losses, a general fan efficiency η' is adopted in which K_D and K_K are combined in the term K_{DL} giving

$$\eta' = 1 - \left(\frac{K_R}{K_{th}} + \frac{K_S}{K_{th}} + \frac{K_P}{K_{th}} + \frac{K_{DL}}{K_{th}}\right) \qquad (15.1)$$

In the case of in-line fan applications, $K_{DL} = K_D$ and $\eta' = \eta_T$

15.1.1 Rotor Loss

Substituting for β_m from Eq. (8.19), for midspan values, in Eq. (10.5) gives

$$\frac{K_R}{K_{th}} = \left[\frac{C_{DP} + C_{DS}}{C_L}\left(\frac{\Lambda}{x} + \frac{x}{\Lambda} - (\epsilon_s - \epsilon_p) + \frac{\Lambda(\epsilon_s - \epsilon_p)^2}{4x}\right)\right]_{MS} \qquad (15.2)$$

Assigning nominal values of 0.8, 0.012, and $0.018C_L^2$ to C_L, C_{DP}, and C_{DS}, respectively, the lift/drag ratio adopted is 34. The latter peaks at approximately the preceding value of C_L (Fig. 9.3). At high Re reductions in the profile drag will increase the ratio to 40 or more.

15.2 FAN EFFICIENCY

Increases in ϵ_s and ϵ_p produce higher and lower rotor efficiencies, respectively, as expressed in Eq. (15.2).

15.1.2 Stator Losses

For the recommended geometries of Chapter 11, the loss coefficients are approximated by

$$\frac{K_S}{K_{th}} = \left[\frac{\Lambda}{x}(0.032 + 0.010\,\epsilon_s)\frac{\epsilon_s}{\epsilon_s + \epsilon_p}\right]_{MS} \quad (15.3)$$

and

$$\frac{K_P}{K_{th}} = \left[\frac{\Lambda}{x}(0.025 + 0.024\,\epsilon_p)\frac{\epsilon_p}{\epsilon_p + \epsilon_s}\right]_{MS} \quad (15.4)$$

where the linear variations are obtained from Figs. 12.3 and 12.4, and the last term in each expression adjusts the loss to the correct K_{th} for the fan in question in accordance with Eqs. (12.10) and (12.11).

15.1.3 Downstream Loss

The alternative values of K_{DL} for in-line and exhaust fans are given in Eqs. (13.2) and (13.4). The corresponding loss in fan efficiency is

$$\frac{K_{DL}}{K_{th}} = \frac{\Lambda\,K_{DL}}{[2x(\epsilon_s + \epsilon_p)]_{MS}} \quad (15.5)$$

15.2 FAN EFFICIENCY

Estimates of fan efficiency as a function of Λ can be obtained when arbitrary values are assumed for lift/drag ratio, boss ratio, swirl coefficient, and downstream losses. Since rotor-straightener units are the most efficient, this fan type has particular appeal in the present context.

It can be shown that ϵ_{s_b} attains a maximum value of unity approximately, with respect to rotor and stator design limits. This value is therefore selected in relation to these high-efficiency studies. The plots of η versus Λ for five boss ratios, and lines of constant K_{DL}, are displayed in Figs. 15.1 to 15.5.

The influence of boss ratio on the peak efficiency is negligible for zero downstream loss.

A typical set of curves is presented in Fig. 15.6 for the prerotator–rotor design case, with an increased constant value of the swirl coefficient, namely, 1.2.

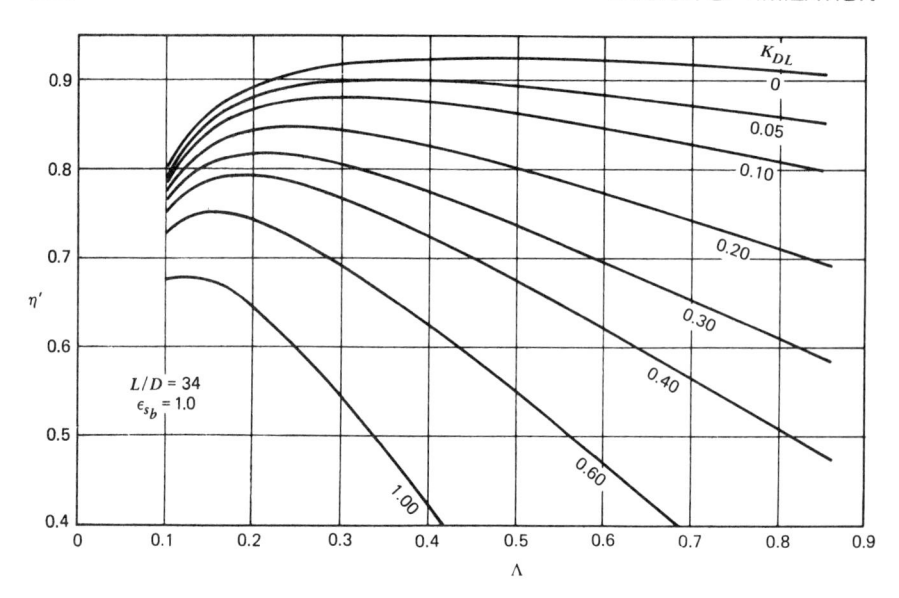

Figure 15.1 Rotor-straightener unit efficiency for 0.4 boss ratio.

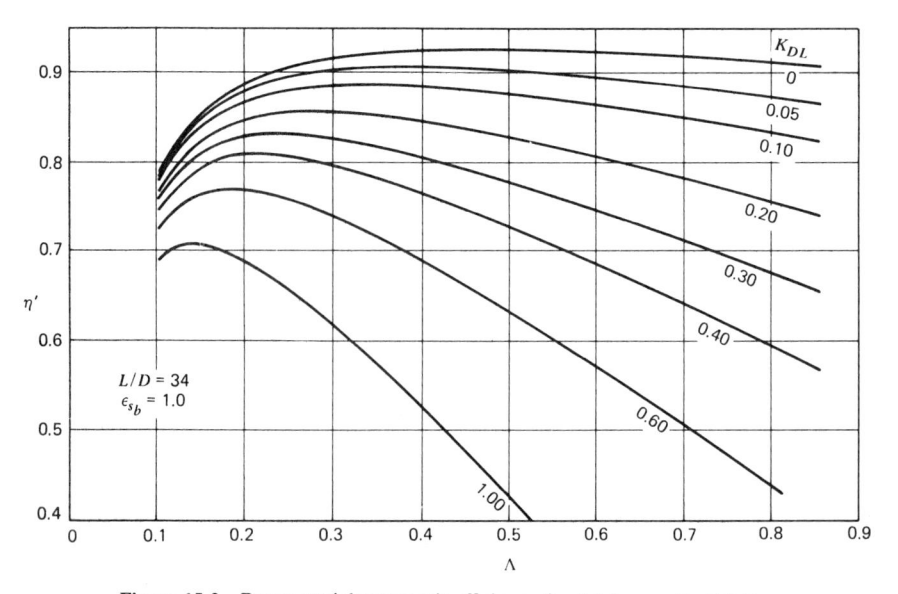

Figure 15.2 Rotor-straightener unit efficiency for 0.5 boss ratio [15.1].

15.2 FAN EFFICIENCY

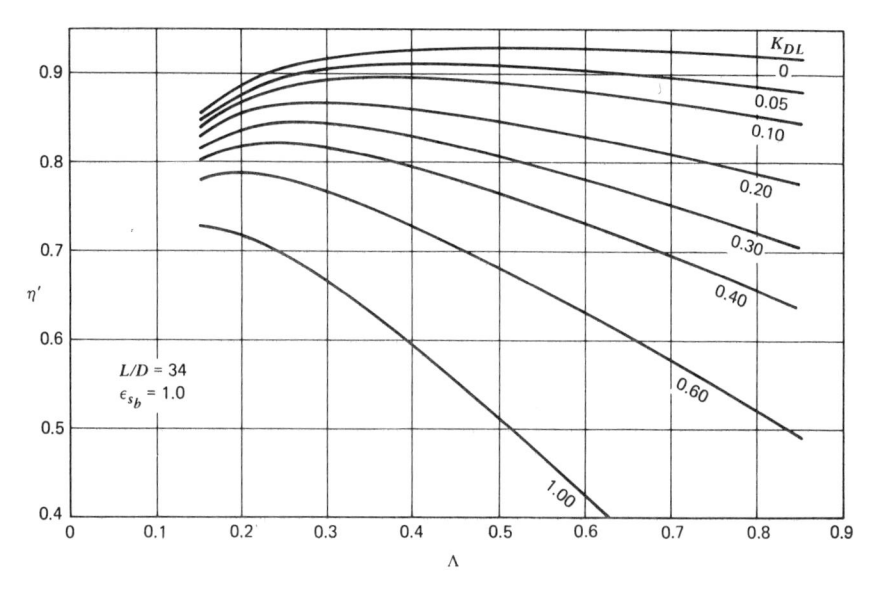

Figure 15.3 Rotor-straightener unit efficiency for 0.6 boss ratio.

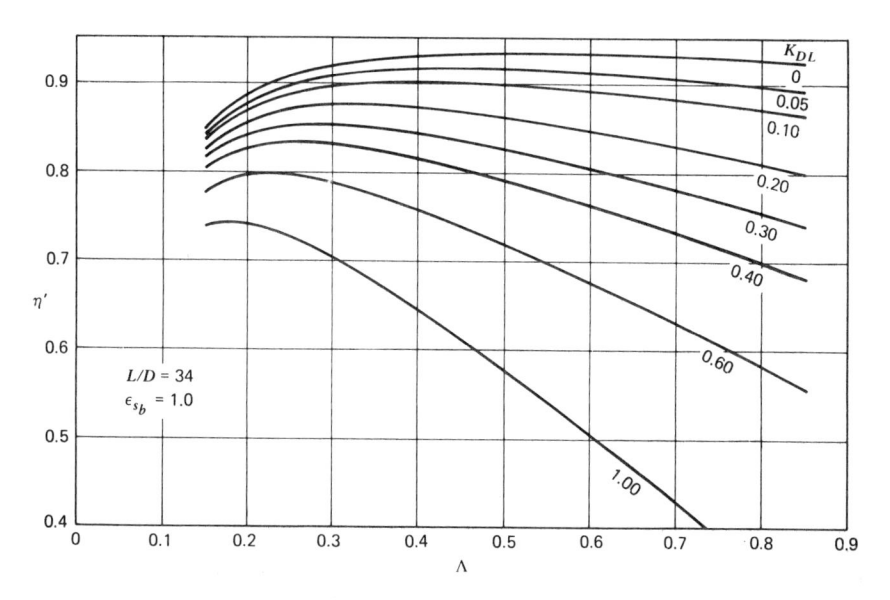

Figure 15.4 Rotor-straightener unit efficiency for 0.7 boss ratio [15.1].

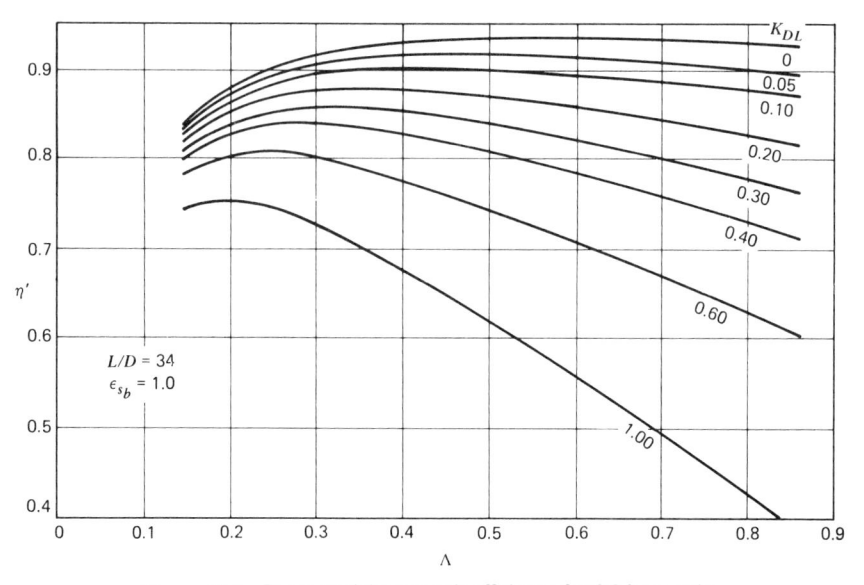

Figure 15.5 Rotor-straightener unit efficiency for 0.8 boss ratio.

At large Reynolds numbers the efficiencies will exceed the values graphically presented by 1 to 2%.

15.3 NONDIMENSIONAL PRESSURE RISE

The blading total pressure rise coefficient for $K_{DL} = 0$ is obtained from the relations

$$K_{bl} = \eta' K_{th} \quad \text{and} \quad K_{th} = \left[\frac{2(\epsilon_s + \epsilon_p)}{\lambda}\right]_{MS}$$

The K_{bl} versus Λ relationships for various boss ratios and a K_{DL} of zero are shown in Fig. 15.7; by subtracting K_{DL}, the required K_T or K_{IT} can be obtained.

15.4 DESIGN OPTIMIZATION

The efficiency curves suggest a fair degree of latitude in relation to the design choice of Λ for high efficiency. However, optimum values of Λ provide a design objective and permit a graphical display of dimensional pressure rise data.

15.4 DESIGN OPTIMIZATION

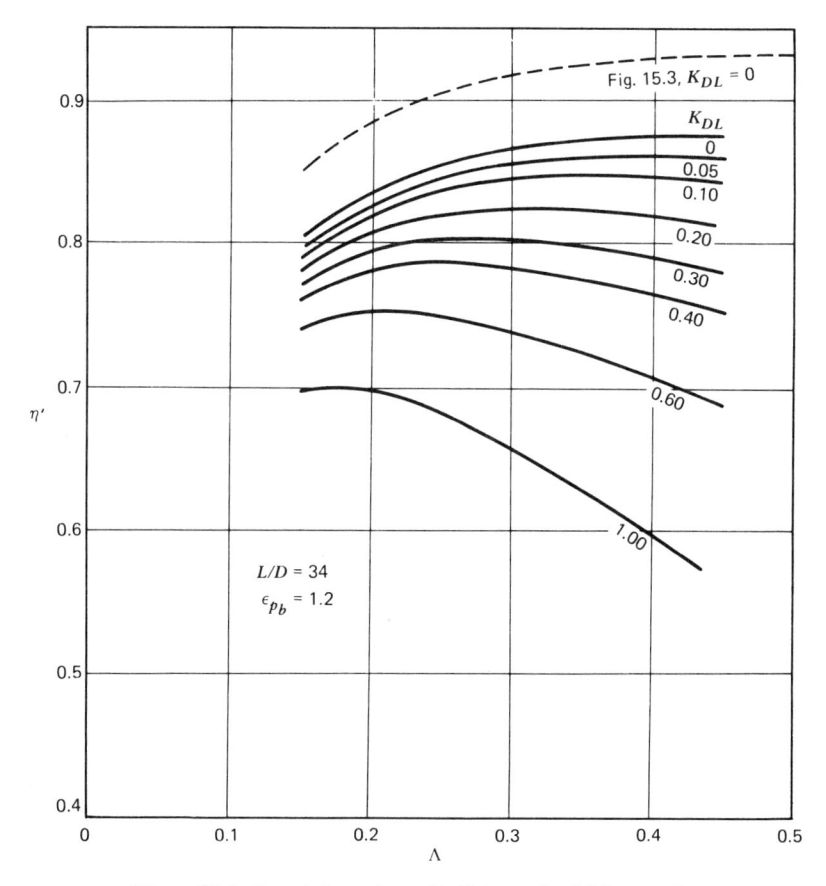

Figure 15.6 Prerotator-rotor unit efficiency for 0.6 boss ratio.

Retaining the preceding assumed values for lift/drag ratio and swirl coefficient, and inserting the various component losses of Eqs. (15.2) to (15.5) in Eq. (15.1), the following expression is obtained for any given set of parameters

$$\eta' = 1 + C_1 - \frac{C_2}{\Lambda} - C_3 \Lambda \tag{15.6}$$

Differentiating with respect to Λ and equating to zero

$$\Lambda_{\text{opt}} = \sqrt{\frac{C_2}{C_3}} \tag{15.7}$$

The Λ_{opt} versus K_{DL} relationships for lines of constant boss ratio values are presented in Figs. 15.8 and 15.9, for the alternate stator arrangements.

270　　　　　　　　　　　　　　　　　　　　　　　　DESIGN OPTIMIZATION

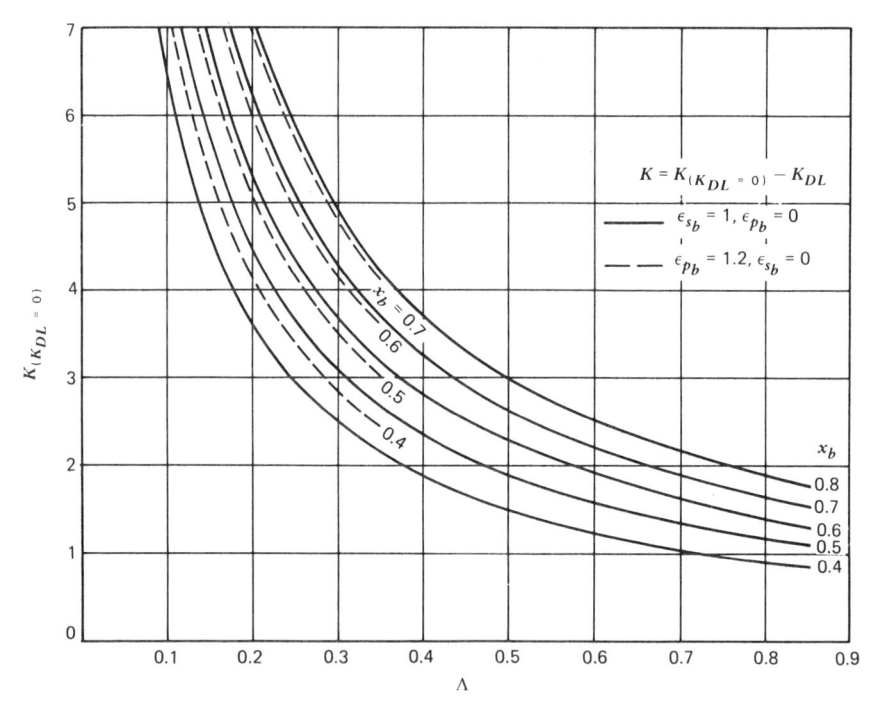

Figure 15.7 Fan pressure coefficients for specified design parameters.

Values of K_{bl} corresponding to Λ_{opt} can be read off Fig. 15.7, enabling either K_T or K_{IT} to be calculated for the relevant K_{DL}.

15.5 DIMENSIONAL DATA, OPTIMIZED UNITS

The use of the foregoing optimized design parameters in developing a simple graphical display of dimensional characteristics is now outlined, bearing in mind the assumption that ϵ_{s_b} is unity.

15.5.1 Fan Pressure

The dimensional useful total pressure rise of in-line or exhaust units of the rotor-straightener type is given in Fig. 15.10, for a tip speed of 100 m/s and a density of 1.2 kg/m³. The relationships displayed on Figs. 15.7 and 15.8 were combined to provide the preceding data. At any fixed nondimensional duty condition the fan pressure is directly proportional to air density, and to the square of tip speed. Hence the quantity ΔH_T, or ΔH_{IT}, can readily be deter-

15.5 DIMENSIONAL DATA, OPTIMIZED UNITS

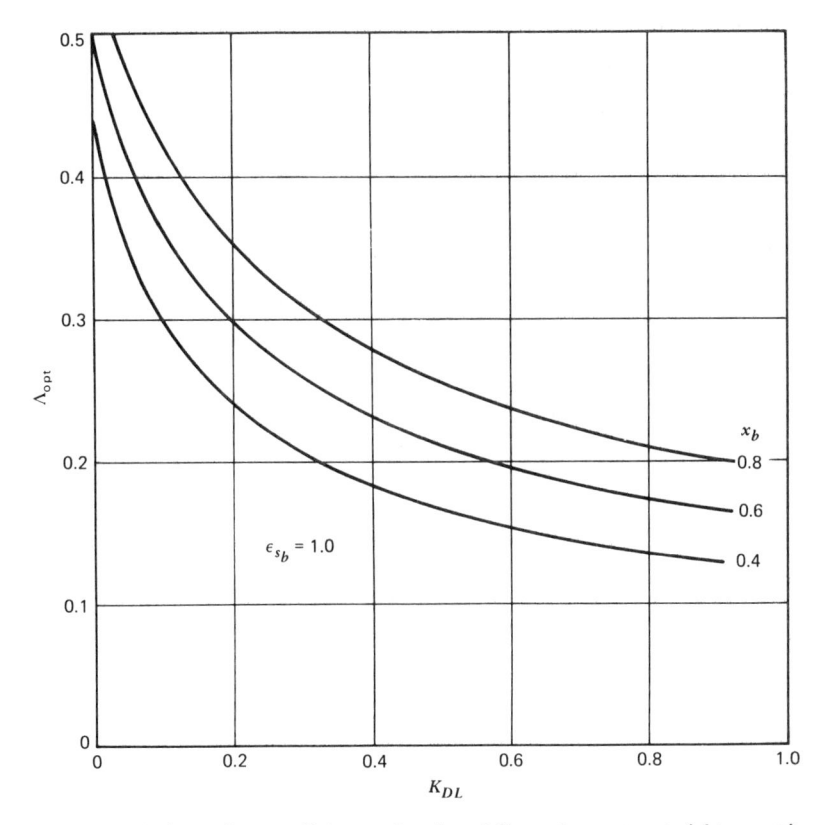

Figure 15.8 Optimum flow coefficient as function of K_{DL} and x_b, rotor-straightener units.

mined for differing combinations of these variables once x_b and K_{DL} are known or selected.

The influence of boss ratio on pressure rise is clearly demonstrated by Fig. 15.10. The equally pronounced effects of downstream loss deserve an explanation.

For a given boss ratio, the optimum Λ reduces rapidly with increasing K_{DL}, augmenting the design value of K_T or K_{IT} (Fig. 15.7). Hence the data of Fig. 15.10 are derived from different pressure rise coefficient quantities.

Doubling the design flow coefficient will halve the design K_{th}. However, pressure rise is a function of Λ^2 and hence the end result is an increased fan blading pressure capability for a given fan speed. When the downstream losses can be spread over a number of rotor stages, increased design Λ represents good practice, necessitating the implementation of cascade design methods.

The degree of reaction is reduced by design increases in Λ. As a result the swirl momentum becomes a substantial proportion of the potential total

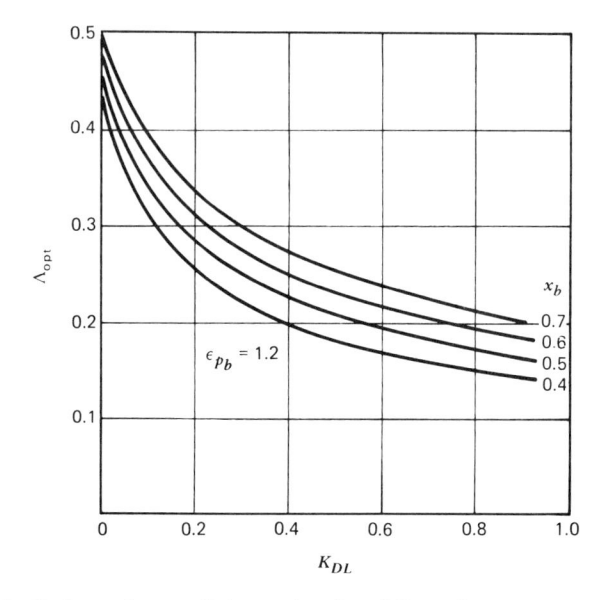

Figure 15.9 Optimum flow coefficient as function of K_{DL} and x_b, prerotator-rotor unit.

pressure rise and must be efficiently recovered as static pressure in the straighteners.

A major difference between fan and compressor design may be explained in terms of the preceding features. In the latter the discharge losses become of little consequence. The choice of 50% reaction for compressor blading ensures a high total pressure rise per stage while limiting tip speed in order to avoid shock-induced blade separation and possible stall.

When designing the optimum multistage unit, K_{DL} will be minimal, allowing larger Λ to be used in design. However, the total pressure rise per stage will be subject to the "work-done" factor and to a possible forced reduction in ϵ_{s_b} from unity in order to restrict blade solidity (Fig. 9.12).

In conclusion, the optimum fan can be described as a unit in which the rotor and stator blading are tailored to the design duty situation and from which all surfaces of a nonaerodynamic nature are eliminated. However, it is important to select the correct overall geometric properties, as a first step. The influence of the downstream losses on these selections is self-evident.

15.5.2 Flow Quantity

The most useful velocity for relating fan diameter and volume flow rate is the one ahead of the nose fairing, designated by U_i. For a given tip speed of 100 m/s and the various Λ_{opt} values, the curves of Fig. 15.11 have been derived. The volume flow rate will, of course, vary linearly with tip speed.

15.6 PRACTICAL ASPECTS

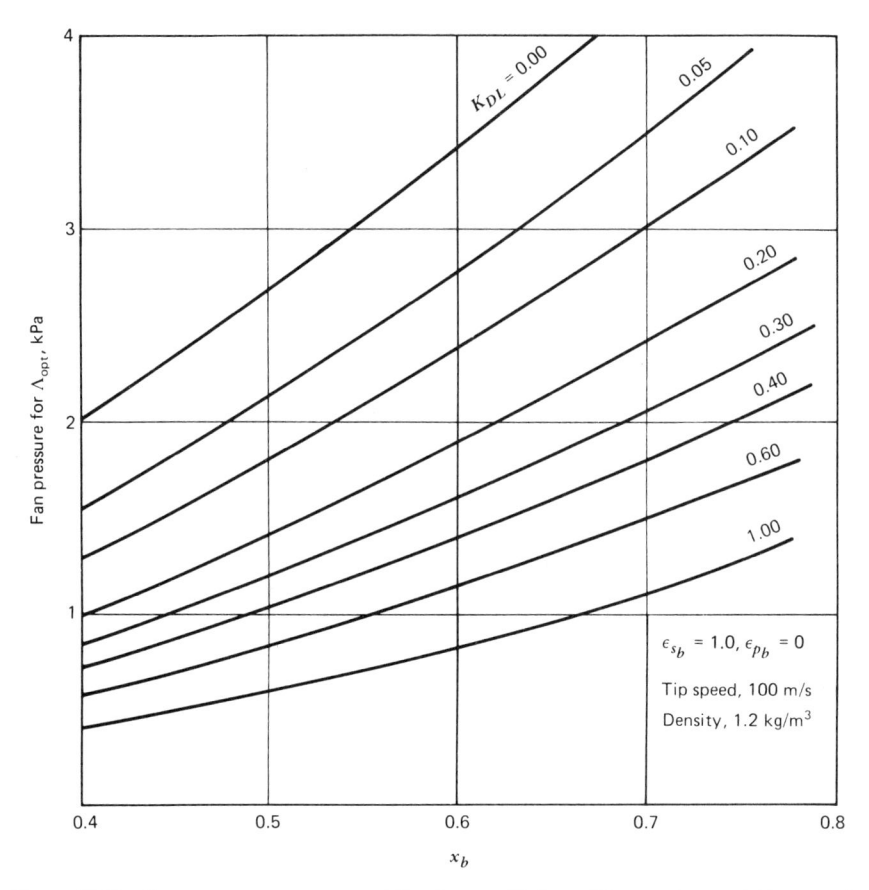

Figure 15.10 Fan pressure versus x_b and K_{DL} for specified conditions, rotor-straightener unit.

It will be noted that at the lower boss ratios and fixed tip speed there is very little change in U_i for the normal well-designed exhaust fan, where $K_{DL} \simeq 0.2$. When an upper limit is placed on tip speed, for mechanical and noise reasons, an approximate design value of U_i is obtained. Hence fan diameter roughly determines volume flow rate while fan pressure is mainly fixed by boss ratio. Therefore transfer of U_i to Fig. 15.12 permits a tentative fan diameter to be selected. The rotational speed then follows from the relationships of Fig. 15.13.

15.6 PRACTICAL ASPECTS

The recommended initial approach to an optimum design exercise is to establish K_{DL} (Figs. 13.3 and 13.4) and from Fig. 15.10 to select a boss ratio that will give the required pressure duty for a selected tip speed. The value of U_i can be obtained from Fig. 15.11 for the chosen set of variables. The diameter and rotational speed are read off Figs. 15.12 and 15.13, for the required volume flow rate. Matching all parameters will require a minimum of trial and error.

Except for research equipment it is seldom possible to design fans that possess all the above optimum properties. However, since the efficiency peaks are relatively flat with respect to Λ, and boss ratio has a minor influence on efficiency for small K_{DL}, a good degree of design adjustment can be made without incurring a performance penalty. Trial-and-error procedures are recommended in obtaining an acceptable combination of major design parameters.

The foregoing optimization study has application to "off-the-shelf" equipment of good pressure rise and high-efficiency capability [15.2]. There is a marked change in Λ_{opt} with K_{DL}, particularly at small K_{DL} (Fig. 15.8). Since this type of equipment may be used in either the in-line or exhaust role, a compromise value of Λ is suggested as the Λ versus η curves are relatively flat for small K_{DL} near the peak efficiency point. An in-line fan has a typical $K_{DL} \simeq 0.1$, whereas this parameter for an exhaust fan will vary between 0.20 and 0.35, depending on the area ratio of the discharge diffuser (Figs. 13.3 and 13.4).

In endeavoring to restrict the number of fan types and sizes it is suggested that two boss ratios and a fan size increment be selected. When the former are 0.552 and 0.69, a given boss diameter is common to both fan types when one fan is 25% larger than the other. Flow coefficients of 0.28 and 0.32 for the two types, respectively, would ensure good pressure rise capability at efficiencies of around 90% for the in-line cases. When a different set of requirements exists, the overall design parameters are readily established in terms of these optimization studies.

Considerations other than peak efficiency may govern many design decisions. For example, a low-pressure-rise fan of lesser rotor blade solidity and swirl, and higher rotational speed may constitute a more acceptable design. Electric motors with fewer poles are smaller and cheaper and the reduced swirl is less of an aerodynamic problem to remove.

Operational considerations may dictate a direct drive from a constant speed electric motor. When the speed suggested by the optimization studies falls in between motor speeds, fixed by pole numbers, the higher speed is usually the preferable one; this avoids blade root overloading, since the local swirl coefficient on which the preceding studies are based is close to the limit. An increase in boss ratio must accompany the alternative lower-speed selection in order to avoid exceeding the blade root design limits.

Since the preceding Λ_{opt} values are in excess of flow coefficients normally encountered in current industrial practice, the fans have a lower tip speed for a given duty requirement. Hence a measure of compatibility between aerodynamic, mechanical, structural, and acoustic matters exists. The extra cost

15.6 PRACTICAL ASPECTS

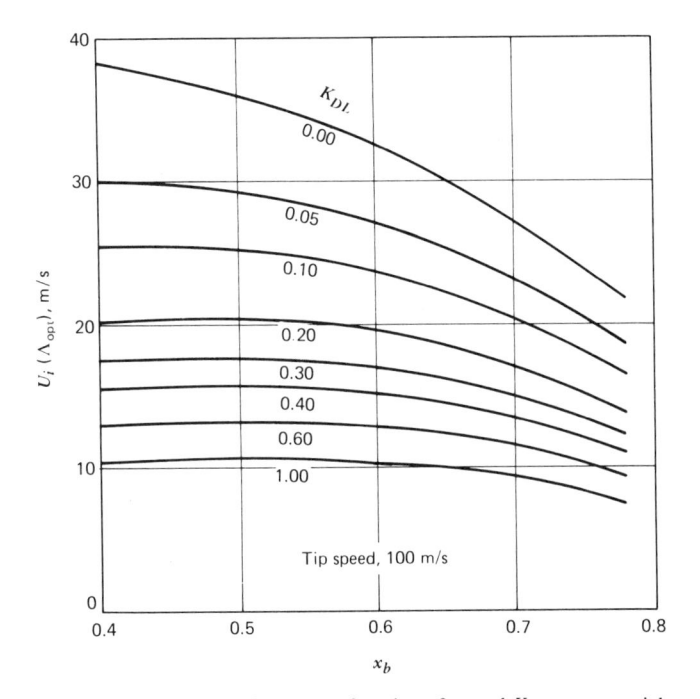

Figure 15.11 Inlet duct velocity for Λ_{opt} as function of x_b and K_{DL}, rotor-straightener unit.

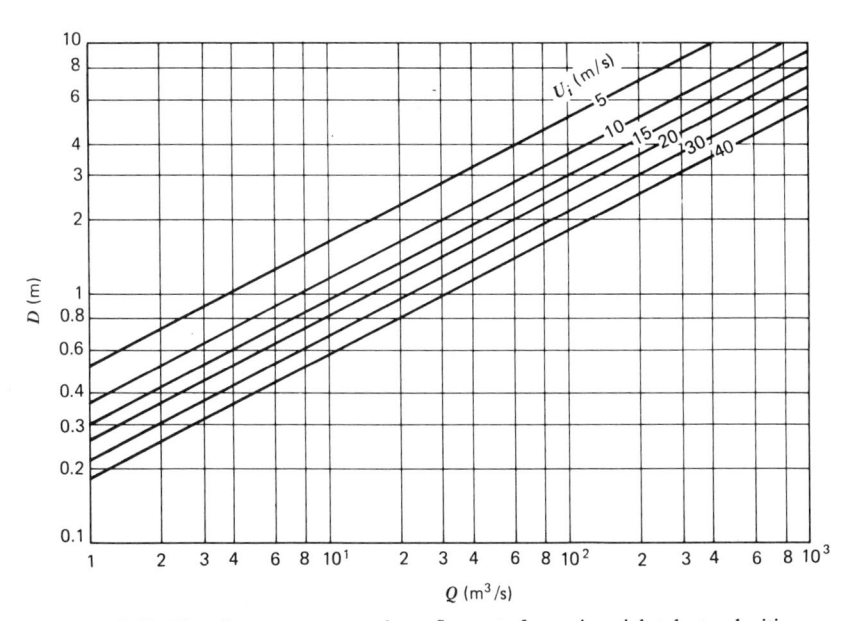

Figure 15.12 Fan diameter versus volume flow rate for various inlet duct velocities.

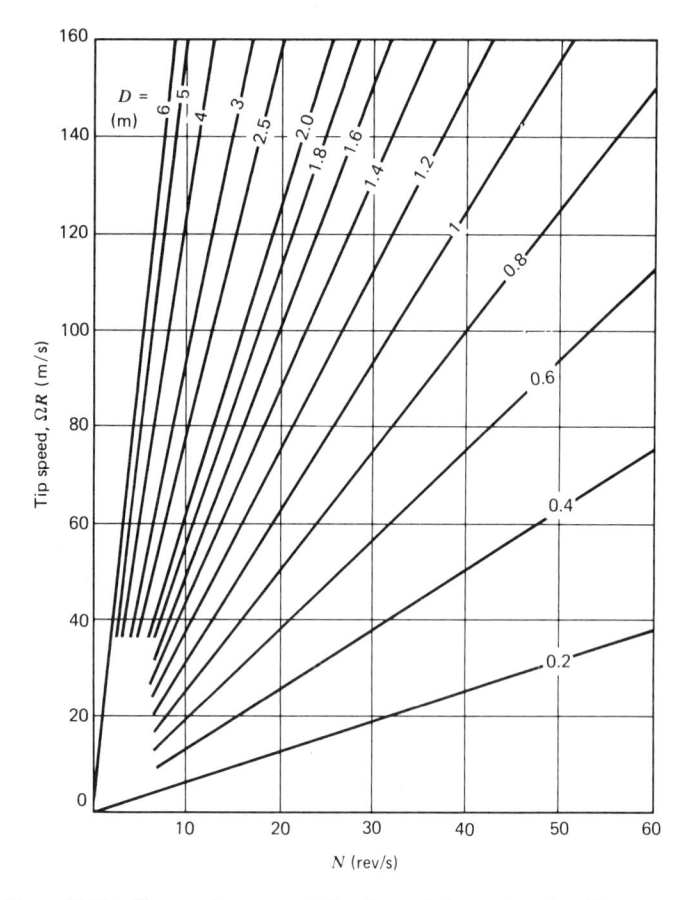

Figure 15.13 Tip speed versus rotational speed, for various fan diameters.

of a well-designed unit of greater blade planform area and slower speed motor is soon recovered in reduced running expenses for continuously operated equipment. The current tendency to undersize exhaust fans absorbing large powers, on a competitive price basis, is clearly in conflict with the desire for optimization, resulting in operational penalties.

REFERENCES

15.1 R. A. Wallis, Optimization of axial flow fan design, *Inst. Eng. Aust.*, **MC4** (1), 31–37, May 1968.

15.2 R. A. Wallis, Design procedures for optimal axial flow fans, *Inst. Eng. Aust., Nat. Conf. Publ. No. 76/12*, Hobart, December 1976.

CHAPTER 16

Fan Materials, Mechanics, and Noise

The following data are directed specifically at alerting the fan designer to the many interface problems that are of importance in achieving the best overall design solution. Fortunately, there is little incompatibility between good aerodynamic design and optimum solutions in the associated disciplines.

General background information in these interactive areas is readily available in standard textbooks. However, the special requirements of fans demand that the interface problems and relevant research findings be discussed here.

16.1 MATERIALS

16.1.1 Environmental Factors

The environment in which the fan is to operate should be one of the first considerations affecting material and manufacturing method selection. For instance, in a furnace system the material should have a high melting point, low creep, and resistance to oxidation. The construction method should avoid stress concentrations, particularly as metal expansions and contractions due to temperature cycling will increase the tendency toward fatigue failure. The high Ni-Cr alloy steels are favored for such installations. When cambered plates are used, extra root strength and stiffness can be obtained by inserting a tapered undersurface gusset plate to match the concave curvature (Fig. 16.1). The ensuing surface irregularities should have an inconsequential influence on the blade aerodynamic properties. Similar type gussets on both outer surfaces of a double-skinned airfoil-shaped blade will result in drag and pressure rise penalties.

It is not possible to design all duct/fan systems to be free of changing load conditions. In some cases this may involve "shock" air loading of the

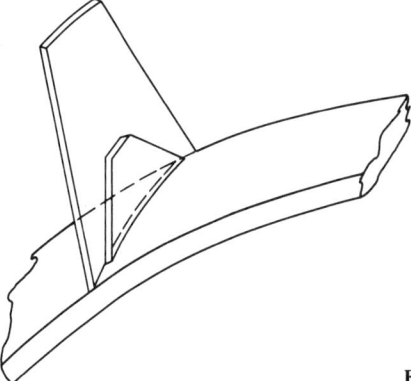

Figure 16.1 Strengthening and stiffening gusset.

blades; extra stiffness and strength must be designed into the unit. However, the most common causes of "shock" loading are gross turbulent inlet conditions and fan stalling. This latter operational condition is often present but can remain unrecognized. The continuing ignorance about partial or complete stall properties is perpetuated by some fan manufacturers who present catalogue data on the complete volume/pressure characteristics without cautionary notes. In most cases sufficient ruggedness, based on experience, is built into the equipment to ensure its reliability.

Blades and vanes may have to operate in moisture- and/or dust-laden flows and hence are liable to water and dust erosion problems. A review of erosion mechanisms and the resistive properties of the normal range of fan materials is presented in [16.1].

In brief, the major variables are dust concentration, hardness, size, shape, velocity, and the strike incidence with respect to the target material. The physical properties of the target will determine the subsequent erosion experienced.

Erosion is of either a ductile or brittle variety where each type is associated with a specific incidence range. The wear rate is a maximum at approximately 25° where the ductile mechanism is active. As the incidence approaches 90°, the brittle fracture process progressively assumes dominance.

The wear rate (mg/g of dust) is proportional to velocity raised to a power that varies within the range 2.0 to 2.3; this relationship is independent of the erosion mechanism for nonbrittle target materials. Exponents of 3 have been recorded for brittle erosion on brittle surfaces.

Hard crystalline dusts such as silica, silicon carbide, and alumina are most destructive. Quartz, which is a particular type of silica abundantly present in nature, provides a ready source of erosion. On impact, the sharp corners of freshly broken crystalline dusts have a much greater destructive power than impacting faces or edges. Rounded noncrystalline dust particles tend to peen metal surfaces at the higher speeds.

16.1 MATERIALS

Aerodynamic drag forces will modify dust trajectories in regions of rapidly changing air flow direction. This is particularly true for dust particles below 10 μm in size. The actual incidence of these smaller particles relative to the blade nose surface is consequently less than the apparent one. The percentage of particles striking the surface is progressively reduced as dust size and the related incidence are decreased.

Experiments show that wear rate rises rapidly with particle dimension, reaching a plateau for sizes above approximately 50 μm. The threshold, as defined in [16.2], is the particle size at which this plateau commences.

The more normal definition of threshold is the condition below which erosion is arrested. Attempts have been made to relate this threshold to particle size, velocity, or kinetic energy. However, an important related variable that has not received much experimental attention is the apparent dust incidence. Experience has shown a substantial threshold improvement with reducing incidence [16.3].

There are a few reports of wear occurring for particles down to 3 μm in size. Hence dust size in a general sense is an inappropriate threshold criterion although some experiments on axial flow compressors have demonstrated zero wear for particle sizes less than 10 μm.

Extensive testing with dust at the incidence for maximum erosiveness has established kinetic energy as the most universal measure of erosion threshold; the target materials were steel and aluminum.

Hence for common dusts at roughly equal specific weights and size, it is possible to fix an approximate velocity at which erosion ceases, for the maximum wear incidence. For steel and aluminum this velocity has a mean value of 45 m/s when dry silicon dust having medium and maximum particle sizes of 35 and 100 μm, respectively, is active. Confirmation of this figure was obtained for an aluminum-bladed rotor exhausting dusty air from an ore-crushing station in an underground mine. The blade leading-edge profile was of a type that facilitated a high wear rate.

As foreshadowed earlier, the trajectory and strike rate changes that accompany small apparent dust incidences will lead to increased threshold velocities. Hence in dusty situations the designer should make a special effort to ensure a smooth inlet flow and the avoidance of blade surfaces permitting dust strikes at incidences promoting peak erosiveness.

Restrictions on airfoil camber, especially in the prime wear region of the leading edge, will result in a minimum of wear and blade speed increases of up to 90 m/s for the previously specified dust before erosion becomes a problem [16.3]; however, the dust must enter the blading annulus in a smooth and controlled manner.

Recent experiences with a large mine fan having a tip speed of 110 m/s show considerable promise. The blade leading edge has been carried forward by 3% to a very small nose radius that unites the upper and lower surface extensions. These flat extensions meet the original airfoil contours along tangential planes. The center line of the wedge-shaped extension must approximate the direction of the oncoming stream relative to the blade. Hence erosion is restricted to the thin extreme leading edge, being mainly of the brittle type. The active dust quantities are consequently a very small percentage of the total throughput; this promises longer blade life. The decrease in $C_{L\ max}$ must be allowed for in design but since a 'shock-free' local incidence is provided, drag will remain low ensuring high efficiency. When this local incidence condition is departed from, increased wear on one surface will sharpen the leading edge and operational efficiency will fall. In other words, the bandwidth of high-efficiency operation will be restricted. This indicates the importance of individual design attention to each blade element shape in erosive airstreams.

The lower blade surface in the vicinity of the trailing edge constitutes an erosion area of secondary importance. As the blade camber increases, the dust to surface incidence and the erosion rate both become greater.

Model tests to obtain quantitative material-loss data on actual turbomachinery are reported in [16.2] and [16.4], the former being concerned with the severe problem faced by the operators of induced-draft, axial flow fans in power station boiler installations.

The two most common blading materials for large fans are steel and aluminum. The weight loss per unit dust mass is approximately identical for both metals, which leaves steel with an advantage of nearly 3 in respect to volume loss. Annealing, tempering, and age hardening of these metals has no substantial influence on the wear rate. This feature is in contradistinction to the abrasion process in which wear is directly proportional to surface hardness. (Abrasion is defined as the rubbing action of an abrasive material over a surface, under applied pressure.)

The erosive wear rate is reduced by metal carbide surface coatings, but the thinness of the flame-applied layers greatly restricts their practical value. The layer eventually erodes away, but in the process undesirable surface roughness is left; the latter affects aerodynamic performance [16.5].

High-alloy white cast irons possessing densely dispersed metal carbides in a tough matrix material represent a possible solution for leading-edge protection under extreme conditions. The high cost of such inserts may, however, restrict their use. On the other hand, the development in Japan of a coating process, designated SIGMA, is apparently showing considerable promise [16.2].

Composite nonmetallic blading materials have relatively poor erosion resistance [16.1].

When elevated temperature and oil mist are nonexistent, resilient materials such as special rubbers or polyurethanes can be used as coatings to resist erosion. The integrity of the underlying blade material can be maintained by regular coating replacements.

Many exhaust fans expel a corrosive and moisture-laden airstream. Provided erosion is not a complementary problem, suitable epoxy paints are available. However, metal surfaces capable of withstanding erosion, but not

16.1 MATERIALS

corrosion, may have to be covered with a material resistant to both these mechanisms. Resilient materials have been the most successful in this regard. However, because fan performance is sometimes severely compromised by these surface coverings, aerodynamic allowance should be made in the design or fan selection exercises for this event.

16.1.2 Material Selection

In addition to the foregoing considerations, the weight, strength and cost of the material have to be taken into account. For any particular application, the blade attachment details will have some bearing on the choice of material. For instance, with a small, fixed-pitch rotor the cheapest solution consists of a cast or molded integral boss/blading assembly. Steel or aluminum are used where strength is required. Molded-foam plastics provide an adequate and cheap unit for low-powered units in protected environments. However, large fans necessitate the blades being cast or fabricated as separate items. Fixed-pitch blades of the latter variety are often welded to the boss and then stress relieved. Cast or molded blades are normally clamped into the boss, or supported in a trunnion-bearing setup for variable-pitch operation.

Large benefits accrue from blade weight reductions. A lighter hub follows, making for dead-weight reductions on the drive shaft and lesser vibration when out-of-balance forces develop. In addition, the "flywheel" effects on starting and stopping problems are reduced. The blades must, however, retain sufficient strength and rigidity.

Wind tunnel fans have been constructed with blades possessing a fiberglass skin and a stiffness-adding foam plastic filling. Starting torque and braking force are both substantially reduced; increased response to wind speed controls is an additional benefit.

Because of its weight advantage over cast or forged steel, aluminum is a common blade and hub material. Although the weight loss due to erosion is comparable to that of steel, the volume loss is, of course, greater. The corrosion- and erosion-resistant properties of cast aluminum alloys with high silicon content (e.g., U.S. Federal Specification QQ-A-601E, A356 with T6 heat treatment), in certain instances could lead to its selection as the best overall solution.

Contrary to popular belief, the normal varieties of stainless steel are subject to corrosion in certain acidic, dust-laden air conditions [16.3]. Blades possessing skins of this material quickly develop a series of pinhole flaws, which by admitting moisture into the interior provide a potential vibration situation. Special corrosion-resistant alloy steels are costly.

Blades with a spar, ribs, and sheet metal skins are favored for large, variable-pitch fans. Weight is kept to a minimum and the ribs assist in maintaining blade profile. The leading edge can be of solid construction, forming a secondary spar. Air tightness is essential in avoiding water ingression, with attendant corrosion and balance problems.

Laminated wood construction, of an integral hub and blade type, is now seldom utilized. However, separate laminated or molded wooden blades are in use, having the advantage of high profile shape accuracy. The surface is frequently protected against moisture by a plastic sheath covering.

The recent advent of production of carbon and boron fibers in commercial quantities is expected to encourage a greater use of fiber-reinforced molded blades. Their cost is currently the main deterrent.

Blade life and replacement costs are two important factors when dealing with fans in adverse environments; these should be considered when selecting the blade material and construction.

16.2 MECHANICAL DESIGN, STRENGTH, AND STIFFNESS

16.2.1 Steady Loads

In general, the normal mechanical stressing and balancing procedures for rotating machinery apply. For steady loading conditions the blade experiences centrifugal and bending loads. The blade deflection under bending load at the design fan pressure duty can be taken into consideration when arranging the centroids of the various sections to lie in a radial line, thus relieving blade root stresses. Since the center of aerodynamic pressure is offset from the section centroid, there exists a turning force about the root shaft fixing, for adjustable or variable-pitch blades. This moment is of small magnitude and is normally neglected in blade stressing. A more significant moment [16.6] is due to centrifugal forces acting on individual blade sections, arising from the fact that the main axes of inertia do not coincide with the axis, and plane, of rotation (Fig. 16.2). The moment is one that tends to turn the blade to a larger stagger angle.

This blade shaft turning moment assumes major importance in relation to variable- (while running) pitch blades. Two diametrically opposed balance weights clamped to the blade shank, inside the hub, and orientated at approximately 90° to the blade chord, are required in order to reduce the actuator and holding brake forces.

The turning moment on a blade is given in [16.6] as

$$M = \rho_m \frac{\omega^2}{2} \int \sin 2\gamma (J_2 - J_1) dz \quad \text{kg-cm} \qquad (16.1)$$

where ω is the angular velocity, ρ_m is material density, γ is the angle specified in Fig. 16.2, and J_1 and J_2 are the main inertial moments referred to the I–I and II–II axes.

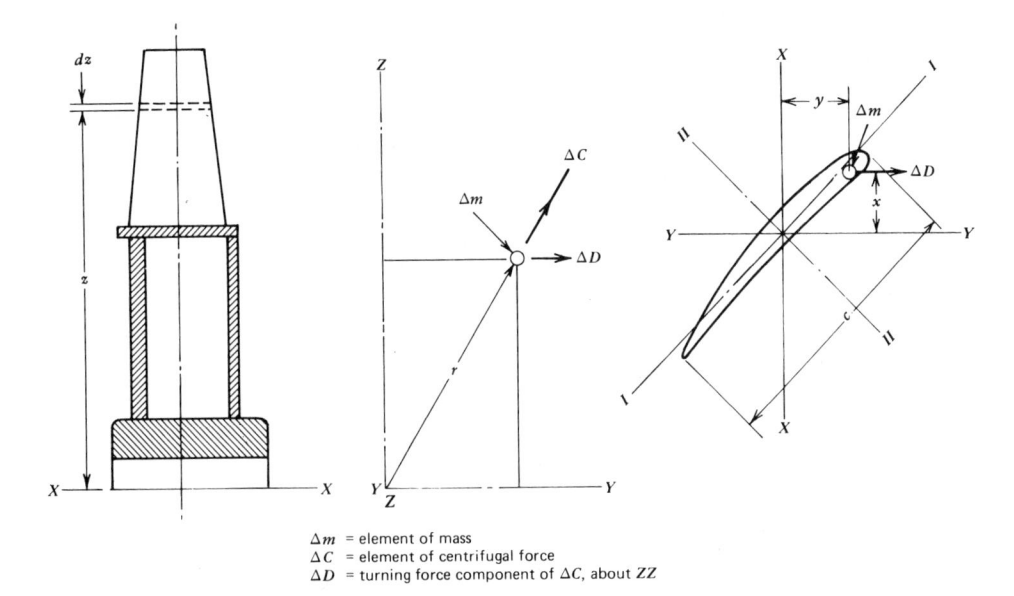

16.2 MECHANICAL DESIGN, STRENGTH, AND STIFFNESS 283

Δm = element of mass
ΔC = element of centrifugal force
ΔD = turning force component of ΔC, about ZZ

Figure 16.2 Origin of blade turning moment. (Adapted with permission of Springer-Verlag from G. Peters [16.6].)

For the solid blade, as defined in Fig. 16.3, the moment is given approximately by

$$M = \rho_m \frac{\omega^2}{2} \int \sin 2\gamma c^4 \eta \, dz \quad \text{kg-cm} \tag{16.2}$$

where η is read from Fig. 16.3.

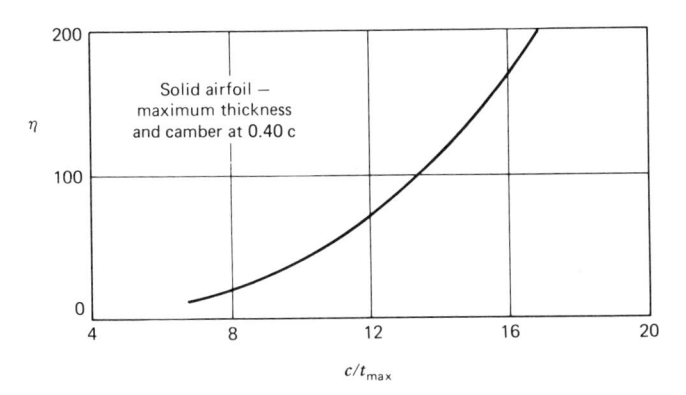

Figure 16.3 Characteristic number for turning moment calculation. (Adapted with permission of Springer-Verlag from G. Peters [16.6].)

In practice, a factory test run with a pair of opposing blades is a wise precaution as weight and clamping-angle adjustments are normally required when selecting the optimum combination. With variable-pitch blades the angle γ at a given cross section is not a fixed quantity, and hence residual torque is present with pitch change. The turning moment due to aerodynamic loading may also assume importance in this rotational balance exercise.

The small shear stress that the bending moment produces is normally of little consequence, particularly when the design usually carries a factor for unsteady load conditions.

A brief summary of stressing procedures is presented in [16.7]; for greater detail [16.8] and [16.9] can be consulted. However, neither of the latter two references indicate a factor for unsteady loads, which all fans will experience in stall.

16.2.2 Unsteady Loads

It is advisable to increase the steady load bending stress by an arbitrary factor, to cater for unsteady air loads. The magnitude of the fluctuating load is often unknown and hence the factor must be based on actual experience. A value suggested in relation to axial flow turbines is given in [16.7] as 75% of the steady bending stress. Impact loading is potentially greater for fans that possess poor inflow conditions and may enter stall. A factor of from 2 to 3 is probably required in these circumstances. When the direction of rotation is reversed for emergency operation (e.g., in countering a mine fire), the fan will often run in a partially or completely stalled state.

A rotor blade cutting the wake from an upstream body, or bodies, will experience aerodynamic load changes, the magnitude of which will depend on wake velocity defects (Section 6.10); the defects for streamlined bodies are a function of downstream distance, gap/chord ratio, and inlet turbulence (Fig. 6.32). In the case of fans with "steady inflow" conditions, the axial separation between the rotor and the trailing edge of the wake-producing bodies should be at least half the chord length of the latter. This will not alter the pressure rise through the fan unit (Section 11.6).

However, the preceding separation must be increased in the presence of highly turbulent inflow conditions to the upstream vanes. The velocity defect decay rate is retarded, making one chord length more appropriate. Noise reduction studies, however, establish blade gap s as a relevant variable, as discussed later.

The most serious consequences arise when the unsteady aerodynamic loads force the blade into resonance. There are two aspects to this problem. In the first instance, when rotor blades cut the wakes of evenly spaced prerotator vanes, or support fairings, at a rate that coincides with the blade natural frequency, or its harmonics, fatigue failure can occur. It is therefore of great importance to establish the natural frequency of the blades. For solid blades an approximate calculation can be made according to the method of

16.2 MECHANICAL DESIGN, STRENGTH, AND STIFFNESS

[16.10]. However, an experimental determination on a stationary blade/hub assembly is preferable and should be mandatory; the stiffening effect of centrifugal force is in most cases small but should not be ignored. In parallel with noise studies, the number of stators or fairings can then be selected. Radial struts confined to a specific sector and providing a bearing support structure must be spaced so that their wakes cannot force the rotor blades into resonant motion.

The instantaneous unsteady blade load can be reduced when the wakes of upstream bodies are not parallel to the blade leading edge at the moment of contact. Prerotators with radial leading edges will possess nonradial wakes when inducing a free vortex flow. Longitudinally aligned, streamlined support vanes can, if required, be inclined at approximately 10° or slightly more to the radial direction, where the vane tip is advanced in the direction of fan rotation.

The natural frequency of rotor blading should be disclosed with all fan catalogue data. A particular rotor may have its support system modified in a manner that heightens the chances of blade failure. The preceding information is vital to the informed engineer who desires to check the operational viability of the selected fan. A prescribed method for measuring the natural frequency of direct drive sheet metal fans commonly used for air conditioning equipment applications is now available in an ASHRAE standard.[†]

The second aspect of blade failure also demands a knowledge of natural frequency. Blades with low natural frequency lack stiffness and are prone to suffer self-excitation. In the various studies of this phenomenon, mechanical damping is assumed to be much less than the aerodynamic variety. Whitehead in [16.11] relates the excitation to changes in blade lift. The associated circulation can only change if vorticity is shed into the wake. It is therefore assumed that the vorticity varies sinusoidally with distance downstream of the trailing edge, and since the wake is swept downstream at the downstream velocity V_2, the wavelength of the fluctuations is given by $2\pi V_2/\omega$ where ω is the angular frequency of the vibration. The nondimensional frequency parameter

$$\chi_2 = \frac{\omega c}{2\pi V_2}$$

is obtained when the wavelength is related to the blade chord. It is shown that the critical value of χ_2, below which bending flutter occurs, is a function of gap/chord ratio, air deflection angle through the blade, and stagger angle.

This type of flutter requires a phase difference between the vibrating motion of one blade and the adjacent ones [16.11]. In aircraft engine compressors this motion is prevented by snubbers located at an appropriate spanwise position; these restrict blade vibratory movements. Large-chord fan blades of adequate stiffness are currently being fitted to a modern large jet engine, increasing engine efficiency.

[†]ASHRAE standard 87-1 method of testing dynamic characteristics of propeller fans, 1982.

The stiffness of industrial fans can be increased without the weight restrictions imposed by aircraft applications and hence the use of blade tip shrouds, tie wires, or rods is unnecessary for correctly designed blades.

The foregoing discussion relates to unstalled blades with moderate turbulence inputs. Stalled blades can also suffer self-excitation, and hence protection from blade flutter under this condition should also be provided. According to [16.11] the limiting value of χ will usually override that for the unstalled case.

When the turbulence level is high, the blades should be designed to withstand a measure of random excitation [16.11,16.12].

Large axial flow fans of considerable pressure capability are currently in service. Since the attainment of a large pressure rise demands high relative blade velocity, the frequency parameter χ tends to be lessened. In addition, the larger blade spans make for smaller natural frequencies, further reducing the value of χ. As large fans are obviously expensive, blade failure is a serious matter. In three instances of blade failure known to the author, the problem has been solved in two of these by increased blade root thickness, and in the third by a continuous shroud ring.

Blades satisfactory for normal turbulence levels can exhibit flutter when upstream large-scale separated or "lumpy" flow is present. The blade natural frequency or its harmonics is believed to lie within the frequency range of the associated large-scale vorticity. The flutter that occurred in each of the preceding three failure cases resulted from operation in periods when random inflow conditions were present.

The fan stall and random flow cases are not amenable to theoretical treatment and hence the designer must depend on data obtained from extensive practical experience. According to [16.11] the only known nondimensional group that appears reasonable for correlation purposes in the stalled fan case is the frequency parameter. The argument is centered on damping considerations.

The critical value of χ will obviously vary, depending on detailed blade features and on the nature and character of the trigger phenomena. The mechanism by which the blade is self-excited and the role played by aerodynamic damping are not currently amenable to analytical determination. To provide guidance for a common fan blade type, relevant details about a 14-bladed 6.1-m-diameter fan that suffered blade flutter are presented. The tapered 10% thick Clark Y blades were of a skinned construction type, with root and tip chords of 0.85 m and 0.37 m, respectively; the corresponding blade speeds were 47 and 94 m/s. The blade length was 1.52 m and blade root stagger angle 55°. The natural frequency of the original 10% chord thick blade was 22 Hz, which was raised to 32 Hz by increases in blade thickness toward the blade root, where the value is now 13% chord. The original blade spar failed at the first change in diameter, outboard of the blade root. The spar showed three zones of fractured cross-sectional area, the first being rust coated, the second a new fatigue failure, and the third an overload break.

16.2 MECHANICAL DESIGN, STRENGTH, AND STIFFNESS

The installation (Fig. 17.2) experienced two periods of high-velocity winds, resulting in separated duct inflows. Self-excitation under normal operating conditions was nonexistent despite a prerotator wake-cutting frequency of twice the fundamental one. Manufacturer's calculations indicated a natural frequency of 32 Hz but prior to failure no experimental confirmation was undertaken that would have disclosed the error, indicating a potential first harmonic problem. The location of blade fracture tends to suggest that second-mode vibrations were the destroying medium.

The order of magnitude of the natural-frequency parameter calculated at midspan is approximately unity. This number is well in excess of the critical range, 0.27 to 0.33, reported in [16.11] and [16.12] for stall-induced self-excited vibration.

Experimental data from [16.12] show a strong tendency for second flexural modes to be excited in compressor blading when subjected to heavy buffeting. It is therefore not surprising that a combination of forces due to wake excitation and random flow features should cause failure at this elevated natural frequency parameter. The highest value for a rotating stall failure (aluminum) recorded in [16.12] is 0.44, but an authority suggesting an upper limit of 0.7 is quoted.

Increased blade stiffness and hence reduced blade amplitude provides an adequate solution to this troublesome phenomenon. Limiting the blade tip amplitude under buffeting load conditions prevents the self-excitation mechanism, at present unknown, from establishing itself. If the critical amplitude were known and the buffet forces predictable, it would be possible to suggest more accurate design guidance. However, for the present the designer must check out the design in terms of a frequency parameter.

16.2.3 Rotor Balance

Whenever factory facilities are available, fans should be dynamically balanced before delivery. Rotors possessing out-of-balance masses in various axial planes can be statically balanced by an added mass in one axial plane. However, when run at speed, an unbalanced couple will make itself evident, increasing drive shaft stresses and vibration levels. Alternatively, dynamic balancing can be done in the field.

16.2.4 Other Design Considerations

The design of the boss, drive shaft, and other mechanical features will follow normal procedures [16.7, 16.8, 16.9] but allowance should be made for significant unsteady aerodynamic and out-of-balance forces; the latter may be present during the fan life and are often due to uneven erosive wear, a broken blade, or mud buildup. The duct wall surrounding the blade tips must be stout enough to resist fracture in the case of blade failure, thus containing the potentially dangerous debris.

288   FAN MATERIALS, MECHANICS, AND NOISE

16.3   NOISE

In achieving a minimum fan noise level, it is important to design, install, and operate the fan in accordance with good aerodynamic principles. For instance, the fan unit of Plates 2 and 3, at the design blade pitch angle and one diameter removed from the fan casing, registered an overall sound pressure level of 75 dB with a peak of 71 dB in the 150 to 300 octave band. The tip speed was 94 m/s and the power 675 kW. Electric motor hum constituted part of this emitted noise. At a distance of 100 m, the high frequencies had been dissipated and the low-frequency components were barely discernible in the quiet surrounding environment. There was a similar measure of noise control in the fan unit of Plates 4 and 5; it was possible to hold conversations at normal speech levels in the immediate vicinity of the installation, which would have been impossible with the units they replaced.

The preceding illustrations are just two of a number of "quiet" fans

Plate 2.   Downcast mine ventilation fan, Mount Isa. (By courtesy of Mount Isa Mines Ltd., Australia.)

16.3 NOISE

Plate 3. Maintenance provisions for downcast fan, Mount Isa. (By courtesy of Mount Isa Mines Ltd., Australia.)

Plate 4. Two-stage auxiliary mine ventilation fan, Mount Isa. (By courtesy of Mount Isa Mines Ltd., Australia).

Plate 5. Assembly of two-stage auxiliary fan components. (By courtesy of D. Richardsons & Sons, Melbourne, Australia).

designed or specified by the author. The commonly held belief that fans are inherently noisy, and consequently must be either tolerated or provided with additional noise control equipment, is obviously out of step with the evidence. Excessive noise can usually be interpreted as indicative of some aerodynamic design flaw and should be treated as such; it should not be ignored. A striking example of the consequences that can accompany a failure to put the correct interpretation on a grossly excessive model noise level is available from the Hinkley Point nuclear power station fan and gas ducting failures [16.13]. Aerodynamic design modifications at the model stage would have resulted in great savings of money and manpower.

In the past the emphasis has been on producing fans of low capital cost for industrial use. Aerodynamic crudity has consequently been accepted by the customer. It is therefore assumed by the latter that the emitted noise is normal; where noise must be controlled, the fitting of attenuators is becoming standard practice. However, the total cost of the installation often exceeds that of a "quiet" fan installation, of good aerodynamic and mechanical design. Attenuator maintenance and additional space requirements must also be considered in the cost comparison.

16.3.1 Design Objectives

By applying the aerodynamic principles outlined in Section 16.2.2 with respect to minimizing unsteady aerodynamic forces, the main noise abatement objectives are achieved. The requirements are as follows:

1. Design the upstream duct system to be free of highly disturbed and swirling flows.

16.3 NOISE

2. Provide a smooth flow into the fan annulus by a suitable outer-wall "bellmouth" and by fitting an adequate nose fairing or spinner.
3. Design the upstream vane or support struts to possess low drag and to be free of flow separation, ensuring that their thin trailing edges are at least one half vane or strut chord upstream of the rotor leading edge (see Section 16.3.2 for qualifications).
4. Ensure that the rotor blades are properly matched, in twist, chord, and camber, to provide efficient and low drag operation at the design condition; also, small tip clearances are most desirable [16.14].
5. The straightener vanes or streamlined support struts must be of good low drag design, free of flow separation, and located at least half a rotor blade chord downstream of the blade trailing edge (see Section 16.3.2 for qualifications).
6. The number of rotor blades and stator vanes must be unequal and preferably should not possess a common factor. In addition, the product of the stator vane number and rotor rotational speed must not equal rotor blade natural frequency. Support struts must be subject to the same requirements; when restricted to a sector, the preceding conditions must be met on the basis of the blade speed and the period between adjacent struts.
7. An unseparated flow condition immediately downstream of the fan blading is desirable.

Interaction with resonant acoustic modes related to duct geometry is discussed later.

16.3.2 Fan Noise Sources

Departure from the preceding design features can lead to unnecessary and damaging noise levels. To set a basis for discussing this matter, noise fundamentals will be briefly discussed.

Fan noise is of a broadband variety on which is superimposed a discrete frequency(s) of negligible or major importance. The resultant sound power is usually associated with contributions from both boundary layer turbulence and wake turbulence each of which increases with relative blade velocity, raised to different powers. However, a single index, namely, 5.5, is commonly applied when predicting industrial fan noise. A range of values from 4.5 to 6 is reported in [16.14], depending on the operating-point location on the fan characteristic curve. This must indicate a change in the relative contributions from the preceding two noise sources, and others.

For below-standard fan and duct equipment the rule for predicting noise power from data pertaining to other units will be inappropriate. However, flow and noise control, as exercised when meeting the conditions previously outlined, will reduce the number and strength of noise sources and improve prospects for prediction reliability.

The mechanics of rotational noise responsible for the discrete frequency component(s) are outlined in [16.15]. The pressure field created by a blade at incidence creates a transverse wave form that is repeated at each adjacent blade. These pressure patterns, with a wavelength equal to the gaps, constitute spinning modes; field strength diminishes exponentially with upstream distance for mode frequencies below the cutoff value [16.15]. Both the blade chord and gap are important parameters in determining the upstream field strength. The first fixes the value from which decay commences and the second determines the rate at which it proceeds.

In the absence of solid bodies upstream of the rotor, such as stators or support struts, the noise intensity rapidly decays with distance, provided the tip Mach number is subsonic [16.15]; the latter condition is always satisfied in the case of fans and hence this noise-producing mechanism may assume secondary importance, provided acoustic resonances due to duct geometry are avoided.

Upstream bodies possess wakes that constitute a velocity nonuniformity entering the rotor disk. As the axial spacing is reduced the wake velocity defect increases (Fig. 6.32) and, in addition, the aft body surfaces enter and interfere with the blade "potential flow" pressure and velocity fields. Cyclic lift changes accompany both these flow features. According to [16.15] these separate events produce effects that have an identical common structure in respect to spinning modes.

Stator/rotor interaction, depending on vane and blade numbers, will produce a varying number of spinning modes related to a fixed blade passing frequency or its harmonics. If a particular interference pattern has fewer lobes than the number of rotor blades, it must rotate faster than the shaft to generate their passing frequency. For example, an eight-blade rotor interacting with a six-vane stator row will possess a $8 - 6 = 2$ lobe pattern spinning at $8/(8 - 6) = 4$ times rotor speed. When the numbers are, respectively, 8 and 9, a one-lobe pattern will be present; this lobe will spin in a direction opposite to rotor motion at a speed eight times rotor speed. Hence the prospect of supersonic mode motion in the outer blade regions, together with consequential noise propagation rather than decay, must be expected in the latter instance. A rapid increase in the radiated sound power, when the acoustic mode begins to rotate supersonically, is cited in [16.14]. This adds weight to the recommendation that the source be reduced by a generous axial separation between blading rows. Blade and vane numbers should never be identical and preferably they should avoid possessing a common factor.

The wake-induced noise source can in most cases be controlled by providing for an axial gap that is at least 50% of the upstream vane chord. However, when the flow is highly turbulent this distance should be doubled on account of the reduced rate of wake decay (Fig. 6.32).

In seeking to minimize the interaction effect of the prerotators on the rotor potential flow field, the rotor blade gap and chord become the relevant parameters. An axial spacing of at least one chord, and blade and vane

16.3 NOISE

numbers chosen to give a high mode number in order to reduce radiation efficiency, are the recommendations of [16.14]. In compressors the blade and vane chords, and the blade gaps, are comparable in magnitude. However, fans often possess lower solidity blading and hence the preceding suggestion needs explanation. Blading with a large gap/chord ratio will not possess a continuous circumferential wave motion; this circumstance could arise for ratios in excess of 3. Therefore expressing axial spacings in terms of blade gap, a dimension that determines the exponential upstream noise decay, could lack relevance. Hence the recommendation of [16.14] can be accepted, subject to experimental verification, when the term *chord* is replaced by *blade chord*.

The suggestion of [16.14] regarding blade and vane numbers, with respect to radiation efficiency, must be considered in relation to mechanical, strength, and blade/vane aspect ratio requirements. Increased rotor blade numbers and hence smaller chords are one way of achieving the stated objective, but decreasing blade stiffness will eventually lead to blade vibration and failure in flutter. Provided the axial spacing is generous, this facet of noise propagation is of minor importance, provided the upstream stators or other bodies are free of flow separation.

Interaction between the rotor blades and the downstream stator vanes has received much less research attention than the former problem. Provided the axial spacing meets the preceding wake-induced noise control requirement of half to one rotor blade chord, the noise is minimal. The cyclic pressure pulses are of lesser magnitude due to the reduced relative air velocity associated with the stator. Reducing the axial spacing will, however, lead to increased noise and possible vibration problems.

As indicated earlier, a considerable proportion of fan-generated noise is of the broadband variety. This can be traced to general turbulence associated with boundary layer and free shear layers, and to vortex shedding from the trailing edges of all blading and immersed bodies. The presence of a nonuniform velocity condition upstream of the rotor will produce cyclic changes in lift, and consequently the variations in blade circulation will lead to increased noise due to vortex action.

The preceding descriptions relate to noise sources that accompany an acceptable standard of aerodynamic fan design. A large number of variations to these cases occur when bluff bodies exist immediately upstream of a rotor or when a fan is partially, intermittently, or fully stalled. A stalled fan can usually be identified by a low-frequency lack of steadiness in the noise level, commonly called woofing.

Methods of predicting fan noise are discussed and referenced in [16.14].

16.3.3 Acoustic Coupling

Fan noise can be substantially increased when resonance with one of the duct acoustic modes is inadvertently established. Simple longitudinal wave motion, commonly known as organ piping, is of particular interest.

The geometry of any duct/fan system will determine the nature and natural frequency of the longitudinal wave system. Because of axial area changes and curved surfaces, the natural frequency may be modified from the normal one, because of curved wave fronts. This wave motion will exist in all ducts possessing air movement, with duct turbulence providing the energy. However, when a strong discrete excitation source of identical frequency exists, for example, blade rotation, the noise level will be dominated by this greatly augmented contribution.

Transverse acoustic waves are alternatively known as cross or higher-order modes. As before, they are determined by duct/fan geometry and exist in the absence of a resonant exciting source. However, when the latter is present serious consequences are possible.

Cyclic interference between upstream stators and rotor blades is one source that can excite acoustically produced noise, as illustrated by the Hinkley Point gas system failure [16.13]. Noise was grossly excessive and the damaged stators were subjected to intense oscillating pressures. The higher harmonics of blade passing frequency were quite prominent. A dramatic reduction in noise accompanied stator removal.

The experimental studies of [16.16] established the magnitude of the normal fluctuating aft surface pressures present on stators that penetrate the potential flow field of the rotor blades, when the axial clearance is inadequate. Relevance to the Hinkley Point failure, where the clearance was relatively small, is claimed.

Transverse modes in a cylindrical or annular duct can be disposed either circumferentially or radially [16.17]. Further data on duct modes are available in [16.18]. In any given situation a number of possible arrangements exist [16.17, 16.19]; this emphasizes the need for avoiding or weakening the noise source by aerodynamic design measures.

A combination of longitudinal and higher-order wave fronts occurs at a sudden expansion in the duct/fan system; truncation of diffusers and tail fairings are two examples of accepted aerodynamic practice but the possibility of resonance cannot be ignored.

The cross modes may be induced into resonance by vortex shedding in the wake of bluff bodies. The shedding frequency is normally given by the Strouhal number but research has now shown that the frequency can change its value over limited speed ranges, adjusting to excited acoustic modes [16.20, 16.21]. The former reference includes an example where acoustic modes were excited between axially aligned stationary vanes.

The complexity of the vortex shedding and its interrelation with the transverse mode mechanisms highlight the need for greater design care. Fortunately in most cases noise problems are avoided when the rules of Section 16.3.1 are followed.

Acoustic model testing is advisable for large and vital installations, such as mine ventilation and power generation fans. For example, the inlet duct arrangement for the fan illustrated in Fig. 17.2 was modeled to establish its

16.3 NOISE

longitudinal acoustic mode characteristic. Unexpected and serious noise problems in an environmental or structural sense could require expensive and physically difficult corrective measures. In the past this mode has received the greater attention because of its frequent prominence in duct and fan systems. However, the more unpredictable higher modes might be expected to come under increasing model test scrutiny for large and important fan installations. The validity of acoustic model testing is supported by sound fundamental reasoning. The state of the art has now advanced to the point where the results of such tests may be satisfactorily assessed.

The strength of the pressure fluctuations associated with organ pipe resonance can be illustrated by a large copper smelting furnace of the fluidized-bed type. The resistance and weight of the bed failed to prevent large and unacceptable bed oscillations. Insertions of a discontinuity in the long compressor output duct line detuned the system. The destructive power of transverse modes is evident from [16.13].

16.3.4 Noise Control Devices

An increase in fan pressure rise capability will eventually make larger tip speeds necessary and hence result in unacceptable noise levels unless noise-attenuating devices are employed. The point at which they are introduced will depend on the maximum permissible noise for the surrounding environment. This is usually specified in terms of a noise criteria (NC) or noise rating (NR) value. The value represents the permissible sound pressure level at 1200 and 1000 Hz, respectively. The latter has been adopted by the International Standards Organization [16.22] and is presented in Fig. 16.4. These curves take account of the much greater tolerance of the human ear to lower-frequency sound, without hearing loss.

The noise requirement(s) in terms of noise rating (NR) will be specified at some location(s) and distance from the fan casing. The readings at mid-octave frequencies are then required to lie below the NR curve specified. In a free field, noise intensity will attenuate according to an inverse distance squared law but reflections and other factors can substantially modify this relation.

High-frequency sound attenuates rapidly with distance, in an open atmosphere. However, this is not true of low frequencies. The problem of low frequencies is carried over into the area of noise control devices. It is perhaps fortunate that our tolerance to such sounds is high. The higher-frequency noise signature of well-designed and installed axial flow fans, as opposed to centrifugal units, is often a distinct advantage. (The lower speed of the latter and the presence of some flow separation at blade exit for most commercial units are together responsible for this characteristic). Provided the low-frequency components are within the noise rating limits, the higher-frequency broadband noise is amenable to attenuation by traditional perforated duct surface treatments.

296 FAN MATERIALS, MECHANICS, AND NOISE

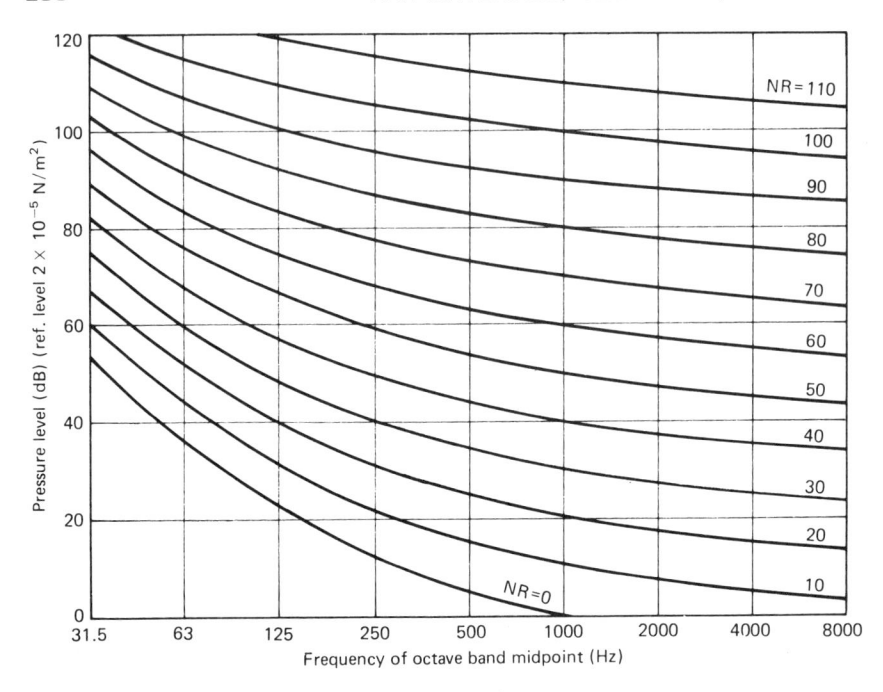

Figure 16.4 Noise rating curves.

This principle dominates the noise control measures currently in use. A length of cylindrical ducting incorporating a central faired body, both featuring porous surfaces backed up by absorbing materials, is attached to the fan casing flange(s). These units are specified in terms of decibel attenuation for given octave bands. For the range 1000 to 2000 Hz, the sound power attenuation claimed by manufacturers is approximately 25 dB; this value is reduced as frequency decreases but some effectiveness is retained at 125 Hz. This refers to noise being transmitted along the duct and not to external levels of fan noise. The fan casing because of its thickness and rigidity will give an overall sound power attenuation of the order of 25 dB.

The flow resistance of the preceding device can constitute a significant proportion of the fan pressure capability, for low-pressure-rise units. An initial attempt to minimize attenuator resistance is outlined in [16.23]. This development arose out of a simple suggestion by the author concerning a possible change in the acoustic coupling mechanisms. Fans with large boss ratios ($\simeq 0.7$) lend themselves to the outer duct wall treatment illustrated in Fig. 16.5. It was postulated that with the blading obscured in line-of-sight vision, the propagation down the duct of wave motions related to blade passing frequency, or its harmonics, would be greatly hindered. (Since the outlet area is approximately equal to the annulus area, the diffusion process in a real-life application is delayed and a downstream diffuser must be added.)

16.3 NOISE

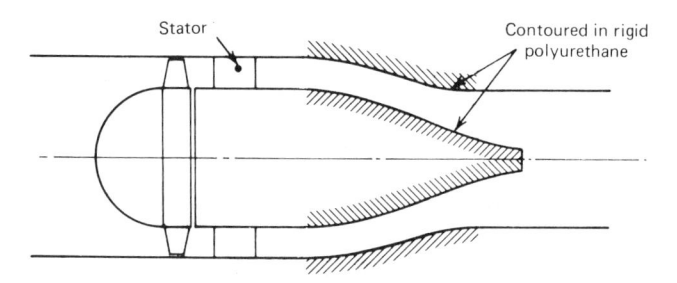

Figure 16.5 Low-resistance noise attenuation system.

The original parallel duct exhibited a peak at blade passing frequency; this feature was absent for the modified duct shape. Further work is required to verify the generality of this result.

Roberts and Luxton [16.23] recognized the longitudinal area changes as an established attenuation principle, for ducts in general. Experiments confirmed this fact for ducted fans, resulting in an overall downstream sound power reduction of 5 dB (measured in the duct); that is, sound power was cut by two-thirds. Attenuation was achieved at all frequencies, with the major gains at frequencies of 1000 Hz and above. At certain frequencies attenuations approaching 10 dB were obtained.

A theoretical study of sound transmission through annular ducts of varying cross section is available in [16.24], providing a means for estimating the effect of area changes and boundary layer growth on acoustic modes.

In tackling a noise attenuation problem the self-induced noise of the control device, and other mechanical and electrical induced noise sources, must be taken into account. For example, a central porous "bullet" will increase flow velocity and possess its own aerodynamic noise-producing characteristics. Silencers of rectangular cross section, and fitted with porous skinned splitters, are often manufactured with little regard for aerodynamic detail; their self-noise can be significant. When a fan is located in a noisy neighborhood, achieving a satisfactory noise level may depend solely on discrete frequency sound suppression.

A discerning and knowledgeable approach to fan noise control is therefore strongly recommended. Fans that are marginally above the required noise level should not be fitted with a device capable of providing high attenuation, at the expense of increased fan pressure duty. It should be remembered that the large potential noise reductions quoted by the makers are dependent on the existence of a high initial sound level, that is, one well in excess of the self-induced noise of the device.

The integrated aerodynamic/acoustic approach of [16.23], in a fully developed form, should provide limited noise control without excessive fan pressure penalties. The possible suppression of the discrete frequencies could produce substantial noise attenuation when strong acoustic coupling is inadvertently present. Extension of the principle to smaller boss ratios is desirable.

A resonator can be designed that absorbs acoustic energy within a narrow band frequency zone. Such selective devices are, however, of limited use, as in most practical cases the noise is spread over a wide range of low frequencies. Fortunately, the troublesome frequencies associated with quality axial flow fans are sufficiently high to respond to other attenuation devices.

Attenuator design is adequately treated in the available literature, for example, [16.25] and [16.26].

A more recent method of noise reduction is described in [16.27]. It consists of a microphone pickup that actuates the generation of sound of opposite signature, thus causing a cancellation of part of the sound power. An experimental study [16.28] has demonstrated reductions of from 16 to 20 dB in the frequency band 30 to 650 Hz.

16.3.5 Noise Measurement

Noise is measured by instruments that record sound pressure levels in decibels, according to the equation

$$\text{SPL (dB)} = 20 \log_{10} \frac{p}{p_{\text{ref}}}$$

where p = rms sound pressure
$p_{\text{ref}} = 2 \times 10^{-5}$

where sound pressure is in Pascals. The sound pressure level can be read for each octave band or added automatically to give an overall decibel reading. An overall dB(A) value can also be obtained where the instrument introduces a correction for the less worrying low-frequency noise, using the standard A-weighting procedure. Most noise levels are quoted in dB(A) units.

The sound power level is likewise expressed in dB, being given by

$$\text{PWL (dB)} = 10 \log_{10} \frac{W}{W_{\text{ref}}}$$

where W = sound power
$W_{\text{ref}} = 10^{-12}$

where sound power is in watts. The performance of most attenuator systems is expressed in sound power terms.

Under ideal conditions (e.g., in an anechoic chamber), sound power level can be established from an average SPL reading, namely,

$$\text{PWL (dB)} = \overline{\text{SPL}} + 20 \log_{10} x + 10.99 - \Delta$$

where x is the distance separating the instrument from the source and Δ is a correction for atmospheric pressure and temperature when these differ from the standard values. The distance should be sufficient to ensure a spherical wave front in an ideal free field. In most practical laboratory test arrangements, however, the room volume and reverberation time must be measured and corrections inserted into the equation. The reader should consult [16.26] or acoustical texts for detailed information.

The results from standardized laboratory test procedures are of limited value when application to industrial installations is attempted. Disparities in inlet flow quality and ducting assemblies can modify noise production and propagation phenomena. In addition, self-generated noise due to the fan loading method, usually the throttling type, and the use of auxiliary fans for overcoming the resistance of flow measuring and acoustic devices for small low-pressure axial equipment, will create additional problems. Code test noise ratings should therefore be interpreted in an informed manner. When the self-generated noise of components extraneous to the delivered product is 10 dB (or more) less than the total measured noise level, their contributions can be ignored.

The in-duct measurement of fan noise is the preferred method [16.29]. The microphone must be well faired and possess a low self-noise level.

The study of discrete tones requires a frequency analyzer that filters out all sound except for any selected narrow frequency band.

Anechoic or reverberant chambers are used in fan noise research. The former effectively cancel all reflected sound and facilitate the study of noise sources. The provision of practical anechoic wedges and an air source create design problems; these have been successfully resolved in a large U.S. test chamber [16.30].

Reverberant chambers are constructed of thick concrete walls and ceiling, internally finished to possess hard smooth surfaces. The walls are of unequal length and at different angles to one another. Providing the air source without affecting the reverberant nature of the chamber is the main design problem. Since no noise energy can escape, the noise absorption must equal the source supply. A reverberation chamber provides a uniform isotropic sound field, some distance from the source, by multiple random reflections from walls or movable surfaces. The relation between SPL and PWL units can be readily established from [16.31, 16.32, 16.33]; these also specify standard test procedures.

REFERENCES

16.1 R. A. Wallis, Dust erosion of fan materials, *Mech. Chem. Eng. Trans. I. E. Aust.*, **MC11** (1, 2), 33–39, 1975.

16.2 H. Sugano, M. Nakajimar, and N. Yamaguchi, Factors affecting blade ash-erosion in axial induced draft-fans, *ASME 80-JPGC/Pwr-7*, 1981.

16.3 R. A. Wallis, Aerodynamics and dust erosion of large axial flow fans, *Mech. Eng. Trans. I. E. Aust.*, **ME2** No. (1), 1–6, 1977.

16.4 G. Grant and W. Tabakoff, Erosion prediction in turbomachinery resulting from environmental solid particles, *J. Aircraft*, **12**(5) 471–478, 1975.

16.5 P. W. Alexieff, Research and development on axial fan blade erosion. *Combustion*, pp. 12–16. July 1979.

16.6 G. Peters, Durch Fliehkraft hervorgerufene Torsionsspannung in Axialschaufeln von Turbomaschinen, Konstruktion, 5. **12**, 419–420, 1953.

16.7 J. H. Horlock, *Axial Flow Turbines*, Butterworths, London, 1966.

16.8 R. Jorgensen (Ed.), *Fan Engineering*, Chap. 11, Buffalo Forge Co., Buffalo, N.Y., 1961.

16.9 W. C. Osborne, *Fans*, Pergamon, London, 1966.

16.10 P. Baur, Natural frequencies of unshrouded cantilever blades, *The Engineer*, pp. 701–706. Oct. 1964.

16.11 D. S. Whitehead, Aerodynamic aspects of blade vibration, *Proc. Inst. Mech. Eng.*, **180** (Part 3, I), 1965–66.

16.12 E. K. Armstrong and R. E. Stevenson, Some practical aspects of compressor blade vibration, *J. Roy. Aero. Soc.* **64**(591), March 1960.

16.13 W. Rizk and D. G. Seymour, Investigations into the failure of gas circulators and circuit components at Hinkley Point nuclear power station, *Proc. Inst. Mech. Eng.* **179**(21), Part I, 627–673, 1964–65.

16.14 B. D. Mugridge, Noise characteristics of axial and centrifugal fans as used in industry–a review, *Shock and Vibration Digest*, **II**(1), Jan. 1979.

16.15 J. M. Tyler and T. G. Sofrin, Axial flow compressor noise studies, *Soc. Automot. Eng. Trans*, **70**, 309–332, 1962.

16.16 R. Parker, Pressure fluctuations due to interaction between blade rows in axial flow compressors, *Proc. Inst. Mech. Eng.* **183**(7), Part 1, 153–163, 1968–69.

16.17 P. M. Morse and K. U. Ingard, *Theoretical Acoustics*, McGraw-Hill, New York, 1968.

16.18 P. Doak, Excitation, transmission and radiation of sound from source distributions in hard walled ducts of finite length (1) The effects of duct cross-section geometry and source distribution on space-time patterns, *J. Sound and Vibration*, **31**(1), 1–72, 1973.

16.19 R. Parker, Discrete frequency noise generation due to fluid flow over blades, supporting spokes and similar bodies, *ASME 69-WA/GT-13*, 1–13, 1969.

16.20 R. Parker and D. C. Pryce, Wake excited resonances in an annular cascade: fan experimental investigation, *J. Sound and Vibration*, **37**(2), 247–261, 1974.

16.21 M. C. Welsh, R. Parker, and S. A. T. Stoneman, The effect of induced sound on the flow around a rectangular body in a wind tunnel, Seventh Australasian Conf. Hydraulics and Fluid Mechanics, 275–278, 1980.

16.22 International Standards Organization Acoustics: Assessment of noise with respect to community response, R 1996–1971.

16.23 A. W. Roberts and R. E. Luxton, Acoustic isolation of fans in ducts, *Proc. Inst. Mech. Eng.*, **181**, Part 1, 948–965, 1966–67.

16.24 A. H. Nayfeh, D. P. Telionis, and S. G. Lekoudis, Transmission of sound through annular ducts of varying cross-section, *AIAA J.* **13**(1), 60–65, Jan. 1975.

16.25 L. L. Beranek, *Noise and Vibration Control*, McGraw-Hill, New York, 1971.

16.26 C. M. Harris, *Handbook of Noise Control*, McGraw-Hill, New York, 1957.

16.27 M. A. Swinbanks, The active control of sound propagated in ducts, *J. Sound and Vibrations*, **27**, 411–436, 1973.

16.28 R. F. La Fontaine and I. C. Shepherd. An experimental study of a broadband active attenuator for cancellation of random noise in ducts, *J. Sound and Vibration* 91(4) Dec. 1983 (provisional).

16.29 ASHRAE, Method of testing sound power radiated into ducts from air moving devices, Standard 68-1978.

REFERENCES

16.30 J. A. Wazyniak, L. M. Shaw, and J. D. Essary, Characteristics of an anechoic chamber for fan noise testing, ASME, Paper 77-GT-74, 1977.

16.31 International Standards Organization, Precision methods for broadband sources in reverberation rooms, ISO/3741.

16.32 International Standards Organization, Precision methods for discrete frequency and narrow band sources in reverberation rooms, ISO/3742.

16.33 International Standards Organization, Precision methods for anechoic and semi-anechoic rooms, ISO/3745.

CHAPTER 17
Fan Applications

The information presented in this chapter, based on actual experience, should aid designers faced with the problem of selecting in any instance the most suitable fan type.

17.1 DUTY-RELATED FAN TYPES

The axial flow fan, in one or other of its various forms, can be designed for efficient operation at any point in the entire fan pressure duty range.

Fan sizes advance according to an arithmetic progression. Under the Imperial unit system each fan is 25% larger than the previous one. This arrangement has limited the number of fan sizes, but in most cases a suitable fan to carry out a specific duty can be chosen. The International Standards Organization has recently put forward a metric series in which the above interval is halved, with a special option for halving this increment once again [17.1]. The listed sizes range from 100 to 2000 mm. In practical terms, it is suggested that the 25% interval be adopted, giving 250, 315, 400, 500, 630, 800, 1000, 1250, 1600, and 2000 as the major fan sizes. However, where space limitations or other special circumstances arise, the ISO recommendation of $12\frac{1}{2}$% should be given first consideration when selecting an intermediate fan diameter.

Fan applications can be divided roughly into three categories, in terms of flow coefficient and pressure rise capability. These are as follows:

1. $\Lambda < 0.20$. These low-pressure-rise fans usually possess a small boss ratio (≈ 0.4) and are frequently operated as exhaust fans, without stator vanes.
2. $\Lambda \approx 0.20$ to 0.40. These units normally have boss ratios between 0.5 and 0.7 and incorporate stator vanes; they are capable of substantial pressure rise, gaining from an optimum efficiency design approach.
3. $\Lambda > 0.40$. Fans in this category are usually of the multistage in-line type, since the single-stage total pressure coefficient capability is re-

17.2 ROTOR-ONLY UNIT

stricted. Compactness and a possible tip speed reduction are associated features.

A study of fans in the second category [17.2] indicated that the number of "off-the-shelf" fan types could be limited by selecting two boss ratios and designing these units to achieve the highest nondimensional pressure rise, and efficiency, in their respective classes (see Chapter 15). The larger boss ratio is 1.25 times the smaller and hence this boss diameter for a given fan equals the smaller boss diameter relative to the next fan size. Some economy of hub and center fairing components can be thereby achieved.

With the available variables of boss ratio, fan diameter, and rotational speed or adjustable rotor blade pitch, a wide range of duties can be covered efficiently by the preceding two design possibilities.

High efficiency is normally assumed for units possessing rotors fitted with airfoil section blading. However, unless the design and installation of the equipment are fully optimized and the fan is well matched to the actual operational duty, this belief could be ill founded. On the other hand, a well-designed unit utilizing single-thickness sheet metal blading can attain very good efficiency when operating in the designed duty region. Producing a good blade shape can be more expensive when three-dimensional curvature is indicated. The required blade stiffness may result in a metal thickness that requires the construction of a special die for pressing.

A simple computational procedure is presented in [17.3] that permits the use of two-dimensional curvature but results in curved leading and trailing edges. Hence the centroids of the various spanwise stations are radially displaced. Special attention has to be given to root fixing. The preceding procedure is confined to constant chord and blade camber, on cylindrical surfaces.

Any deviation from these restrictions to conical blade surfaces, with varying chord and camber, would require calculations on a trial-and-error basis. In any case, this technique for avoiding a three-dimensional curvature need has limitations for free vortex flow rotors with zero preswirl. The ensuing rate at which twist is changing with blade span (the major cause of sheet metal blade three-dimensionality) accelerates with reducing boss ratio. Hence to restrict leading- and trailing-edge curvatures the fan must have a relatively large boss ratio and/or be of an arbitrary vortex flow type.

A prerotator-rotor unit design, presented in Appendix A, illustrates an approximate procedure for developing blades and vanes out of sheet material, rolled to cylindrical and conical contours, respectively.

17.2 ROTOR-ONLY UNIT

Induced draft cooling tower fans are almost exclusively of this type. The discharge diffuser is normally short and of restricted area ratio, thus allowing increased wall angles. Since the installation is a low-pressure-rise one, swirl angles will be correspondingly small; this flow rotation can prove beneficial to diffuser performance when careful design and development techniques are employed.

The reversing fan used in many furnace and drying kiln applications possesses symmetrical blades with flat, wedge, or contoured cross section. Symmetrical, axially aligned stators upstream and downstream of the rotor will only be effective for limited degrees of swirl; this condition is associated with low-pressure-rise duties and small flow coefficients. However, the resistance of furnace and kiln equipment is usually appreciable, making the rotor-only configuration the enforced one. The loss in efficiency due to swirl is normally unimportant (provided the pressure duty is attained), as it represents an additional low-grade heating source. However, the swirl should not be allowed to persist in closed circuit equipment, thereby affecting fan inlet flow.

Optimized equipment in the preceding two categories usually involves a degree of arbitrary vortex flow in the blade design. This is related to design problems associated with small boss ratios. Suggestions for dealing with these are contained in Appendix A.

The fan industry is presently producing a relatively large proportion of its axial flow units as "tube-axial" equipment, as demanded by a capital-cost-conscious market. However, increasing power costs must in the near future force the user and fan industries to revise their thinking on stator use.

17.3 ROTOR-STRAIGHTENER UNIT

Correctly designed units of this type are potentially the most efficient. As a result, the preceding axial flow fan type is employed wherever the power requirements are high, for example, in relation to large mine ventilation and boiler plant aspiration equipment. Because of the inherent flow stability of the unit it is the common choice for wind tunnel applications.

This blading arrangement is particularly suited to multistage units, as the stators remove, for all practical purposes, an appreciable range of swirl angles resulting from changing duty conditions.

Associated with the use of stators is the need for a suitable inner termination. The normal fluted motors used in direct drive assemblies have no stator attachment capability, and hence a motor casing component must be provided. Motor cooling is then achieved in the manner illustrated in Fig. 17.1.

When dealing with large powers and solids transport, external air must be ducted in and out of a sealed motor compartment. Hollow stator vanes, or special streamlined fairings, are required as airflow passages. The cooling air flow is maintained by an auxiliary fan, the motor cooling fan, and/or static pressure differentials. The cooling air can be rejected to the duct airstream when the static pressure of the latter is less than the motor compartment

17.3 ROTOR-STRAIGHTENER UNIT

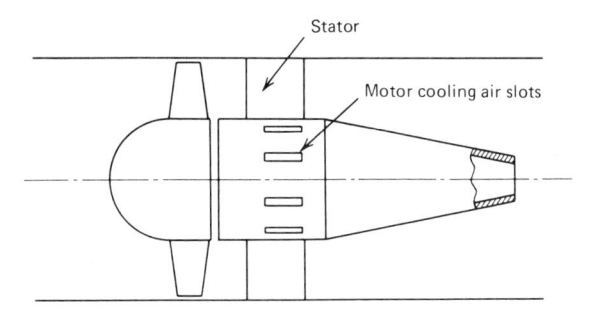

Figure 17.1 Air exchange system for motor cooling.

pressure. Chamber and/or stator partitioning may be required to separate the ingoing and outgoing airstreams. A large fan of this type is illustrated in Plate 6; the unit features an expanding centerbody diffuser.

Straighteners normally consist of sheet material of a single, cambered and twisted-plate type or of a double-skin airfoil variety. The former are simpler to produce but carry reduced performance and increased noise penalties;

Plate 6. Upcast mine ventilation fan featuring 4:1 diverging centerbody diffuser, Mount Isa. (By courtesy of Mount Isa Mines Ltd.).

these are particularly evident when the fan is operating well away from its design duty.

17.4 PREROTATOR-ROTOR UNIT

In applications where the axial dimension of the blading unit is limited, the preceding unit can provide an efficient, high-pressure-rise solution. For a given duty the solidities of both the stator and rotor blading are less than those for the previous fan type, and the rotor blade setting with respect to the plane of rotation is also less. Provided an adequate gap between the blading rows exists, to minimize noise, an extremely compact unit can be designed. Also, both the nonlinear twist change and total twist for a rotor blade are significantly less than those for the previous blading arrangement, using free vortex flow design procedures. As illustrated in Appendix A, this latter feature can be exploited to simplify the construction procedure to allow for sheet metal blading of two-dimensional curvature.

For a given rotational speed the flow acceleration induced by the prerotators increases the relative rotor blade velocity and gives the potential for significant gains in fan unit pressure rise, as opposed to the rotor-straightener unit. However, there is a moderating influence on this increment due to drag increases arising from the blade passage effects discussed in Section 9.3.4.

The use of untwisted, symmetrical inlet vanes of variable incidence as a means of fan duty control on rotor-only installations is an efficient method of achieving modest changes in the fan characteristic. However, outside a restricted range of vane incidence angles, departure from free vortex flow conditions, and eventual flow separation from the vanes, will lead to a lowering of overall efficiency.

17.5 PREROTATOR-ROTOR-STRAIGHTENER UNIT

The characteristic curve of a rotor-straightener unit can also be changed by the addition of the foregoing variable incidence inlet vanes, with similar provisos. At the other end of the range fixed, axially oriented, symmetrical vanes are occasionally used in the dual role of residual swirl removers and structural supports for electric motors, bearings, fairings, and so on. From the aerodynamic design viewpoint, these units belong to the previous two categories.

Units in which both prerotators and straighteners contribute to the design duty of the fan are uncommon. Their use is confined mainly to installations with special requirements. For example, the 6.1-m-diameter, 900-kW vertical mine ventilation fans at Mount Isa Mines Ltd. lend themselves to such a solution [17.4]. The blading arrangement is identical for both the upcast and downcast units (Figs. 17.2 and 17.3, Plates 2 and 3, pages 288–289).

17.5 PREROTATOR-ROTOR-STRAIGHTENER UNIT

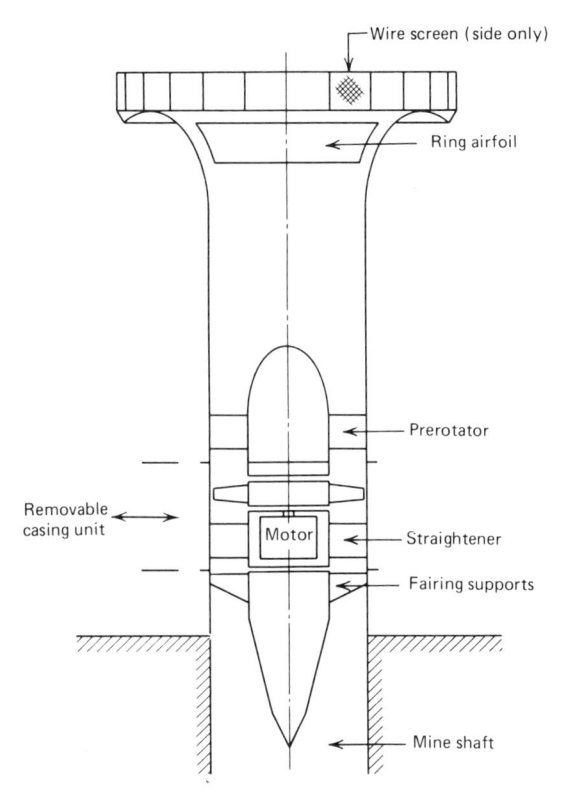

Figure 17.2 Vertical 6.1-m-diameter downcast mine ventilation fan, Mount Isa Mines Ltd.

Since these fans are sited in housing areas the need to minimize noise led to a limit on tip speed. This in turn results in a relatively large total swirl, $(\epsilon_p + \epsilon_s)$ at the blade root. In addition, structural support for the upstream casing and enclosed motor in the upcast fan case necessitates the use of a considerable number of faired struts. Therefore the use of prerotators in the combined structural/aerodynamic role represents an acceptable solution. In the downcast case, the straighteners fulfill a similar motor support purpose. The requirement for identical blading arrangements in relation to both fan types, with its attendant gains of rotor interchangeability and lower capital and maintenance costs, highlighted the advantages of the dual stator arrangement.

At the rotor design condition, 40% of the total swirl is induced by the prerotators. However, since the fans are of a variable rotor blade pitch type, the straighteners are designed for swirl amounts in excess of the 60% remainder, to cover the full range of likely mine duty requirements.

For the downcast fan unit, the prerotators provided a flow smoothing and orientation device under windy atmospheric conditions, thus minimizing rotor blade stresses.

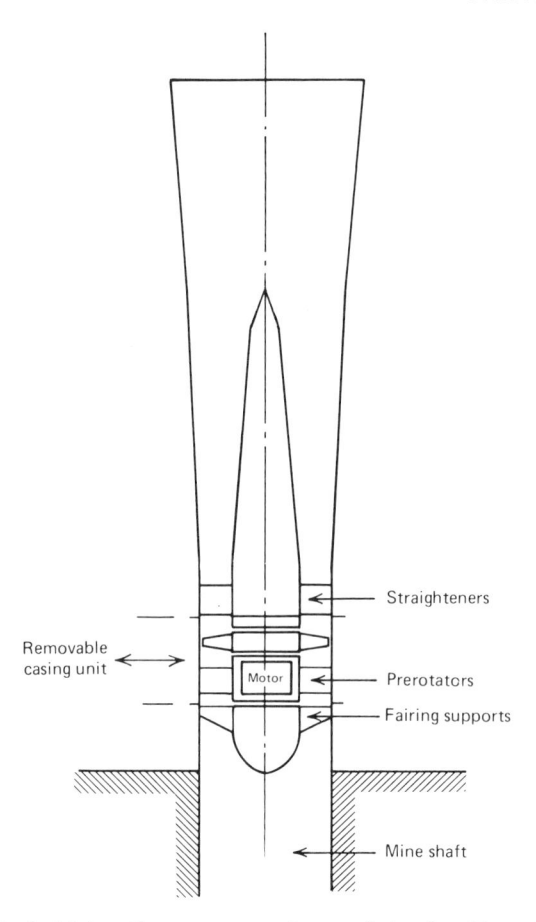

Figure 17.3 Vertical 6.1-m-diameter upcast mine ventilation fan, Mount Isa Mines Ltd.

The flow accelerating action of the prerotators can be used to obtain a greater sensitivity for volume flow rate measurement. The pressure differential between upstream wall tappings and those at a selected position on the vane surfaces represents a more adequate quantity than that obtained when the downstream ring is located in the annulus wall, for boss ratios of 0.5 or less. This eliminates the need for expensive contractions, or nozzles.

On a subsequent unit designed for changed duty requirements, the preswirl had to be increased to 50% of total design swirl, owing to speed and fan sizing constraints.

The preceding blading arrangements carry slight efficiency penalties, but overall these are relatively small.

Limits on rotor blade root loading actually fix the amount of total swirl, $(\epsilon_p + \epsilon_s)_b$, achievable in design. The increase over the design value for the rotor-straightener case is significant but not substantial (Chapter 9).

17.7 MULTISTAGE UNIT

The rather detailed description of these mine fans illustrates the type of solution that evolves when all design and operational requirements are permitted to interact.

17.6 CONTRAROTATING UNIT

This compact multistage unit is currently used in regard to high-pressure-rise applications. However, excessive noise from existing commercial versions has constituted a major acceptability problem, particularly when the rotors are adjacent to one another. In the latter case the noise has a peak of approximately twice the frequency of a single rotor; the resultant frequency and intensity are both critical from a hearing-loss viewpoint.

There are three different arrangements of fan unit, namely,

1. Motor-rotor/rotor-motor.
2. Rotor-motor/motor-rotor.
3. Rotor-motor/rotor-motor.

Since the motors of the first arrangement can be supported on axially aligned plate systems, aerodynamic interference is minimized, but rotor proximity maximizes the noise. Tie rod mounting of the motors for the latter two arrangements minimizes the aerodynamic and noise problems associated with the motor support systems; unfortunately, some fans in these categories have foot-mounted motors supported by a bench and plate system. In certain instances the terminal boxes are directly attached to the motors. Excessive noise and poor aerodynamic performance are the inevitable consequences of such crudity, which is condoned on the grounds of capital cost and ease of manufacture.

To function correctly over a wide range of duty requirements, variable speed is required on the downstream rotor if the axial outflow condition is to be maintained. This refinement is justified for wind tunnel fan equipment where swirl-free flow is essential.

17.7 MULTISTAGE UNIT

Despite the current widespread industrial use of contrarotating fans, the author believes that in most instances a more acceptable result can be achieved by staging rotor-straightener units of identical and appropriate design. The motor-cooling aspect is discussed in a previous subsection.

These units have been favorably received by Australian mining industry for use in the underground auxiliary ventilation role. As the duct length is increased, rotor units are progressively added to the total fan assembly. To minimize noise and hence eliminate the need for noise attenuation equipment, fan units operating off four-pole motors (1450 rpm) were developed. The pressure capability is obtained by the increased axial velocity that accompanies a boss ratio of 0.7. However, careful attention must be paid to all aerodynamic design features if excessive losses, potentially associated with high velocities, are to be avoided. Nose and tail fairings for the first and last stages, respectively, are essential. The duct flow passage should be parallel and free of nonaerodynamic features, with the stators providing the structural supports. When the fan assembly is installed as a blower unit, a curved or conical inlet bell is required with the large open-area safety screen attached at the bell inlet (Plates 4 and 5, pages 289–290). Compactness is a feature of this multistage fan equipment.

These fans are in the third category outlined in Section 17.1. The restriction on tip speed calls for an increase in the axial velocity component, hence the higher Λ value. Since the latter velocity is maintained through multiple stages, the losses are mainly skin friction ones. The subsequent loss over the tail fairing is then only a small proportion of the total gain in pressure for the multistage unit.

When designing a multistage unit for a given high-pressure duty, economy and capital cost considerations may dictate a smaller number of stages, with a tip speed increase and reduced Λ. As a consequence, noise will become an increasing problem for which solutions must be found. However, the high blade passing frequency of likely arrangements will simplify the design of satisfactory noise attenuation devices. The ensuing unit could prove more satisfactory from a noise viewpoint than existing less-compact centrifugal fans that presently cover this portion of the fan pressure range in the broad industrial scene. However, a crude approach to such a development would, in the author's opinion, have limited or little success and could lengthen the time for eventual acceptance by industrial clients.

With multistage units a "work done" factor is required for use in determining the design total pressure rise coefficient (Section 23.1). This constitutes an arbitrary correction for velocity nonuniformity influences.

17.8 CONCLUSION

Broadly speaking, the last two of the three duty categories listed previously favor a "free vortex" design approach, whereas fans in the first category may for practical reasons require an "arbitrary vortex" design approach. The fans in this latter category cover a wide range of cooling and unitized equipment applications. Unfortunately, the restricted upstream and downstream flow spaces, in certain cases, make application of the design equations and methods presented here of limited value. For example, a motor vehicle cooling fan will experience substantial radial velocity components and large-scale secondary tip flows, when unshrouded. However, the guidance provided here should lead to improved fan installation layouts and design development procedures (see Appendixes C and D).

REFERENCES

17.1 ISO Standard, *General Purpose Industrial Fans: Circular Flanges—Dimensions* (draft ISO/TC117/SC4).
17.2 R. A. Wallis, Design procedure for optimal axial flow fans, *Inst. Eng. Aust. Nat. Conf. Publ. No. 76/12*, Hobart, December 1976.
17.3 R. A. Wallis, Sheet metal blades for axial flow fans, *Engineering* (London) *171*. 681–683, 1951.
17.4 K. E. Mathews, et al. Development of the primary ventilation system at Mount Isa, Australasian *Inst. Min. Metall. Proc. No. 222*, pp. 1–61, 1967.

CHAPTER 18
Design Examples

The four free vortex design exercises have been chosen with a view to covering the full range of design possibilities outlined in the previous chapter. The first of these commences with dimensional duty requirements and proceeds through to the completion of the aerodynamic design. All designs are subject to mechanical design scrutiny with respect to strength and vibration characteristics. A spanwise redistribution of blade thickness or a rearrangement of blade numbers and chords while maintaining the calculated solidities are two measures that can be taken readily in correcting any difficulty.

In the latter three designs the calculations start with the requirements being expressed as nondimensional coefficients and conclude with the blading dimensions being expressed as a function of fan radius. Conversion of these data into an actual dimensional situation is indicated at the completion of the second design.

All designs are considered to have small tip clearances; hence no adjustments to fan pressure and efficiency are required.

A prerotator–rotor-type design is the subject of a design example in Appendix A.

18.1 LOW-PRESSURE FAN

The specification is for a forced-draft cooling tower fan with a capacity of 13 m³/s. The sum of the total pressure losses within the tower and at tower discharge is 180 Pa, at standard density. Direct coupling to a 50-Hz, four-pole motor is required.

For the purposes of the preliminary design study it is assumed that the fan incorporates a diffuser, of area ratio 2, ahead of an abrupt discharge into the tower. From Fig. 13.3 for a η_D of 0.86, the K_D is 0.1. The discharge loss will be a percentage of the discharge velocity pressure (see Fig. 4.30) and since this loss will be inseparable from others associated with tower flow, it is

18.1 LOW-PRESSURE FAN

included in tower resistance, which is defined as the tower inlet total pressure with reference to ambient static pressure at discharge.

It is proposed to use stators of the modified NPL type supporting a cylindrical casing enclosing the motor on the downstream side; the cylinder is to terminate at diffuser exit. The order of the fan and boss diameters is quickly established by rough calculations. The boss diameter will not exercise a dominant influence and hence is kept constant in the design preliminaries. A constant blading efficiency for rotor and stator of 0.85 will be assumed in these initial studies; the rotational speed is taken to be 24 rev/s.

Inspection of Tables 18.1 to 18.3 shows a preference for design B, as the efficiency is higher and the swirl coefficient at the blade root is within the swirl-removing capability of the modified NPL-type stators. The flow coefficient Λ is also a reasonable figure for this fan type. The smaller fan would require an increase in the diffuser area ratio, to match the efficiency of B, and to possess similar tower inlet flow conditions. Increasing the boss diameter of C would reduce ϵ_{s_b}, but the diffuser losses would increase with velocity.

The closest standard fan size to design B is 1.125 m (Section 17.1), and hence the new parameters are as follows:

Fan diameter = 1.125 m
Boss diameter = 0.4 m (x_b = 0.356)
Rev/s (N) = 24 (ΩR = 84.8 m/s)
$\bar{V}_a$ = 14.98 m/s ($\frac{1}{2}\rho \bar{V}_a^2$ = 134.6 Pa)
Λ = 0.177
ΔH_{bl} = 180 + (0.1 × $\frac{1}{2}\rho \bar{V}_a^2$) = 193.5 Pa
K_{th} = (193.5/134.6)/0.85 = 1.69

(The estimated efficiencies will be slightly less than those calculated for design B.)

Table 18.1 Determination of Main Design Parameters

Design	A	B	C
$2R$	1.0	1.1	1.2
$2r_b$	0.4	0.4	0.4
x_b	0.4	0.364	0.333
$\bar{V}_a = Q/\pi R^2(1 - x_b^2)$	19.7	15.77	12.93
$\frac{1}{2}\rho \bar{V}_a^2$	233	149	100
$\Delta H_{DL} = (0.1 \times \frac{1}{2}\rho \bar{V}_a^2)$	23	15	10
$\Delta H_{bl} = (180 + \Delta H_{DL})$	203	195	190
$K_{th} = (\Delta H_{bl}/\frac{1}{2}\rho \bar{V}_a^2)/\eta_{BL}$	1.03	1.54	2.24
$\Omega R = 2\pi r \times$ rev/s	75.4	82.9	90.5
$\Lambda = \bar{V}_a/\Omega R$	0.261	0.190	0.143

314   DESIGN EXAMPLES

Table 18.2   Check on Blade Root Loading

Design	A	B	C
λ_b	0.653	0.522	0.429
$\epsilon_{s_b} = \dfrac{K_{\text{th}} \lambda_b}{2}$	0.336	0.402	0.481
$(C_L \sigma)_b$ Fig. 8.9	≃0.4	≃0.4	≃0.4

Table 18.3   Efficiency Estimation

Design	A	B	C
$\lambda_{\text{MS}}(x_{\text{MS}} \simeq 0.7)$	0.373	0.271	0.204
$(\epsilon_s)_{\text{MS}} = \dfrac{K_{\text{th}} \lambda_{\text{MS}}}{2}$	0.192	0.209	0.229
$\gamma K_R/K_{\text{th}}$ (Fig. 10.1)	2.9	3.7	4.8
$\gamma = C_L/(C_{D_P} + C_{D_S}) \simeq 30$ (Fig. 9.4)	30	30	30
K_R/K_{th}	0.097	0.123	0.160
K_S/K_{th} (Fig. 12.3)	0.013	0.009	0.008
$\eta_{\text{BL}} = 1 - [(K_R + K_S)/K_{\text{th}}]$	0.890	0.868	0.832
K_D/K_{th}	0.097	0.065	0.045
$\eta_T = \eta_{\text{BL}} - K_D/K_{\text{th}}$	0.793	0.803	0.787

Table 18.4   Blading Design Procedure

x	0.356	0.45	0.6	0.8	1.0
λ	0.497	0.393	0.295	0.221	0.177
$\epsilon_s = 1.69\lambda/2$	0.420	0.332	0.249	0.187	0.150
$\tan \beta_m = (1 - \tfrac{1}{2}\epsilon_s \lambda)/\lambda$	1.802	2.379	3.265	4.431	5.575
β_m	61.0	67.2	73.0	77.3	79.8
$\cos \beta_m$	0.485	0.388	0.293	0.220	0.177
$C_L \sigma = 2\epsilon_s \cos \beta_m$	0.408	0.257	0.146	0.082	0.053
$\tan^{-1} \beta_1 = 1/\lambda$	63.6				80.0
$\tan^{-1} \beta_2 = (1 - \epsilon_s \lambda)/\lambda$	57.9				79.7
$\beta_1 - \beta_2$	5.7				0.3
C_L	0.85	0.74	0.65	0.62	0.70
σ	0.480	0.349	0.224	0.132	0.076
$nc/R = 2\pi\sigma x$	1.074	0.988	0.846	0.661	0.478
c ($n = 4$, $n/R = 7.111$)	0.151	0.139	0.119	0.093	0.067
α (12% Clark Y, Fig. 6.14)	3.4	2.3	1.4	1.0	1.9
$\xi = \beta'_m - \alpha$	57.6	64.9	71.6	76.3	77.9
$\phi = 90° - \xi$	32.4	25.1	18.4	13.7	12.1

18.1 LOW-PRESSURE FAN

A considered choice of root and tip values of lift coefficient was made in Table 18.4 and the respective chords were established. A linear chord distribution was assumed from which the intermediate values of C_L were calculated. The rather moderate values of C_L toward the tip occur as a result of practical considerations with respect to chord and planform. Stiffness has to be given special attention, since the fan may be subject to gusty atmospheric conditions. Since the required flow-turning angles are small, a flat undersurface airfoil of low camber has been selected. A Clark Y section of 12% thickness to chord ratio has been chosen. A stiffness check on the resulting blade must be carried out after making a decision on blade material and construction.

The flat undersurface is the reference chord line from which all blade angles are measured.

The stators will consist of symmetrical airfoil sections, axially aligned, and attached to a cylindrical casing surrounding the motor. When the chord remains constant along the span, design is concentrated at the vane root where $\epsilon_{s_b} = 0.42$.

The required $(\beta_1 - \beta_2)$ for the stator root is therefore approximately 23°. Selecting a solidity of unity entails a C_L of 0.82 (Fig. 11.3). However, when Fig. 6.29 is consulted it is clear that for low stagger angles and the preceding solidity, an interference factor of approximately 0.7 is present. Hence $C_{L_i} = 1.17$ and with an isolated airfoil lift curve slope of 0.1 per degree of incidence ($m = 5.7$) an air incidence angle of 11.7° is required. The geometric angle from Fig. 10.11 is given by

$$\tan^{-1} \beta_{m_s} = \frac{\epsilon_s}{2} = 0.21$$

where $\beta_{m_s} = \alpha = 11.9°$.

(The solidity of a cambered airfoil doing the same amount of air turning would, according to the design recommendations of Fig. 9.5, approximate 0.7.)

Hence in view of the preceding checks a 12% thick symmetrical airfoil with a root solidity of unity is adopted (see Section 11.4). Therefore,

$$nc = 1.258$$

when $n = 5$ and $c = 0.252$ m. This chord dimension will provide adequate axial support for the central casing within which the motor is mounted.

Assuming a unit efficiency of 0.80

$$W = 180 \times 13/0.80 = 2.925 \text{ kW}$$

18.2 CAMBERED PLATE FAN

The design parameters adopted for this exercise were suggested in Section 15.6 in relation to a rotor-straightener configuration. On completion of the nondimensional design development, in-line and exhaust fan applications are discussed in dimensional terms.

The fan unit parameters selected are as follows: $\Lambda = 0.28$, $K_{th} = 3.6$, and $x_b = 0.552$.

In Table 18.5 the solidity at the blade root is based on Fig. 9.12 and the interference factors of Fig. 6.29, since the corresponding data for cambered plates are unavailable. As the blade camber distribution has been arranged to provide smooth inlet flow at the blade leading edge, the present design is satisfactory, but stall will occur at a lesser pressure duty than that for faired sections. The choice of camber is made on the basis of Eq. (9.3) and Fig. 6.27, where C_{L_i} is considered the relevant lift quantity. Twelve blades provide a satisfactory aspect ratio ($\simeq 2$) that, together with the 7% blade root camber, ensure good blade stiffness. A three-dimensional blade surface curvature is required for this design.

The increased stagger angle adopted in Table 18.6 follows from the recommendation of Section 11.5. Three-dimensional surface curvature is present in the preceding vane design.

Adopting $x = 0.8$ as the mean station for efficiency calculations, then the estimate is as follows:

Rotor loss, $\quad \dfrac{\gamma K_R}{K_{th}} = \dfrac{\lambda}{\cos^2 \beta_m} = 2.613$

$$\dfrac{K_R}{K_{th}} = 0.118$$

where $C_L = 0.8$, $C_{D_P} = 0.020$ and $C_{D_S} = 0.025\, C_L^2$.

Stator loss, $\quad \dfrac{K_S}{K_{th}} = 0.014 \times 1.5 = 0.021 \quad$ (Fig. 12.3)

Blading efficiency, $\quad \eta_{BL} = 0.861$

Conversion of the preceding design into dimensional terms can be achieved by the following equations:

$$\Delta H = \tfrac{1}{2} \rho K_{th} \eta \Lambda^2 (\Omega R)^2, \quad \text{Pa} \tag{18.1}$$

$$Q = \pi R^2 (1 - x_b^2) \Lambda \Omega R, \quad \text{m}^3/\text{s} \tag{18.2}$$

18.2 CAMBERED PLATE FAN

Table 18.5 Rotor Blade Design Procedure

x	0.552	0.6	0.7	0.8	0.9	1.0
λ	0.507	0.467	0.400	0.350	0.311	0.280
$\epsilon_s = 3.6\lambda/2$	0.913	0.841	0.720	0.630	0.560	0.504
$\tan \beta_m = (1 - \tfrac{1}{2}\epsilon_s \lambda)/\lambda$	1.516	1.721	2.140	2.542	2.935	3.319
β_m	56.6	59.8	65.0	68.5	71.2	73.2
$\cos \beta_m$	0.551	0.502	0.423	0.366	0.323	0.288
$C_L \sigma = 2\epsilon_s \cos \beta_m$	1.006	0.845	0.610	0.461	0.361	0.291
$\beta_1 - \beta_2$ [Eqs. (8.28), (8.29)]	16.4	12.6	7.5	4.9	3.3	2.5
C_L selected	0.75	0.77	0.81	0.80	0.78	0.75
σ	1.340	1.097	0.753	0.576	0.463	0.388
$nc/R = 2\pi\sigma x$	4.648	4.136	3.312	2.895	2.618	2.438
C_L/C_{L_i} (Fig. 6.29)	0.69	0.84	1.0	1.0	1.0	1.0
C_{L_i}	1.08	0.92	0.81	0.80	0.78	0.75
b/c (%) selected	7.0	6.3	5.4	4.75	4.3	4
θ [Eq. (6.9)]	31.7	28.5	24.4	21.5	19.5	18.1
α (Fig. 6.24)	4.7	3.6	3.2	3.7	3.8	3.9
$\xi = \beta_m - \alpha$	51.9	56.2	61.8	64.8	67.4	69.3
$\phi = 90° - \xi$	38.1	33.8	28.2	25.2	22.6	20.7
$c/R(n = 12)$	0.388	0.345	0.276	0.241	0.218	0.203
$\cos \beta_1/\cos \beta_2$	0.66					0.87

For the preceding design these expressions reduce to functions of rotational speed, fan radius, and efficiency, the latter depending on the fan type and downstream duct components. Hence at standard density

$$\Delta H = 6.685 R^2 \eta N^2, \quad \text{Pa} \quad \text{and} \quad Q = 3.843 R^3 N, \quad \text{m}^3/\text{s}$$

where N = rev/s

Table 18.6 Straightener Vane Design Procedure

x	0.552	0.6	0.7	0.8	0.9	1.0
ϵ_s	0.913	0.841	0.720	0.630	0.560	0.504
s/c (Fig. 11.5)	0.52	0.62	0.84	1.04	1.20	1.32
θ (Fig. 11.6)	52.0	50.2	47.0	43.6	40.5	38.0
b/c (%) $= \tfrac{1}{2}\tan \theta/4$	11.5	11.1	10.4	9.6	8.9	8.4
ξ (Fig. 11.6)	16.0	15.0	12.2	10.5	9.0	8.0
C_L (Fig. 11.5)	0.87	0.96	1.14	1.25	1.28	1.28
$nc/R = 2\pi x/(s/c)$	6.670	6.081	5.236	4.833	4.712	4.760
$c/R(n = 11)$	0.606	0.553	0.476	0.439	0.428	0.433
$\xi_{\text{mod}} = 1.5 + \xi$	17.5	16.5	13.7	12.0	10.5	9.5

An in-line fan with a tail fairing enclosed within a cylindrical duct has a downstream area ratio of 1.44; for a η_D of 0.90 (Fig. 4.30), $K_D = 0.05$ (Fig. 13.3). Hence

$$\eta_T = 0.861 - 0.05/3.6 = 0.847 \text{ and FTP} = 5.66R^2N^2, \text{ Pa.}$$

Assuming an area ratio of 2.5 and a η_D of 0.85, then the ensuing exhaust fan will possess a K_{DL} of 0.29 (Fig. 13.4), giving

$$\eta_{IT} = 0.861 - \frac{0.29}{3.6} = 0.78 \quad \text{and} \quad \text{FITP} = 5.21R^2N^2, \text{ Pa}$$

The preceding design volume flow rate and fan pressure expressions are illustrated in Figs. 18.1 and 18.2 for rotational speeds of 16 and 24 rev/s and

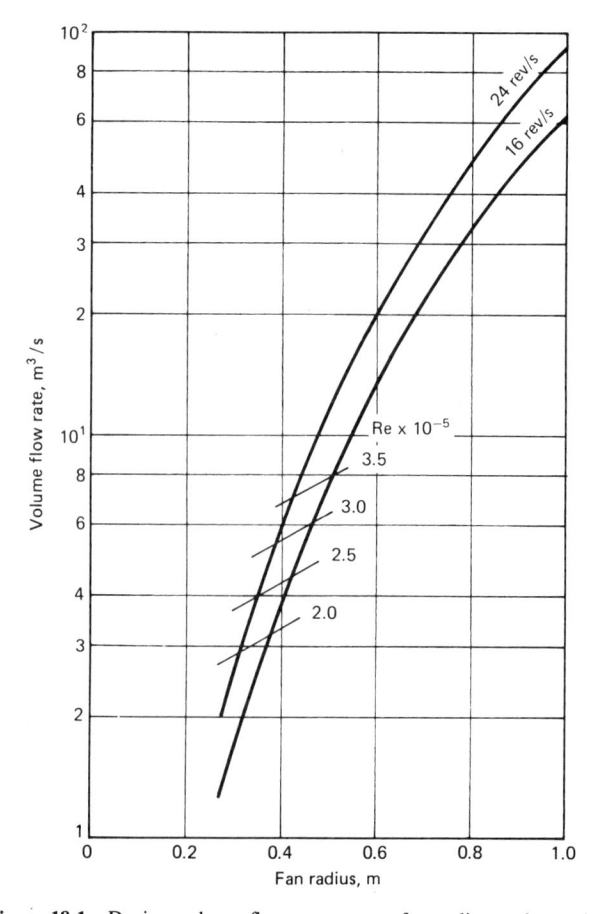

Figure 18.1 Design volume flow rate versus fan radius and speed.

18.3 ROTOR-STRAIGHTENER UNIT WITH LARGE BOSS RATIO

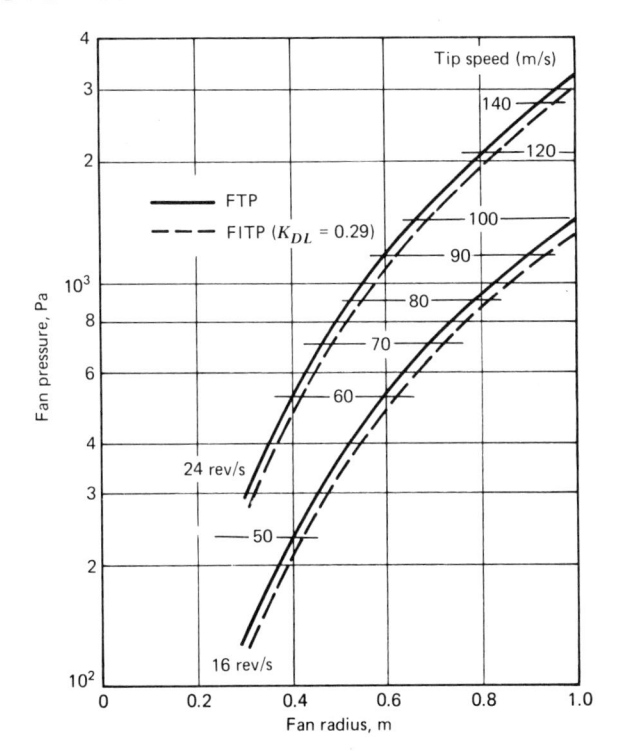

Figure 18.2 Design fan pressures versus fan radius and speed.

radii less than 1 m. For any selected value of R the blade and vane chords follow from the tabulated ratios calculated previously.

The crossplot of blade tip Reynolds number, based on tip speed, given in Fig. 18.1 provides some guidance with respect to the region below which performance can be expected to diminish (see Section 10.2.2). Although it is of limited relevance to this design, this exercise should not be omitted for faired airfoils.

Tip speed is also an extremely significant parameter and hence it is included in Fig. 18.2.

18.3 ROTOR-STRAIGHTENER UNIT WITH LARGE BOSS RATIO

This design will illustrate higher-pressure-rise units, of restricted tip speed. It represents the larger boss ratio unit suggested in the optimal studies of Section 15.6.

The selected design parameters are $\Lambda = 0.32$, $K_{th} = 4.2$, and $x_b = 0.69$. F-series airfoils with 2% leading-edge droop have been selected in the rotor design.

Root values of solidity were selected from Fig. 9.12 in the design procedure of Table 18.7.

The stator blading is based on the C4 thickness form (see Appendix D for section coordinates). The design data are given in Table 18.8.

The blading efficiency estimate based on the $x = 0.85$ station is as follows:

Rotor loss, $\quad \dfrac{\gamma K_R}{K_{th}} = \dfrac{\lambda}{\cos^2 \beta_m} = 2.30$

$$\dfrac{K_R}{K_{th}} = 0.068$$

where $C_L = 0.86$, $C_{D_P} = 0.012$, $C_{D_S} = 0.018 \, C_L^2$

Straightener loss, $\quad K_S/K_{th} = 0.015$ (Fig. 12.3)

Blading efficiency, $\quad \eta_{BL} = 0.917$

Conversion to dimensional terms, and for specific applications, can be

Table 18.7 Rotor Blade Design Procedure

x	0.69	0.73	0.80	0.85	0.9	1.0
λ	0.464	0.438	0.400	0.377	0.356	0.320
$\epsilon_s = 4.2 \, \lambda/2$	0.974	0.920	0.840	0.792	0.748	0.672
$\tan \beta_m = [1 - \tfrac{1}{2}\epsilon_s \lambda]/\lambda$	1.668	1.823	2.080	2.257	2.435	2.789
β_m	59.1	61.3	64.3	66.1	67.7	70.3
$\cos \beta_m$	0.514	0.481	0.433	0.405	0.380	0.338
$C_L \sigma = 2\epsilon_s \cos \beta_m$	1.002	0.885	0.728	0.642	0.568	0.454
$\beta_1 - \beta_2$ [Eqs. (8.28), (8.29)]	15.4	12.6	9.3	7.6	6.3	4.5
C_L selected	0.74	0.79	0.85	0.86	0.85	0.82
σ	1.354	1.120	0.857	0.747	0.668	0.554
$nc/R = 2\pi\sigma x$	5.870	5.137	4.308	3.989	3.778	3.481
C_L/C_{L_i} (Fig. 6.29)	0.73	0.84	1.0	1.0	1.0	1.0
C_{L_i}	1.01	0.94	0.85	0.86	0.85	0.82
θ [selection based on Eq. (9.2)]	26	23.5	20	18.5	17	15
α_{od} (F-series)	4.2	4.0	4.0	4.5	4.7	5.0
α (2% droop) Eq. (6.13)	3.1	2.9	2.9	3.4	3.6	3.9
$\xi = \beta_m - \alpha$	56.0	58.4	61.4	62.7	64.1	66.4
$\phi = 90° - \xi$	34.0	31.6	28.6	27.3	25.9	23.6
c/R ($n = 20$)	0.294	0.257	0.215	0.200	0.189	0.174
c/R ($n = 24$)	0.245	0.214	0.180	0.166	0.157	0.145
$\cos \beta_1/\cos \beta_2$	0.65					0.81

18.4 ROTOR-STRAIGHTENER UNIT WITH LARGE Λ

Table 18.8 Straightener Vane Design Procedure

x	0.69	0.73	0.8	0.9	1.0
ϵ_s	0.974	0.920	0.840	0.748	0.672
s/c (Fig. 11.5)	0.46	0.52	0.63	0.78	0.94
θ (Fig. 11.6)	53.8	52.3	50.3	47.7	45.1
ξ (Fig. 11.6)	17.4	16.4	14.7	12.9	11.2
C_L (Fig. 11.5)	0.8	0.86	0.96	1.09	1.2
$nc/R = 2\pi x/(s/c)$	9.425	8.821	7.979	7.250	6.684
c/R ($n = 23$)	0.410	0.384	0.347	0.315	0.291

made in a manner similar to that used in the previous design example. Calculations based on Eq. (18.1) for a given tip speed show this fan to have a 62% greater *design* pressure rise capability than the prior unit.

The thickness/chord ratios can be adjusted to suit the strength, stiffness, and inertial requirements of the proposed installation. Consideration should be given to the matters discussed in Section 9.6. Within the limits 7 to 13%, no adjustment to the preceding aerodynamic design is required.

18.4 ROTOR-STRAIGHTENER UNIT WITH LARGE Λ

This variety of fan is best suited to multistage units of the in-line type, in which category blower fans are included. The discharge loss is consequently avoided and the diffusion loss over the tail fairing is a once-only event. The maintenance of smooth flow throughout the unit is very important.

The parameters chosen for the following design example, namely, K_{th} = 2.1, $\Lambda = 0.6$, and $x_b = 0.7$, will from Eq. (18.1) for given blading efficiency and tip speed give a 75% total pressure rise increase over the previous design for the first stage blading. The blading efficiency according to Fig. 15.4 (K_{DL} = 0) increases slightly and hence the comparison is valid. With multiple stages the "work done" factor must be applied (Fig. 23.1), but the influence of K_D will diminish. For this fan $K_D \simeq 0.1$ (Fig. 13.3), which, from Fig. 15.4, reduces η' by 4.5% for a single-stage unit.

Both the rotor and stator blading for this unit will consist of C4 airfoils (i.e., possess zero nose droop).

The choice of blade numbers from Table 18.9 will depend mainly on practical considerations such as root fixing details and boss depth. However, aerodynamic matters should not be neglected. For instance, a decreasing number of blades will reduce the aspect ratio and eventually lead to an efficiency decrement. On the other hand, a greater number could result in the blades possessing a higher drag because of below Re_{crit} operation and physical problems. The decision must be weighted in terms of all these factors.

Table 18.9 Rotor Blade Design Procedure

x	0.70	0.75	0.80	0.85	0.90	1.00
λ	0.857	0.800	0.750	0.706	0.667	0.600
$\epsilon_s = 2.1\lambda/2$	0.900	0.840	0.788	0.741	0.700	0.630
$\tan \beta_m = (1 - \tfrac{1}{2}\epsilon_s\lambda)/\lambda$	0.717	0.830	0.939	1.046	1.149	1.352
β_m	35.6	39.7	43.2	46.3	49.0	53.5
$\cos \beta_m$	0.813	0.769	0.729	0.691	0.656	0.595
$C_L\sigma = 2\epsilon_s \cos \beta_m$	1.463	1.293	1.149	1.024	0.919	0.749
$\beta_1 = 1/\lambda$	49.4	51.3	53.1	54.8	56.3	59.0
$\beta_2 = (1 - \epsilon_s\lambda)/\lambda$	14.9	22.3	28.6	34.0	38.6	46.0
$\beta_1 - \beta_2$	34.5	29.0	24.5	20.8	17.7	13.0
σ (Fig. 9.12)	1.90	1.63	1.39	1.20	1.04	0.82
C_L	0.77	0.79	0.83	0.85	0.88	0.91
$nc/R = 2\pi\sigma x$	8.357	7.681	6.987	6.409	5.881	5.152
s/c	0.526	0.614	0.719	0.833	0.962	1.220
$\theta' = [\beta_1 - \beta_2]/[1 - 0.26 \sqrt{s/c}]$	42.5	36.4	31.4	27.3	23.8	18.2
$\xi' = \beta_1 - (\theta'/2)$	28.2	33.1	37.4	41.2	44.4	49.9
m (Fig. 9.3)	0.26	0.275	0.29	0.30	0.315	0.34
i (Fig. 9.1)	2	2	2	2	2	2
$\theta = [\beta_1 - \beta_2 - i]/[1 - m\sqrt{s/c}]$	40.1	34.4	29.8	25.9	22.7	17.6
$\xi = \beta_1 - i - (\theta/2)$	27.4	32.1	36.2	39.9	43.0	48.1
$c/R (n = 20)$	0.418	0.384	0.349	0.321	0.294	0.258
$c/R (n = 24)$	0.348	0.320	0.291	0.267	0.245	0.215
$\cos \beta_1/\cos \beta_2$	0.67					0.74

Since the stators contribute substantially to the useful total pressure rise by swirl removal, it is quite important for these to be of an efficient type, hence the adoption of the C4 cambered airfoil in Table 18.10.

Blading efficiency is calculated from $x = 0.85$ data, namely,

Rotor loss, $\quad \dfrac{\gamma K_R}{K_{th}} = \dfrac{\lambda}{\cos^2 \beta_m} = 1.48$

$$\frac{K_R}{K_{th}} = 0.051$$

where $C_L = 0.85$, $C_{D_P} = 0.016$, $C_{D_S} = 0.018 \, C_L^2$.

Straightener loss, $\quad \dfrac{K_S}{K_{th}} = 0.027 \quad$ (Fig. 12.3)

Blading efficiency, $\quad \eta_{BL} = 0.922 \quad$ for $\text{Re} \approx 3 \times 10^5$

The overall efficiency for any given multistage arrangement of the preceding blading stage can be readily calculated by introducing the appropriate

18.4 ROTOR-STRAIGHTENER UNIT WITH LARGE Λ

Table 18.10 Straightener Vane Design Procedure

x	0.7	0.75	0.80	0.85	0.9	1.0
ϵ_s	0.900	0.840	0.788	0.741	0.700	0.630
s/c (Fig. 11.5)	0.54	0.62	0.71	0.80	0.88	1.04
θ (Fig. 11.6)	51.7	50.2	48.8	47.4	46.0	43.7
ξ (Fig. 11.6)	16.0	14.8	13.7	12.8	12.0	10.5
C_L (Fig. 11.5)	0.88	0.96	1.04	1.10	1.16	1.24
$nc/R = 2\pi x/(s/c)$	8.145	7.601	7.080	6.676	6.426	6.042
$c/R (n = 23)$	0.354	0.331	0.308	0.290	0.279	0.263

downstream loss coefficient. There will be some multistage efficiency penalty, but this will be less severe than the "work done" factor effect on available pressure rise. The additional drag will be associated with an increase in the secondary flows; with sufficient stages the flow pattern and losses are stabilized.

Motor cooling requires special attention. A special double-ended motor with an enlarged carcase to which the stators are attached is featured in Plate 5. For a standard motor, heat transferred by convection to the casing constituting the inner wall of the fan unit will be dissipated by convection from the casing and stator surfaces. The motor cooling fan will circulate the air within the casing but the resultant convective heat transfer to the casing may reach equilibrium at a temperature in excess of the allowable motor maximum. An air exchange across the casing wall must then be provided through slots. This should be accomplished with the minimum flow disturbance, to be verified experimentally.

CHAPTER 19
Rotor and Stator Analysis

An analytical method of establishing off-design duty performance is an important part of practical fan technology. It permits the construction of characteristic curves for a wide range of blade settings and enables fan stall approximations to be obtained. The procedure is invaluable in relation to the development of adjustable- and variable-pitch rotors and "off-the-shelf" equipment. In addition, from measurements of blade profile, solidity, and stagger angle, the likely performance of a rotor can be estimated in the absence of aerodynamic design data.

The method proposed is centered on mean flow and pressure rise calculations. Provided the design meets free vortex flow conditions, the midspan station is a possible choice. However, when the axial velocity component varies along the span, some other location may be preferable.

At the mean station it is only necessary to calculate ϵ as a function of λ, for a given blade geometry, in order to compute all other variables involved in characteristic curve construction; the estimation of stall point is an exception. The latter requires load limit studies at the blade extremities, normally the blade root.

19.1 MEAN FLOW AND PRESSURE CONSIDERATIONS

The total flow through a fan is given by

$$\text{Volume flow per second} = 2\pi \int_{r_b}^{R} V_a r \, dr \qquad (19.1)$$

The mean axial velocity V_a is therefore

$$\bar{V}_a = \frac{2}{1 - x_b^2} \int_{x_b}^{1} V_a x \, dx \qquad (19.2)$$

19.1 MEAN FLOW AND PRESSURE CONSIDERATIONS

assuming a linear velocity distribution given by

$$V_a = C_1 + nx \tag{19.3}$$

where C_1 and n are constants, and integrating Eq. (19.2),

$$\bar{V}_a = C_1 + \tfrac{2}{3}n \frac{1 - x_b^3}{1 - x_b^2} \tag{19.4}$$

Defining the radial position at which $V_a = \bar{V}_a$ by $x_{\bar{V}_a}$

$$x_{\bar{V}_a} = \frac{2}{3} \frac{1 + x_b + x_b^2}{1 + x_b} \tag{19.5}$$

Alternatively, when the velocity varies according to the relationship

$$V_a = C_2 + m\sqrt{x} \tag{19.6}$$

the position of mean velocity is given by

$$x_{\bar{V}_a} = \frac{16}{25}\left(\frac{1 - x_b^{5/2}}{1 - x_b^2}\right)^2 \tag{19.7}$$

In both cases the mean position is independent of the constants, and of negative and positive velocity gradients.

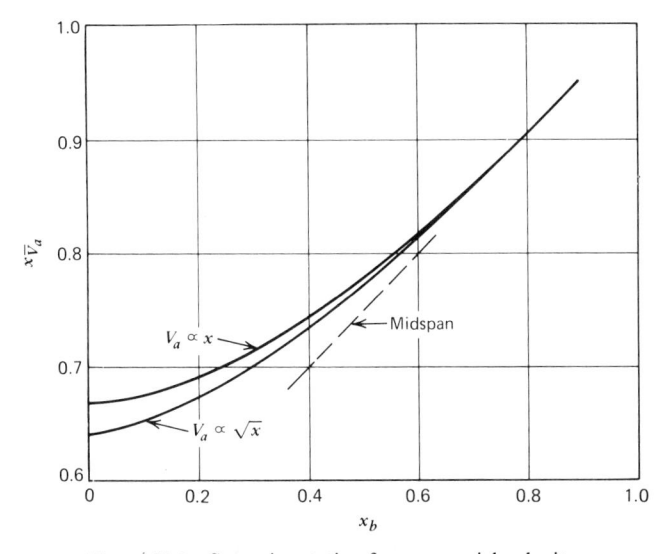

Figure 19.1 Spanwise station for mean axial velocity.

The graphical presentations of Eqs. (19.5) and (19.7) are given in Fig. 19.1. The small distance between the mean velocity locations for the usual boss ratios suggests a relative insensitivity to the actual velocity distribution, which for off-design conditions is normally unknown. The relationship of Eq. (19.6) is approximated in the case of a free vortex fan operating away from design and hence is recommended for use. However, for boss ratios of 0.5 and above, the calculated performance characteristics will not be altered significantly by a midspan station choice.

Swirl distributions that approximate free vortex ones over an appreciable proportion of the blade span may result in two spanwise locations at which the local velocity equals the mean, the velocity distribution being peaked. This matter is discussed further in Appendix A.

The mean total pressure rise is assumed to occur in the same location as mean velocity.

19.2 BLADE ELEMENT ANALYSIS—ROTORS

The general configuration of a rotor with prerotation and downstream residual swirl (Chapter 8) is adopted for this analysis method, ignoring the effect of radial flow components.

The rotor vector diagram of Fig. 8.3 is modified in the manner shown in Fig. 19.2. Alterations to fan duty will produce changes in both the preswirl and axial velocity components; however, ϵ_p, which defines a deflection angle, remains unchanged.

For the zero work case the velocity vector diagram is represented by EBG where

$$\text{Mean resultant velocity} = V_{m_N} = \text{EB}$$
$$\text{Axial velocity} = V_{a_N} = \text{BG}$$
$$\text{Tangential velocity} = \Omega r + (V_{\theta_p})_N = \text{EG}$$

With increasing load the point D will move away from B and hence

$$V_m = \text{ED}$$
$$V_a = \text{CG} = \text{BG} - \text{BC}$$
$$\Omega r + \tfrac{1}{2}(V_{\theta_p} - V_{\theta_s}) = \text{EG} - \text{CD}$$

Expressions for λ and ϵ_s are now developed from

$$V_a = [\Omega r + (V_{\theta_p})_N] \cot \beta_N - \left[\frac{(V_{\theta_p})_N - \tfrac{1}{2}(V_{\theta_p} - V_{\theta_s})}{\tan \psi} \right]$$

Dividing by Ωr

$$\lambda = (1 + \epsilon_p \lambda_N) \cot \beta_N - \frac{\epsilon_p \lambda_N - \tfrac{1}{2}\lambda(\epsilon_p - \epsilon_s)}{\tan \psi} \tag{19.8}$$

19.2 BLADE ELEMENT ANALYSIS—ROTORS

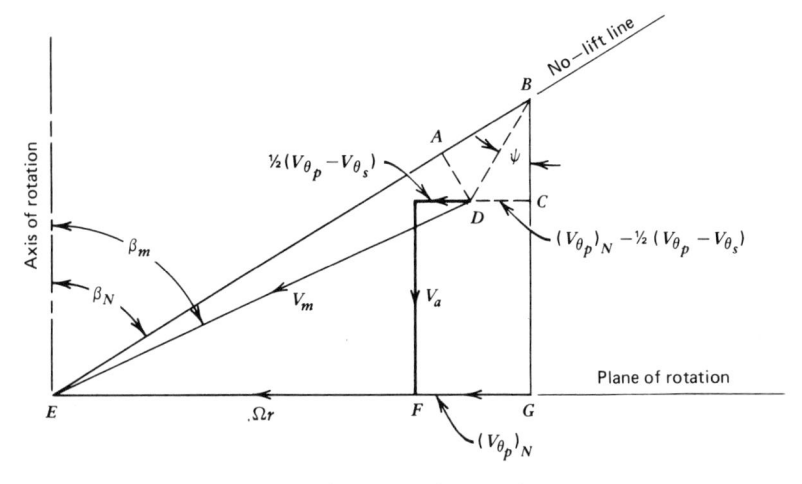

Figure 19.2 Velocity vectors for rotor blade element.

where

$$\frac{(V_{\theta_p})_N}{\Omega r} \frac{\overline{V}_{a_N}}{\overline{V}_{a_N}} = \epsilon_p \lambda_N$$

and

$$\lambda_N = (1 + \epsilon_p \lambda_N) \cot \beta_N = \frac{1}{\tan \beta_N - \epsilon_p} \tag{19.9}$$

Equation (19.8) can be rewritten as

$$\epsilon_s = 2 \tan \psi \left[\frac{\lambda_N}{\lambda} - 1\right] - \epsilon_p \left[\frac{2\lambda_N}{\lambda} - 1\right] \tag{19.10}$$

in which ψ is the only unknown being a function of blade element load capability.

Assuming that the lift coefficient can be expressed as a simple function of incidence angle, then

$$C_L = m \sin (\beta_m - \beta_N) \tag{19.11}$$

where m is a constant and $\beta_m - \beta_N$ is the incidence relative to the no-lift chord line. For airfoils of low to moderate camber, of faired shape, and not subject to multiplane interference (see Section 19.4), $m = 5.7$. The values that apply to cambered plate blade sections of varying camber are presented in Fig. 6.26.

An expression for $\sin(\beta_m - \beta_N)$ is obtained as follows:

$$\begin{aligned} AD &= V_m \sin(\beta_m - \beta_N) \\ &= BD \sin(\beta_N - \psi) \\ &= \frac{DC \sin(\beta_N - \psi)}{\sin \psi} \end{aligned}$$

Substituting the velocity vector of Fig. 19.2 for DC

$$V_m \sin(\beta_m - \beta_N) = [(V_{\theta_p})_N - \tfrac{1}{2}(V_{\theta_p} - V_{\theta_s})] \frac{\sin(\beta_N - \psi)}{\sin \psi} \quad (19.12)$$

Momentum considerations are satisfied when Eqs. (8.37) and (19.11) are equated. Rearrangement gives

$$\sin(\beta_m - \beta_N) = \frac{2(\epsilon_p + \epsilon_s) \cos \beta_m}{m\sigma} \quad (19.13)$$

Substitution in Eq. (19.12) produces the expression

$$\tan \psi = \frac{\sin \beta_N}{\dfrac{4(\epsilon_p + \epsilon_s)}{m\sigma[\epsilon_p(2\lambda_N/\lambda - 1) + \epsilon_s]} + \cos \beta_N} \quad (19.14)$$

where $(V_{\theta_p})_N / V_a = \epsilon_p \lambda_N / \lambda$.

Introducing Eq. (19.14) into Eq. (19.10) gives

$$\begin{aligned} \epsilon_s &= \frac{2 \sin \beta_N [\lambda_N/\lambda - 1]}{\dfrac{4(\epsilon_p + \epsilon_s)}{m\sigma[\epsilon_p(2\lambda_N/\lambda - 1) + \epsilon_s]} + \cos \beta_N} - \epsilon_p\left(\frac{2\lambda_N}{\lambda} - 1\right) \\ &= \frac{2 \sin \beta_N (\lambda_N/\lambda - 1) - 4\epsilon_p/m\sigma - \epsilon_p \cos \beta_N (2\lambda_N/\lambda - 1)}{4/m\sigma + \cos \beta_N} \end{aligned} \quad (19.15)$$

In the absence of prerotation Eq. (19.10) reduces to

$$\epsilon_s = 2 \tan \psi \left[\frac{\cot \beta_N}{\lambda} - 1\right] \quad (19.16)$$

Eq. (19.14) reduces to

$$\tan \psi = \frac{\sin \beta_N}{4/m\sigma + \cos \beta_N} \quad (19.17)$$

and Eq. (19.15) reduces to

$$\epsilon_s = \frac{2 \sin \beta_N (\cot \beta_N/\lambda - 1)}{4/m\sigma + \cos \beta_N} \qquad (19.18)$$

The desired relationship between λ and ϵ_s has now been established that, in the absence of substantial radial flow, is generally applicable to each and every blade element. Axial velocity at a given radius will not remain constant within the rotor stage when radial flows are present; the average of inlet and outlet velocity is required in establishing λ.

19.3 APPLICATION OF λ AND ϵ ESTIMATES—ROTORS

The major use of these data is in producing characteristic curves of pressure rise versus volume flow rate and related efficiency estimates. Provided the rotor blades and boss are free of flow separation, calculations based on the mean radius (Fig. 19.1) will provide satisfaction, irrespective of the design method. Since the mean radius is relatively insensitive to the sign and magnitude of the spanwise axial velocity gradients, there is no need to average upstream and downstream quantities.

The mean theoretical total pressure rise is calculated from

$$K_{th} = \left[\frac{2(\epsilon_p + \epsilon_s)}{\lambda}\right]_{x\bar{V}_a} \qquad (19.19)$$

Since the actual total pressure rise across the rotor blading is given by

$$K_r = K_{th}\left(1 - \frac{K_R}{K_{th}}\right) \qquad (19.20)$$

the rotor efficiency loss must be estimated from Eq. (10.2), namely,

$$\frac{K_R}{K_{th}} = \frac{C_{Dp} + C_{Ds}}{C_L} \frac{\lambda}{\cos^2 \beta_m}$$

where guidance in the selection of drag data is obtained from Chapter 10, and β_m and C_L are established, respectively, from Eqs. (8.19) and (8.37).

The importance of the foregoing no-separation proviso is self-evident when one gives consideration to the assumptions on which the preceding design equations were developed.

General use of the λ versus ϵ_s relationships for the attainment of straightener design data in relation to variable-pitch rotor units, and for stall prediction, is desirable. Provided the inlet flow is reasonably uniform and the design is of the free vortex type, radial flow effects will be insignificant;

hence the relation $\lambda = \Lambda/x$ will hold. However, the local values of ϵ_s when converted to k_{th} may display a spanwise gradient of theoretical total pressure rise for off-design and variable-pitch duty conditions. The greater these gradients become, the more chance there is of encountering radial flows. When the latter become appreciable, accuracy can only be improved by taking cognizance of the need for flow coefficients based on the average local axial velocity or by acknowledging that the λ used in calculation corresponds to a lesser or greater Λ than the one at which the fan performance was established. With experience the problem can be overcome.

Calculations of ϵ_s and C_L at both blade extremities will enable these quantities to be plotted as functions of Λ and blade stagger setting angles. The use of these data for straightener design, and for the study of rotor and stator stall probabilities, is discussed in Chapter 20.

19.4 MULTIPLANE INTERFERENCE—ROTORS

High blade solidity will decrease the lift curve slope from its isolated airfoil value. Hence m must be adjusted when blade conditions at the blade root and mean stations demand.

For fans of optimum Λ design (Chapter 15), it is not normal for the mean station to be subject to lift interference. However, adjustment to m for blade root calculations will be required.

As the design Λ is increased progressively, multiplane interference effects will become evident at the mean blade station.

The interference factor (C_L/C_{L_i}) is a function of both the zero lift angle and lift curve slope. However, reasonable accuracy is attained in the normal duty range when the factor is assumed to operate on lift curve slope alone. (Experimental cascade data indicate that changes in the former are small.) Therefore the required m is given by

$$m = \frac{5.7 C_L}{C_{L_i}}$$

where C_L/C_{L_i} is obtained from Fig. 6.29.

The general design situation in terms of $(\beta_1 - \beta_2)$, and for nominal conditions, is displayed on Fig. 19.3. Airfoil cambers of less than 8% c ($\theta = 36°$) will be suitable for most design requirements and hence the preceding interference factor approach can be implemented in the vast majority of analysis exercises.

Turning angle is not a unique function of C_L, even for the isolated blading case. As the stagger angle decreases, the turning angle is increased and hence the airfoil camber must be adjusted. In the limit the turning and camber angles become relatively large; flow separation from the rear of the suction surface is prevented by pressure field modifications imposed by the

19.4 MULTIPLANE INTERFERENCE—ROTORS

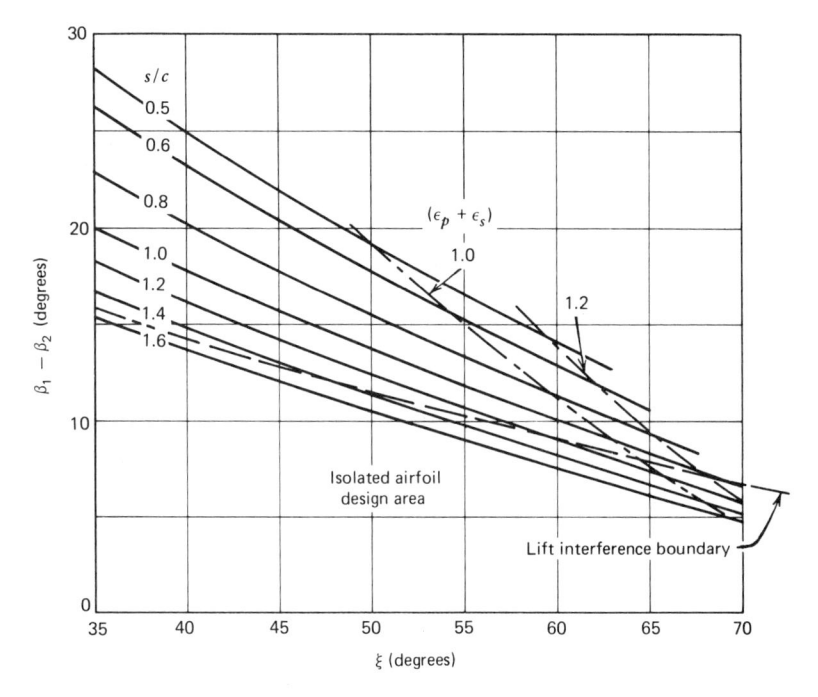

Figure 19.3 Diagram showing approximate lift interference boundary, based on Howell expression [Eq. (6.15)] [6.1].

adjacent blade, in a high-solidity arrangement. The changed pressure field is reflected in C_L reductions.

From the preceding explanation the interrelationship between blade solidity and camber is self-evident. Provided the nexus recommended in the enclosed design techniques is approximately maintained, the preceding procedures will produce reliable results.

NACA 65-series blower sections can be fitted into this general analysis framework when the equivalent camber angle is obtained from Fig. 6.12.

When the air-turning angle exceeds 28° the blade camber angle requirement will be greater than that provided by an 8% c airfoil. In these relatively rare circumstances the analytical method advocated in [19.1] provides an alternative approach.

By plotting all deflection data in the manner illustrated in Fig. 19.4, Howell obtained a relatively unique curve. The angle of incidence, i^*, is the angle at which the nominal deflection $(\beta_1 - \beta_2)^*$, defined as $0.80(\beta_1 - \beta_2)_{max}$ is obtained. A mean value of i^* for the high camber conditions with which we are concerned is, according to Fig. 9.1, approximately 2°. The incidence change $(i - i^*)$ can be replaced by $\beta_1 - \beta_1^*$.

When design calculations are available giving nominal deflection data, new β_1 and $(\beta_1 - \beta_2)$ data can be generated from Fig. 19.4. These can then be translated into λ and ϵ_s by use of the following relations, when ϵ_p is either known or zero, namely,

$$\tan \beta_1 = \frac{1 + \epsilon_p \lambda}{\lambda} \qquad (8.28)$$

$$\tan \beta_2 = \frac{1 - \epsilon_s \lambda}{\lambda} \qquad (8.29)$$

Since $(\beta_1 - \beta_2)$ seldom exceeds 28° at the mean station, the main use for the preceding data is with respect to root stall prediction.

In the absence of design calculations, blade geometry measurements of solidity, stagger, and camber angle will allow estimates of β_1 and $(\beta_1 - \beta_2)$ to be obtained for nominal design conditions.

$$\xi = (\beta_1^* - i^*) - \frac{\theta}{2} \qquad (9.11)$$

$$\theta = \frac{(\beta_1 - \beta_2)^* - i^*}{1 - 0.26 \sqrt{s/c}} \qquad (9.10)$$

The occasions when the latter exercise is justified will be extremely limited. In the absence of design data this work should be carried out and assessed for relevance by an experienced fan aerodynamicist.

19.5 PREROTATOR ANALYSIS

The air-turning performance of this stator type, expressed by ϵ_p, remains constant with changing fan duty. When the aerodynamic design calculations are unavailable, the outlet angles can be calculated from Eq. (11.14), namely,

$$\theta = \frac{\beta_2 - i}{1 - 0.19 s/c}$$

where vane solidity and camber are known from measurement at each spanwise station. For uniform swirl-free inlet flow and free vortex conditions

$$\epsilon_p x = \text{constant}$$

where $\tan \beta_2 = \epsilon_p$.

The vane outlet angle becomes the inlet angle β_1 in rotor blade element analysis.

Variable-pitch inlet vanes are used to change the fan characteristic curve and power requirements. Cambered airfoil vanes designed with twist to achieve free vortex flow can be used in this role, but this arrangement is most unusual in present-day practice. Instead symmetrical, untwisted vanes are the most common type. Provided the angle is not excessive, a reasonable estimate of swirl angle can be obtained from Chapter 11. In these circumstances a loss of accuracy must be expected in rotor analysis as the vane angles are advanced and vane flow separation eventually occurs. The radial flow component also increases with departure from free vortex conditions. A combined analytical and experimental approach to this problem is advocated when building up design and operational experience relative to this fan type.

The loss in fan unit efficiency due to prerotators is estimated from

$$\frac{K_P}{K_{th}} = \left(\frac{C_{DP} + C_{DS}}{C_L} \frac{\lambda}{\cos^2 \beta_{m_p}} \frac{\epsilon_p}{\epsilon_p + \epsilon_s} \right)_{MS} \quad (12.11)$$

adopting the procedures outlined in Chapter 12 and using λ and ϵ_s as determined from rotor analysis.

19.6 STRAIGHTENER ANALYSIS

It is apparent from Fig. 19.4 that β_2 remains reasonably constant for loadings less than the design and is associated with a progressive trend toward underturning as the load approaches the stall. However, provided the stators remain unstalled, the latter is of little consequence, since the related energy is negligible.

The main interest in straightener analysis is related to stator ability to handle rotor outlet swirl without stalling and to establish loss estimates for use in fan efficiency predictions.

For swirl coefficient values greater than 0.5, the cascade design method defines the geometric vane properties required in obtaining a flow-turning angle that is 80% of the maximum or stall value. Since β_2 for the vane outlet angle approximates zero

$$\beta_{1,\text{stall}} \simeq 1.25 \, \beta_{1,\text{design}}$$

Because of secondary flow influences it is recommended that the more conservative ratio of 1.2 be adopted. Because of the large vane root design values of β_1, this ratio does allow adequate latitude. For instance, when a vane element is designed for 45° of turning ($\epsilon_s = 1.0$), the maximum allowable becomes 54° ($\epsilon_s = 1.38$). High solidity and a large camber angle are of assistance is this regard.

When cambered plates are used as straighteners, a further degree of conservatism is required in assessing the possibility of vane stalling. However, their performance is normally quite good, requiring only small adjustments

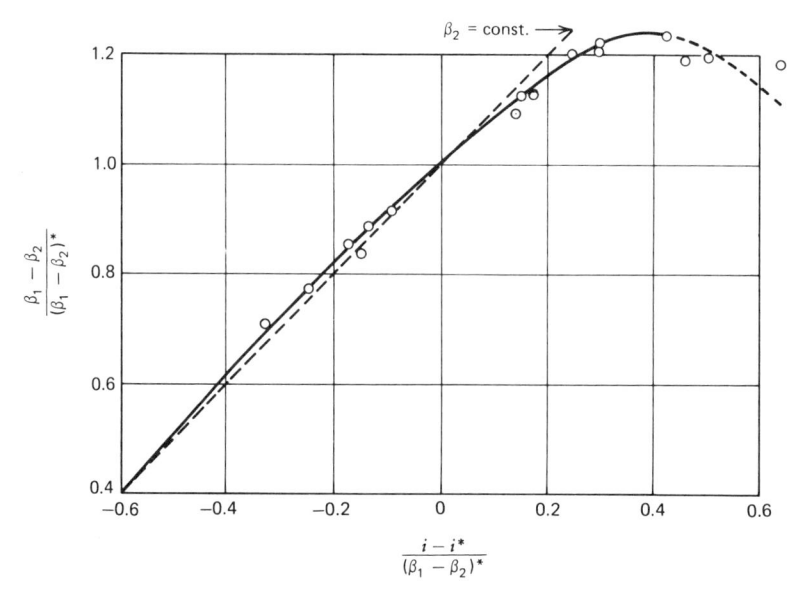

Figure 19.4 Flow deflection data for cascade airfoils. (Taken from A. R. Howell [19.1] and reproduced with the permission of the Controller of Her Majesty's Stationery Office.)

that, in the absence of published data, must be based on individual experience.

When the stator design calculations are available their ability to remove all swirls up to the predicted rotor stall duty(s) can be readily assessed once ϵ_s has been established from rotor analysis, at the three stations, namely, root, tip, and mean. In the absence of design information, a reference β_1 must be established at each required spanwise station by use of Eq. (11.10), namely,

$$\theta = \frac{\beta_1}{1 - 0.26\sqrt{s/c}}$$

This relationship is assumed to hold when the outlet vane angle is equal to the deviation angle δ, as given by

$$\delta = 0.26\theta\sqrt{s/c} \qquad (9.1)$$

making β_2 equal to zero.

Finally, the fan efficiency loss due to straighteners is estimated from

$$\frac{K_S}{K_{\text{th}}} = \left(\frac{C_{DP} + C_{DS}}{C_L} \frac{\lambda}{\cos^2\beta_{m_s}} \frac{\epsilon_s}{\epsilon_s + \epsilon_p}\right)_{\text{MS}} \qquad (12.10)$$

where the procedures outlined in Chapter 12 are used. Alternatively, use may be made of Fig. 19.5.

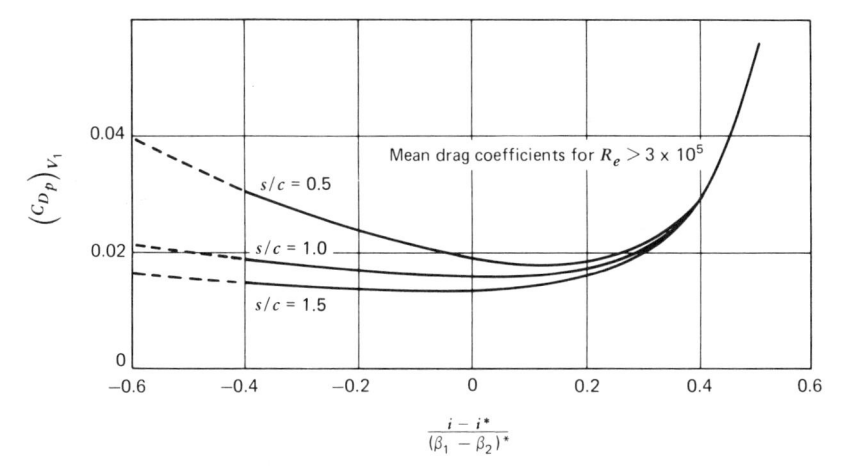

Figure 19.5 Drag coefficient versus flow deflection for cascade airfoils. (Taken from A. R. Howell [19.1] and reproduced with the permission of the Controller of Her Majesty's Stationery Office.)

19.7 OVERALL FAN UNIT PERFORMANCE

For the general blading arrangement adopted in this chapter the overall fan unit efficiency is given by

$$\eta_T = \frac{K_{th} - K_R - K_P - K_S - K_D}{K_{th}} \qquad (14.7)$$

where K_D estimates are obtained from data given in Chapter 13.

The overall total pressure rise coefficient is then obtained from

$$K_T = \eta_T K_{th}$$

for the various values of Λ.

Since reliable drag data are available in the design range for carefully designed units, the analytical performance predictions will achieve design accuracy. However, for operational conditions well away from the design duty point this high accuracy may not be maintained, because of gaps in our knowledge.

REFERENCE

19.1 A. R. Howell, The present basis of axial flow compressor design. Part 1: Cascade theory and performance, Gt. Britain Aero. Research Council, *ARC R & M 2095*, 1942.

CHAPTER **20**
Examples of Fan Analysis

The procedures developed and outlined in the preceding chapter are extremely useful in the design of variable-pitch fan units. To demonstrate this feature the mine ventilation fan depicted in Fig. 17.2 is analyzed for a range of off-design duties, the computations being compared with actual test data.

Two alternative methods for analyzing high-solidity fans possessing multiplane interference at the midspan station are also investigated and compared with experiment.

20.1 VARIABLE-PITCH FAN ANALYSIS

Comprehensive aerodynamic studies and model testing preceded the manufacture of the large vertical fan units installed at Mount Isa Mines Ltd. during 1965.

The rotor blading initially consisted of 10% thick Clark Y sections since the turning angles were generally small and the desire for a wide band of high-efficiency operation was dominant. Cambered C4 sections were selected for the prerotator and straightener vanes.

The midspan reference stagger angle of the rotor blade is 66° and since $\alpha_N = -4.5°$, $\beta_N = 61.5°$. Blade solidity is 0.455 and with zero multiplane interference the lift curve slope is 5.7. The Reynolds number approximates 2×10^6. The prerotators were designed to give a swirl coefficient of 0.264 at midspan.

For the reference pitch setting, the common computational data for use in Table 20.1 are $\tan \beta_N = 1.842$, $\sin \beta_N = 0.879$, $\cos \beta_N = 0.477$, $4\epsilon_p/m\sigma = 0.407$, $\lambda_N = 0.634$ [Eq. (19.9)], $\epsilon_p \cos \beta_N = 0.126$, $4/m\sigma + \cos \beta_N = 2.019$.

The efficiency loss due to the prerotators is estimated in Table 20.2.

Estimates of straightener resistance and the related fan unit efficiency losses for vanes with midspan design values of 0.5 and 1.0 for ϵ_s and σ,

20.1 VARIABLE-PITCH FAN ANALYSIS

Table 20.1 Rotor Analysis at Reference Blade Setting Angle

Λ	0.20	0.25	0.30	0.35
λ	0.267	0.333	0.400	0.467
$2 \sin \beta_N (\lambda_N/\lambda - 1)$	2.416	1.589	1.028	0.629
$\epsilon_p \cos \beta_N (2\lambda_N/\lambda - 1)$	0.472	0.354	0.273	0.216
ϵ_s Eq. (19.15)	0.761	0.410	0.172	0.003
$\epsilon_p + \epsilon_s$	1.025	0.674	0.436	0.267
$K_{th} = 2(\epsilon_p + \epsilon_s)/\lambda$	7.68	4.05	2.18	1.14
$\tan \beta_m = [1 - \frac{1}{2}(\epsilon_s - \epsilon_p)\lambda]/\lambda$	3.497	2.930	2.546	2.272
β_m	74.04	71.16	68.56	66.24
$\cos \beta_m$	0.275	0.323	0.366	0.403
$C_L \sigma = 2(\epsilon_p + \epsilon_s) \cos \beta_m$	0.564	0.435	0.319	0.215
C_L	1.239	0.957	0.701	0.473
C_{Dp} (Fig. 6.16, Re = 3×10^6)	0.017	0.012	0.009	0.008
$C_{Ds} = 0.015\, C_L^2$	0.023	0.014	0.008	0.004
$\gamma = C_L/(C_{Dp} + C_{Ds})$	31.0	36.8	41.2	39.4
$\gamma K_R/K_{th} = \lambda/\cos^2 \beta_m$	3.53	3.19	2.99	2.88
K_R/K_{th}	0.114	0.087	0.073	0.073
η_R	0.886	0.913	0.927	0.927

respectively, are presented in Table 20.3 where K_D is taken to be 0.05, from Fig. 13.3.

The C_{Dp} versus ϵ_s relationship displayed in Fig. 20.1 represents an adequate measure of profile drag for fan unit efficiency estimation purposes. It is based on Fig. 19.5 and the midspan design value of ϵ_s for the relevant solidity. Because of the high Re the conversion of $(C_{Dp})_{v_1}$ to C_{Dp} has been ignored.

Estimation of fan total efficiency and dimensional characteristic data are available from Table 20.4.

Similar calculations for other blade settings result in the plotting of estimated performance data in Fig. 20.2. The efficiency contour is the consequence of a smoothing process when η_T is plotted against Q, for each blade setting.

The estimates are compared with test data for the actual installation where FTP is taken as the difference between the mean total pressure just

Table 20.2 Prerotator Analysis at Reference Blade Setting Angle[a]

Λ	0.20	0.25	0.30	0.35
$\gamma K_P/K_{th} = \lambda/\cos \beta_{m_p} \cdot \epsilon_p/(\epsilon_p + \epsilon_s)$ Eq. (12.11)	0.076	0.143	0.266	0.507
K_P/K_{th}	0.003	0.005	0.009	0.016

[a] Midspan data: $\epsilon_p = 0.264$, $\tan \beta_{m_p} = \epsilon_p/2$, $\beta_m = 7.5°$, $\cos \beta_{m_p} = 0.992$, $\sigma = 0.77$, $C_L = 0.68$ (Fig. 11.3), $C_{Dp} = 0.015$ (Fig. 12.2), $C_{Ds} = 0.015\, C_L^2$, $\gamma = 31$.

Table 20.3 Efficiency Loss Due to Straighteners and Fairing

Λ	0.20	0.25	0.30	0.35
ϵ_s (rotor analysis)	0.761	0.410	0.172	0.003
$\tan \beta_{m_s} = \epsilon_s/2$	0.381	0.205	0.086	
β_{m_s}	20.86	11.59	4.92	
$\gamma K_S/K_{th} = \lambda/\cos^2 \beta_{m_s} \cdot \epsilon_s/(\epsilon_s + \epsilon_p)$ Eq. (12.10)	0.227	0.211	0.159	
C_L (Fig. 11.3), $\sigma = 1$	1.42	0.80	0.33	
C_{D_P} (Fig. 20.1)	0.055	0.016	0.020	
$C_{D_S} = 0.018 C_L^2$	0.036	0.012	0.002	
$\gamma = C_L/(C_{D_P} + C_{D_S})$	15.6	28.6	15.0	
K_S/K_{th}	0.015	0.008	0.011	(0.02)
K_D/K_{th}	0.007	0.012	0.023	0.044

downstream of the tail fairing, and atmospheric pressure. Hence the fan "unit" incorporates losses associated with the inlet duct, and devices, for minimizing the adverse effects of winds. For light wind conditions the loss component was measured as 7% of $\frac{1}{2}\rho \bar{V}_a^2$. In addition, the rotor blades possess square-cut root sections for which no allowance is made in the preceding computations. When consideration is given to normal difficulties in carrying out accurate on-site testing, particularly the effect of varying wind strength, the agreement is considered good.

Computations of K_{th} versus Λ at the blade root result in relatively small changes away from the mean values presented earlier. Hence radial flows are substantially absent for off-design duty conditions.

The ϵ_s and C_L data for the blade root are useful for studying straightener design and rotor stall, respectively.

The design C_L at the blade root is 0.68, which when multiplied by a solidity of 1.238 gives a loading factor of 0.84. Stalling from the blade root occurred at a calculated C_L of 0.95 for this design pitch, representing a factor of 1.18, or a 40% increase. The square-cut blade root probably precipitated a slightly earlier fan stall.

Table 20.4 Dimensional Data for Reference Blade Setting Angle[a]

Λ	0.20	0.25	0.30	0.35
η_T, Eq. (14.7)	0.861	0.888	0.884	0.847
K_T	6.61	3.60	1.93	0.97
$\bar{V}_a = 94.2\Lambda$	18.84	23.55	28.26	32.97
$Q = 21.89 \bar{V}_a$	412.2	515.3	618.4	721.5
$\frac{1}{2}\rho \bar{V}_a^2 (\rho = 1.2 \text{ kg/m}^3)$	213.0	332.8	479.2	652.2
FTP $= K_r \cdot \frac{1}{2}\rho \bar{V}_a^2$	1408	1198	925	633

[a] $D = 6.096$ m, annulus area $= 21.89$ m², and $N = 4.92$ rev/s.

20.1 VARIABLE-PITCH FAN ANALYSIS 339

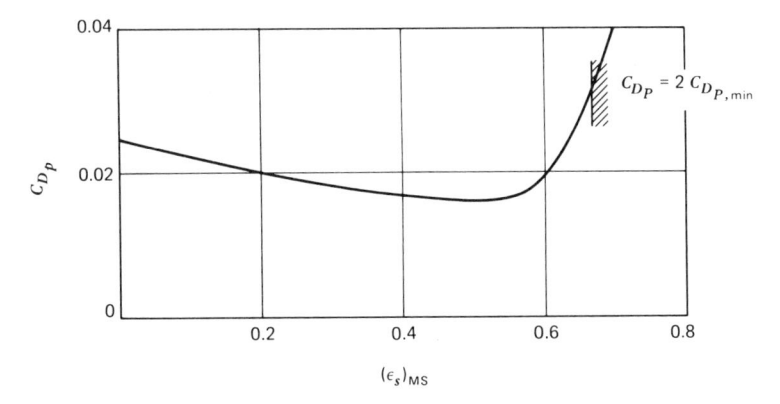

Figure 20.1 Postulated drag characteristic for stators in fan analysis exercise.

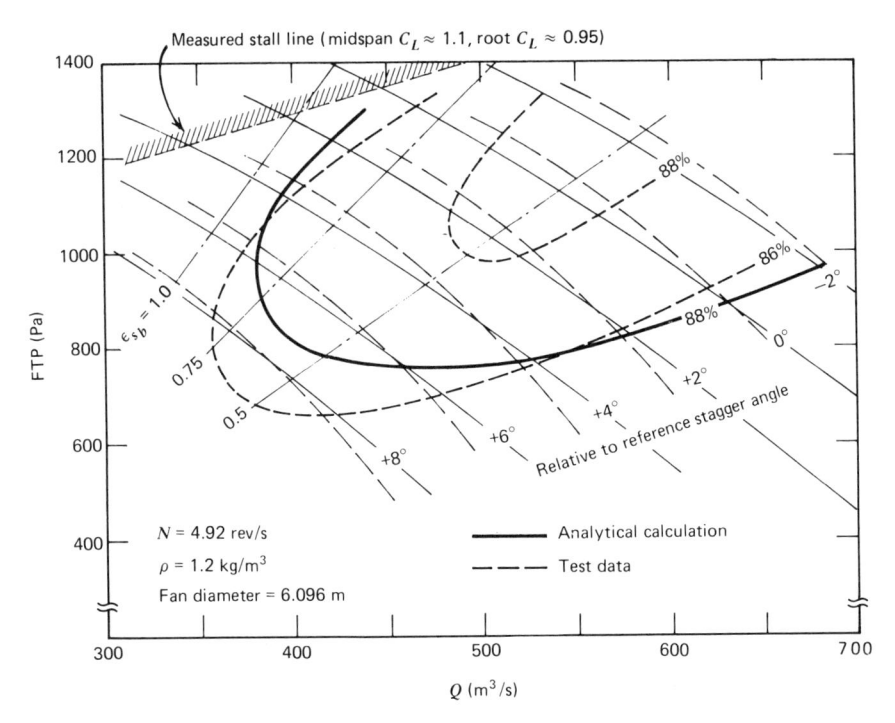

Figure 20.2 Variable-pitch mine ventilation fan performance data. (By courtesy of Mount Isa Mines Ltd., Australia.)

The preceding result is relevant only to clean and smooth 10% Clark Y high root solidity blading operating at flow coefficients of approximately 0.2. The many factors that influence stall are discussed elsewhere.

A measure of judgment is required when designing straighteners for a variable-pitch fan. The lines of constant ϵ_{s_b} superimposed on Fig. 20.2 indicate the nature of the problem. Stator stalling in the top left-hand corner of the performance chart, ahead of rotor stall, will be present. The selected design value of $\epsilon_{s_b} = 0.75$ represents a compromise solution, favoring the more probable higher-pressure duty range. (Redesigned rotor blading has been successful in raising the operational fan pressure on equipment of the type illustrated in Fig. 17.3.)

20.2 ANALYSIS OF ROTOR-STRAIGHTENER UNIT OF LARGE Λ

The off-design performance of the fan designed in Section 18.4 is calculated by the two alternative methods discussed in the previous chapter.

Midspan design data for use in Table 20.5 are $\theta = 25.9°$, $\xi = 39.9°$, $\sigma = 1.2$, $s/c = 0.833$.

The no-lift angle from Fig. 6.7 is $-6°$ and C_L/C_{L_i} from Fig. 6.29 is 0.67, resulting in a lift curve slope (m) of 3.82. Hence $\beta_N = 33.9°$.

Off-design performance calculations by the Howell method yield the listed data of Table 20.6.

When the K_{th} obtained by both methods are plotted against Λ, their mean lines are virtually coincident. Since the multiplane interference data of Fig. 6.29 were obtained from an integration of actual test data on isolated and cascade airfoils, with optimum circular-arc camber lines, the preceding coincidence at the design point could reasonably be expected. However, the agreement at off-design duties suggests a degree of latitude with respect to the relationships of Fig. 6.29.

Applying the drag data of Fig. 19.5 for efficiency estimation purposes, it is first necessary to convert the drag coefficient to one based on V_m. From Fig. 6.4 the relationship

$$\frac{V_m}{V_1} = \frac{\cos \beta_1}{\cos \beta_m}$$

is obtained.

Table 20.5   Rotor Analysis by the Lift Interference Factor Method[a]

Λ	0.75	0.70	0.65	0.60	0.55	0.50	0.45
λ	0.882	0.824	0.765	0.706	0.647	0.588	0.529
$2 \sin \beta_N (\cot \beta_N/\lambda - 1)$	0.769	0.899	1.055	1.236	1.451	1.708	2.023
ϵ_s [Eq. (19.18)]	0.452	0.528	0.620	0.726	0.852	1.003	1.188
K_{th} [Eq. (8.17)]	1.02	1.28	1.62	2.06	2.63	3.41	4.49

[a] Design data: $\sin \beta_N = 0.558$, $\cos \beta_N = 0.830$, $\cot \beta_N = 1.488$, $4/m\sigma + \cos \beta_N = 1.703$.

20.2 ANALYSIS OF ROTOR-STRAIGHTENER UNIT OF LARGE Λ 341

Table 20.6 Rotor Analysis by Cascade Method

$(i - i^*)/(\beta_1 - \beta_2)^*$	-0.4	-0.2	0	0.1	0.2	0.3	0.4
$(\beta_1 - \beta_2)/(\beta_1 - \beta_2)^*$	0.61	0.82	1.0	1.08	1.16	1.225	1.25
$i - i^*$	-8.3	-4.2	0	2.1	4.2	6.2	8.3
$\beta_1 - \beta_2$	12.7	17.1	20.8	22.5	24.1	25.5	26.0
i	-6.3	-2.2	2.0	4.1	6.2	8.2	10.3
$\beta_1 = \beta_1^* + i - i^*$	46.5	50.6	54.8	56.9	59.0	61.0	63.1
$\lambda = 1/\tan\beta_1$	0.949	0.821	0.706	0.652	0.601	0.554	0.507
β_2	33.8	33.5	34.0	34.4	34.9	35.5	37.1
$\tan\beta_2$	0.669	0.662	0.675	0.685	0.698	0.713	0.756
$\epsilon_s = 1/\lambda - \tan\beta_2$	0.385	0.556	0.741	0.849	0.966	1.092	1.216
K_{th}	0.81	1.35	2.10	2.60	3.21	3.94	4.80
Λ	0.807	0.698	0.600	0.554	0.551	0.471	0.431

[a] Design data: $i^* = 2°$, $\beta_1^* = 54.8°$, $(\beta_1 - \beta_2)^* = 20.8°$.

The required C_{Dp} is therefore given by

$$C_{Dp} = (C_{Dp})_{V_1}\left(\frac{\cos\beta_m}{\cos\beta_1}\right)^2$$

where $\tan\beta_m = \frac{1}{2}[\tan\beta_1 + \tan\beta_2]$ [Eq. (8.31)].

The design rotor efficiency deduced in Table 20.7 is a little less than that calculated in Section 18.4 because of the higher C_{Dp}. The value used in design for the stated Re is believed to be the more realistic.

The stator can be analyzed in a manner similar to that outlined for the previous example.

Analytical studies of the root and tip stations indicate little appreciable variation of K_{th} along the blade for off-design conditions, suggesting no significant radial flows. Hence a constant axial flow component can be assumed in the stall analysis.

Table 20.7 Rotor Efficiency from Cascade Method

$(i - i^*)/(\beta_1 - \beta_2)^*$	-0.4	-0.2	0	0.1	0.2	0.3	0.4
$(C_{Dp})_{V_1}$ (Fig. 19.5)	0.022	0.019	0.017	0.016	0.017	0.021	0.030
$\tan\beta_m$	0.862	0.940	1.046	1.109	1.181	1.259	1.364
$\cos\beta_m$	0.758	0.729	0.691	0.670	0.646	0.622	0.591
C_{Dp}	0.027	0.025	0.025	0.024	0.027	0.034	0.050
$C_L\sigma = 2\epsilon_s \cos\beta_m$	0.584	0.811	1.024	1.138	1.248	1.358	1.437
C_L ($\sigma = 1.2$)	0.49	0.68	0.85	0.95	1.04	1.13	1.20
$C_{Ds} = 0.018 C_L^2$	0.004	0.008	0.013	0.016	0.020	0.023	0.026
$\gamma = C_L/(C_{Dp} + C_{Ds})$	15.8	20.6	22.4	23.8	22.1	19.8	15.8
$\gamma K_R/K_{th} = \lambda/\cos^2\beta_m$	1.65	1.54	1.48	1.45	1.44	1.43	1.45
K_R/K_{th}	0.104	0.075	0.066	0.061	0.065	0.072	0.092
η_R	0.896	0.925	0.934	0.939	0.935	0.928	0.908

Computation of the blade root value of cos β_1/cos β_2 indicates the wall stall criterion for thin boundary layers (Section 9.5) is reached when the blade root flow deflection is 1.15 times the design quantity. This occurs when Λ is approximately 0.475, and hence fan stall can be considered imminent for lesser flow coefficients.

An analysis of the fan illustrated in Plate 4 provides a check on method accuracy. A unique curve of K_{th} versus Λ results from both computational methods, and this curve is confirmed by actual test results when equal work output and efficiency from each of the two rotors is assumed (Fig. 20.3). This fan provides a more severe test of the multiplane interference method of analysis, since $C_L/C_{L_i} = 0.45$ and $\alpha_N = -9°$ for a midspan blade solidity of 1.8. The probable design values of β_1 and $(\beta_1 - \beta_2)$ were calculated from the geometric quantities of θ and ξ, by the method outlined in Section 19.4. The experimental stall point is given in Fig. 20.3 and this corresponds to a midspan flow deflection that is 20% greater than the computed design value. This percentage is a little lower than that suggested for the preceding worked example, partly because of the more severe wall flow conditions of this two-stage unit. These stall recommendations are for good leading-edge and general surface conditions and above-critical Re. Allowance for other circumstances must be made on the basis of experience.

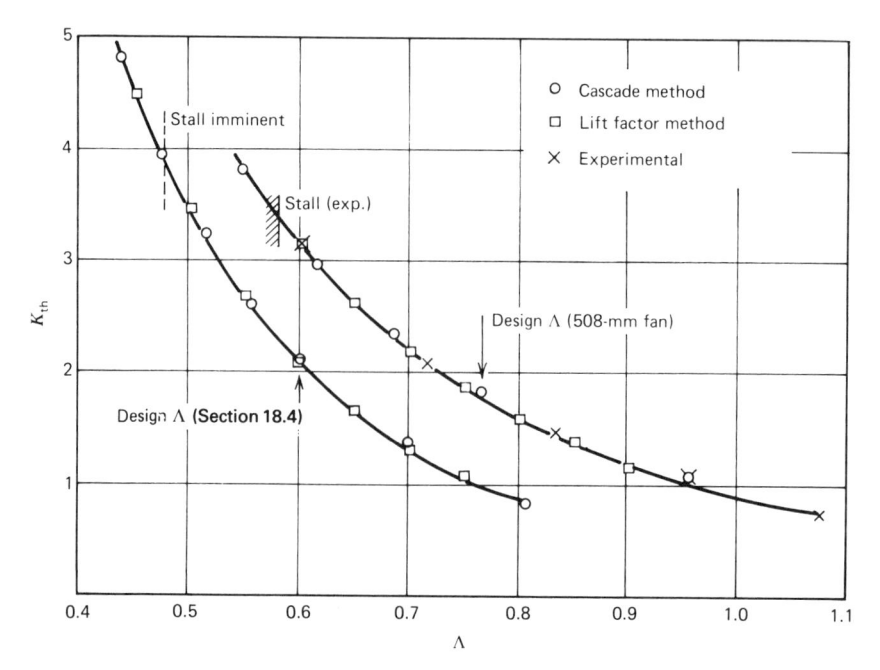

Figure 20.3 Analysis of two fans designed for large flow coefficient Λ. (Data for 508-mm unit by courtesy of D. Richardson & Sons, Melbourne, Australia.)

20.3 ANALYSIS OF OTHER FAN UNITS

The methods outlined here have been used with success in a wide range of equipment designed in accordance with the free vortex flow principle. Of particular interest is the smooth and accurate link between isolated and cascade design data for airfoils with a circular-arc camber line.

In low-pressure-rise fans of moderate blade root solidity, the fan stall will be associated with a higher C_L than that pertaining to the first fan analyzed. However, the type of airfoil used can vary greatly and hence the designer must rely on personal text experience when predicting stall.

Fans designed by the arbitrary vortex flow method of Appendix A can be analyzed approximately by the foregoing method, provided the blade root value of downstream axial velocity remains above zero.

CHAPTER 21
Fan Testing—Commercial

The current policy of presenting all aspects of fan technology in a sound aerodynamic manner is continued in the present instance. The many complex issues that currently beset the fan industry are not insoluble, provided the problems are approached correctly. Although test codes cannot dictate design procedures or grade their merits, test specifications can be drafted that may provide the customer with additional useful information and incentive from the selection viewpoint.

National technology levels are normally reflected in their test codes. In relation to fans, most would agree that the present situation is unsatisfactory. The difficulty appears to be in reaching agreement on improved methods, particularly on the international level. However, when emphasis is placed on airflow rather than equipment features, satisfactory progress in the upgrading of current test codes can be achieved.

21.1 SYSTEM EFFECT FACTORS

Commercial laboratory-type testing provides the user with "benchmark" data with which to assess the capability and suitability of equipment in regard to the intended duty. When the upstream ducting is designed on good aerodynamic lines and the fan inlet flow is free of swirling, nonuniform flow conditions, then the fan performance can be directly equated to the total pressure loss in the duct system. The performance of adjacent downstream duct components will be modified by poor fan outlet flow quality, and this may increase duct resistance above the normal value for such a component.

The code authority for the United States and Canada, namely, the Air Moving and Conditioning Association (AMCA), has accepted the fact that many installed fans suffer substantial performance penalties because of poor inlet flow quality. When applying laboratory test data to a proposed installa-

21.2 FAN PRESSURE

tion, adjustment by the introduction of System Effect Factors (SEF) is recommended [21.1]. It is acknowledged that the procedure lacks accuracy, but it is claimed that the correction data are the best available.

However, no distinction between fan types is made. Since the fluid processes by which total pressure is added are different for axial and radial flow fans, the influence of poor inlet flows cannot be identical. The axial unit, which relies entirely on velocity vector changes, will be more affected by a swirl vector than the radial variety, which possesses a large pressure component resulting from centrifugal action. The majority of SEF data has been established from radial flow fans.

In its *Fan Application Manual* [21.1] AMCA highlights some of the worst installation features and suggests alternatives. However, this is often a question of relative crudity, as a SEF correction to fan performance can still be required.

The SEF concept is carried through into the field test document of [21.1]. Because of the large number of variables in an air-moving system layout, adoption of a standard method of field testing has not been considered feasible and hence the document takes the form of a guide. In contrast, field tests in some European countries are covered by code specifications.

The use of System Effect Factors can be avoided by adherence to sound aerodynamic design practice. The duct design and development data presented here, when used in the correct context and within the stated limits, will produce factors of unity.

Greater use should be made of model testing in cases of special design difficulty, as advocated in [21.1]. For example, fans in compact heat exchanger units may be subject to substantial radial flow curvatures at fan inlet and outlet, even when the inlet flow is free of separation. These curvatures will affect fan performance; hence experimental studies on a full- or smaller-scale model are essential in the development of a fan/duct arrangement of satisfactory and known aerodynamic performance (see Appendix C).

21.2 FAN PRESSURE

Fan pressure must always be considered in terms of total pressure. Static pressure measurements are valid only when used in establishing or estimating mean total pressure at a chosen duct or fan station.

The first fan codes were formulated in the 1920s in both the United States and Europe. It is believed that the term *fan static pressure* (FSP) arose out of a test method for centrifugal fans [21.2] where, for blower configurations, a static pressure measurement in the outlet duct was taken. This quantity relative to atmospheric pressure, which is defined as "fan static pressure," plus the mean velocity pressure at the measurement station, is equal to the total pressure rise across the fan. However, the flow is accelerated from rest into the rotor inlet, resulting in a negative static pressure; hence the static pressure rise across the fan rotor is not given by fan static pressure. The continuing use of this latter quantity seems to be enshrined in this early tradition, and despite an early attempt [21.2] to put the emphasis on fan total pressure, the industry continues to give precedence to FSP.

The problem has been compounded by the usefulness of FSP in relation to exhaust fan equipment, where this quantity is equal to the total pressure loss in the duct system. When the system resistance expressed as a K in terms of the dynamic pressure in the ducting upstream of the fan is large, the static and total pressure losses, which differ by only 1.0, are close to equal and hence FSP has a degree of practical if not correct significance to the user. However, under different circumstances the errors can be large and crucial, particularly for axial flow fans.

The first step toward a rational system is to subdivide fans into two major classes, namely, in-line and exhaust equipment. The correct pressures for each of these units are *fan total pressure* (FTP) and *fan inlet total pressure* (FITP), respectively. This eliminates FSP in relation to in-line fans and replaces it in the exhaust case by FITP.

In-line and exhaust fans are defined as units possessing useful and zero downstream work components, respectively. When the exhaust unit incorporates a downstream diffuser, the combined assembly constitutes the exhaust fan.

Since a rotor unit may be used in either of the two preceding configurations, some conversion between FTP and FITP is required. The total pressure rise through an exhaust fan is equal to outlet total pressure (OTP) minus inlet total pressure (ITP), where OTP is given by the mean dynamic pressure associated with a good-quality outlet flow.

For separation and swirl-free outlet flows the only matter needing special consideration is in relation to different downstream treatments, when these exist. Adjustments to the outlet total pressure (OTP) for differences in duct resistance and velocity pressure introduced by diffusers or other downstream components are necessary on the basis of calculations or measurements. However, when these flow conditions are not met within reasonable bounds, separate test programs are the recommended alternative. It is an advantage to have test duct components common to both methods.

In seeking a general test solution for all fans irrespective of outlet flow quality, the existing codes have defined the *fan velocity pressure* (FVP) as that based on the mean hypothetical velocity at the outlet flange; this is calculated on gross area, neglecting the presence of internal bodies and surfaces. Unlike the previous proposal, this conversion method carries no provisos in respect to outlet flow quality. Hence in correct aerodynamic terms this procedure in the general sense is erroneous and unsound. Some fans possess outlet flows of gross nonuniformity, high turbulence, and large fluctuating amounts of swirl velocity. These flow features present no measurement or interpretive problems for exhaust fan testing, but for in-line fans the situation is different, as discussed later.

21.3 LABORATORY TESTING CONSIDERATIONS

Therefore the proposal is for testing in relation to two distinct fan types; data conversion from one to the other is subject to an outlet flow quality check.

21.3 LABORATORY TESTING CONSIDERATIONS

There are three categories of required testing, namely,

1. *Specific.* A test carried out to determine the precise characteristics of one particular fan. The test conditions of fan speed and air density should approximate those pertaining to the installed fan.
2. *Type.* A test made on a unit that is intended to represent a production run of a particular size and model of fan.
3. *Geometrically similar.* A test made on one particular size of fan that is intended to generate performance data for geometrically similar units of varying size, speed, and operational air density.

Testing methods are not normally a function of the preceding requirements, although the detailed data presentations may differ.

Since in-line and exhaust fans both have upstream ducting, suitable inlet conditions are not difficult to arrange. A blower fan qualifies as an in-line fan when a flared inlet and controlled air approach path to the rotor blading is assured; these are required for efficiency in addition to test reasons. Unducted diaphragm-mounted propellor-type fans are excluded from these test categories, although chamber test methods can be adjusted to suit.

The measurement planes have to be located in regions where the accuracy limits of normal instrumentation are not exceeded. On the inlet side, the station must have axial separation from the nose fairing in order to avoid local stream curvature effects, which induce radial static pressure gradients.

The exhaust fan requires no outlet measurement station, since the kinetic energy of the discharge flow is not a useful work component; its influence on fan performance is fully reflected in the power measurement. Since all turbulence and velocity components are eventually dissipated, the ambient atmospheric pressure is the final outlet total pressure and hence becomes a reference pressure.

In the in-line fan case, the outlet flow quality will have a pronounced effect on the measurement plane position. Flow separation from motor supports (Fig. 13.5) or similar obstructions will always give rise to a turbulent swirling flow. Hence outlet flow quality can be qualitatively discussed in terms of swirl only; this may be of a steady or unsteady character.

Since a fan is tested over a wide duty range, an amount of negative or positive swirl will be present in all cases, even for fans designed for zero swirl at the duty point; rotor-straightener units will minimize such quantities (Section 19.6). Provided the swirl does not exceed approximately 10° within the transverse plane, the radial static pressure gradient will be virtually zero, and hence a piezometer ring will record mean static pressure.

Downstream of the motor, for the unit illustrated in Fig. 13.5, three eddies will initially be present, ruling out the prospect of a satisfactory measuring station close to the outlet flange. However, these eddies will coalesce further downstream to form a single vortex of a steadier and more stable variety. This condition is usually accompanied by a return to velocity profiles of a more regular type. The measurement plane specified in existing test codes is governed by the latter feature.

Swirl remains the outstanding question with regard to fan performance determination. It leads to pitot-static tube errors when the instrument is axially aligned and is responsible for radial static pressure gradients. In addition, swirl is normally considered a nonuseful part of fan output. The codes resolve the instrument problem by adding a flow straightener upstream of the measurement plane. The OTP of the fan is credited with the calculated losses in the test duct assembly, from fan outlet to the measuring station. This calculation is for swirl-free, fully developed pipe flow, at the appropriate Reynolds number. The additional leading-edge separation loss incurred by the straightener when dealing with swirling flows at large magnitude is tacitly assumed to approximate the eradicated swirl momentum, and hence no attempt is made to measure this loss increment.

The user receives the ensuing data in good faith, not realizing that he may have acquired, in terms of flow, a different fan from the assembly-tested one. The problem of excessive swirl, which has been countered by the manufacturer, is now in the court of the user, who may be ill equipped to deal with it in potentially critical circumstances. Incorporating the air straightener into the merchandise and making its losses integral with others in the fan assembly would remove basic aerodynamic opposition to the use of this "fix" device.

Recognition of the swirl properties as part of the fan performance characteristics is recommended. Since swirl and velocity readjustment require approximately the same test duct length, a common test station results. The combined yawmeter and total pressure tube of Fig. 22.10 will permit these flow characteristics to be established for fans with swirls exceeding 10°. The radial static pressure variation can be obtained with a hooked tube (Fig. 22.11) oriented along the yawmeter established flow paths. Conservation of angular momentum is the basic principle underlying the preceding assumption, namely, that the swirl characteristics measured some distance downstream are representative of the fan properties. Negligible and significant errors will accompany single and multisource vortices, respectively. This procedure should be confined to circular ducts, as the rectangular duct walls will modify and place some restraint on the vortex flow pattern.

It is not practicable and barely possible to map the outlet (flange) flow of fans similar to that shown in Fig. 13.5. A reduction in equipment standards will inevitably lead to increased uncertainty in respect to test data and opera-

21.3 LABORATORY TESTING CONSIDERATIONS

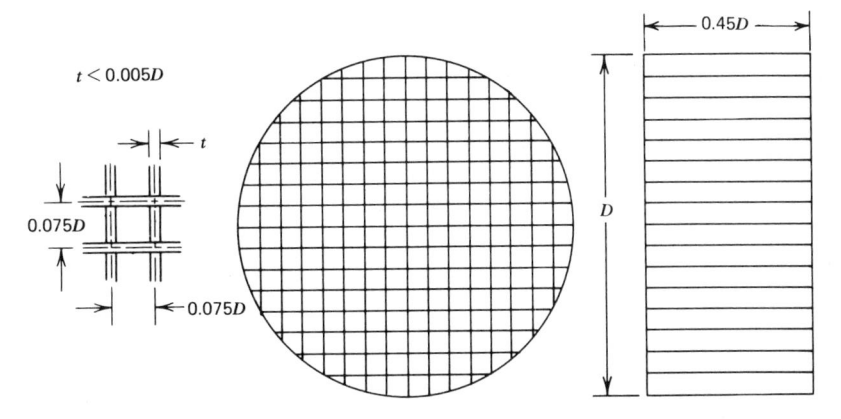

Figure 21.1 Airflow straightener, [21.3]. (Reproduced by permission of AMCA and ASHRAE from AMCA standard 210-74, published by the Air Movement and Control Assoc. Inc. Arlington Heights, Ill., USA.)

tional properties. Hence in the overall context the errors arising from the suggested test procedures should be an acceptable risk. The advantage is that the user now has test data that are relevant to the unit purchased.

The preceding suggested development is particularly important when related to axial flow fans. Their limited K_{th} capacity makes them more vulnerable than the centrifugal type. However, the swirl effects on downstream duct components will remain independent of fan type.

Logically, the length of downstream test ducting should be determined by outlet flow quality. However, the codes have to establish a "standard" length of duct covering all fan equipment. A length of 8.5 duct diameters of straight parallel tubing upstream of the measuring station is specified in the AMCA document [21.3]; in BS 848 [21.4] the requirement is for 5 duct diameters. Each code specifies a flow straightener of a completely different type (Figs. 21.1 and 21.2).

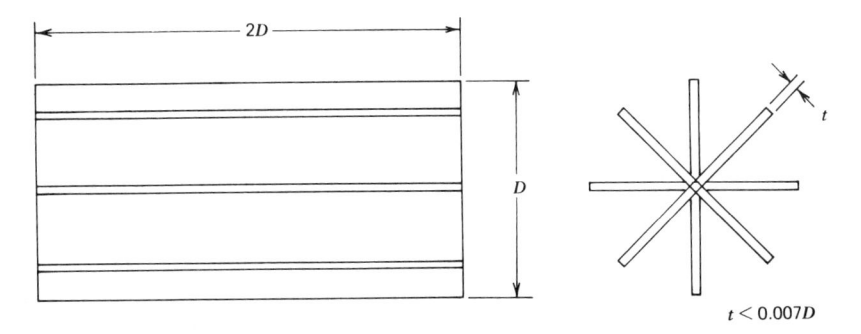

Figure 21.2 Airflow straightener, [21.4]

The flow straightener used in the British code originated in Belgium, being developed in relation to the International Standards Organization's fan code project [21.5]. The device by permitting radial flow movements enables a more rapid return to uniform flow conditions than does the AMCA version. This is reflected in the shorter duct specified.

The outlet duct length should logically be made conditional on flow quality for large-scale equipment where available finance and test space are both strictly limited. This could remove the need for model testing in certain instances.

Circular test ducts provide greater rigidity and simplify test procedures.

21.4 LABORATORY TEST METHODS

In principle, laboratory test methods should be centered on computer techniques, be economic in regard to equipment costs and laboratory space, and be of a limited number to enhance the comparative value of performance data for alternative makes and types of fan.

The AMCA Standard lists 10 test methods from each of which both FTP and FSP data can be obtained. In eight of these an auxiliary fan is required as a load control device, and a means of overcoming the loss in the venturi and nozzle-flow-measuring systems at the higher-volume flow rates, when fan pressures are low. Seven of these methods require a settling chamber of relatively large dimensions, particularly when the fan under test is located upstream.

Three of the test methods call for pitot-static traverses, the remainder relying on static pressure measurements at various stations. The latter data are suitable for computer collection and processing, resulting in a graphical display of performance.

The two test methods that are independent of auxiliary fans feature the test fan at either the inlet or outlet of the test duct setup. In the latter case, fan loading is applied with a throttle at the test duct inlet, and hence a duct length of 8.5 diameters is required from the inlet to the measuring station.

A multiplicity of test methods, all differing from the AMCA ones, is presented in BS 848 [21.4]. Once again both FTP and FSP are obtained with a single test procedure. Hence it is reasonable to assume that in the preceding codes one test method could represent the optimum, with a minimum number of alternative procedures to deal with special circumstances, such as small volume flow rates and double entry fans.

The inlet test duct of Fig. 21.3 illustrates a desirable combination of a controlled entry flow with swirl remover, a volume flow rate measuring arrangement of low loss (Fig. 21.4), and a loading device (Fig. 21.5) that ensures a uniform cross-sectional loss in total pressure. The sliding-plate system for resistance control must, however, possess excellent hole uni-

21.4 LABORATORY TEST METHODS

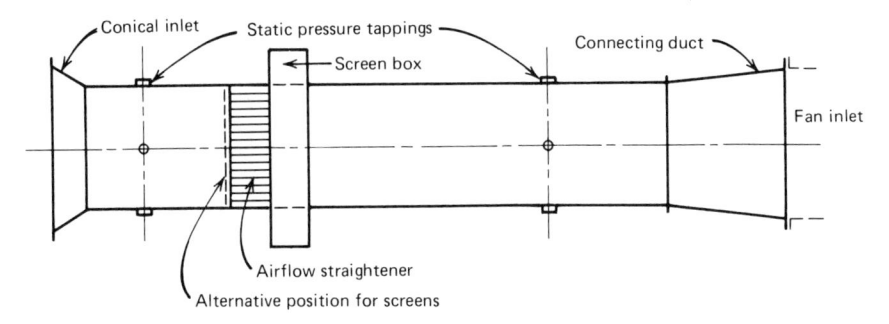

Figure 21.3 Inlet ducting arrangement.

formity and be fitted with a fine-threaded screw drive. Alternatively, wire screens may be inserted in the box or upstream of the air straightener.

A minimum length of test ducting is a consequence of the flow control designed into the test duct assembly. Calibration of the flow- and pressure-measuring systems is desirable but not mandatory, since information from Fig. 21.4 should normally be acceptable.

The stated objective of computerized collection and reduction of test data is also achieved, since traversing techniques are eliminated with this inlet assembly.

Single-entry exhaust fan tests can be carried out with the preceding test assembly. When the inlet flow to a double-entry centrifugal fan is bificated, the test assembly should be attached per medium of a transformation element ahead of the bification. For comparison of performance with axial

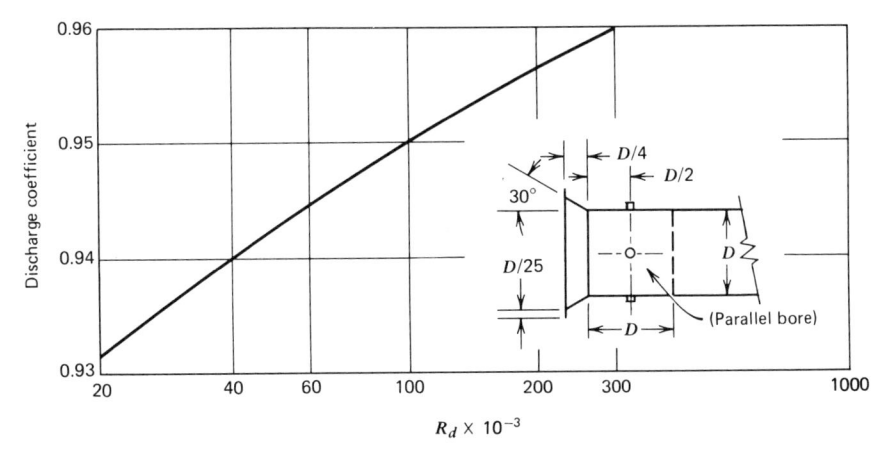

Figure 21.4 Conical inlet for volume flow measurement, [21.4]. (Reproduced by permission from BS 848: Part 1: 1980, published by the British Standards Institution, 2 Park Street, London W1A 2BS, from whom complete copies may be obtained.)

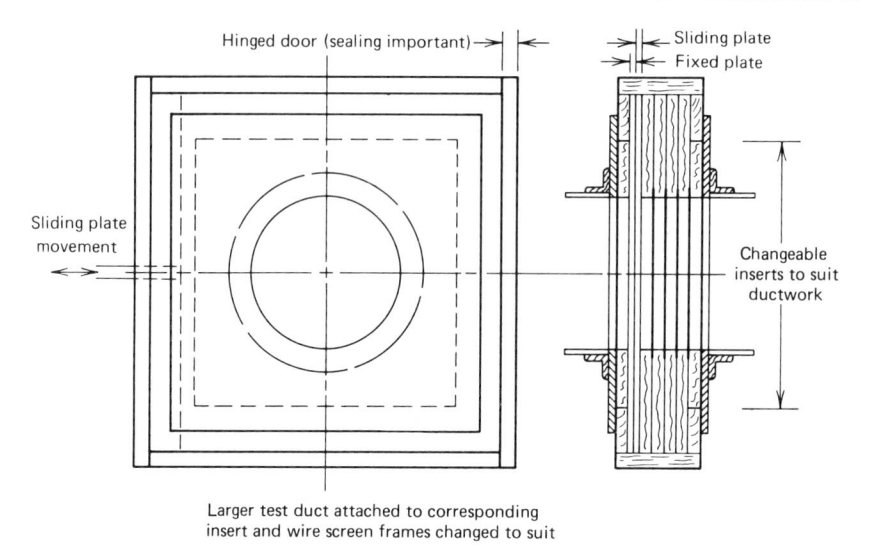

Figure 21.5 Schematic illustration of combination box for wire screen and sliding perforated plates.

units, and single-entry centrifugal fans, the bification ducts and inlet boxes must be included in the exhaust fan assembly.

The preceding compact inlet-flow-measuring and fan-loading system can be retained for in-line fans. Provided the fan unit possesses good-quality outlet flows, a minimum length of outlet duct equipped with a piezometer ring produces the required data without recourse to pitot-static or yawmeter traverses (Fig. 21.6).

When the fan outlet flow quality is below standard, the downstream length of test ducting must be increased and traversing techniques adopted. Measurements of yaw angle, total pressure, and static pressure are required at specified radial locations. From these the resultant velocity vectors are es-

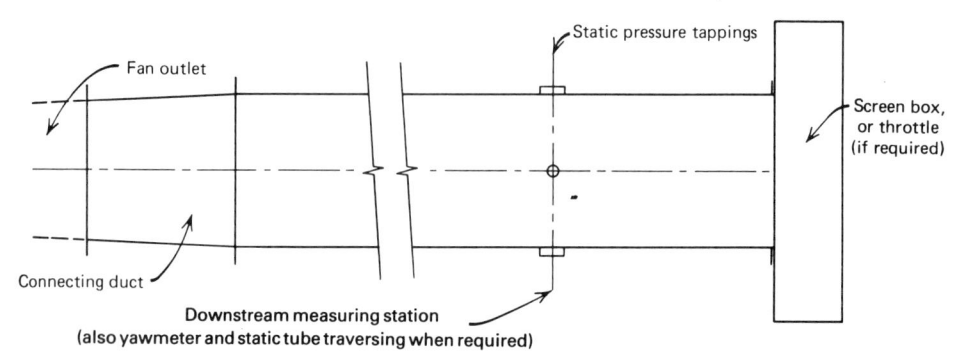

Figure 21.6 Outlet test ducting arrangement.

21.4 LABORATORY TEST METHODS

tablished for resolution into axial and tangential components. The former are the basis for volume flow rate determination.

The mean velocity pressure of the tangential components represents a nonuseful contribution to total pressure and must be subtracted from the mean measured total pressure in the process of establishing fan performance.

When laboratory space does not permit the retention of the inlet duct assembly, a short length of parallel duct with a flared inlet may be substituted. A suitably designed screen box can be transferred from the inlet duct to the point of discharge.

In presenting the test data, the angle of mean swirl and the mean swirl velocity pressure should be included as functions of volume flow rate. These quantities assume great importance for low-pressure units of the type illustrated in Fig. 13.5. The customer will increasingly become aware of the qualitative and quantitative value of this additional information.

Additional test methods should be reserved for special circumstances such as when the velocity pressures are too small for accurate flow rate determination. Employing a test method in which the flow is speeded up through a venturi, nozzle, or orifice plate will normally require the aid of an auxiliary fan.

The preferred selection is an inlet chamber method. (The chamber diameter is about 60% of that of an outlet chamber rig for a given fan size and type.) Selected alternatives from the AMCA and BS test codes are illustrated in Figs. 21.7 and 21.8. The former incorporates the flow-measuring system within the chamber and requires greater chamber length.

Exhaust fan tests require an entry duct of specified form (Fig. 21.9). The in-line fan must have a similar inlet and an outlet duct of suitable length, as discussed previously. However, in cases where low velocities make swirl determination difficult, the OTP for FTP calculation might be assumed to

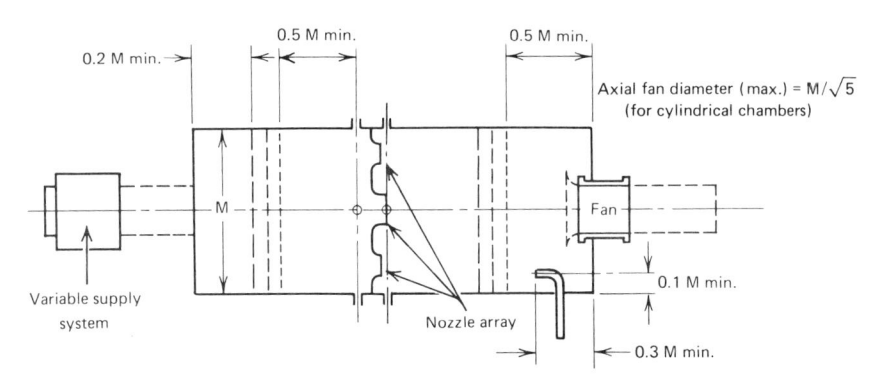

Figure 21.7 Inlet chamber test equipment, [21.3]. (Reproduced by permission of AMCA and ASHRAE from AMCA Standard 210-74, published by the Air Movement and Control Assoc. Inc. Arlington Heights, Ill., USA.)

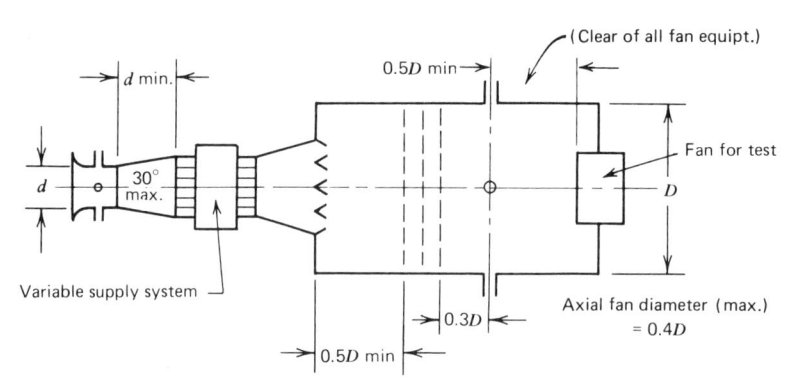

Figure 21.8 Inlet chamber test equipment, [21.4]. Reproduced by permission from BS 848: Part 1: 1980, published by the British Standards Institution, 2 Park Street, London, W1A 2BS, from whom complete copies may be obtained.)

equal the sum of the static and velocity pressures at the downstream station. In the case of low-flow, high-pressure blowers the errors are negligible, but for fans at the other end of the pressure scale the approximate nature of the test results should be highlighted.

Large manufacturing companies with a wide range of fan sizes to test would probably consider the chamber method as the optimum one for general use. A chamber sized for the largest fan can then be used for all fans of lesser diameter. The large laboratory space requirement and high capital cost can usually be justified in terms of workload and operational scale. However, since no special attention has to be given to the open return flow, a more fruitful use of floor space can be achieved.

Provided the specified minimum chamber-to-fan-inlet area ratio condition is met, double-entry fans of the *blower* type can be tested by this method. However, exhaust or in-line equipment with inlet boxes could require an

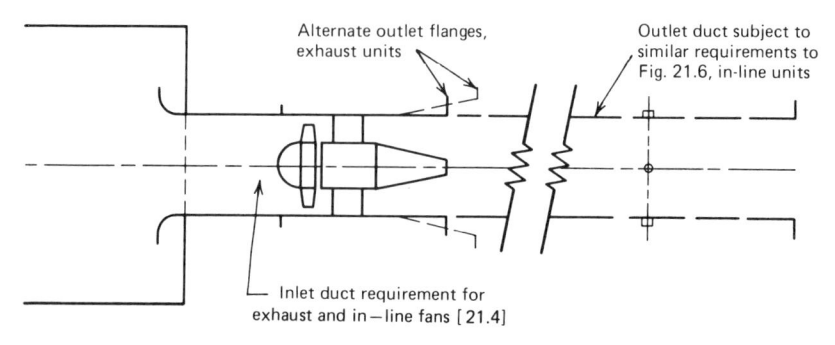

Figure 21.9 Axial fan arrangement, inlet chamber test.

outlet chamber method or a test arrangement similar to Fig. 21.3, with the latter providing the most practical and cost-effective solution.

As implied earlier, laboratory tests are not normally performed on large-scale equipment. When model tests are required, the fan size and experimental conditions should be selected with a view to obtaining data at Reynolds numbers clearly above the critical range (for axial fans see Section 10.2.2).

In the case of geometrically similar units, the nondimensional test results obtained under working conditions that ensure an above Re_{crit} situation should not be extrapolated downward to equipment operating in or below the critical range. Since Re_{crit} is dependent on many design and operational factors, accurate predictions cannot be made. However, when correct design and installation procedures are followed in relation to axial fans, Section 6.9 provides good guidance.

The test codes allow little or no scope for extrapolation to Reynolds numbers less than the test value. However, many models with speed control can be tested progressively through the critical range, thus enabling fan characteristics to be predicted for a wider range of geometrically similar fan units.

Measurable air compressibility effects will be present for fan pressures above 2.5 to 3 kPa. The complex problem is resolved in the fan codes by the use of a compressibility factor based on polytropic compression: this is considered more accurate than the previously used isentropic compression factor.

21.5 TEST CERTIFICATION

The foregoing rationalization of test methods has resulted in slight modifications to selected AMCA and BS test methods. However, all codes call for strict adherence to their specified procedures, which are meticulously presented in the relevant documents. Hence the suggested modified procedures must remain at present as voluntary options.

In the United States and Canada, the AMCA issues licenses to Approved Laboratories that operate in accordance with AMCA standards. The AMCA also tests fan equipment and carries out research in its own laboratories. Central test authorities exist in a small number of European countries, whereas in Australia laboratory registration for industry in general is given by the National Association of Testing Authorities.

The requirements for an AMCA license are enunciated in specific detail in [21.6]; NATA registration is granted on the basis of favorable reports from competent assessors. In Britain each fan company assumes a test responsibility, with the National Engineering Laboratory carrying out the supportive research into related code matters. However, a National Testing Laboratory Accreditation Scheme for general industry, under the control of the National Physical Laboratory, was inaugurated at the end of 1981.

Information on practices in other countries has not been specifically sought, but naturally large and well-developed test laboratories and authorities exist throughout Europe and in other parts of the world.

21.6 ON-SITE TESTING

With the approaching energy shortage, greater attention is being given to the installed effectiveness and efficiency of fan equipment. However, the break with the old tradition, where the emphasis was on compactness and capital cost, is slow. An upgrading of available design data and reeducation are urgently required.

A detailed analysis of fan acquisition for the mining industry is given in [21.7]. The procedures recommended can be extended to industry in general, provided acceptance of the basic principles by all concerned can be obtained. The building construction industry with its many diverse viewpoints and areas of responsibility is one in which consensus is extremely difficult to achieve.

Briefly, the ideal approach to fan acquisition is for a wide-ranging and expert study of various equipment alternatives, before completing the job specification and seeking tenders. In addition to indicating any special features desired, the document should detail the on-site test requirements. When the preceding study highlights a knowledge gap or area of uncertainty, provision should be made for model testing either by the contractor or other qualified authority.

The implementation of these principles within the Australian mining industry is reported in [21.8]; the same principles and standards have been successfully applied in other sectors of Australian industry. The minimal cost increase is often recovered within 6 months in reduced running expenses. On occasion, the initial study may result in considerable simplification of operational and maintenance requirements, with very large savings. Personal experience with a variety of industrial fan arrangements has shown that it is possible to gain millions of dollars by using better equipment and eradicating lost production time.

Both the AMCA [21.1] and BS [21.4] fully acknowledge the vital importance of collaboration between supplier and user in achieving improved installation design and satisfactory test conditions. In South Africa there is now a realization that improved cooperation between the interested parties is very desirable [21.9].

The design information available herein on contractions, corners, and short diffusers must result in equipment assemblies of comparable compactness to current installations. Hence the codes should offer some incentive to designers making full use of these data. For instance, a Class A test procedure that specified flow quality requirements at the measurement stations, and the fan inlet, would be suitable for inclusion in any national code. Provided these conditions were met, the actual placement of measuring planes

21.8 PRESENTATION OF FAN PERFORMANCE DATA

and points could and would vary greatly with installation features. A strong inducement to most design and test teams to meet Class A standing would subsequently be created.

In addition to flow quality, prime importance should be placed on establishing duct resistance and fan duty in terms of total pressure, by measurement or summation.

The use of vane anemometers in Class A on-site test programs is not recommended. Experience has revealed substantial discrepancies between measurements taken with different makes and sizes of instrument, despite valid wind tunnel calibration certificates. Overestimation of flow quantity is the usual fault.

21.7 PACKAGED EQUIPMENT

The development of low-noise, high-efficiency fans for car cooling systems, air conditioners, air heaters, and similar unitized equipment involves testing the fan in a similar environment to the final assembly.

In normal circumstances only the volume flow rate and fan power are required to be determined. (An approximate estimate of system and fan losses is required in establishing the order of the K_{th} for design.) The chamber methods of Figs. 21.7 and 21.8 can be used for measuring the former by adjusting the auxiliary fan to give an identical inlet total pressure condition to that experienced by the actual installation. Discharge may be either to the atmosphere or to a duct system the resistance of which can be simulated by an adjustable throttle. The development of a suitable fan can be greatly speeded by this method, which ensures accuracy in capacity determination. A great variety of duct/fan assemblies can be accommodated, the only requirements being to preserve inlet and outlet geometries and to maintain inlet and outlet total pressure conditions.

For large equipment a scale model may be required in developing a suitable item of equipment.

The efficiency of small motors can vary greatly with type, make, and care in manufacture and assembly. Hence electrical input measurements will not permit accuracy in fan power determination. A simple rope or prony brake system can readily be produced for motor calibration. In very small motors, the power output remains relatively insensitive to input current. However, motor speed reduces quickly with load; hence this variable may provide a more convenient measure of fan power, after motor calibration.

21.8 PRESENTATION OF FAN PERFORMANCE DATA

The fan codes specify the procedures that must be followed in obtaining graphical data, of which Fig. 21.10 is typical. For a variable geometry fan with adjustable stator or rotor blades, the normal presentation method is

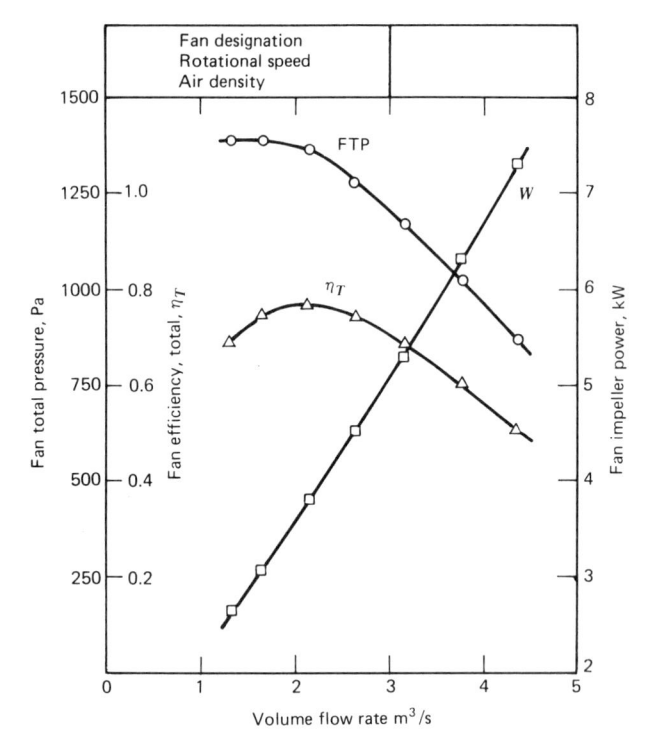

Figure 21.10 Typical fan characteristic curves.

illustrated in Fig. 20.2. The efficiency contours represent mean values of the measured quantities at different pitch settings.

In displaying type test data the actual experimental points may be omitted. It is important to qualify such data with regard to manufacturing tolerances, surface finish, and Reynolds number trends.

The appropriate test data presentation for geometrically similar fans is illustrated in Fig. 21.11, utilizing nondimensional coefficients. From these simple relationships the desired performance curves can be constructed for any combination of fan size, speed, and air density. For reasonable agreement between the predicted and actual fan performance, strict nondimensional similarity in construction is essential. For example, blade shapes—particularly in the vicinity of the leading edge, where burrs, casting dressing, and other malformations may be present after manufacture—should be closely controlled; blade settings should be held within $\pm\frac{1}{4}°$ of the specified value; and a nondimensional similarity should exist in respect to surface roughness and to blade clearances at root and tip. Some adjustment to the performance curves, on the grounds of Reynolds number considerations, will be required.

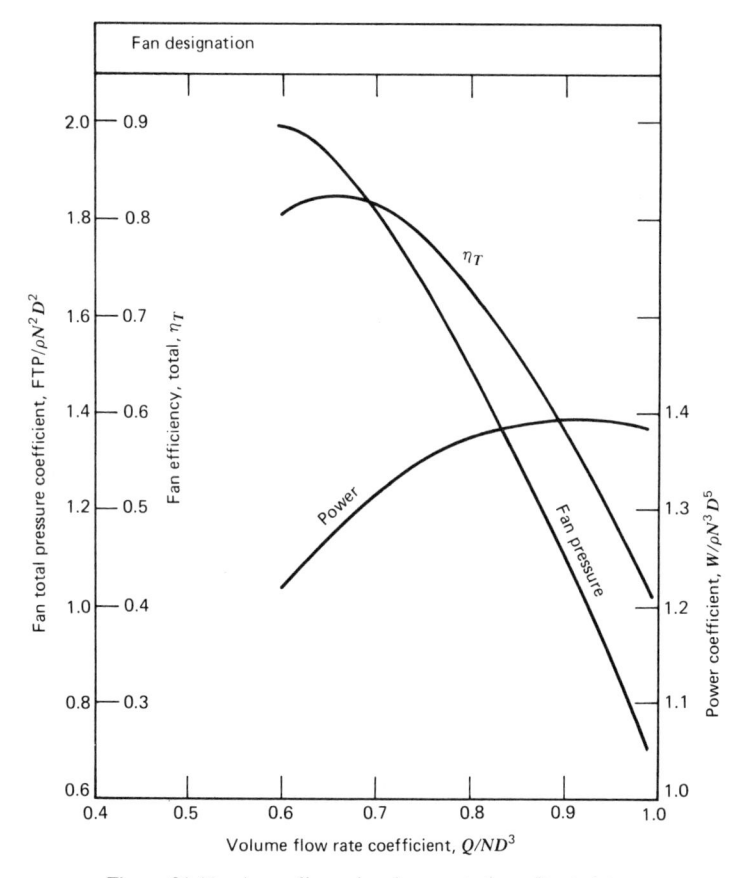

Figure 21.11 A nondimensional presentation of test data.

Making accurate adjustments for the latter raises problems as Reynolds number effects are closely related to leading-edge shape and surface roughness; these geometric properties in industrially produced equipment vary from fan to fan. Provided the operating Reynolds number is above the critical range, as determined during the model test program, the degree of uncertainty should be within acceptable limits. Complete dynamic similarity (see Section 2.4.1) is usually unattainable.

A number of the foregoing recommended features are under consideration with respect to a draft Australian test specification [21.10].

REFERENCES

21.1 Air Moving and Conditioning Association, *Fan Application Manual:* Fans and Systems Publ. 201-73, Troubleshooting Publ. 202-72, A Guide to the Measurement of Fan-System Performance in the Field, Publ. 203-76, Arlington Heights, Ill., 1972–1976.

- **21.2** Institution of Heating and Ventilating Engineers, *Report of the Fan Standardization Committee appointed by the Institution of Heating and Ventilating Engineers*, London, June 1927.
- **21.3** Air Moving and Conditioning Association, Laboratory methods of testing fans for rating, *AMCA Standard 210-74*.
- **21.4** British Standards Institution, Fans for general purposes. Part 1: methods of testing performance, *BS 848: Part 1*, 1980.
- **21.5** International Standards Organization, "Air performance test methods of industrial fans using standardised airways" (in preparation).
- **21.6** Air Moving and Conditioning Association, Certified ratings program air performance, *AMCA Publ. 211-80* and *Publs. 3 to 5*, Arlington Heights, Ill., 1980.
- **21.7** R. A. Wallis, Primary ventilation in mines: Some factors affecting the acquisition of new fan equipment. *Proc. Eighth Commonwealth Mining and Metallurgical Congress*, 1965, pp. 1487–1494.
- **21.8** N. Ruglen, Requirements for on-site testing of axial flow fans, *Proc. Australasian Inst. Min. Metall. No. 245*, March 1973, pp. 17–24.
- **21.9** M. N. Harrison et al., Application of the thermomatic fan test method to the testing of main upcast mine fans (and discussion), International Mine Ventilation Congress, Johannesburg, 1975, pp. 81–91.
- **21.10** Standards Association of Australia, "Determining the performance characteristics of industrial fans" (in preparation).

CHAPTER 22
Fan Testing—Developmental Research

22.1 GENERAL DISCUSSION

A thorough understanding of fan flow features is essential in dealing with design problems, in developing special fan equipment, and in resolving troublesome flow problems. This knowledge cannot be acquired without experimentation of a wide-ranging nature.

A rotor-stator unit designed for free vortex flow, and installed in a well-designed and matched duct system, will perform in a predictable and satisfactory manner. The amount of testing required is either minimal or nil. When the volume flow rate and fan pressure have to be regularly monitored, for example, in ventilation and power generation systems, instrument calibration tests are required.

However, the need for experimentation is greatly increased when the blading produces an arbitrary vortex flow; these design development needs are discussed in Appendixes A to C.

The four segments of fan testing recommended here are as follows:

1. Visual studies to qualitatively establish flow directions, regions of disturbed flow, and possible areas of separated flow from rotating and stationary surfaces.
2. Determination of volume flow rate and mean total pressure, at selected reference locations.
3. Traverses upstream and downstream of a particular blade row to determine the radial distributions of velocity, total pressure, and flow direction.
4. Power determination.

The basic features of these procedures will be outlined prior to discussing their use in solving problems.

Techniques for measuring boundary layers and static pressures on rotating blade surfaces were considered to be outside the present developmental research scope.

22.2 VISUAL STUDIES

Most experimental fan rigs feature a duct wall panel(s) composed of Perspex placed in the desired location(s). However, unless specially requested in the job specification, this facility is unavailable in commercial equipment. When the on-site test program is planned at the job specification stage, it is wise to consider fitting two adjacent windows, one for introducing light and the other for viewing; the fans featured in [22.1] possess this facility. Windows are sometimes fitted retrospectively to commercial installations that possess flow problems.

Assuming the means for viewing exists, whether directly or through windows, smoke or tufting can provide much useful information to the experienced observer. An important aspect of this work is to identity flow patterns, and in particular the location, extent, and growth of duct and blading separation, with changing duty conditions. Surface-attached tufts, a tufted fine-wire grid at a duct cross section, a tufted surface-attached post, and a wandering tufted wand all provide useful data. In the rotor case, goose feathers (from a soft feather duster) provide a high-drag, low-density, and highly flexible tuft material, minimizing centrifugal force effects; stroboscopic lighting is essential. All tufting must be neatly surface-attached with tape or glue (no uplifted edges). The wand should be of adequate stiffness to prevent whipping in the airstream but the tip to which the tuft is attached will require a size reduction. The tuft length should match the application. For example, on duct and blading surfaces the length will be in the range 1 to 5 cm, depending on boundary layer conditions, and will approach 15 cm for tufted wands in large-scale free shear flows; on occasion cloth strips, which resist a tendency to knot around the wand tip, are preferable for highly turbulent flows. Cotton and wool threads are the two most common tuft materials.

Coatings that indicate surface air movement are a useful aid, but considerable experience and aerodynamic knowledge are essential in correctly interpreting the traces produced, particularly on rotating surfaces. The interaction between the radial static pressure gradient and centrifugal forces that act on air particles in boundary layer and separated flow must be adequately assessed. Inadvertent duct or fan surface coatings of mud or dust often provide an initial clue.

A greater use of visual studies is strongly advised. When the detailed flow pattern is known, the type and extent of testing can be defined, together with the most appropriate instrumentation. For example, using a standard pitot-static tube to measure velocity in a swirling flow can produce gross errors. A tuft of suitable length and flexibility attached to the instrument head will enable the tube to be approximately aligned with the stream direction, for minimum error. In addition, the identification, location, and extent of a separated flow by visual means will avoid the taking of surface static pressure and pitot-static readings of uncertain and misleading value.

22.3 VOLUME FLOW RATE AND MEAN TOTAL PRESSURE 363

Visual studies can also be employed in troubleshooting exercises. It is sometimes unprofitable to undertake extensive test programs on installations that have failed to give satisfaction. A limited visual investigation will often establish the source and general nature of the problem, opening the way to remedial measures and subsequent complete test programs. Often the studies will involve unpleasant and difficult conditions for test personnel, but the rewards are substantial. Instrument readings without an actual, physical picture of the troublesome flow feature can produce negative thinking, documenting a difficulty without necessarily providing guidance for subsequent fruitful action.

Where duct or blade surfaces are hidden from view, the use of fixed mirrors and special lighting should be seriously considered.

The interpretation of tuft motions is an art that develops as the investigator gains experience.

22.3 VOLUME FLOW RATE AND MEAN TOTAL PRESSURE DETERMINATION

Design development demands more information than that available from code testing. In studying the spanwise blade loading features, a test rig similar to that illustrated in Fig. 22.1 is essential. This item of equipment ensures relatively uniform inlet and outlet flows and provides a parallel annular passage upstream and downstream of the blading for traverse experiments. When the drive motor is located upstream of the fan assembly, it becomes necessary to fit a contracting annulus of the type illustrated in Fig. 22.2; fan load is then applied on the downstream side.

The open-return air circuit results in some degree of nonuniformity in fan entry flow. Corrective measures include honeycomb air straighteners and inlet screens. Alternatively, a large, low-porosity screen of perforated metal or nylon cloth placed upstream and enclosing the inlet area (Fig. 22.3) ensures a relatively swirl-free and steady inflow.

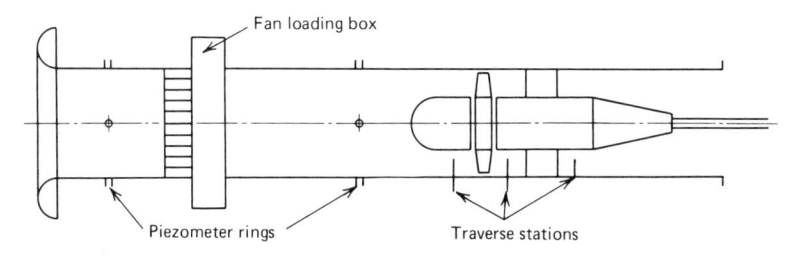

Figure 22.1 Suitable test rig for blading research studies.

364   FAN TESTING—DEVELOPMENTAL RESEARCH

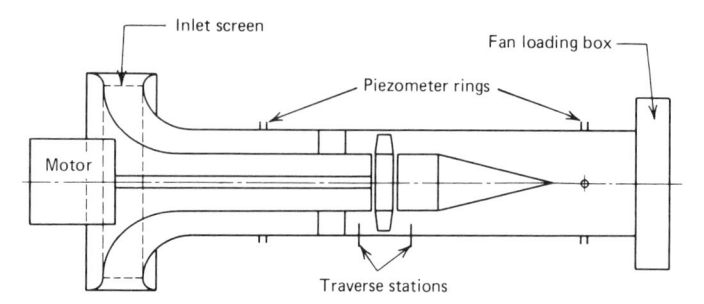

Figure 22.2 Alternative test rig for blading research studies.

One of the main sources of nonuniformity is the ground vortex that results from large-scale angular momentum being intermittently converted into the intense small-scale variety. The screen of Fig. 22.3 eliminates this phenomenon. A device that is successful in certain circumstances is illustrated in Fig. 22.4; the center of the radiating arms should be coincident with the original vorticity axis. Dimensions are established by trial-and-error procedures.

Volume flow rate is the major reference parameter in fan developmental studies. This quantity is often determined from calibrated surface static tappings. The internal duct surface in the vicinity of the tappings should be free of waviness within a distance of 20 hole diameters and the hole geometry should conform to the instructions of Fig. 22.5. There are a number of ways in which the four tappings can be interconnected to obtain a mean static pressure reading. The method illustrated in Fig. 22.6 is gaining acceptance following the research findings of [22.2].

The pressure differential with reference to atmospheric pressure, namely, ΔP, enables the duct volume flow rate to be established from a single pressure reading. It is important to ensure that the inlet equipment remains clean, undamaged, and unchanged at all times. For any given flow rate, two pitot-static tube traverses on diameters 90° apart provide data for calculating

$$\text{Volume flow rate per quadrant} = \frac{2\pi}{4} \int_0^R ur\, dr \quad (22.1)$$

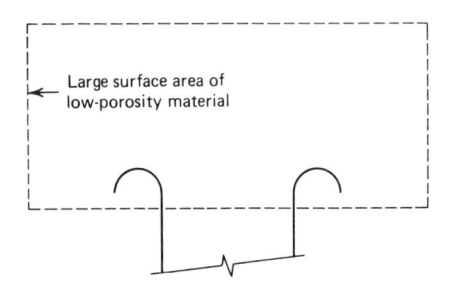

Figure 22.3 Inlet-flow smoothing device.

22.3 VOLUME FLOW RATE AND MEAN TOTAL PRESSURE

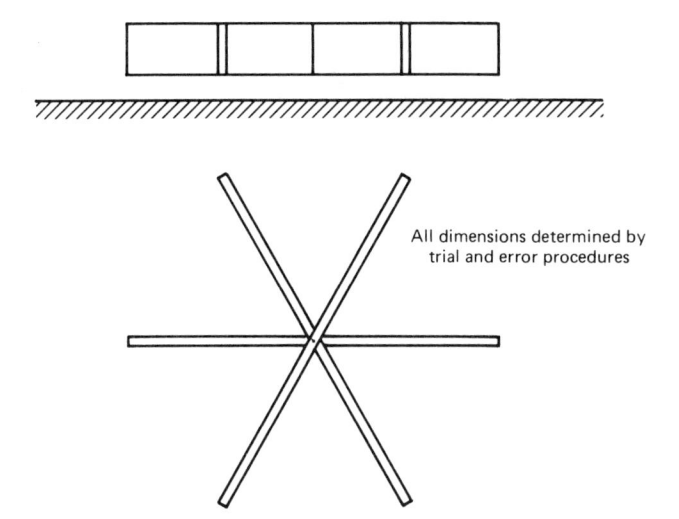

Figure 22.4 Suggested device for ground-effect vortex control.

By summing all quadrant flows, the total volume flow rate Q is obtained for the measured pressure differential ΔP at a density ρ. Repeating the process for four or five flow rates will provide data for a straight line graph of Q versus $\sqrt{\Delta P/\rho}$, provided the inlet is free of Re effects. The slope of the line then determines the constant C_c in the equation

$$Q = C_c \sqrt{\frac{\Delta P}{\rho}} \tag{22.2}$$

which is then used in all subsequent calculations. A trip wire may be required on the bellmouth entry in achieving Re insensitivity.

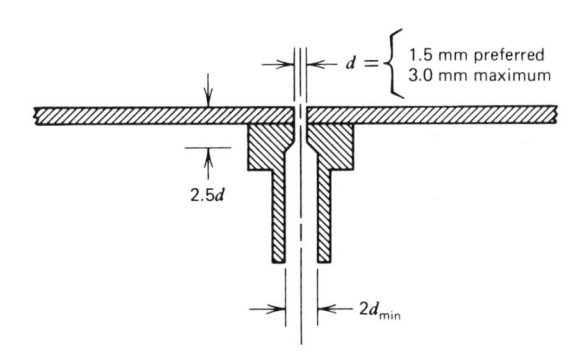

Figure 22.5 Surface static pressure tapping.

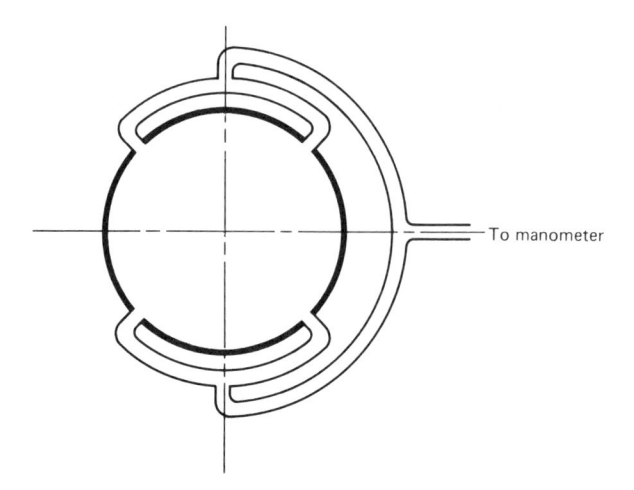

Figure 22.6 Recommended method for connecting surface static pressure tappings, [22.2].

The mean velocity at any chosen duct cross section is established once the local area is known.

In the graphical integration of Eq. (22.1) it is important to map the outer wall boundary layer flow as accurately as possible. This requires smaller spacing intervals as the wall is approached.

The corresponding calibration procedure for the rig shown in Fig. 21.2 is similar, with the inner limit in Eq. (22.1) now r_i, not zero. The inner wall boundary layer should also be traversed at small intervals. Most rigs of this kind possess a wire screen at the inlet that must be cleaned regularly. However, the small total pressure loss across the screen (caused by low inlet velocities) ensures a reasonable degree of test insensitivity to minor dust buildups on the screen.

An alternative calibration method consists of using a standard orifice plate setup to which the equipment to be calibrated is attached upstream. This technique is especially suited to small-scale test equipment.

The use of an orifice plate in the test rig as a flow-measuring device is acceptable, provided it is located downstream of the fan, thus avoiding fan inlet flow disturbances. Since the resulting arrangement is unlikely to conform to the requirements of the British Standard, 1042, Part 1, 1964, (on orifice plates), calibration of the interchangeable orifice plates can be attained by the traverse technique described earlier. Because the orifice plate will constitute an appreciable fan pressure load, this measuring system will normally be restricted to the higher-pressure-rise fans. The ensuing relationship is

$$Q = C_0 d^2 \sqrt{\frac{\Delta P}{\rho}} \tag{22.3}$$

22.4 RADIAL TRAVERSES

where d is the orifice diameter, C_0 is the orifice coefficient which may vary with Re, and ΔP is the static pressure differential across the orifice.

The piezometric ring device described earlier can be employed in the measurement of mean total pressure, provided the flow is of good quality. Normally this requires the flow to be axisymmetric, possessing less than 10° of swirl. The magnitude of error for nonconforming flows is unknown, but considerable interpretative caution should be exercised in all such cases.

For conforming flows the mean total pressure at a specific cross-section is given by

$$H = P + \tfrac{1}{2}\rho \bar{U}^2 \tag{22.4}$$

where P is the piezometer ring pressure and $\bar{U}$ is calculated from the local area and the volume flow rate Q as determined in one of the preceding ways.

Alternatively, mean total pressure can be established from traverse experiments.

22.4 RADIAL TRAVERSES

Detailed studies of blade performance require measurements of flow features both upstream and downstream of a particular blade row. The various types of instruments are

1. Pitot-static tubes (Fig. 22.7).
2. Total pressure tubes.
3. Yawmeters.
4. Hot-wire anemometers.

The sensitivity of two types of pitot-static tubes to flow direction is indicated in Fig. 22.8. When an open-ended tube is used solely for total pressure

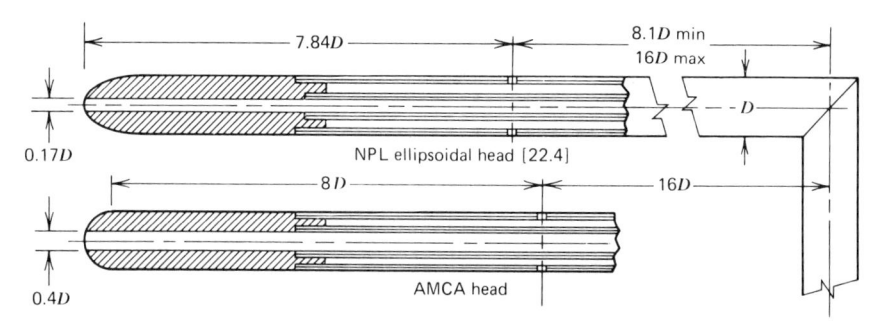

Figure 22.7 Alternative pitot-static tubes.

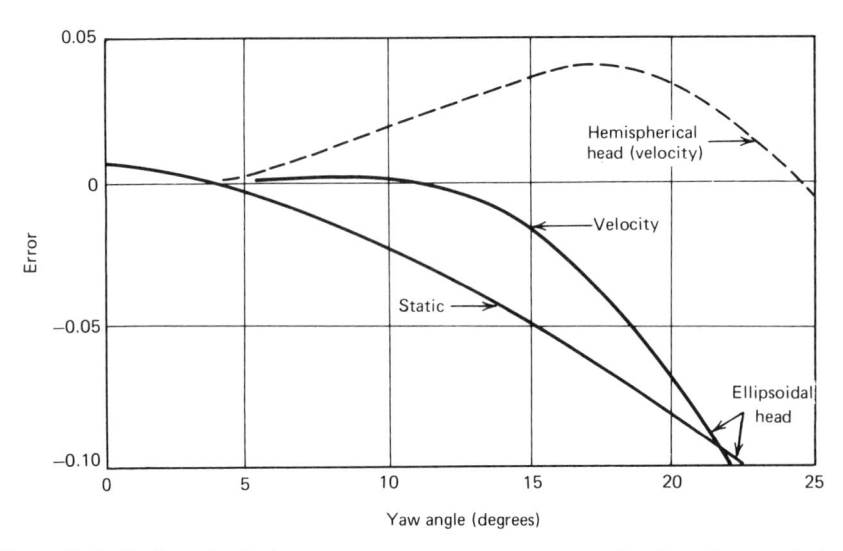

Figure 22.8 Static and velocity pressure measurement errors as fraction of true velocity pressure. (Adapted from D. W. Bryer and R. G. Pankhurst [22.4] and reproduced with the permission of the Controller of Her Majesty's Stationery Office)

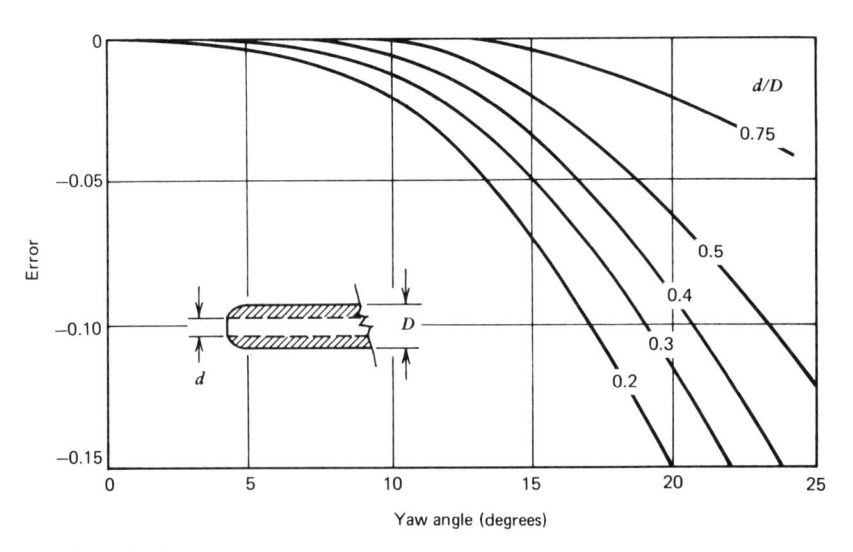

Figure 22.9 Total pressure measurement errors as fraction of true velocity pressure. (Adapted with permission of Pergamon Press from E. Ower and R. C. Pankhurst [22.15]).

22.4 RADIAL TRAVERSES

measurement, the hole to tube diameter ratio can be increased, with beneficial results (Fig. 22.9) from a tolerance viewpoint.

Flow direction can be determined with a variety of yawmeters. The two most popular in fan research usage are the combination transverse-cylinder and Cobra instruments (Fig. 22.10). The instruments are rotated until equal pressure is obtained in the two side orifices. The center orifice then registers total pressure. The angle of inclination is measured by a protractor fixed to the tube stem.

By keeping a three-hole Cobra tube aligned in the axial direction it is possible with special calibration techniques to obtain yaw, velocity, and total pressure information [22.3]. A calibration wind tunnel and the use of a computer are two additional requirements.

A hot-wire anemometer (Fig. 22.10) can be used in a similar manner to measure yaw angle and velocity, although it possesses a reduced sensitivity to yaw at low angles. When comparison was made between this instrument and the above Cobra tube arrangement on a fan rig, satisfactory agreement was achieved [22.3].

The three-hole cylindrical yawmeter can also be used to obtain an approximation of flow velocity. The side holes specified in Fig. 22.10 register a pressure that is close to the local static pressure, when pressure balanced. However, since the surface static pressure gradient is at a maximum and the pressure is subject to change with Reynolds number and stream turbulence, measurement precision cannot be claimed. It is important to calibrate the instrument in a wind tunnel of comparable stream turbulence, and over the appropriate Re range, in order to establish the K factor. Hence,

$$\tfrac{1}{2}\rho V^2 = K\Delta P_y \qquad (22.5)$$

where ΔP_y is the differential between the center and side hole pressures. However, the hooked rotatable static tube of Fig. 22.11 is the preferred instrument. Direct calibration of the transverse tube in the exact environment then becomes possible.

The existence of gradients in total pressure and hence velocity dictates the use of small-scale instruments. The associated sluggish response of water-type manometers, and micromanometers, to this small-bore equipment has increased the use of electrical gear, such as the Furness micromanometer, where the latter is regularly calibrated against a water-filled Betz or similar-type micromanometer; electrical instruments often exhibit drift. The use of computers for data storage and reduction is aided by electrical test outputs.

Details of instrument construction, calibration, limitations, and usage can be obtained from [22.4].

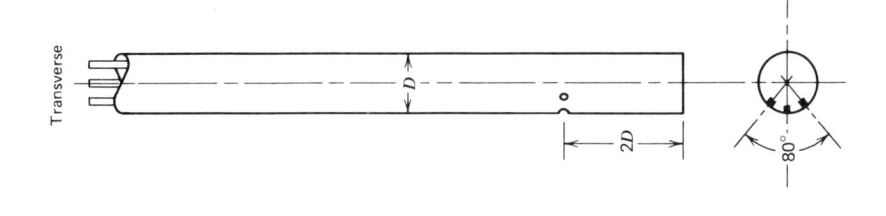

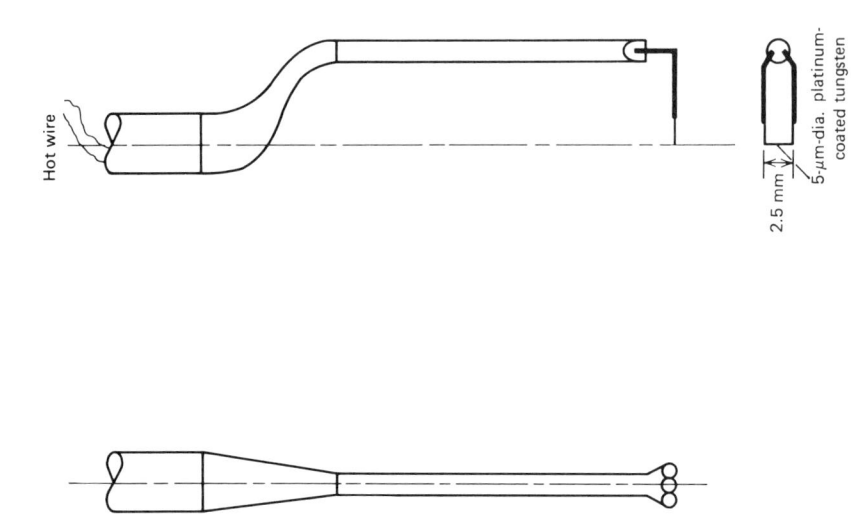

Figure 22.10 Three yawmeter types.

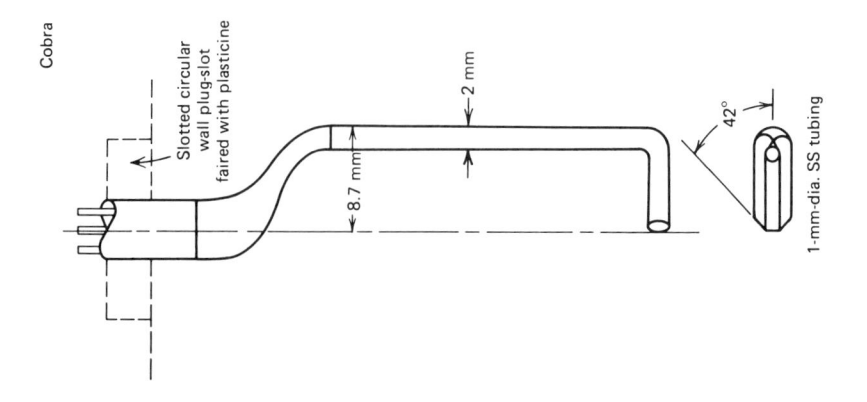

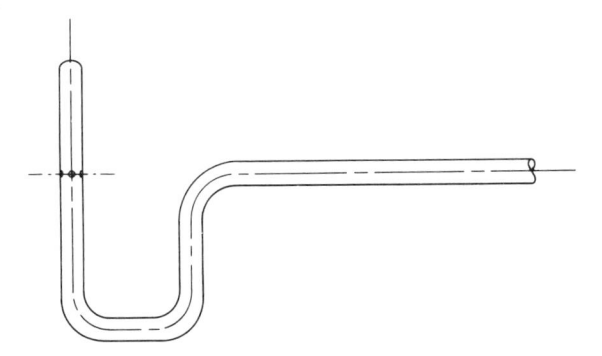

Figure 22.11 Hooked static pressure tube head.

22.5 POWER DETERMINATION

Most laboratory test rigs feature a common power source that drives the rotor shaft through a transmission-type dynamometer. The swinging arm and torsion variants are the types usually favored. In the case of the former it is important to have a drive motor that is balanced on the drive shaft axis. When some imbalance exists, it is essential to return the arm to its calibrated horizontal position. The sensitivity and accuracy both deteriorate as the mass out-of-balance of the driver source is increased.

Torsion-type equipment has either electrical or mechanical torque-indicating outputs. The latter, which is the most common and more reliable, requires an optical reading of the angular deflection of a rod, tube, or coil spring. A simple and relatively cheap unit of the spring type is illustrated in Fig. 22.12. Although a direct drive variant could be designed, it would need to be more rugged to absorb the high starting torque on a rubber overload stop.

When driving through a V-belt, the load can be applied gradually by a belt-tensioning device. Either a stroboflash or a slotted plate with additional illumination is suitable for viewing the spring deflection. A range of springs is desirable, for different power ranges.

The torsion dynamometer may be calibrated either statically by weights hung on an attached arm or dynamically by means of an absorption unit. The friction block or rope type is suitable for small power calibrations, but electric power generation or hydraulic absorption units should be considered whenever the power is large. A linear relationship between torque and deflection is essential for accuracy.

When rotor shaft power is obtained from motor input data, the motor calibration procedure should be carefully controlled. The identical instruments used in the motor calibration tests should, where possible, be used in

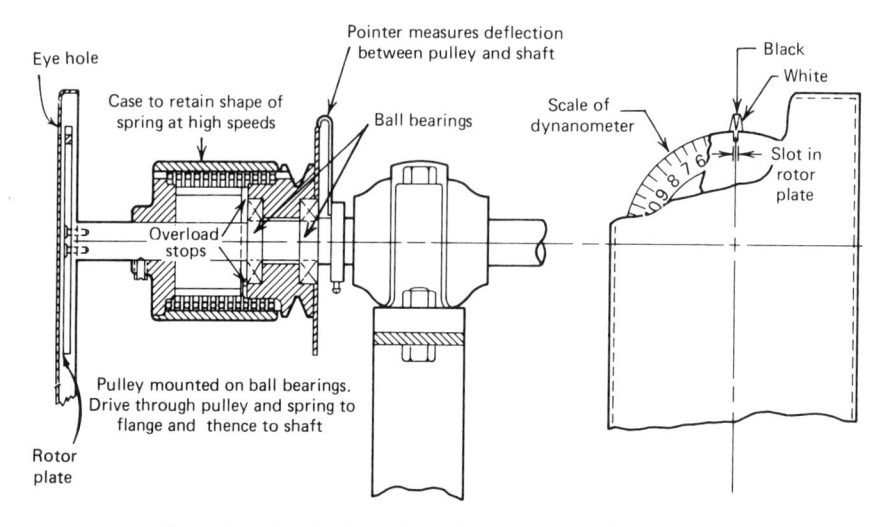

Figure 22.12 Simple torsion-style transmission dynamometer.

the fan experiments. In any case, the instruments should be calibrated before and after a fan test program, against a standard reference unless experience indicates a lesser calibration frequency.

The latter procedure is not suited to the smaller electric motors, since self-heating tends to destroy calibration sensitivity to motor output load. Motor speed slip provides a better measure, but high accuracy is unlikely to be achieved by this technique.

22.6 AXIAL AND CIRCUMFERENTIAL LOCATION OF TRAVERSE STATIONS

The stations should be located in parallel duct passages, to avoid mean axial velocity gradients. In particular, a traverse should not be taken in close proximity to the trailing edge of the rotor blades or the straightener vanes. The studies of [22.5] and [22.6] indicated errors of the type illustrated in Fig. 22.13. Because of the flow complexity [22.7], [22.8] in regions of wake-chopping phenomena, no precise explanation can be offered. Downstream distances of close to half the blade chord are recommended.

One radial traverse location is sufficient downstream of a rotor, since an axisymmetric flow situation is assumed to apply. A number of radial traverse locations have to be provided downstream of stator vanes, since the wake locations are fixed.

22.7 MEAN VALUE CALCULATIONS FROM TRAVERSE READINGS

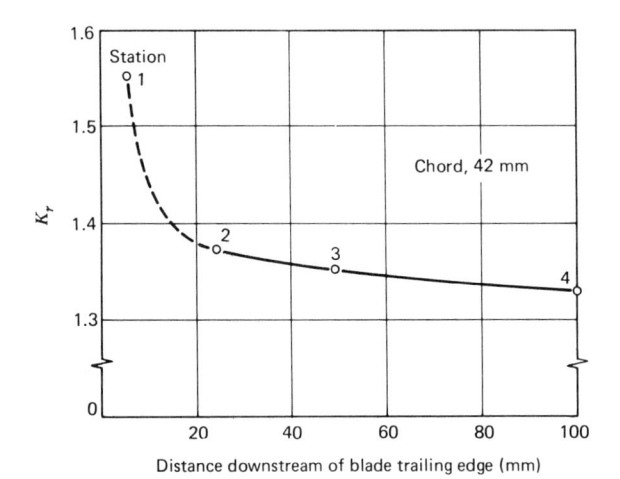

Figure 22.13 Measured total pressure rise versus downstream distance, cambered plate fan.

22.7 MEAN VALUE CALCULATIONS FROM TRAVERSE READINGS

According to [22.9], the more precise method of averaging individual total pressure readings to obtain a mean value should be based on mass. However, in experimental fan work the mean values obtained from area averaging continue to be used.

22.7.1 Mean Velocity

The total volume flow rate is given by Eq. (19.1),

$$Q = 2\pi \int_{r_b}^{R} V_a r \, dr$$

and mean axial velocity by Eq. (19.2),

$$\bar{V}_a = \frac{2}{1 - x_b^2} \int_{x_b}^{1} V_a x \, dx$$

22.7.2 Mean Total Pressure Rise

When Δh is the local difference in total pressure across a rotor blade row, the mean total pressure rise ΔH is given in similar terms to Eq. (19.2), namely,

$$\Delta H = \frac{2}{1 - x_b^2} \int_{x_b}^{1} \Delta h \, x \, dx \qquad (22.6)$$

or

$$K = \frac{2}{1 - x_b^2} \int_{x_b}^{1} kx \, dx \qquad (22.7)$$

The mean total pressure downstream of a stator row requires the evaluation of a double integral, expressed by

$$H = \frac{n_s}{\pi(1 - x_b^2)} \int_{x_b}^{1} \int_{0}^{2\pi/n_s} hx \, d\theta \, dx \qquad (22.8)$$

where h is the local total pressure and n_s is the number of stator vanes. It is assumed that the flow in all sectors is identical; hence the various traverses are confined to one-stator vane spacing. The total pressures are with references to atmospheric pressure.

As an alternative, a plot such as that in Fig. 22.14 can be divided into a large number m of equal areas to each of which a value of h is ascribed. The mean value is then

$$H = \frac{\Sigma h}{m} \qquad (22.9)$$

22.7.3 Rotor Torque

The angular momentum change across a rotor blade row can be equated to the rotor shaft torque. Hence an integration of the swirl properties will give a torque value that closely approximates that obtained by more usual means.

As a development of equations presented in Chapter 8, it can be shown that for the general case

$$T_c = \int_{x_b}^{1} \frac{2(V_{a_1} + V_{a_2})(V_{\theta_s} + V_{\theta_p})x^2 \, dx}{\bar{V}_a^2} \qquad (22.10)$$

The torque then follows from Eq. (8.35), namely,

$$T = T_c \tfrac{1}{2} \rho \bar{V}_a^2 \pi R^3$$

22.8 TRAVERSE EXPERIMENT OBJECTIVES

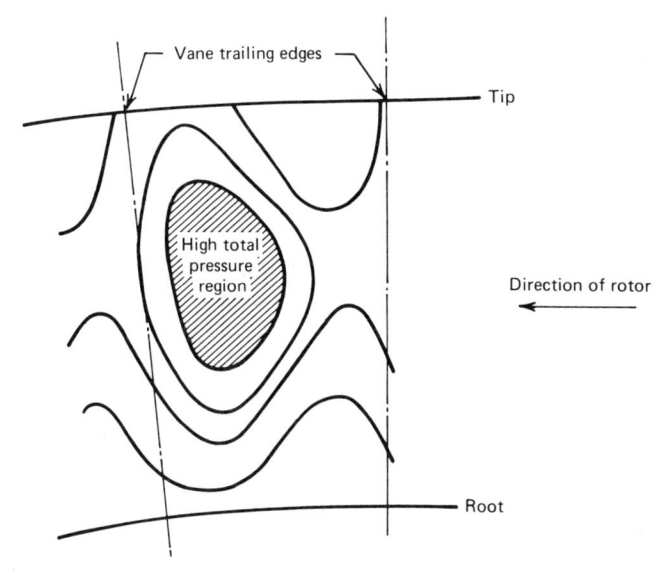

Figure 22.14 Typical total pressure contours aft of stators, [22.6]

22.7.4 Power from Yawmeter Traverses

Shaft power is calculated from Eq. (14.13), namely,

$$W = 2\pi TN$$

where N is shaft speed in rev/s when using the foregoing T value.

22.8 TRAVERSE EXPERIMENT OBJECTIVES

A major objective is in proving, or improving, the mathematical design methods available. The validity of the free vortex method for low-solidity fans was established a long time ago. More recently, justification was required for the proposed unified blade design technique, where cascade interference factors are introduced for use in the inner blade span region. The results of this work are most encouraging but a continuing experimental effort is desired in order to study the interference factor data of Fig. 6.29, in a completely general context. The "axial velocity ratio" (Chapter 23) is important in this respect, particularly for multistage units.

Publications relating to traverse experiments on fans possessing an arbitrary vortex-type blading are meager. Since a sizable proportion of current "off-the-shelf" fans is in this category, the recent experimental studies of Appendix B make a most important contribution to axial flow fan technology. Distributions of the nondimensional axial, tangential, and radial velocity components are displayed for rotors possessing a 0.38 boss ratio. The design conclusions drawn from these data are discussed in the Appendix.

In a previous publication [22.10], experimental traverses are available for two 0.69 boss ratio rotors of arbitrary vortex flow design. The distributions of axial velocity, total and static pressure, and swirl are reproduced in Figs. 22.15 to 22.18, showing good agreement with the design estimates.

The two sets of experimental results at opposite ends of the boss ratio range promote a real sense of confidence in the design methods expounded. The recommendations regarding design limitations on some of the variables should also add to a sense of technical security.

For free vortex flow rotors, the design assumption of constant total pressure rise across the rotor is approximately attained for design conditions, as illustrated in Figs. 22.19 to 22.21 for three completely different fan types [22.1], [22.5], and [22.11]. The corresponding swirl coefficient distributions are available in Figs. 22.22 to 22.24.

In the variable-pitch fan, the square cutting of the blade root for flow reversal reasons has undesirable consequences that are particularly evident

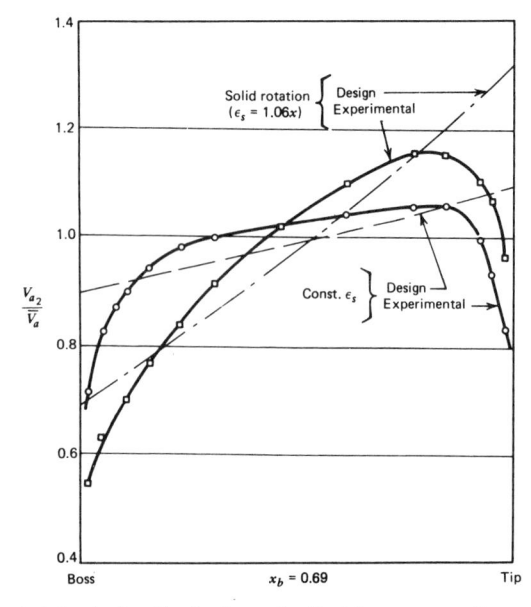

Figure 22.15 Axial velocity distributions aft of arbitrary vortex flow rotors, [22.10].

22.8 TRAVERSE EXPERIMENT OBJECTIVES

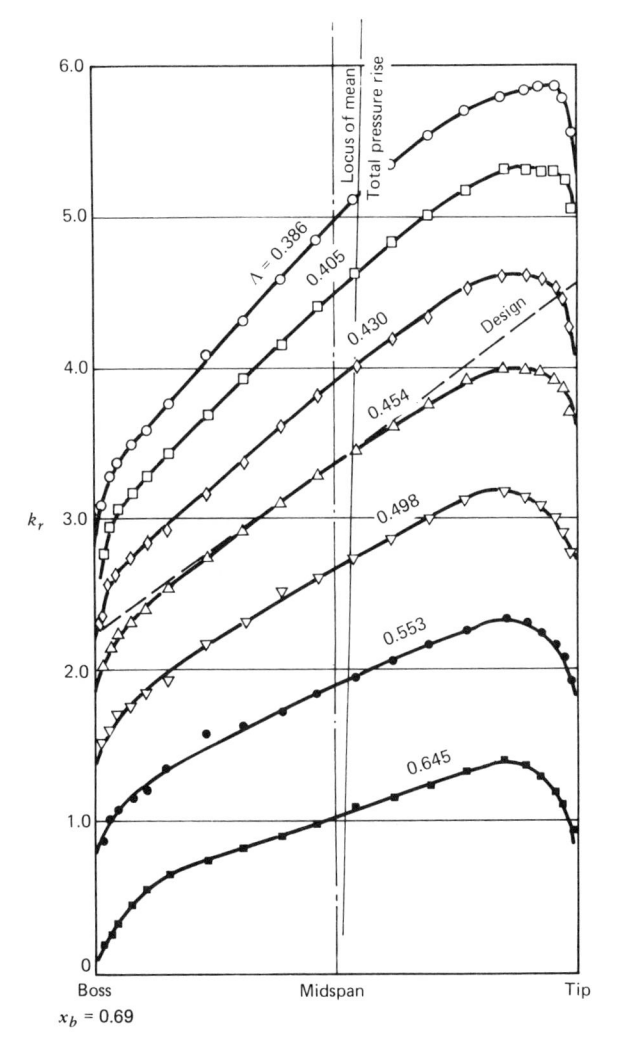

Figure 22.16 Total pressure rise distributions aft of arbitrary vortex flow rotors, [22.10].

as the stall is approached. Large secondary flows are present, as evidenced by the swirl coefficient distribution; the total pressure distributions of Fig. 22.19 also reflect an increasing flow disturbance region. (A change in flow reversal procedures has now permitted a substantial reduction in root clearance.) The slightly higher total pressure rise at the blade tips is associated with the thick boundary layer that results from the inlet treatment devices. Overall performance data for the unit are presented in Fig. 19.2.

Care was taken to seal the blade roots of the optimized research fan [22.11]. Prior to the wall stall onset, the inboard swirl properties indicate

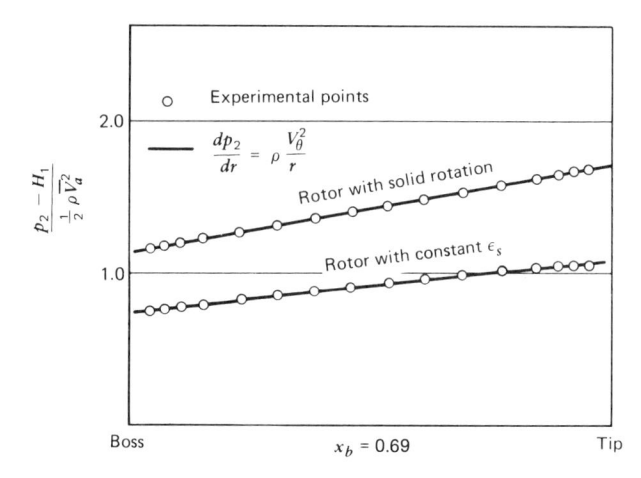

Figure 22.17 Static pressure distributions aft of arbitrary vortex flow rotors, [22.10].

secondary flows of small spanwise extent. The corresponding secondary flows for the cambered plate fan [22.5] exist over a greater percentage of blade span.

A feature of particular interest in relation to the latter fan is the local tip stall and the associated change in slope of the total pressure rise curve (Fig. 22.20). This is due to a static pressure rise associated with the centrifugal forces acting on the local air mass. As the stall progresses, the air mass involved becomes greater, resulting in fan pressures that exceed the stall pressure, for units of large stagger angle.

The use of swirl data in torque and power measurement is illustrated in Fig. 22.25 [22.12].

Since the induced tangential velocities are related to the rotor torque, a means exists for establishing the power losses due to secondary and other three-dimensional flows. The meager amount of relevant experimental data

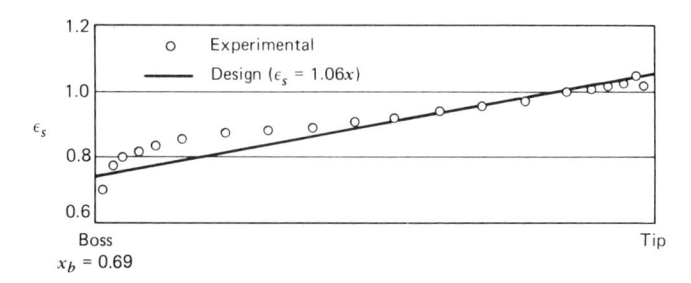

Figure 22.18 Swirl distribution aft of arbitrary vortex flow rotor, $\Lambda = 0.454$, [22.10].

22.8 TRAVERSE EXPERIMENT OBJECTIVES

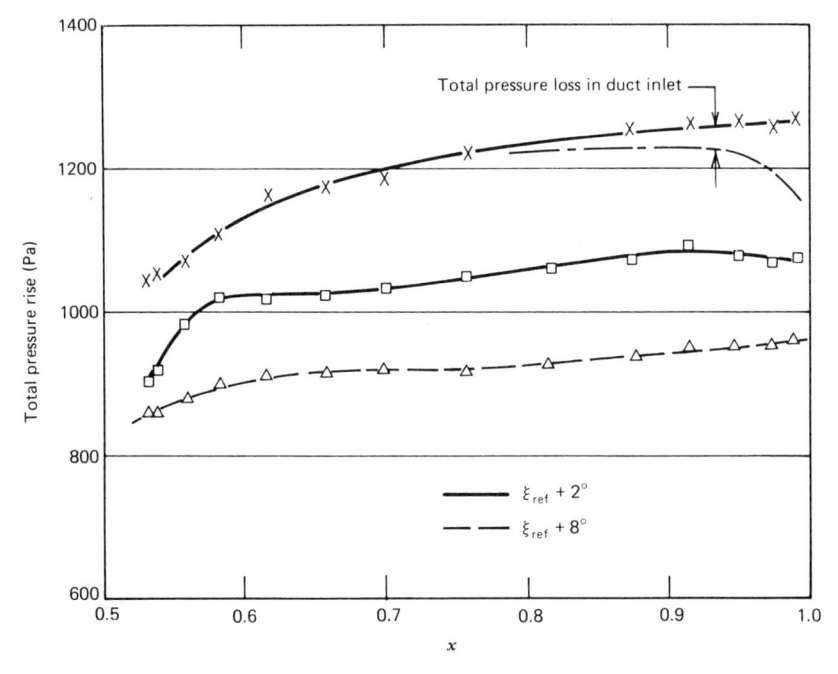

Figure 22.19 Total pressure rise distributions for 6.1 m diameter rotor, [22.1].

has discouraged studies of this nature. However, the present lack of definitive data with respect to these design problems could be largely overcome by systematic yawmeter-based research. The test work should be carried out at various Reynolds numbers, and for a range of nondimensional design pressure duties, on equipment of advanced design.

A nondimensional performance curve for the optimized fan [22.11] is presented in Fig. 22.26 and compared with calculations based on the analytical method of Chapter 19. The close agreement between the design and experimental data is representative of all well-designed fans.

The deviation angles (Fig. 22.27) for this latter fan were deduced from the yawmeter readings. At a given duty, the spanwise variation of δ is contained in a bandwidth of approximately 1°. The recommendation of Eq. (9.2), namely, a design value of $\delta = 10°$, is based on this result.

The mean of four radial traverses, in which the swirl downstream of a prerotator vane row was established, is presented in Fig. 22.28. This illustrates the accuracy of the prediction equation for relatively small amounts of swirl, the only case studied experimentally.

The adequacy of the straightener design method can be gauged from Figs. 22.29 and 22.30 [22.13] for the optimized rotor design case discussed previously. Blade wakes were avoided with one local exception.

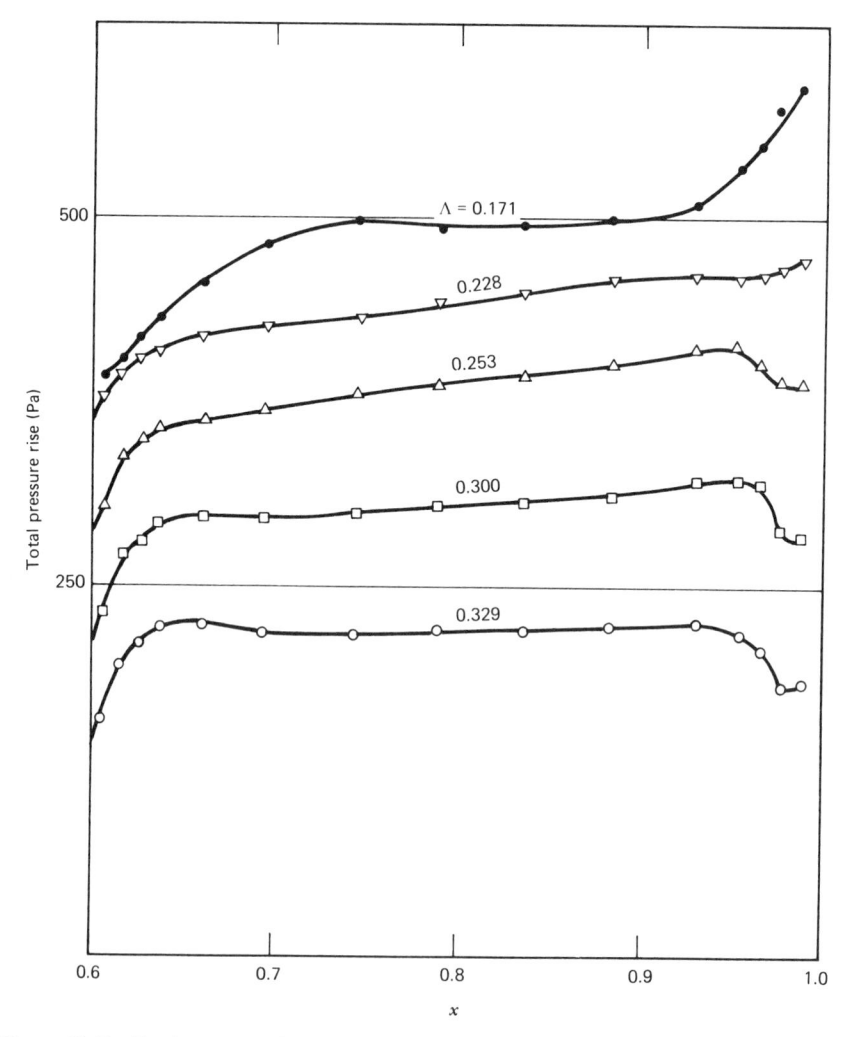

Figure 22.20 Total pressure rise distributions for cambered plate rotor, design $\Lambda = 0.294$, [22.5].

From the preceding discussion it is clear that total pressure and yaw traverses, particularly the latter, can be of great assistance in the development of advanced and efficient equipment. This is best illustrated in relation to the foregoing variable-pitch fans, where techniques normally confined to the laboratory were employed in the field to obtain valuable information for use in later design developments.

22.8 TRAVERSE EXPERIMENT OBJECTIVES 381

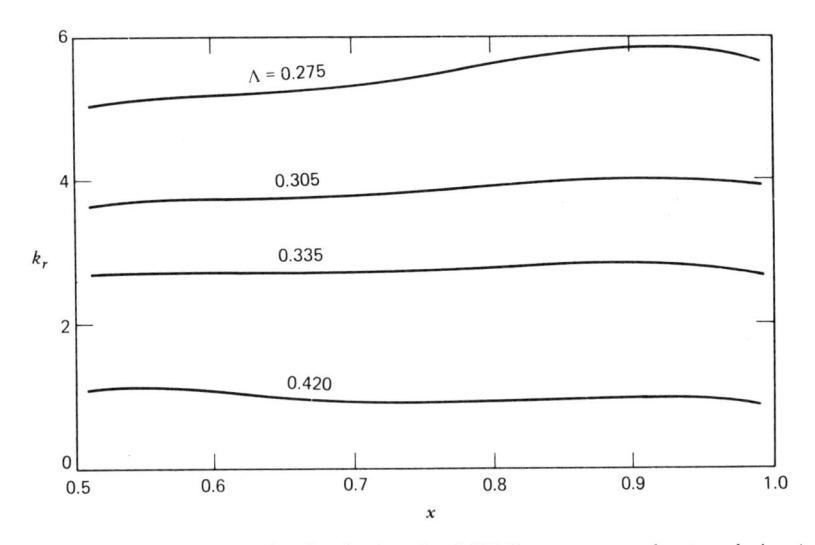

Figure 22.21 Total pressure rise distributions for 0.915-diameter research rotor, design $\Lambda = 0.34$, [22.11].

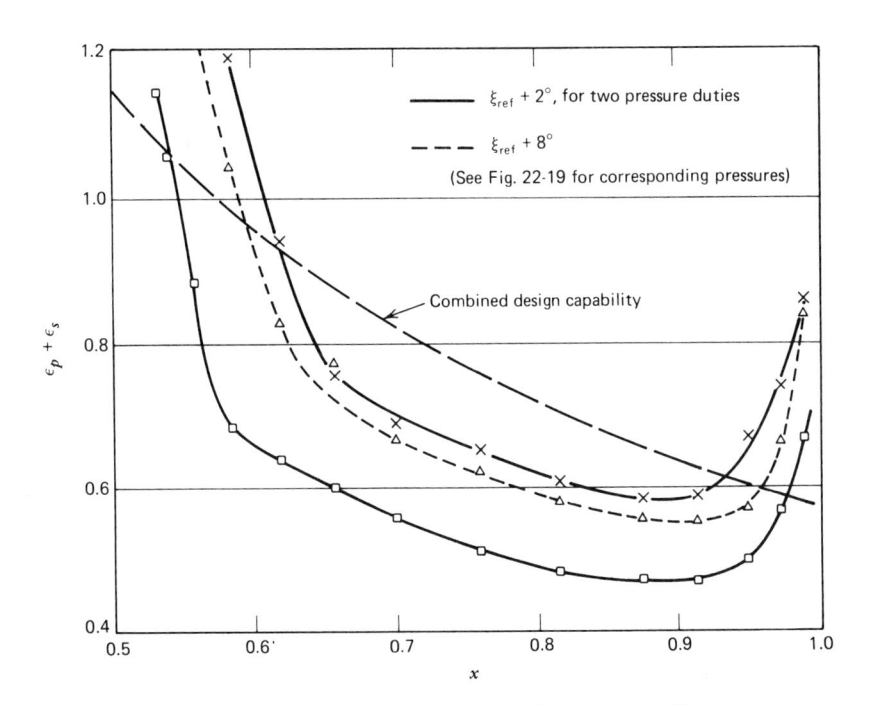

Figure 22.22 Swirl distributions for 6.1-m-diameter rotor, [22.1].

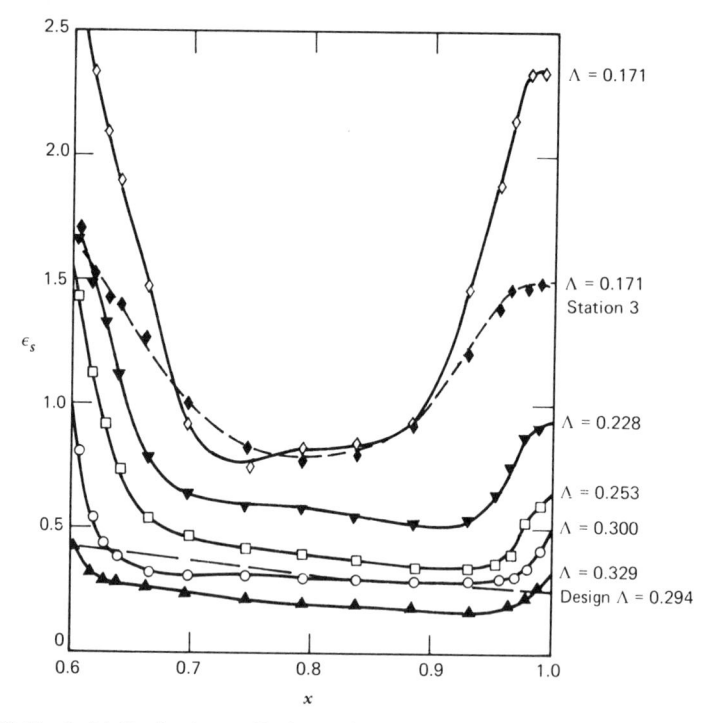

Figure 22.23 Swirl distributions at Station 1, for cambered plate rotor (see Fig. 22.13), [22.5].

22.9 DIFFUSER TESTING

Diffusers are an important element in many fan assemblies and hence demand accurate and careful performance testing.

A test rig used in discharge diffuser studies is illustrated in Fig. 22.31 and Plate 7. An adequate length of parallel annulus has been provided between the trailing edge of the straightener vanes and the diffuser inlet. This ensures the existence of a suitable station for the static pressure measuring piezometer ring. (At diffuser inlet there is a local surface flow acceleration as the air is deflected down the diffuser.)

Experimental data are reduced according to the approximate expressions given by Eqs. (13.5) and (13.6), respectively.

$$C_p = \frac{p_2 - p_1}{\frac{1}{2}\rho \bar{V}_a^2} \tag{13.5}$$

$$\eta_D = \frac{C_p}{1 - \left(\frac{A_1}{A_2}\right)^2} \tag{13.6}$$

22.9 DIFFUSER TESTING

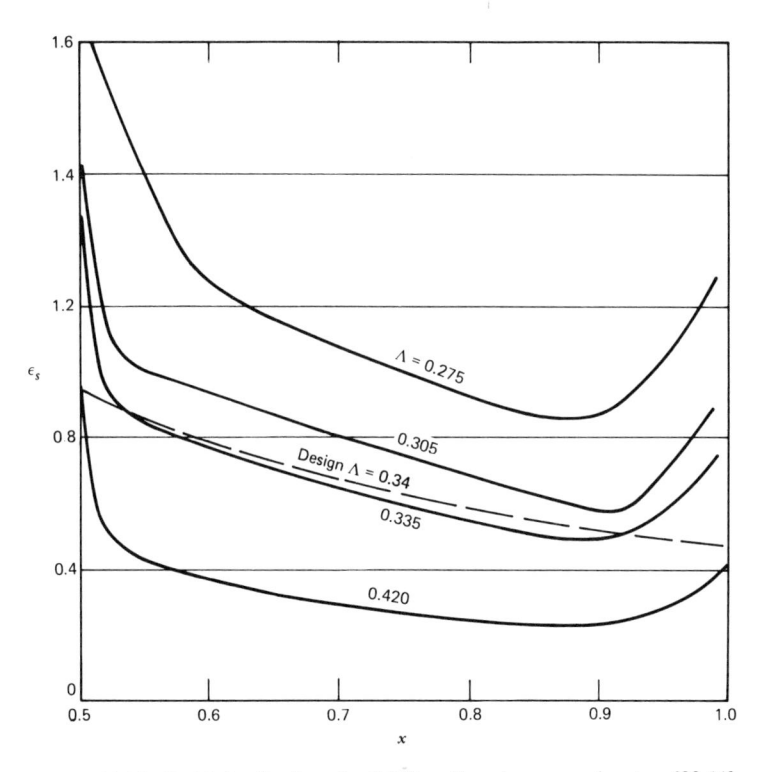

Figure 22.24 Swirl distributions for 0.915-m-diameter research rotor, [22.11].

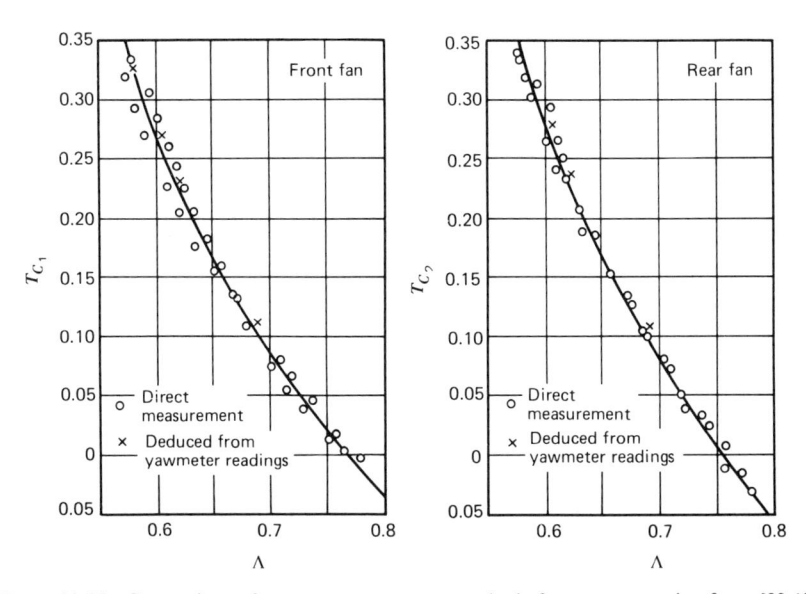

Figure 22.25 Comparison of torque measurement methods for contra-rotating fans, [22.12].

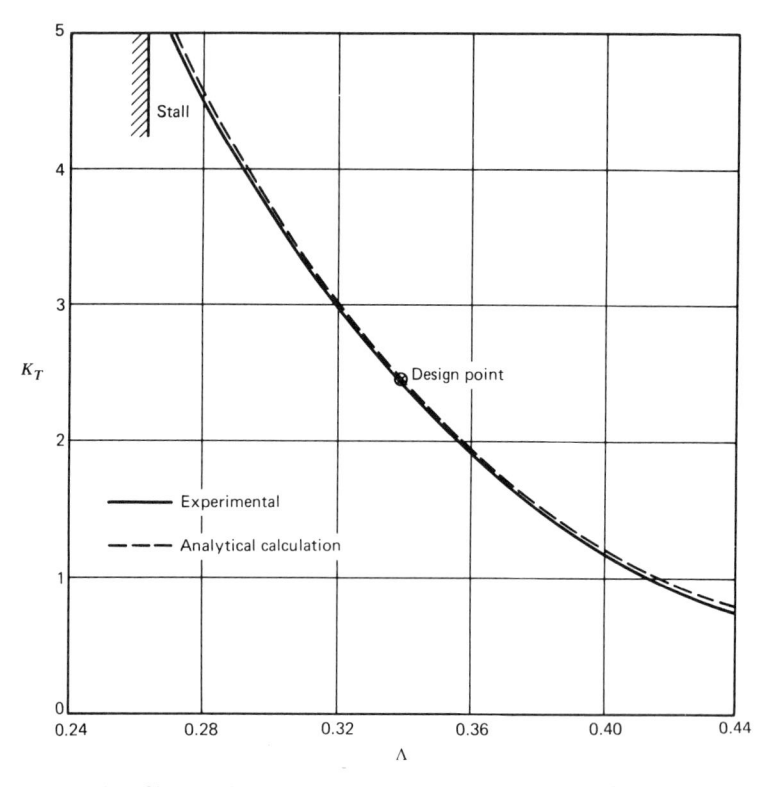

Figure 22.26 Characteristic curve for 0.915 m diameter research fan unit, [22.11].

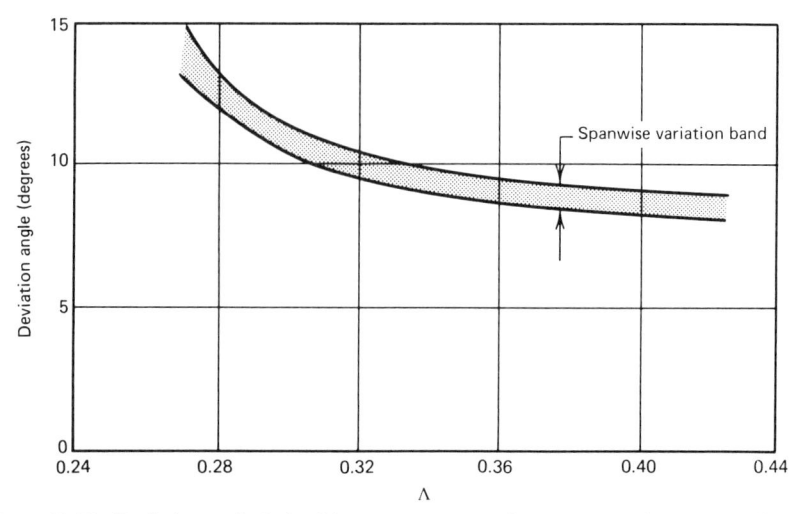

Figure 22.27 Deviation angle derived from yawmeter readings versus Λ, for 0.915-m-diameter research rotor, [22.11].

22.9 DIFFUSER TESTING

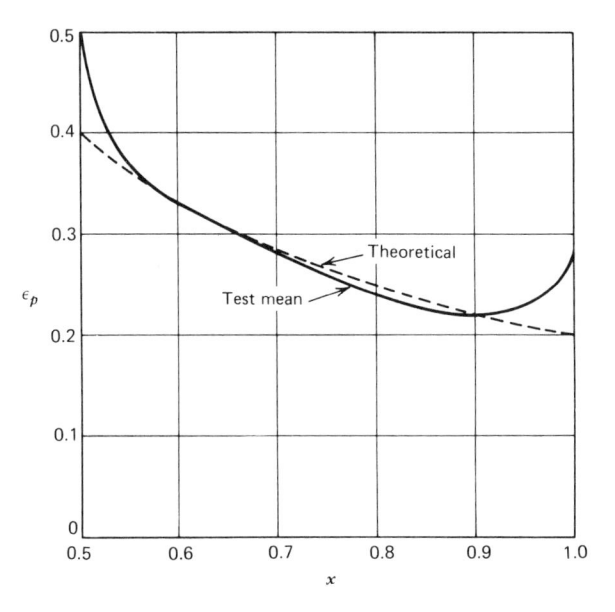

Figure 22.28 Prerotator performance for 6.1-m-diameter fan unit, [22.1].

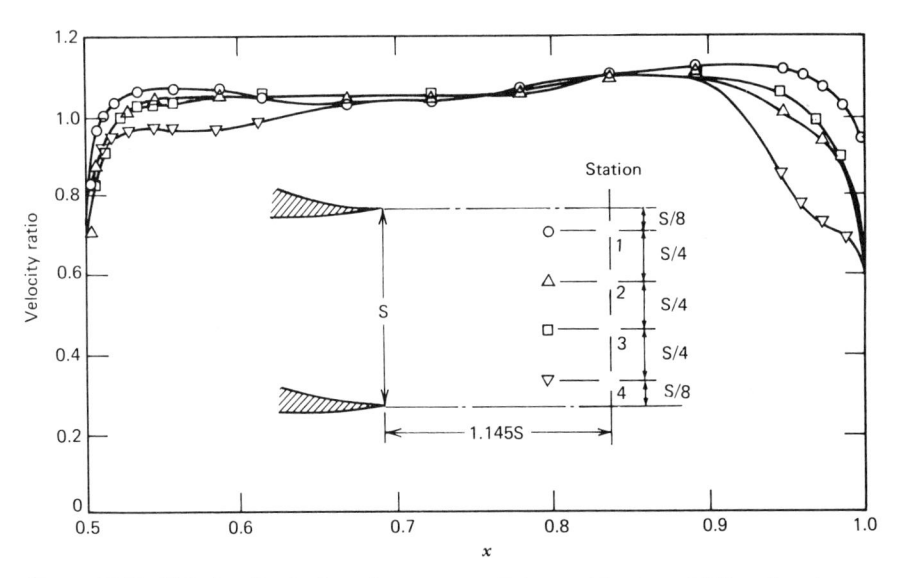

Figure 22.29 Velocity distributions downstream of the straighteners, 0.915-m-diameter research fan unit, design duty, [22.13].

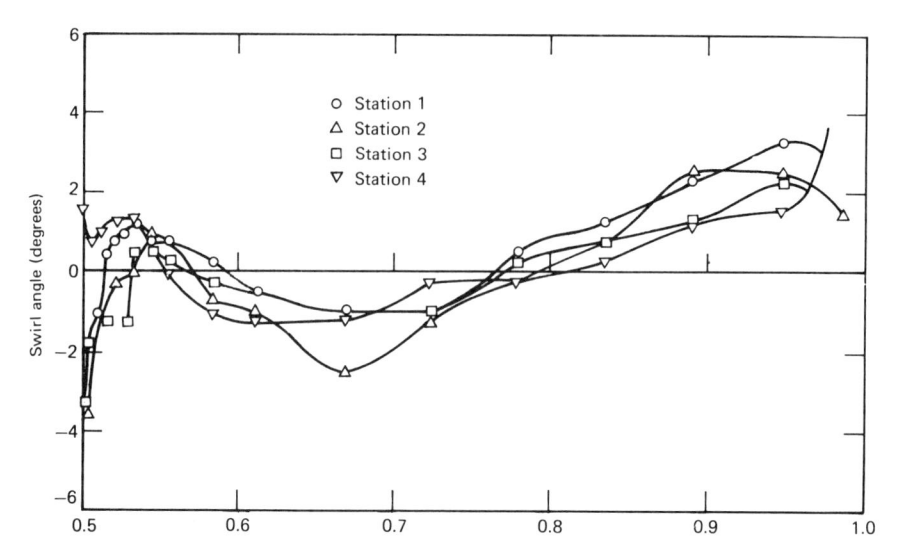

Figure 22.30 Swirl distributions downstream of the straighteners, 0.915-m-diameter research fan unit, design duty, [22.13].

Alternatively, the mean total pressure at diffuser inlet, relative to atmospheric pressure, represents the overall total pressure loss, composed of diffuser resistance plus discharge velocity pressure. The coefficient K_{DL} follows from a knowledge of the mean inlet velocity pressure [Eq. (13.3)].

Because of continuing static pressure regain in the parallel duct downstream of the diffuser termination, for in-line fan arrangements, the location

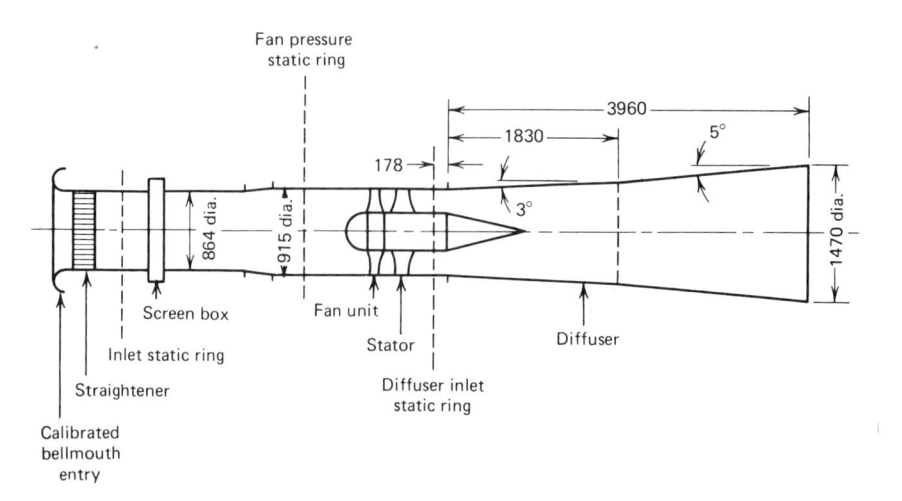

Figure 22.31 Fan and diffuser research facility, [22.13].

22.10 TROUBLESHOOTING

Plate 7. Research fan and diffuser test rig. (By courtesy of CSIRO Division of Energy Technology, Melbourne, Australia.)

of the station at which p_2 is measured must be recorded. The mean total pressure difference between stations 1 and 2 represents the diffuser loss.

22.10 TROUBLESHOOTING

With the exception of fan and duct systems, each of good aerodynamic design but poorly matched, the major portion of work in this class is related to isolating design faults. Insufficient flow and excessive noise are probably the two most common complaint areas, and these problems often arise because of flow separations. The various sources of such flows have been well covered in other chapters; hence, in the present context, only the recommended test approach will be outlined.

Initially, the visual study techniques outlined previously should be undertaken. For instance, a tufted rod traversed downstream of a rotor will provide a good preliminary guide to blade flow conditions. When the fan is partially or wholly in stall, the swirl angles will be excessive and the flow highly turbulent. The inlet and outlet flows can be qualitatively observed with tufts and a decision reached on whether meaningful flow and pressure measurements can be obtained. The type of instrument and its location, in the less severe cases, can then be selected.

A preliminary study may reveal an incompatibility between the fan and duct system because of an underestimate of duct losses. Reliable test data on system total pressure loss at a measured volume flow rate are essential in achieving a solution. The nondimensional system resistance can then be calculated. A discussion of the model approach to this test exercise is contained in Chapter 5. A badly stalled fan in the actual system assembly could produce abnormal flow patterns and hence higher than normal total pressure losses. Model tests can avoid this problem and, in addition, provide an accurate measurement of volume flow rate for establishing the nondimensional resistance.

In solving the particular problem, the aerodynamicist has to select from one or more of the following alternatives:

1. Modifications to, or replacement of, the fan unit.
2. Fan installation alterations.
3. Reduction of system pressure losses, by duct changes or redesign.

In extreme cases, replacement of the entire assembly by redesigned equipment is required for complete satisfaction. The author has recommended one such accepted solution in relation to a large acid cooling tower installation [22.14].

When accuracy is sought in full-scale investigations, vane-type anemometers should not be used, as these have a tendency to overestimate velocity. Velometers used strictly in accordance with instructions, and calibrated before and after a test program, are more reliable. Experience has shown that velometer calibration, as determined by the manufacturer, has a poor lifespan, despite the continuing serviceability of the instrument.

REFERENCES

22.1 K. E. Mathews et al. Development of the primary ventilation system at Mount Isa, *Australasian Inst. Min. Metall. Proc. No. 222*, pp. 1–61, 1967.

22.2 K. A. Blake, The design of piezometer rings, *J. Fluid Mech.* **78**, 415–428, 1976.

22.3 J. H. Perry, Calibration and comparison of Cobra probe and hot wire anemometer for flow measurements in turbomachinery, *CSIRO Div. Mech. Eng. Tech. Rep. TR1*, 1974.

22.4 D. W. Bryer and R. C. Pankhurst, *Pressure-probe methods for determining wind speed and flow direction*, Nat. Phy. Lab., Dept. Trade and Ind., London, 1971.

22.5 R. A. Wallis, Performance of sheet metal bladed ducted axial flow fans, Aust. Dept. Supply, Aero. Research Labs, *ARL Aero. Rep. 90*, 1954.

22.6 R. Fail, Tests on a high solidity engine cooling fan in the R.A.E. full scale fan testing tunnel, *Gt. Britain RAE Rep. Aero. 2068*, 1945.

22.7 L. H. Smith, Jr. Wake dispersion in turbomachines, ASME *J. Basic Eng.*, pp. 688–690, Sept. 1966.

22.8 B. Reynolds and B. Lakshminarayana, Blade loading and spanwise effects on wake characteristics of compressor rotor blades, *J. Aircraft*, **19**, 97–103, Feb. 1982.

22.9 J. L. Livesey and T. Hugh, 'Suitable mean values' in one-dimensional gas dynamics, *J. Mech. Eng. Sci.*, **8**, 374–383, 1966.

22.10 A. Kahane, Investigation of axial flow fan and compressor rotors designed for three-dimensional flow, *NACA Tech. Note 1652*, 1948.

REFERENCES

22.11 J. H. Perry, A study of the design and off-design flow characteristics of a single stage axial flow fan, Div. of Mech. Eng. CSIRO, 1973 (unpublished report).

22.12 J. F. M. Scholes and G. N. Patterson, Wind tunnel tests on contra-rotating fans, Australian Council for Aeronautics, *Rep. ACA 14,* 1945.

22.13 R. A. Wallis, Annular diffusers of radius ratio 0.5 for axial flow fans, Div. of Mech. Eng., CSIRO, *Tech. Rep. 4,* 1975.

22.14 L. Aa et al. A cooling tower for use in zinc electrolysis, Thermodynamics Conf. Inst. Eng., Aust., August 1970.

22.15 E. Ower and R. C. Pankhurst, *The Measurement of Air Flow* Pergamon, Oxford, 5th ed., p. 51, 1977.

CHAPTER 23
Review of Design Assumptions and Limitations

23.1 EFFECT OF NONUNIFORM INLET VELOCITY

A dominant feature of an axial flow fan is its ability to improve rather than accentuate spanwise total pressure gradients on account of nonuniform, swirl-free inlet flow. Blading designed for uniform inlet flow, in accordance with the free vortex method, will possess an excess of incidence for inlet velocities less than the mean value, and vice versa. Consequently, the local total pressure added will parallel the spanwise variation in incidence, and hence lift coefficient, resulting in a mean fan total pressure rise that on most occasions approximates the one desired.

When the flow just upstream of the nose fairing is of the fully developed pipe flow type, the velocity distribution will possess nonuniformity of the magnitude illustrated in Fig. 3.5 for $H = 1.3$. However, because of the contraction induced by the fairing, the nonuniformity will be progressively decreased with increases in boss ratio. In the case of small boss ratios the inboard local velocities will be substantially greater than the outboard ones, thus increasing the pressure rise capability in these areas of reducing blade velocity; this is another favorable aspect of axial flow fan operation.

A rotor operating in a fully developed annulus flow, or downstream of initial rotor stages in a multistage unit, will experience a different situation to the one just discussed. This results in a work output that is less than the one computed on the basis of flow uniformity. Many attempts have been made to put this phenomenon on a firm design foundation. A "work done" factor that is a function of the number of rotor stages, or resultant flow distortion, has been widely used in the past. An empirically developed relationship [23.1] for multistage units is presented in Fig. 23.1.

23.1 EFFECT OF NONUNIFORM INLET VELOCITY

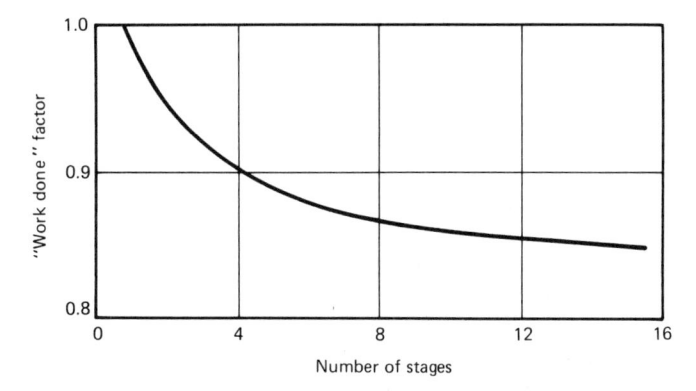

Figure 23.1 "Work done" factor versus blading stages. (Reprinted by permission of the Council of the Institution of Mechanical Engineers from A. R. Howell and R. P. Bonham [23.1] in Proceedings, No. 163, 1950.)

An artificially created approach to the fully developed annulus flow case is reported in [23.2] for spoiler-induced boundary layer thickening on both walls. The subsequent losses in pressure rise and efficiency were small. In these tests and those of [23.3], the rotor tended to reenergize the boundary layer. However, the increased energy input in the blade-end regions is in [23.3] considered to be nonuseful from a fan pressure rise point of view, since tip and secondary losses tend to absorb a large proportion of this input.

Comparative tests on three alternative compressors designed for (a) constant axial velocity, (b) the actual axial velocity distribution, and (c) increased total pressure rise toward the two blade extremities demonstrated a higher efficiency for the initial variant [23.4]. In the second design, the downstream velocity distribution showed a deterioration, whereas some improvement in this regard was achieved with the third type of blading. The efficiency loss in this latter instance was traced to stator flow separations brought about by the large swirl angles encountered at the stator extremities.

The axial velocity ratio (AVR) concept represents an attempt to improve the accuracy of rotor blade design—for example, [23.5], [23.6], and [23.7]. Because of the flow speed-up effect that accompanies the rapid boundary layer growth on inner and outer walls, the inlet axial velocity component is less than its outlet value. The AVR is defined as the ratio of these two velocities. This necessitates an approximate deviation angle correction given in [23.6] by

$$\delta = \delta^1 + 10(\text{AVR} - 1) \qquad (23.1)$$

where δ^1 is the value for AVR = 1.

In addition to the foregoing acceleration effect, the secondary flows that accompany the turning of the end wall boundary layers also impose a change in blade outlet air flow direction that is more pronounced toward the blade extremities [23.8].

However, the preceding developments have their greatest value in the interpretation and analysis of experimental data. In addition, they have increased relevance to axial flow compressor design. The isolated airfoil design method, in conjunction with multiplane interference factors, provides an adequate and sufficiently accurate fan rotor design procedure with few exceptions. Knowledge of the deviation angle is, however, required in straightener design. Nevertheless, an angle correction on the basis of Eq. (23.1) is unwarranted as small residual swirls have negligible influence on fan efficiency or downstream duct losses. In addition, the design value of AVR cannot be estimated in normal circumstances.

The "work done" factor (Fig. 23.1) is recommended as an adequate design variable for multistage fans.

$$K_{th} = \frac{\text{fan pressure rise coefficient}}{\text{fan efficiency} \times \text{"work done" factor}} \quad (23.2)$$

However, modifications to these tentative and conservative design suggestions may be required when relevant experimental multistage fan information becomes available.

23.2 VORTEX FLOW CONSIDERATIONS

A large number of fans currently produced are in the arbitrary vortex flow category. Small boss ratios, constant chord, and blades with little or no twist are features normally associated with this fan type. Low efficiency is not an uncommon event. These units are often the result of experimental ad hoc development.

Advantages of the free vortex flow design approach are ease and accuracy of design and the assurance of a high operating efficiency, particularly when stators and a high effectiveness diffuser (with uniform inlet flow) are elements of the assembly.

As demonstrated in Appendix A, preswirl offers some assistance in dealing with blade root loading problems. The recent work described in Appendix B now removes the need for the tentative attitude adopted in respect to the rotor-only design exercise for Appendix A.

The experimental work described in Appendix B has shown that the radial velocities associated with arbitrary vortex flow fans are of a minor nature, provided the downstream axial velocity is greater than zero at the blade root. Since a fair proportion of industrial fans operate at duties beyond this latter duty point, there exists a practical requirement for a method that will predict fan performance in this extended arbitrary vortex flow regime. At present there is no analytical treatment capable of dealing with the associated complex flows.

23.3 FLOW DISTORTION AND ASYMMETRY

An approximate ad hoc procedure for dealing with the problem is presented in Section B.5.

23.3 FLOW DISTORTION AND ASYMMETRY

Relaxation of the assumption of axisymmetric flow through the blading annulus gives rise to problems that can only be resolved experimentally, for each particular fan installation. Examples of the penalties involved are presented in [23.9] and [23.10] for two different installations possessing upstream segmental-type blockage. Initially with increasing amounts of obstruction the losses in fan pressure rise and efficiency are small, prior to the rapid deterioration phase being encountered. Surging will almost certainly be present for high-pressure-rise units, and the low-pressure-rise machine may be rendered impotent. The practice of placing gate or butterfly valve arrangements upstream of fans for control purposes should be questioned in view of the preceding consequences. Hence the preceding fan inlet condition is of little concern to the informed duct and fan designer.

Distorted flows arising from upstream air-turning circumstances are, however, unavoidable in many instances. For example, a fan taking air from the free atmosphere will be subjected to varying wind conditions, which in the most severe case will be at right angles to the fan axis. For small fans some form of shielding from cross winds is often feasible. However, in the case of large vertical ventilation fans supplying downcast mine air, the inlet duct requires special attention, as outlined in Section 4.3. Provision of a reasonable length of inlet duct, plus the calming effect of prerotators, complete the low-loss fan inlet treatment illustrated in Fig. 4.20.

Inlet boxes (Section 4.3) will introduce flow distortions that are associated with acceptable penalties when the box is correctly proportioned and a relatively large acceleration is present as the air enters the blading annulus.

In recent times there has been an increasing emphasis on matters associated with installation effects on fan performance [23.12]. In these papers it is acknowledged that, with poorly designed ductwork, the axial flow fan will sustain losses that are not necessarily of the same magnitude as those suffered by centrifugal units. Inlet flow conditions are far more important than those associated with ducts on the outlet.

However, the ducting components involved in the experimental work of [23.12] reflect present day practice. In some instances, the equipment items need updating in order that system effect factors of approximately unity may be achieved. More attention has to be focussed on the flow features that the various ducting components possess.

The large scatter of experimentally determined system effect factors presented in [23.12], and the opposing views expressed, emphasize the seriousness of the situation. By allowing swirl to persist in fan inlet flows, there is little hope of ever getting consistent system effect factors. This is made clear in all fan test code methods, by the various measures that are taken to eliminate inlet swirl.

The tangential velocity components associated with swirl are an integral part of axial flow fan design. Therefore the scale, rotational direction(s), and random nature of the vorticity which makes up the conglomerate we loosely refer to as swirl, must have a serious effect on axial flow fans. Non-uniformities in the axial component of the inlet flow, from a fundamental viewpoint, will potentially be far less important, provided the fan inlet duct runs full. Duct components that eliminate or reduce the effect of swirl on axial flow fans do exist.

Common standardized ducts components are used in both radial and axial flow fan installations. However, very little thought is given to their suitability as far as the latter fan type is concerned. As a result, an understandable bias has been built up against axial flow units, which are quite often noisy and lacking in both performance and efficiency. As discussed in Chapter 21, the pressure rise performances of radial and axial fans result from entirely different flow features.

For instance, the loss in a 4-piece lobster back corner may be of the same order as that of a vaned corner but the latter is more suited to an installation that features a corner upstream of an axial flow fan. The flow difference is highlighted in Figs. 2.9 to 2.11, which diagrammatically show the secondary flow features. The smaller scale vorticity is more uniformly spread, and as a result will tend to dissipate in a shorter downstream distance and hence have less effect on the fan performance. The two large contra-rotating vortices from an unvaned corner will persist right up to the rotor face.

The fan entry flow with a vaned corner will consequently be of improved quality. If placed in an airstream containing solid material, then provision has to be made for periodic cleaning of the vanes. The improved operating efficiency and lower noise level will more than compensate for this inconvenience. In wind tunnel design recommendations, it is accepted that the fan should be at least 2 duct diameters downstream of the compact vaned corner. The compactness of this type of corner is an attractive feature that should be of merit in the confined spaces that designers have to cope with in many industrial installations. Corners of the type illustrated in Fig. 4.18, when installed in circular ducting, have proved most successful in practice. However, one seldom sees their use.

Rapid increases in duct area immediately upstream of a fan will produce flow instabilities and an associated loss in fan performance, and augmented noise problems. Alternating flow separation from first one side and then the other, or rotating separation in a conical duct, can on occasions be present, (Fig. 23.2). Structural and mechanical failures in the fans and ducts are quite commonplace.

Once again there is a need to look at the special requirements of an axial flow fan. Implementation of the special 2 and 3-dimensional devices

REFERENCES

Separated flow may switch from side to side

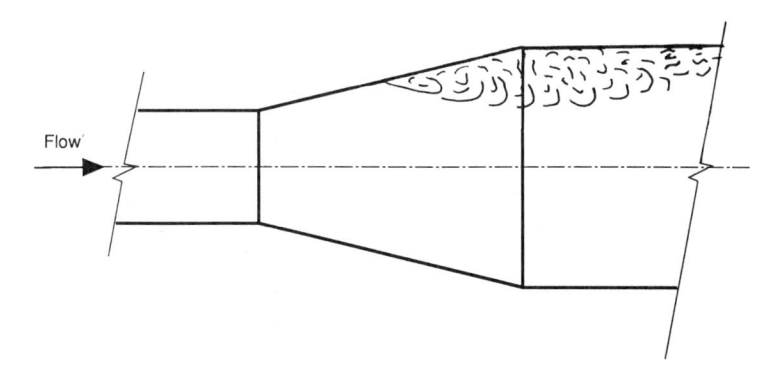

Figure 23.2 Separated flow in a 25 degree included angle diffuser.

presented in Sections 4.4.5 through to 4.4.7 will reduce duct losses and make for a more even spread of a smaller scale turbulence. However, of greater importance will be the beneficial effect on fan operation. Provided a short length of duct that changes the cross-section from rectangular to round, or "calming" duct, is fitted upstream of the fan, a satisfactory operational solution is possible.

Fan air inlets from boxes, plenums, or open spaces can affect fan performance, varying from the ideal to the disastrous, where the grossly disturbed inlet flow sets up extreme vibrations that lead eventually to structural failure. Axial and radial flow equipment alike suffer this consequence. The subject is of equal importance to the above corner and diffuser topics, but the diverse and complicated nature of the problem has restricted any generalized discussion on the subject. However, the types of flow that create these problems are readily identified as "lumpy" eddying flow, and large scale vorticity, created by crossflow asymmetries at inlet.

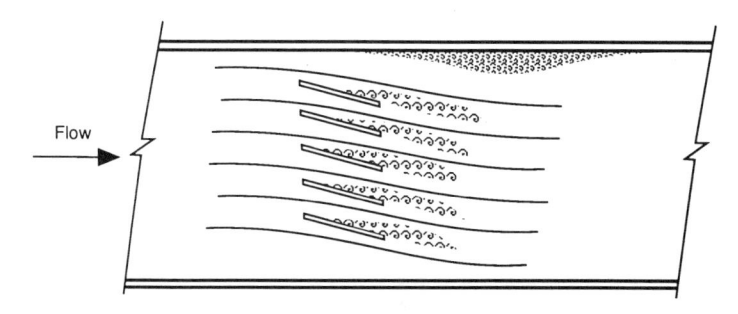

Figure 23.3 Flow through louvre array set at 20 degrees.

The former may be the result of flow separation from a vaned damper system in the inlet box, from poor inlet duct geometry, or from the wakes of adjacent obstacles, the design effort should be aimed at eliminating or minimizing these flows.

All engineers will be conversant with the characteristics of the wake from a bluff body, as they will have seen the flow patterns downstream of a pier in water. However, the same is not necessarily true of flow separation from surfaces. When air meets a free surface such as a vane or louvre at a greater angle than approximately 10 to 15 degrees, respectively, the air will separate completely from the sheltered surface, and conditions analogous to a bluff body will result, where the bluffness dimension increases with the incident flow angle (Fig. 23.3).

The extent to which flow separation can be avoided at entry, with increasing crossflows, will depend on the shape of the inlet to the fan duct, (see Fig. 4.20). Some of the worst examples of poor fan installation are in the chemical and petroleum industries where axial flow fans are used in condenser arrays that lack attention to aerodynamic matters.

The inlet flow to many plenums is often at right-angles to the fan axis. Multiple fans are also on occasions subjected to this flow condition. Depending on the height of the box, the energy associated with stray vorticity in the plane of the box inlet flow can be quite substantial (Fig. 23.4). When drawn into the fan inlet, the tangential momentum will be con-

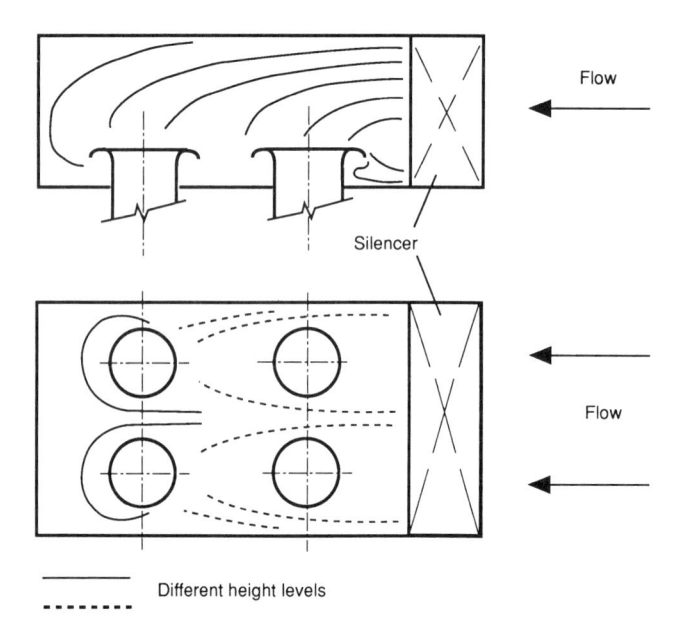

Figure 23.4 Schematic illustration of crossflows in inlet box.

23.4 FAN STALLING

served, resulting in swirling flows of uneven and often considerable strength. Loss of performance and excessive noise are the ensuing penalties. When the box inlet flow is not symmetrical about the fan centers(s), or when the inlet flow is substantially nonuniform, the flow problems will be exacerbated. Splitter plates or other devices, based on model tests, that tend to inhibit these swirling crossflow motions, are recommended.

Fan inlets should be free of any obstructions in their vicinity that will cause swirling and wake flows to enter the fan duct. If after examining all design options the difficulty is unavoidable, a low loss honeycomb device insert in the fan duct to straighten the flow is recommended.

The foregoing briefly outlines the thought processes that are recommended when dealing with axial flow fan installations. It is vital to consider the potential air flow paths in the design stage in order that the effective, efficient, and low noise character of this fan type can be realized. The extra expense is amply repaid in energy savings, and in having a completely reliable and quieter installation. A tufted hand held rod is often the only piece of equipment necessary to demonstrate, to an aerodynamicist, an alarming state of gross air turbulence and flow instability in instances where operational difficulties are experienced.

23.4 FAN STALLING

The high maximum lift coefficients associated with two-dimensional airfoils in wind tunnel tests can never be reproduced on fan equipment. The onset of stall initially occurs at a blade extremity in a region of low velocity and hence higher local C_L. Once local flow separation is present there is an interaction with the adjacent attached regime, with unsettling consequences.

Rotors with high root loading factors tend to stall from this extremity. The sequence of events is that, with increasing pressure load, an orderly streamwise vortex is created on the hub wall, being located in the upper blade surface corner. As the fan stall is approached, the vortex strength grows, and this has an increasing influence on the local blade surface flow. Eventually the blade boundary layer is sufficiently disturbed to experience flow breakdown accompanied by an end to the orderly corner vorticity. Depending on the distributions of chord, twist, and section profile along the blade, the stall can develop rapidly or slowly. Hence although hub flow conditions are a very important factor in fan stall, blade design is also of consequence.

Lightly loaded rotors, of modest root solidity will be less influenced by hub flow and hence higher lift coefficients before stall can be assumed. However, a more sudden stall is likely to occur.

In general, blade aspect ratios of 0.75 or less tend to make for a more gradual stall, since the two end wall secondary flow vortices exert a stabilizing influence on the blade boundary layer; the action is similar to that produced by vortex generators. However, peak pressure may be reduced.

The buildup of mud and dirt on mine ventilation fans may dramatically precipitate stall.

The attachment of blades to a conical or streamlined hub will reduce the wall adverse pressure gradient and hence delay the onset of wall stall and may influence the subsequent fan stall.

The effects of blade setting errors on stall are highlighted in [23.11]. In the author's experience blade twist and setting angle errors in excess of 0.25° for individual blades can increase the degree of surge or stall severity.

In view of the preceding complexities and large range of design possibilities it is not practicable to seek a quantitative solution to stall prediction. However, the design data and recommendations contained herein, when applied correctly and intelligently, should result in equipment free of stall problems.

REFERENCES

23.1 A. R. Howell and R. P. Bonham, Overall and stage characteristics of axial flow compressors, *Proc. Inst. Mech. Eng.*, **163**, 1950.

23.2 E. Boxer, Influence of wall boundary layer upon the performance of an axial flow fan rotor, *NACA Tech. Note 2291*, 1951.

23.3 A. Mager, J. J. Mahoney, and R. E. Budinger, Discussion of boundary layer characteristics near the wall of an axial flow compressor, *NACA Tech. Rep. 1085*, 1952.

23.4 S. J. Andrews, R. A. Jeffs, and E. L. Hartley, Tests concerning novel designs of blades for axial compressors, Parts 1 and 2, Gt. Britain Aero. Research Council, *ARC R&M 2929*, 1956.

23.5 D. Pollard and J. P. Gostelow, Some experiments at low speed on compressor cascades, *ASME J. Eng. Power*, **89**, 427–436, 1967.

23.6 J. H. Horlock, Some recent research in turbo-machinery, *Proc. Inst. Mech. Eng.*, **182**, Part 1, 1967–68.

23.7 S. Soundranayagam, Effect of axial velocity variation in aerofoil cascades, *J. Mech. Eng. Sci.*, **13** (2), 92–99, 1971.

23.8 M. F. Bardon, W. C. Moffatt, and J. L. Randall, Secondary flow effects on gas exit angles in rectilinear cascades, *ASME J. Eng. for Power*, **97**, 93–100, 1975.

23.9 R. C. Turner, J. Ritchie, and C. E. Moss, The effect of inlet circumferential maldistribution on an axial compressor stage, Gt. Britain Aero. Research Council, *ARC R&M 3066*, 1958.

23.10 J. L. Koffman, Fans for traction applications-2, *Diesel Railway Traction*, pp. 87–94, April 1951.

23.11 R. M. El-Taher, Experimental study of the effect of blade setting errors on stalling performance of cascade, *J. Aircraft*, **17**, 313–318, May 1980.

23.12 Installation Effects in Fan Systems, London Conference, Proc. Inst. Mech. Engr. 204. March 1990.

APPENDIX A

Approximate Design Procedures

Designers seeking to optimize a fan unit from an overall technical and capital cost viewpoint may find the free vortex flow design procedure unacceptable. For instance, the inboard swirl velocity required to ensure free vortex flow may be excessive for the boss ratio desired. A modified design procedure giving reduced inboard swirl is therefore desirable.

The aerodynamic features that accompany a reduction in inboard blade setting angles, necessary to achieve the preceding result, are reduced local axial velocities and total pressure rises, combined with a radial outward flow movement. Theoretical attempts to cover these developments in an approximate manner [A.1] have failed to produce a design method that is reliable in all instances. Rather than resort to the more mathematically complex streamline curvature, matrix, or finite element techniques, which are computer based, the author recommends a modified free vortex method for restricted design amounts of arbitrary flow. This permits a ready extension of the foregoing design techniques and data on which good reliability can be placed.

The procedure is illustrated in two design examples. Similar units to the first of these, but of increased pressure duty, have performed satisfactorily when installed in less than ideal equipment layouts.

A.1 FIRST DESIGN EXAMPLE

A compact blading unit is required for a large preheat furnace application. The required total pressure rise across the blading unit is given as $K_{bl} = 4.2$. Assuming the blading efficiency is 0.80, K_{th} is 5.25.

The pressure duty magnitude makes the use of stators mandatory and, for reasons outlined in Section 17.4, prerotators are selected. The ability of

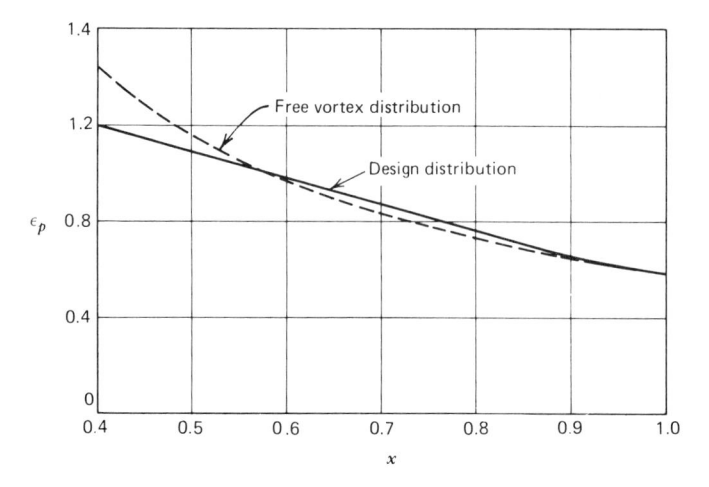

Figure A.1 Modified preswirl distribution for first design example.

prerotators to eliminate residual upstream swirl and present the rotor with a steadier inflow is an additional factor.

Sheet metal of two-dimensional curvature and single thickness will be used in constructing the rotor blades and the prerotator vanes.

The design will be generated from the following main design parameter values:

$$\Lambda = 0.22, K_{th} = 5.25, \text{ and } x_b = 0.4$$

For free vortex flow conditions, the swirl distribution is as given in Fig. A.1. The magnitude of the root coefficient is undesirably high and would lead to blade construction problems. As a consequence the swirl coefficient distribution will be assumed to be straight-lined between 0.65 at $x = 0.9$ and 1.20 at $x_b = 0.4$. The mean fan total pressure rise of the resulting unit is expected to equal the desired value.

In the calculations of Table A.1 the coefficients λ and ϵ_p relate to $\overline{V}_a$ and hence retain their identity, as defined in the Notation. The fan blading efficiency is calculated at mid-span ($x = 0.7$), for local values of λ and ϵ_p

Prerotator loss, $K_P/K_{th} = 0.015 + 50\% = 0.023$

Rotor loss, $\dfrac{\gamma K_R}{K_{th}} = \dfrac{\lambda}{\cos^2\beta_m} = 4.405$

$$\gamma = C_L/(C_{D_P} + C_{D_S}) = 0.82/(0.019 + 0.025\, C_L^2) = 22.9$$

$$K_R/K_{th} = 0.192$$

A.1 FIRST DESIGN EXAMPLE

Blading efficiency, $\eta_{BL} = 1 - (0.192 + 0.023) = 0.785$

This approximates the value initially assumed (0.80); the difference is within the order of accuracy which can be expected from the design and efficiency calculations. Hence recalculation is not essential.

In the selection of the blade variables, σ, C_L and θ, the requirement for a two-dimensional curvature blade shape, of constant radius of curvature, has to be realized. The plate curvature is given by

$$R_{cur}/R = c/R \; 2 \sin \theta/2 \quad (6.10)$$

and the camber/chord ratio by

$$b/c = 0.00221 \; \theta \quad (6.9)$$

When the chord is a linear function of x (straight leading and trailing edges) it can be shown from Eqs. (6.10) and (6.9) that θ and b/c are also approximately linear with x. Finally by small adjustments to blade stagger angle it is possible to make this variable a linear function of x. This adjustment is in the correct aerodynamic sense as the reduced axial velocity resulting from local work reductions will result in an increased stagger angle.

A trial and error process is initiated in seeking a design solution, within the above geometric restrictions. As a first step, a blade taper relationship has to be assumed. Since R_{cur} is a constant, a selection of camber/chord

Table A.1 Rotor Design

x	0.4	0.5	0.6	0.7	0.8	0.9	1.0
λ	0.55	0.44	0.367	0.314	0.275	0.244	0.220
ϵ_p (Fig. A.1)	1.2	1.09	0.98	0.86	0.75	0.64	0.58
$\tan \beta_m = (1 + \tfrac{1}{2}\epsilon_p\lambda)/\lambda$	2.418	2.818	3.215	3.615	4.011	4.418	4.836
β_m	67.5	70.5	72.7	74.5	76.0	77.2	78.3
$\cos \beta_m$	0.382	0.335	0.297	0.267	0.242	0.221	0.203
$C_L\sigma = 2\epsilon_p \cos \beta_m$	0.917	0.729	0.511	0.459	0.362	0.283	0.235
β_1-β_2 (Eqs. 8.28, 8.29)	10.5	7.1	5.1	3.5	2.6	1.8	1.4
$nc/R = 2\pi\sigma x$	2.89	2.74	2.60	2.45	2.30	2.16	2.01
selected: $\sigma = nc/2\pi Rx$	1.15	0.88	0.69	0.56	0.46	0.38	0.32
selected: C_L	0.80	0.83	0.85	0.82	0.79	0.75	0.73
selected: b/c (%)	6.5						4.5
selected: θ	29.5	28.0	26.5	25.0	23.4	21.9	20.4
α (Fig. 6.24)	2.0	2.6	3.2	3.2	3.1	3.0	3.1
$\beta_m - \alpha$	65.5	67.9	69.5	71.3	72.9	74.2	75.2
$\beta_m - \alpha$(linear)	66.4	67.9	69.4	70.9	72.4	74.0	75.5
c/R ($n = 10$)	0.289	0.274	0.260	0.245	0.230	0.216	0.201

ratio at the blade tip will allow the root value to be calculated from the assumed blade taper. A consistent and acceptable set of variables soon emerges from this trial and error method.

In the present instance, emphasis is placed on low drag blades with good lift potential, together with adequate blade stiffness. The degree of camber required in meeting these objectives is greater than for the F-series aerofoils. The selected camber distribution represents a good compromise solution.

The cambered plate centroids for this constructional method will lie along a slightly curved radial line. Attention should be given to this characteristic when selecting plate thickness and designing the root attachment.

The root station will be subject to multiplane interference effects (see Fig. 6.29), but keeping in mind (1) the local nature of this phenomenon, (2) the small effect of this region on overall fan pressure rise, (3) the measure of arbitrary vortex flow already allowed in this region, and (4) the blade pitch setting adjustment required for geometric reasons, lift and incidence corrections are inappropriate.

Assume the midspan chord is 40% of the blade span, that is, 24% of fan radius. Since

$$\frac{nc}{R} = 2.45 \text{ (Table A.1)}$$

$$n = \frac{2.45}{0.24} = 10.2$$

After selecting 10 rotor blades and making the radius R equal to unity, the blade length is 0.6 and the tip and root chords are 0.201 and 0.289, respectively (Fig. A.2).

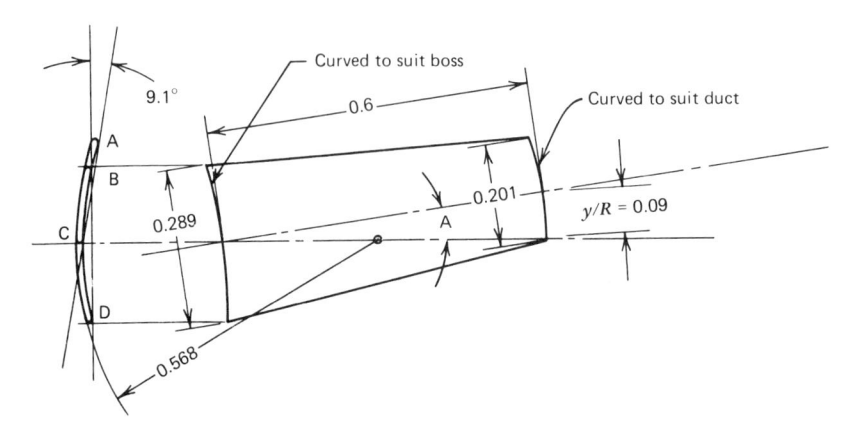

Figure A.2 Rotor blade development from circular cylindrical surface, fan radius = 1.

A.1 FIRST DESIGN EXAMPLE

Blade chords are strictly curved lines about the fan axis, since this is the route traversed by an air particle as it is deflected by the blade around a circumference. For present purposes, however, the blade shape will be calculated on a straight chord assumption.

As a first approximation the cos A correction will be ignored. From Eq. (6.10) and for the tip

$$\frac{R_{cur}}{R} = \frac{0.201}{2 \sin 10.2} = 0.568$$

Multiplying this quantity by the blade twist (9.1°) expressed in radians fixes the value of y/R at 0.09. Hence $\tan A = 0.15$ and $\cos A = 0.989$, making any correction of little practical value.

The extension of the leading and trailing edges around the boss surface will increase the effective blade twist by a small and negligible amount. When attaching the blade to the boss, the blade-setting angle at the midspan is the appropriate reference quantity. In the present instance its value is 70.9°, which can be rounded off to 71°.

The rolled sheets from which the blades are cut should be sufficiently large to ensure that the correct curvature is attained within the blade contour; this also applies to prerotator development.

The prerotator design will be centered on the conical development of a sheet possessing two-dimensional curvature. Hence the radius of curvature will vary in a linear manner between root and tip. Constant chord will be assumed. In this instance the tip determines the design value of chord, with the root having a lesser design requirement (Fig. 11.8).

The ensuing design of Table A.2 is illustrated in Fig. A.3; the leading edge is aligned in the radial direction. The number of vanes (17) avoids a common multiple with the number of rotor blades, for noise and vibration reasons.

Table A.2 Prerotator Design

x	0.40	0.60	0.80	1.0
ϵ_p (Table A.1)	1.2	0.98	0.75	0.58
$\beta_2 = \tan^{-1} \epsilon_p$	50.2	44.4	36.9	30.1
σ (const. c)	1.70	1.133	0.850	0.68
s/c	0.588	0.882	1.176	1.471
$\theta = \beta_2/(1 - 0.19s/c)$	56.5	53.3	47.5	41.8
$c/R = \sigma 2\pi x/n_p$ ($n_p = 17$)	0.251			0.251
$R_{cur}/R = c/2R \sin \theta/2$	0.265			0.352
θ (linear with x)	58.0	52.7	47.3	42.0
Revised R_{cur}/R	0.260			0.350
$\xi = \theta/2$	29.0			21.0

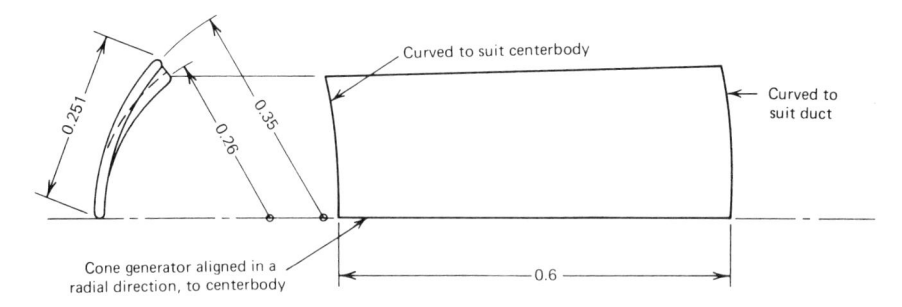

Figure A.3 Prerotator vane development from conical surface, fan radius = 1.

The axial length of the blading unit, in terms of the fan radius, is given by

    Vane tip chord projection = $c \cos 21°$ = 0.234
    Gap (half vane chord, for noise suppression) = 0.126
    Blade root chord projection = $c \cos 66.4°$ = 0.116
                                                        Total = 0.476

Hence the blading unit is very compact, being less than 25% of fan diameter.

A selection of fan diameter, rotational speed, and air density will enable the dimensional fan properties and design duty to be determined.

A.2 SECOND DESIGN EXAMPLE

This example illustrates a rotor design for which the coefficients Λ, K_{th}, and x_b are 0.2, 2, and 0.3, respectively. Units in this design regime are suitable for cooling tower fans operating against low flow resistance.

A preliminary design study along free vortex flow lines indicated blade root design difficulties that were overcome by adopting a straight line swirl coefficient distribution (Fig. A.4). Since the percentage reduction in work output at the blade root exceeds that of the previous example, calculation of the probable velocity distribution is in order. An approximate relationship developed in [A.1] is, for $\epsilon_p = 0$,

$$\left(\frac{V_{a2}}{\overline{V}_a}\right)^2 = \left(\frac{V_{a1}}{\overline{V}_a}\right)^2 + \left[\epsilon_s\left(\frac{2}{\lambda} - \epsilon_s\right)\right] - \left[\epsilon_s\left(\frac{2}{\lambda} - \epsilon_s\right)\right]_{x\overline{V}_a} - 2\int_{x\overline{V}_a}^{x} \epsilon_s^2 \frac{dx}{x} \quad \text{(A.1)}$$

where for uniform inlet flow $V_{a_1}/\overline{V}_a = 1$.

A.2 SECOND DESIGN EXAMPLE

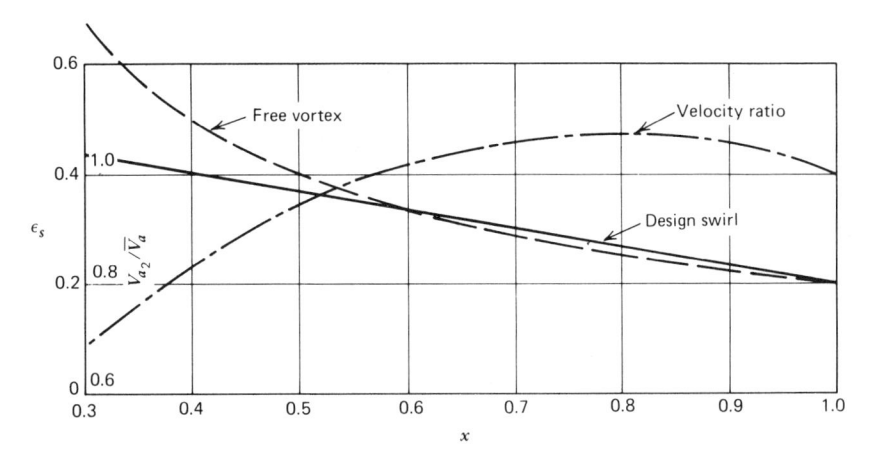

Figure A.4 Spanwise swirl and associated velocity distributions for second design example.

The distribution of ϵ_s, as proposed, is given by

$$\epsilon_s = a + bx \tag{A.2}$$

where $a = 0.543$ and $b = -0.343$. Hence

$$2\int_{x\bar{V}_a}^{x} \epsilon_s^2 \frac{dx}{x} = [2a^2\ln x + b^2x^2 + 4abx] - [2a^2\ln x + b^2x^2 + 4abx]_{x\bar{V}_a} \tag{A.3}$$

Equation (A.1) is the theoretical equilibrium relationship developed from the data of Section 2.5. The selection of a station at which the velocity is estimated to equal the mean reduces the computational effort.

Inspection of Fig. A.4 suggests peak and mean velocity stations at approximately $x = 0.8$ and 0.6, respectively. Selection of the latter value gives a distribution of $V_{a_2}/\bar{V}_a$, which when multiplied by x and integrated in accordance with Eq. (19.2) produces a $\bar{V}_a$ value just under unity. Trial and error establishes $x = 0.57$ as the correct location for mean velocity. The resulting velocity distribution is superimposed on Fig. A.4.

The swirl and flow coefficients at $x = 0.3$ and 0.4 are taken in Table A.3 to be

$$\epsilon' = 2\epsilon/(1 + V_{a_2}/\bar{V}_a), \qquad \lambda' = \lambda(1 + V_{a_2}/\bar{V}_a)/2 \tag{A.4}$$

where $V_{a_2}/\bar{V}_a$ equals 0.685 and 0.836, respectively.

Table A.3 Rotor Design

x	0.3	0.4	0.5	0.6	0.7	0.8	0.9	1.0
λ	0.667	0.500	0.400	0.333	0.286	0.250	0.222	0.200
(λ')	(0.562)	(0.457)						
ϵ_s (Fig. A.4)	0.440	0.406	0.371	0.337	0.303	0.269	0.234	0.200
(ϵ_s')	(0.552)	(0.442)						
$\tan \beta_m$	1.518	(1.958)	2.315	2.835	3.345	3.866	4.388	4.90
β_m	(56.6)	(62.9)	66.6	70.6	73.4	75.5	77.2	78.5
$\cos \beta_m$	(0.550)	(0.455)	0.397	0.333	0.286	0.250	0.222	0.200
$C_L \sigma$	(0.607)	(0.402)	0.294	0.224	0.173	0.135	0.104	0.080
$\beta_1 - \beta_2$	(9.9)	(5.2)	3.4	2.2	1.4	1.0	0.7	0.5
Selected								
nc/R	1.26	1.18	1.10	1.02	0.94	0.86	0.78	0.70
σ	0.669	0.470	0.350	0.271	0.214	0.171	0.138	0.111
C_L	(0.91)	(0.86)	0.84	0.83	0.81	0.79	0.75	0.72
α(10% Clark Y) Fig. 6.15	(5.0)	(4.5)	4.0	3.9	3.8	3.6	3.2	3.0
$\beta_m - \alpha = \xi$	(51.6)	(58.4)	62.6	66.7	69.6	71.9	74.0	75.5
ξ (faired)	52.5	58.5	63.0	66.5	69.5	72.0	74.0	75.5
$c/R(n = 6)$	0.210	0.197	0.183	0.170	0.157	0.143	0.130	0.117

A.2 SECOND DESIGN EXAMPLE

If we take $x = 0.7$ as representative of the mean rotor properties, for purposes of efficiency calculations, then

Rotor loss, $\quad \gamma \dfrac{K_R}{K_{th}} = \dfrac{\lambda}{\cos^2 \beta_m} = 3.497$

$$\gamma = \dfrac{C_L}{C_{D_P} + C_{D_S}} = \dfrac{0.81}{0.010 + 0.018 C_L^2} = 37.1$$

$$K_R/K_{th} \approx 0.10$$

to which must be added a small but significant amount to cover the anticipated higher blade root flow losses. The latter losses will be more substantial in the absence of an aerodynamically designed nose fairing.

In the absence of straightener vanes, the overall efficiency loss is increased by a further

$$\left(\dfrac{\epsilon_s^2}{K_{th}}\right)_{MS} = \dfrac{0.303^2}{2} = 0.05$$

The proposed use of this rotor unit, in an induced cooling tower capacity, influenced to some degree the blade planform selection. When the fan draws air from a plenum-type space, some tip flow disturbances are likely even when care is taken in the design of the duct inlet shape; a slightly conservative C_L is therefore selected at the tip.

The 10% thick Clark Y airfoil chosen has a maximum lift/profile drag ratio in the vicinity of $C_L = 0.85$. This low camber airfoil with its flat undersurface is the one favored for the proposed application as the flow turning angles $(\beta_1 - \beta_2)$ are small.

Adopting straight leading and trailing edges, the root chord follows once C_L decisions at $x = 0.5$ and 1.0 are reached.

Design conditions in the blade root vicinity are considered to be satisfactory. This statement is based on intuitive reasoning rather than on rational argument, since the latter would require the introduction and discussion of three-dimensional flow theories. However, the author would not recommend the use of this design technique for significantly greater blade root load reductions. Such must inevitably increase the outboard velocities, inflating the potential dynamic pressure loss at fan outlet.

The potential efficiency loss due to swirl (≈ 0.05) is small and makes the rotor-only fan configuration an acceptable proposition.

This design is based on a K_{th} of 2, and in order to assess its applicability to actual installations, the probable downstream total pressure losses must be estimated. The available "useful" fan inlet total pressure is then obtained nondimensionally from

$$K_{IT} = \eta_T K_{th} - (\epsilon_s)^2_{mean} - K_{DL}$$

A rotor discharging direct to atmosphere would have a K_{DL} of unity and hence a η_{IT} of approximately 35%. Installing a diffuser of area ratio 1.65 and an effectiveness of 0.80 would result in a pressure recovery factor (C_p) of 0.5 (Fig. 4.27), thus reducing K_{DL} to 0.5 and increasing η_{IT} by 25%; this increases the useful workload by 70%.

It is clear from the preceding discussion that errors of up to 5% in estimating rotor efficiency are not crucial, since the performance of the vitally important discharge diffuser has a much greater influence on η_{IT}. This margin should cover any additional rotor losses incurred in this design exercise.

In developing a diffuser design, some considerations should be given to an increased wall angle to take advantage of the unremoved design swirl at the wall ($\approx 12°$). On the debit side the central core of turbulent mixing, in the absence of a centerbody, will have an unknown influence on diffuser effectiveness, and on the critical wall angle for which unseparated flow conditions can just be maintained. Because of the reduced inboard velocities, an expanding centerbody is the most acceptable type.

The preceding discussion assumes that the drive and support systems are of an efficient aerodynamic nature.

REFERENCE

A.1 R. A. Wallis, *Axial Flow Fans*, Newnes, London, 1961.

APPENDIX B
ARBITRARY VORTEX FLOW DESIGN

A need for definitive design information with respect to this extremely common industrial variety of fan unit is implied in the first edition of this book. The absence of published experimental data that might validate the method of [B.1] led to the present program of work. The objective was to establish limits for the simple design method of [B.1], for the most common type of unit, namely, a rotor with no preswirl.

The ensuing research work was undertaken by the CSIRO Division of Building, Construction, and Engineering. The fan duty selected falls into a common duty area within industry.

The boss ratio chosen was 0.38, a value that is in contrast to the 0.69 figure selected by Kahane [B2], and is therefore representative of the lower end of the useful range. Three different blading shapes were designed for the 610 mm fan, all for the same duty condition. The first featured a tapered cambered plate blade with solidities of 1.42 and 0.34 at the root and tip respectively. The second cambered plate blade shape had corresponding values of 1.02 and 0.39. The last blade shape featured the F-series airfoil with identical camber and chord to the previous blade, and with a leading edge droop of 2 ½% chord.

The second blade shape was modified to facilitate the rotor being die cast in one piece. The reduction in blade solidity at the root was countered by a modest increase in blade camber to give the desired increase in lift coefficient, keeping the product $C_L\sigma$ the same, a case of trading off solidity for C_L (see page 207). Reduced solidity results in a gain in $C_{L,max}$ and this feature was also taken into consideration.

The increased solidity at the tip was the consequence of a constant chord blade, considered desirable from the point of view that the fan would eventually stall out at the tip, and hence an improvement in maximum pressure duty could result from the increase. The design C_L at the tip is therefore lowered. All the design expectations were realized.

The selected test results presented in Section B.3 validate the method of

[B1] as a simple and practical design approach to arbitrary vortex flow fans. Therefore the more exacting design measures mentioned in Section 2.6 are considered unessential.

B.1 DESIGN PROCEDURES

The initial assumption is that the required mean total pressure rise is the same as the local value at the midspan. Hence the first step entails calculating either K_t or K_{it} for in-line or exhaust fans respectively. This requires a knowledge of flow quantity, fan and boss diameters, and air density, in order to evaluate

$$K_T(\text{or } K_{IT}) = \frac{\text{total pressure requirement}}{1/2\,\rho \bar{V}_a^2}$$

This permits a tentative value of K_{th} to be established from Eqs. (14.3) through to (14.5) when a first estimate is made of fan efficiency. Knowing the rotor speed, the flow coefficient can be established at the midspan, and from Eq. (8.17), the corresponding swirl coefficient is determined. A more accurate estimate of fan efficiency can now be made from data presented in Chapters 10, 12, and 13, and a value of β_m from Eq. (8.31). The tests show good estimation accuracy, using the recommended free vortex data and procedures. By iteration, suitable values of efficiency and K_{th} will be arrived at, provided subsequent allowance has been made for tip clearance effects, according to Eqs. (10.9) and (14.10). The experimental studies mentioned above have indicated tip clearance adjustments approaching a 100% increase on these loss values, particularly with respect to the peak fan pressure; excess clearance results in a large reduction in this pressure. This could be due to the local radial velocity component and hence generous additions are advised for similar small boss ratio fans.

This relation

$$\epsilon_s = a + bx$$

represents a practical swirl coefficient distribution and is therefore recommended, although not essential. This swirl distribution governs the spanwise distribution of total pressure rise. Graphing the free vortex distribution and then drawing in the preferred straight line distribution that you desire, as in Appendix A, the values of the intercept "a" on the $x = $ zero line, and "b" the slope, can be determined. The value of ϵ_s at the midspan should be the same as that calculated in respect to free vortex flow.

The nondimensional relationship that determines the gradient of velocity distribution downstream of the rotor is given in Eq. (A.1), where the integral for the above ϵ_s distribution is calculated from Eq. (A.3). This involves making a provisional assumption regarding the spanwise location at which the local V_{a2} will equal the mean value. Since the values of velocity

B.2 DESIGN LIMITATIONS AND RECOMMENDATIONS

must satisfy Eq. (19.2), the values of $V_{a2}/\overline{V}_a$, when integrated graphically with respect to x, must result in a LHS quantity of unity. This requires an iterative process in which successive assumptions are made in Eq. (A.1) regarding the x at which the local $V_{a2}/\overline{V}_a$ is unity.

The design relationships applicable to free vortex flow then apply, provided the above flow and swirl coefficients are replaced by the relationships given in Eq. (A.4). The procedure is illustrated in Table A.3.

B.2 DESIGN LIMITATIONS AND RECOMMENDATIONS

Since the arbitrary vortex flow fan offers a potential increment in maximum pressure rise to that value attainable with the free vortex variety, it is important to have some idea of the likely benefit. The recommended maximum design value of ϵ_s at the midspan is 0.7 for a boss ratio of 0.4, increasing by 0.05 for each 0.1 increase in boss ratio up to 0.7. These figures represent a probable gain in design pressure capability of from 27% to 13% respectively for the boss ratio range stated. In that range the slope "b" will progressively change from a negative value to a positive one.

However, the above estimates may not reflect the magnitude of the increases in peak, as distinct from design fan pressure, for reasons outlined in Section B.3.

In the design exercise for fans with small boss ratios, selecting progressively smaller values of negative slope will result in the downstream axial velocity at the blade root approaching zero and then going negative. It is recommended that values of "b" that result in root velocity ratios less than 0.4 should be avoided. For high boss ratios, blade loading considerations, rather than axial velocity ratio, will determine the allowable "b" values.

The procedure for determining the design distribution of lift coefficient differs from that recommended in the free vortex flow case. Since the axial velocity in a given annulus does not remain constant across the rotor blade, the relationships developed in Section 9.3.3 tend to become irrelevant, particularly for the inboard flows associated with small boss ratios. Nevertheless, the use of the blade interference factors of Section 6.8.6 have been shown to apply with reasonable accuracy. (These factors have been used with success in cambered plate blade designs and hence their universal application to all circular arc cambered sections, and possibly other sections as well, may be assumed in industrial practice).

At the blade root, for small boss ratios and relatively high pressure loads (K_{th} in excess of 4), interference factors will apply. Therefore some guidance is desirable in determining the relationship between C_L and solidity since the recommendations based on constant axial velocity in an annulus are invalid.

When the calculated blade root value of $C_L\sigma$ is approaching 1.2, the recommended procedure is to adjust the solidity so that the design $C_L i_i$

value, in general, is kept below 1.3. This figure is usually associated with camber angles of approximately 36 degrees (8% c). As the solidity is increased there is a greater related lift interference and hence an increase in C_{Li} will be needed for the calculated C_L design requirement.

A normal sequence of events in the case of arbitrary vortex fan stalling is for flow separation to occur first in the vicinity of the blade root, and spread outwards as pressure duty is increased. This is a relatively stable flow situation. The radial velocities are substantially increased and this leads to stronger outboard flow, and a tip stall delay.

Therefore the recommendation concerning spanwise lift distribution is to first establish the C_L at the root, from the above considerations. The calculated product of C_L and solidity then enables root solidity to be established. The spanwise distribution of solidity is then selected, on the basis of constant or tapered chord. The recommended value of the tip C_L should be quite modest, say around 0.6, in order to take advantage of an enlarged increment in peak total pressure rise after the onset of flow separation from the blade root. The intervening spanwise values of C_L can then be determined in the usual manner from $C_L\sigma$.

Before the blade incidences can be determined, the distribution of camber angle must be selected. As the calculation for the inner blade sections of $(\beta_1 - \beta_2)$ can no longer be used as a reliable guide, we have to replace the free vortex recommendation with an alternative. Since the radial flows produce a small effective increase in chord, the percentage camber based on chord is slightly decreased. Hence the selected camber angle for arbitrary vortex blading can be in excess of that recommended for free vortex flow. A modest increase on normal free vortex values is therefore recommended at the blade tip. (A calculation of β_1 and β_2 from the unmodified flow and swirl angles at the blade tip will produce a meaningful answer regarding this value). The value at the blade root will be governed by the ability of the isolated blade section to deliver the C_{Li} required, with a reasonable margin to $C_{L.\max}$. A smooth curved line, of decreasing negative slope towards the tip, will provide a suitable distribution of camber.

The above information is with respect to design limits, and recommendations, for low boss ratio fans in the high K_{th} range. Such fans normally operate in the design Λ range of 0.2 to 0.25. For less extreme design conditions, and for fans of arbitrary vortex design in the high boss ratio range, the appropriate design recommendations for lift and camber values will tend to approach those associated with free vortex flow. As indicated previously, the precentage gains diminish with increasing boss ratio.

B.3 TEST VALIDATION OF DESIGN METHOD

The tests were carried out in a carefully constructed and instrumented rig, described fully in [B.3]. Total and static pressures, and all three velocity components, were measured using a small 4-hole pressure probe. A transmission dynamometer measured the rotor input power.

The fan performance is based on a downstream station located approximately one midspan half-chord from the trailing edge at this location. Measurements at two stations further downstream showed similar characteristics, indicating flow stabilization.

The Reynolds numbers of the tests were 2.2 and 2.5 × 10^5, based on the blade tip speed, for the first and latter two rotors, respectively. Corresponding test Reynolds numbers as low as 1.55 and 1.75 × 10^5 showed no discernable scale effect, a result that was unexpected for the airfoil bladed rotor. However, the lack of scale effect for the cambered plate blades is in accord with the statement in Section 10.2.2. Leading edge droop is effective in reducing Re_{crit}, and hence the 2 ½% chord designed into the airfoil blades may have been partly responsible for the observed insensitivity. (However, a reduction in fan diameter for a commercial application produced a substantial reduction in performance).

The important detailed flow features from a design validation viewpoint are the nondimensional spanwise distributions of the three downstream components of velocity, and of the total and static pressures. The results for the first and last rotor tests are displayed in Figs. B.1 to B.10 for various operating conditions. (The first and second rotors have distributions that are quite similar, with the latter possessing a small increase in peak performance).

The axial velocity generally follows the design prediction except in the inner region, where the retarded flow in the boundary layer is compensated for by a velocity increment above the mean outside the layer. With flow breakaways at the blade root, the changes in distribution are quite radical.

The tangential components of velocity, namely the swirl, also show good agreement with the design estimate. Curves for the off-design conditions illustrate the effect of flow separation from the blade root.

Of particular interest are the values of the radial velocity ratios for which no estimate existed, due to the quasi two-dimensional nature of the design method. The radial velocities are approximately 8% and 10% of the mean axial velocity for the Nos. 1 and 3 rotors respectively. The pressure forces associated with these velocities are less than 1% of the overall pressure forces and hence their neglect in the design process is justified.

The nondimensional total and static pressure distributions of Figs. B.7 to B.10 indicate a close agreement with the design estimates. The curves maintain much the same slope up to the duty at which the downstream axial velocity at the root diminishes to zero.

The characteristic curves and rotor efficiencies shown in Figs. B.11 and B.12 are of special interest. The margin to stall, from the design duty point, is much greater for the airfoil bladed rotor, and this feature is reflected in the spanwise distributions of Figs. B.7 and B.8. Flow improvements in the tip vicinity are believed to be due to the presence of an airfoil

414   ARBITRARY VORTEX FLOW DESIGN

shape, and a small tip clearance. The peak pressure drops away quite sharply with increasing tip clearance.

The fan pressure curves are both slightly in excess of the design duty point.

Despite the inboard flow disturbances, the rotor total pressure efficiencies in both cases remain reasonably high. Peak efficiencies are approximately 86% and 88.5% for rotors Nos. 1 and 3 respectively.

B.4 PERFORMANCE ESTIMATES

An approximate performance curve can be established by the use of midspan data concerning stagger angle, blade chord, blade number, and the no-lift angle of the blade section. This exercise follows an identical procedure to that outlined in Chapter 20 for free vortex flow rotors, where the flow and swirl coefficients are with respect to the mean axial velocity. Once again it is assumed that the mean total pressure rise for the fan is represented by midspan conditions. This is a conservative value as the true mean will normally be slightly higher.

However, the above procedure will lose relevance for fan duties that lie beyond the point at which zero axial velocity at the blade root is reached. A method that gave reasonable answers in the foregoing experimental studies is now outlined.

Since the spanwise distribution of the downsteam axial velocity component varies with changes in fan duty, the procedure involves assuming a new distribution of swirl coefficient at a slightly higher flow coefficient than design. Selecting a Λ from the preceding calculations, the corresponding computed value of ϵ_s at the midspan will provide the point through which the line passes. A slightly more negative slope than the design should be assumed, and the related value of "a" determined.

The design computational procedure is then re-activated, with the axial velocity ratio distribution being determined for the above assumed value of "b". Entering these quantities into the blade design exercise, in a similar manner to Table A.3, and inserting the values of solidity at each spanwise station from the original design data, tentative estimates of the stagger angles are computed. When these agree closely with the actual design values of stagger, at all spanwise stations, the selection of "b" has been correct. Some iteration may be necessary in achieving this goal.

The precedure should be repeated for another two duty conditions. The duties selected should not be too close to the zero velocity case, as a marked computational sensitivity exists in the region. When $V_{a_2}/\bar{V}_a$ at the root is plotted against Λ for the design and three additional points, a line drawn through the points, and extrapolated, will give the approximate value of Λ at which zero velocity occurs.

The terminal point on the characteristic curve, from the computational point of view, has now been approximately established. The curve extension is a matter on which no definitive guidance is offered.

The tedium of the above procedures, and the possibility of human errors when calculating, can be avoided when computer programs are employed. This is an area were they are most valuable since the computer makes light work of iterative procedures.

B.5 AD HOC EXTENSION TO PERFORMANCE CURVE

The test results indicate that with good quality swirl-free inlet flows, airfoil blades, adequate tip chord, and restricted tip clearance, a fan can operate efficiently in the flow regime that follows the appearance of zero axial velocity at the blade root. Provided these conditions are met with reasonable accuracy, then the procedure outlined will provide practical design solutions.

An ad hoc design approach to fans that have to operate in this upper flow regime is to draw in an imaginary characteristic curve, and then design a fan for a duty on this curve that will ensure a positive value of axial velocity at the blade root. Conservative values of C_L should be used in the design process. The lower part of the performance curve is then calculated by the method described in Section B.4 for positive spanwise axial velocities, enabling you to check the slope of the curve against your original estimate. When adjustment is required, a second design will be necessary in order that the desired duty point in the top end of the characteristic curve may be approximately attained.

A characteristic curve is then calculated for the new design, up to the point at which the root axial velocity diminishes to zero. The procedure for estimating this specific point is described in the preceding Section. The extension to the characteristic curve should then pass through the desired duty point, flattening out at the estimated peak usable fan pressure point. (Axial flow fans of moderately low Λ will continue to produce increased levels of fan pressure, due to centrifugal forces, after this usable peak has been passed in the diminishing flow direction. Reduced fan efficiency and flow instability are the related penalties.) The length and shape of the extrapolated part of the curve will constitute the ad hoc component of the design exercise, requiring considerable but not unattainable experimental skills.

However, good accuracy cannot always be expected when employing this procedure. When the length of the computed characteristic curve is short, a goodly measure of practical experience will be essential, firstly, in designing a practical blade plan form with a low C_L distribution and, secondly, in the extrapolation of the likely performance curve. Nevertheless, the method is recommended as it provides the designer with a realistic modus operandi. It is important to ensure that the fan fulfils the design requirements laid down at the commencement of this Section, for best and meaningful results.

B.6 TESTS WITH CONICAL FAN BOSSES

The boss ratio of the airfoil bladed fan was increased to 0.5, so that cone angles of 70 and 50 degrees could be introduced into the boss. This was achieved by the addition of molded polysterene inserts.

Tests were carried out on three boss conditions, namely, taper angles of 0, 10, and 20 degrees. Increasing the taper angle resulted in slightly greater radial velocities along the entire blade span, with minimal increments at the tip. Just prior to the axial velocity at the root reaching a zero value, the inboard radial velocity was just 12% of the mean axial velocity for the largest taper angle. There was a slight delay and improvement in respect to the operational duty at which the zero axial velocity condition was reached.

At the peak total pressure rise, inboard radial velocity ratios of 27 and 40% were recorded for the 0 and 20 degree taper angle cases. Slight increases in the pressure duty were recorded at the low end of the characteristic curve, but the two curves were virtually coincident in the region of peak fan pressure. The curves of spanwise total pressure rise, for these two conditions, gave no indication of complete flow separation from the blades, at the related value of Λ for peak total pressure.

Fan efficiencies were relatively insensitive to the actual taper angle, with peak values of 88%. All the above findings were obtained with no spinner fitted, as the differences in performance with a spinner fitted were minor, an unexpected result.

Therefore, in this instance it can be concluded that conical bosses, with their inherent manufacturing problems, have very little to offer when compared with carefully designed, airfoil bladed, arbitrary vortex flow fans.

B.7 SUMMARY

The rotors tested are broadly representative of common type industrial fans, in a nondimensional duty sense and for small boss ratios. Since the design duty adopted was close to the limiting value, the foregoing establishes the approximate limits of the current design techniques and demonstrates the chief flow features that precede the stall.

A modified free vortex flow method of design is adequate to deal with all rotor designs that have essentially swirl-free inlet flow. Except for excessive tip clearance, the rotor efficiencies are similar to those for free vortex rotors. As the boss ratio approaches 0.7, the gains in pressure potential are reduced and the blading design rules of the free vortex method tend to become more relevant.

The procedure recommended in Appendix A for the small nondimensional pressure duties is very appropriate for such equipment.

REFERENCES

B.1 R. A. Wallis, *Axial Flow Fans*, Newnes, London. 1961
B.2 A. Kahane, Investigation of axial flow fan and compressor rotors designed for three-dimensional flow, NACA Tech. Note 1652, 1948.
B.3 R. J. Downie, M. C. Thompson, and R. A. Wallis, An engineering approach to blading design for low to medium pressure rise axial flow fans, to be published in the International Jnl. of Experimental Heat Transfer, Thermodynamics, and Fluid Mechanics.

418　　　　　　　　　　　　　　　　　　　　ARBITRARY VORTEX FLOW DESIGN

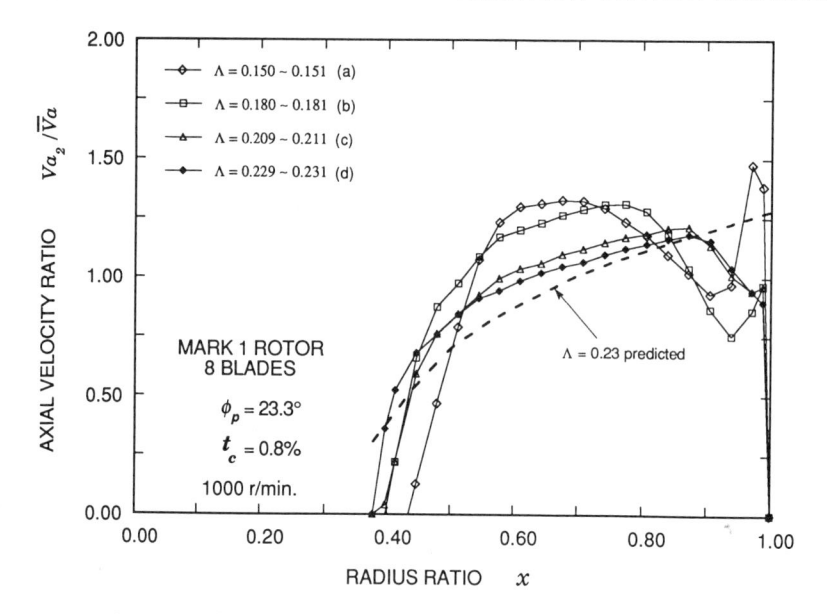

ϕ_p is the angle between the airfoil chord line and the plane of rotation, at the midspan location
t_c is the tip clearance as a % of the blade span

Figure B.1 Downstream axial velocity ratio versus spanwise location.

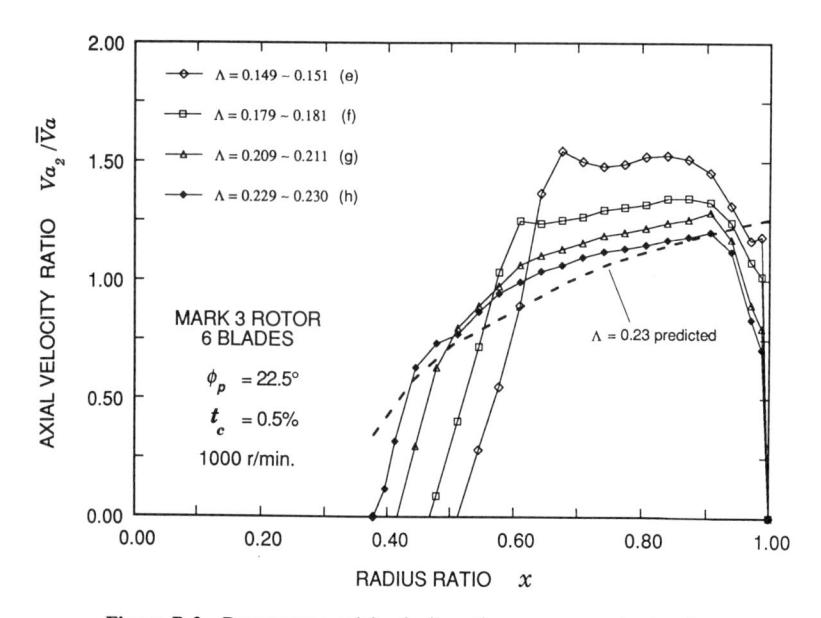

Figure B.2 Downstream axial velocity ratio versus spanwise location.

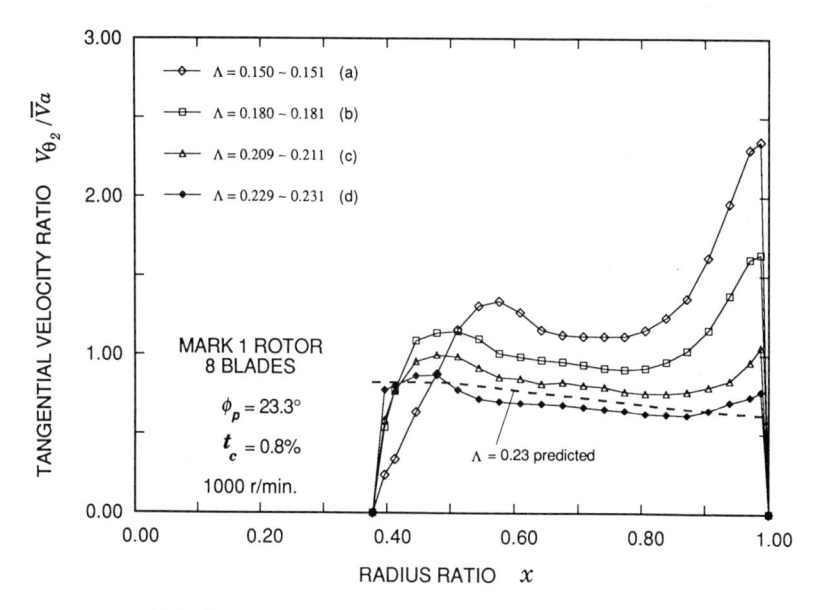

Figure B.3 Downstream tangential velocity ratio versus spanwise location.

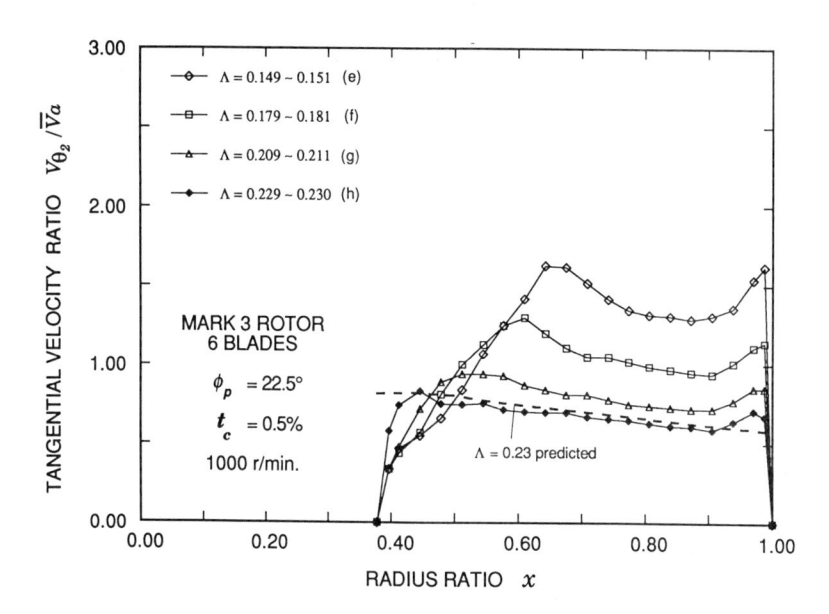

Figure B.4 Downstream tangential velocity ratio versus spanwise location.

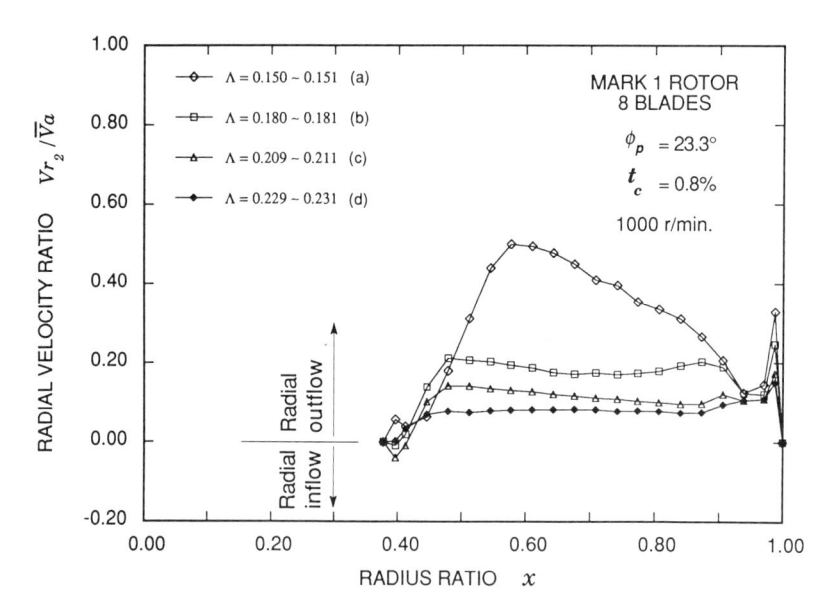

Figure B.5 Downstream radial velocity ratio versus spanwise location.

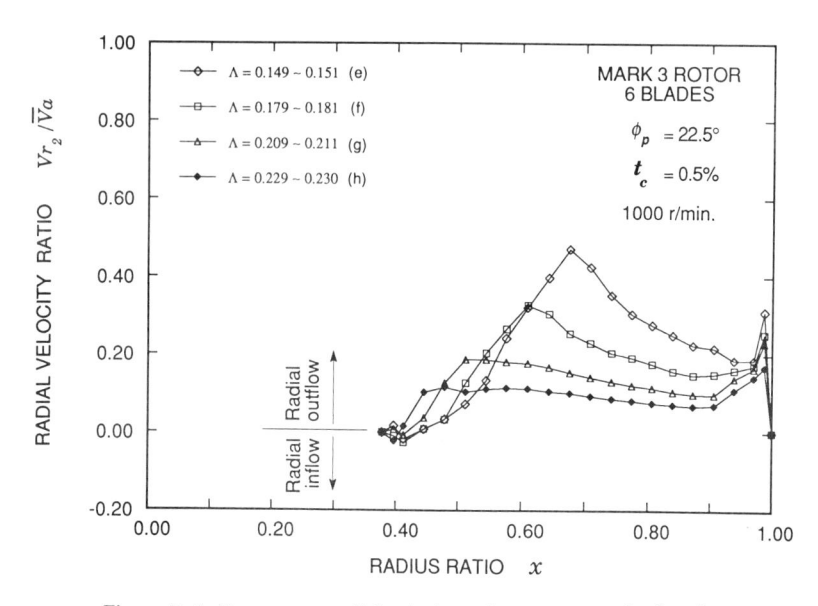

Figure B.6 Downstream radial velocity ratio versus spanwise location.

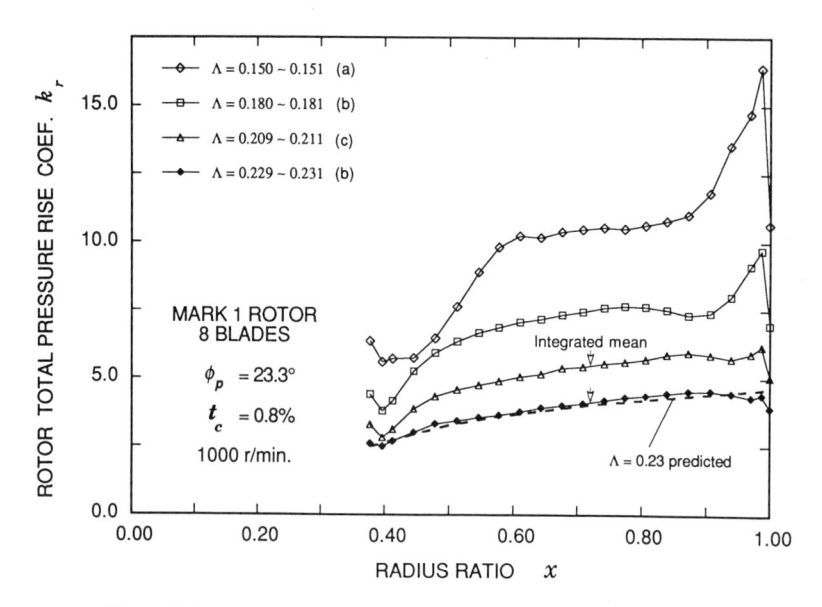

Figure B.7 Spanwise distribution of the total pressure rise coefficient.

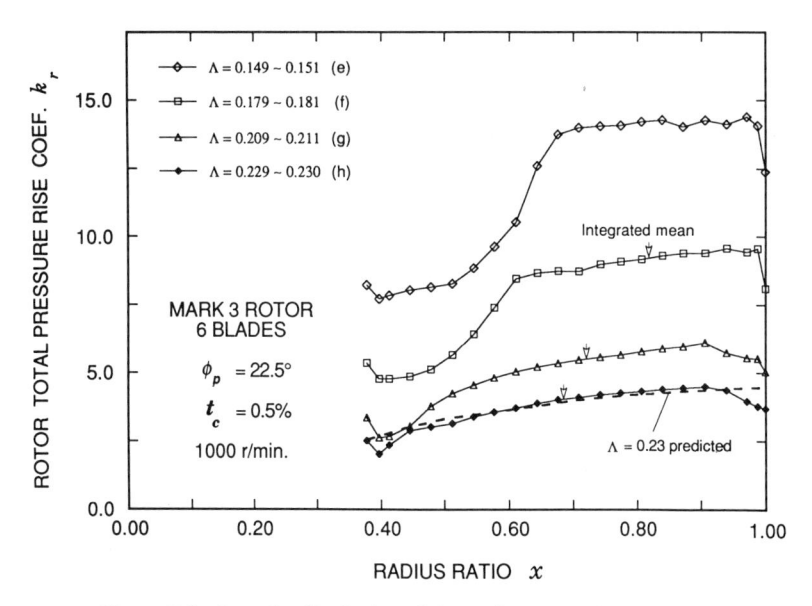

Figure B.8 Spanwise distribution of the total pressure rise coefficient.

422 ARBITRARY VORTEX FLOW DESIGN

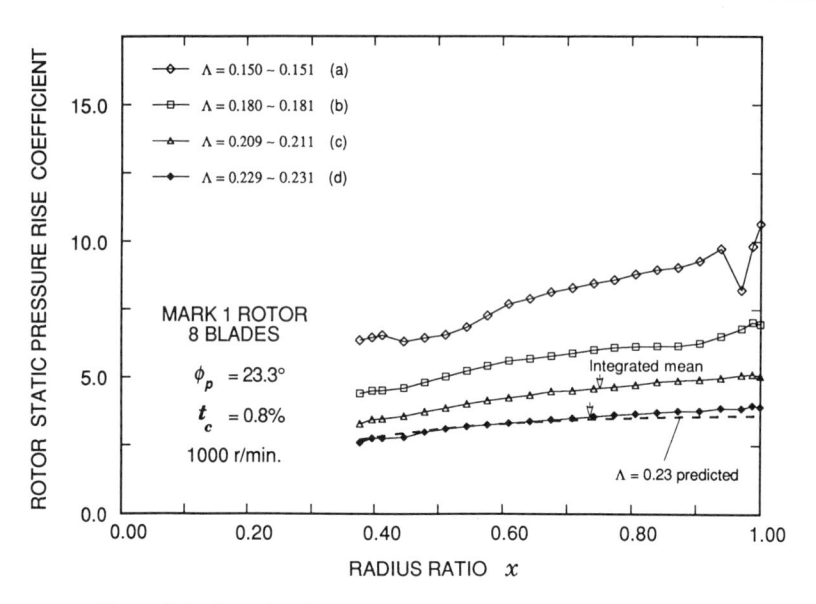

Figure B.9 Spanwise distribution of the static pressure rise coefficient.

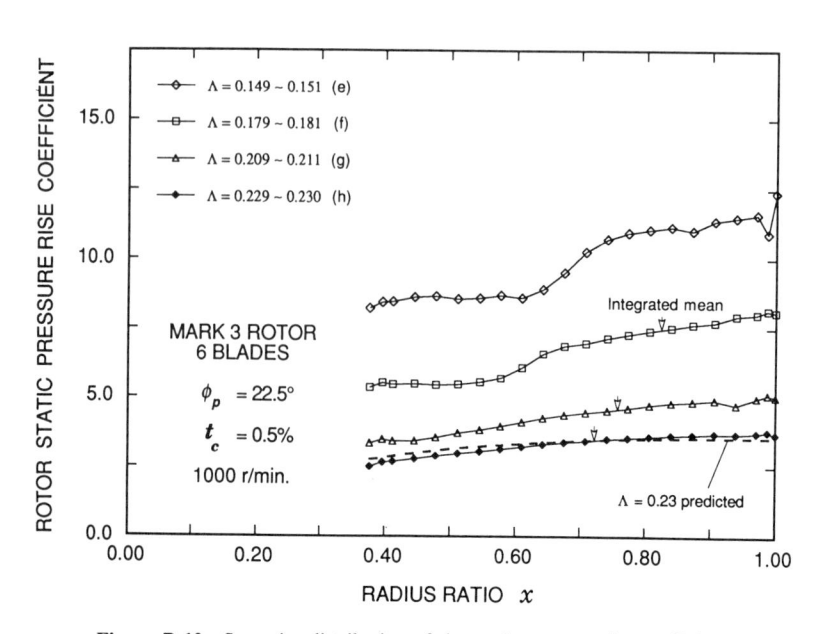

Figure B.10 Spanwise distribution of the static pressure rise coefficient.

423

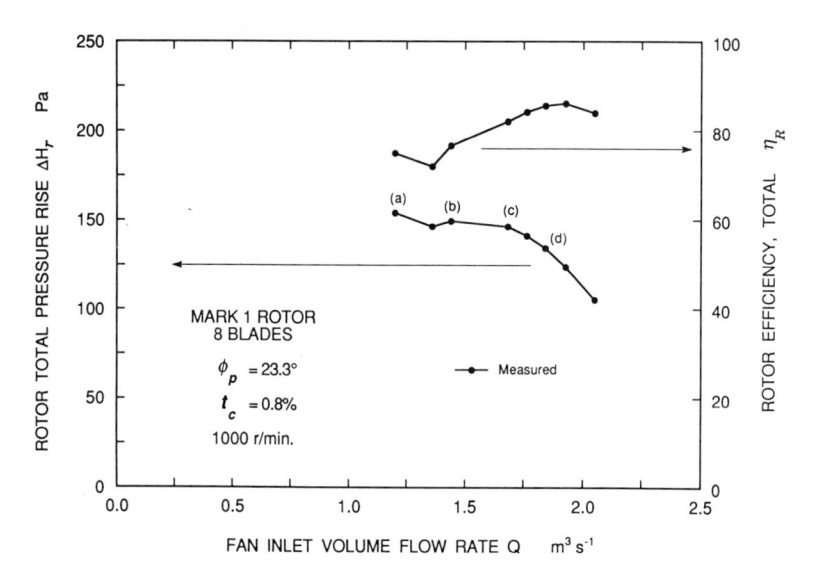

Figure B.11 Total pressure rise versus volume flow rate.

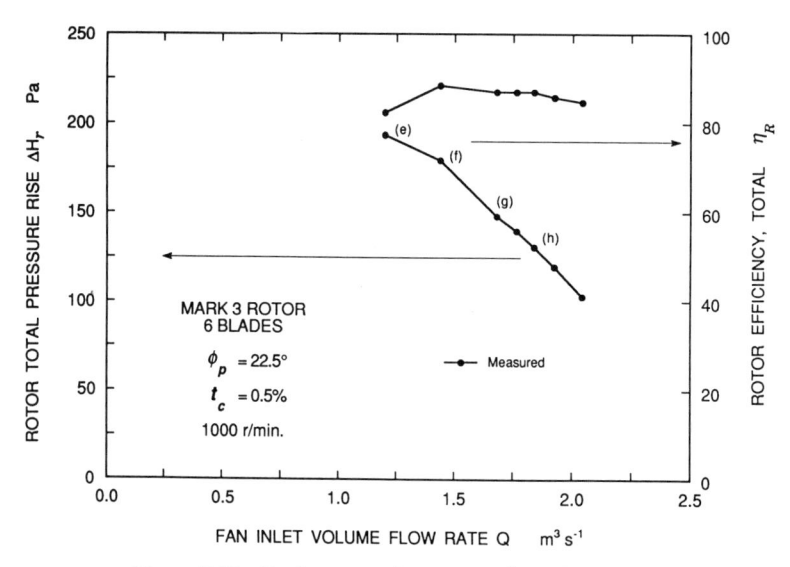

Figure B.12 Total pressure rise versus volume flow rate.

APPENDIX C
Diaphragm-Mounted and Shrouded Fans

Many existing fans in the title category are unnecessarily crude and consequently are inefficient and noisy. The increased use of air conditioning and automatic drive systems in motor vehicles has highlighted the inadequacy of the old-style, unducted automotive cooling fan. Fan pressure rise limits are governed by tip flow recirculation, which degenerates into the "vortex ring state" described in the following appendix. High blade camber, in the hope of achieving a greater pressure rise and flow, is only mildly effective in raising these limits and always results in increased inefficiency and noise. A duct of limited length incorporated in a radiator embracing box is becoming more common on new models of motor cars. When the rotor is engine mounted, large tip clearances must be provided to allow for engine sway. Electric drive systems avoid this problem.

Since the axial dimension of the fan unit is limited, the addition of a diffuser is usually out of the question, resulting in a low overall efficiency.

Other examples of this fan type are found in domestic and industrial heat exchangers for cooling and heating duties and in building ventilation equipment.

Free vortex flow design solutions are normally unsuitable for this fan class. Appreciable nondimensional total pressure rise requirements have to be achieved with restricted blade speeds and small boss ratios. Consequently, the values of ϵ_s and $C_L \sigma$ at the blade root are excessive.

In the circumstances arbitrary vortex flow equipment may provide a practical solution. The introduction of a gradient of total pressure rise along the blade produces an out-of-balance static pressure field, resulting in outward radial flow components. However, there is an obvious limit to the amount by which the root load can be eased by a reduction in blade twist. When elements near the blade root provide an insufficient total pressure rise to meet the downstream static pressure needs, the whole downstream flow field will readjust itself to the prevailing conditions. Although the inner portion of the blade is not stalled in the usual sense, hub flow separation will be present and grow rapidly to occupy a large annular area immediately downstream of the blade trailing edge. The action of the blades is to rotate this flow body, raising local static pressure as a result of centrifugal effects. The subsequent mixing with the higher energy outer flow, through shear stresses, enables such a fan to perform in a relatively stable manner.

Increases in the disturbed flow mass will lessen the useful mean total pressure rise of the unit resulting from energy mixing and swirl factors. The normal fan stall under these conditions will occur at the blade tip. However, because of the centrifugal effect of the entire blade in swirling the air, pressure is not necessarily reduced, although the flow rate is. As the stall condition intensifies, the static pressure normally continues rising to its highest levels. The above flow features are confined to fans with small boss ratios, and design flow coefficients, Λ, that are generally below 0.25.

Fans with untwisted blades operating in this design regime will be subject to these preceding features. Increasing the degree of blade twist will raise the proportion of load being carried by the blade root. Increasing the twist by stages will result in a progressive reduction in the foregoing disturbed flow volume. When a hand held tufted wand indicates a large swirl component, but no substantial disturbed flow in the inboard trailing edge region, at the desired volume/pressure operating point, then the work output of the blade root has reached the minimum acceptable level. The fan total pressure at tip stall will have risen progressively with the associated twist increases.

Although some of these characteristics may not be present in all cases, the above scenario may be taken to represent the general state of affairs. The recent experimental work of Appendix B now enables computational design to move into an area where previously entirely ad hoc developmental procedures were required. However, there is still a low efficiency regime in which the axial velocity component is rarely above zero at any point on the complete characteristic curve. The radial flows will be quite substantial in such cases and hence the unit is not an axial flow fan in the true sense, but rather one that resembles a crude "mixed-flow" variety.

An alternative to the latter fan type should always be sought but, when this option is impractical, developmental procedures based upon test results are the only practical avenue left open to the designer. A fair measure of experience, aerodynamic skill, and detailed flow observations on the "mixed-flow" fan type are required in achieving the best compromise fan unit.

APPENDIX D
Air-Circulating Fans

The function of a free fan is to circulate air in a common air space. This may serve the purpose of achieving uniform mixing and hence steady temperature conditions within the air space. The current of air provided by a free fan is often adequate for the purposes of cooling men or heat-generating equipment; the former relies on evaporative cooling and the latter on heat dispersion within the cooler surrounding air space.

Generally speaking, therefore, the free fan is restricted to one or two specific duties and in consequence design procedures can be simplified. In addition, the aerodynamic specifications to which the fan is designed are relatively flexible. These circumstances are indeed fortunate, as the flow field in the vicinity of the fan blades is a complex one for which no precise analysis is available. Smith [D.1] has, however, managed to develop a design method that is simple and relatively accurate, Useful information on the general characteristics of free fans is also available in [D.1].

Although in what follows no attempt will be made to cover the subject of free fans in a comprehensive manner, data sufficient for the purpose of making an informed approach to design problems will be presented. In particular the design method for near optimum performance as developed in [D.1] will be outlined.

D.1 GENERAL FLOW FEATURES

The fan discharges a jet of air possessing an unrestrained or free boundary. Owing to turbulent shear the surrounding air in contact with this boundary is accelerated as a result of momentum transfer from the jet stream. This process of entrainment continues as the air stream moves further away from the fan. Representative data for such a phenomenon are expressed graphically in Figs. D.1 and D.2. The nominal jet boundaries presented are, of course, in regions of large velocity fluctuations; the lines drawn represent a mean condition. From Fig. D.1 it will be seen that downstream of 8 diameters,

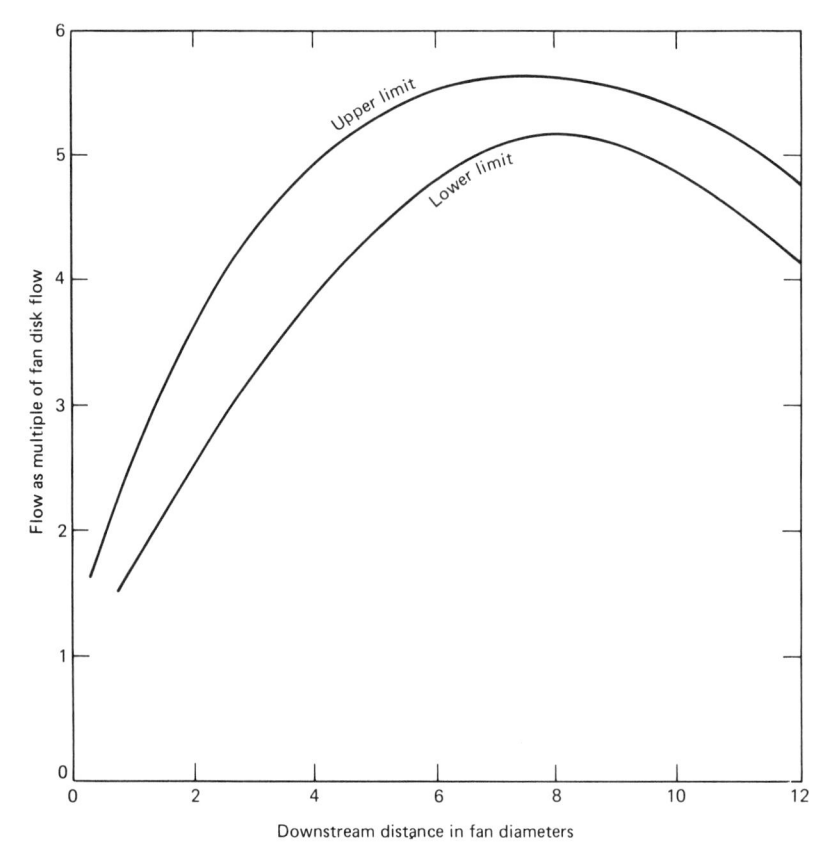

Figure D.1 Jet entrainment data. Test data for twelve fans, 305 to 915 mm in diameter, lie within these boundaries.

where entrainment ceases, there is a net loss of air through the jet boundaries of Fig. D.2.

When the fan operates in a well-defined air space (e.g., in a room), the air will continue to recirculate through the rotor with a frequency determined to some extent by the size of the room. The recirculation flow pattern can be broadly divided into two parts: (1) the general flow in the room and (2) the local flow at the blade tips. In the first case, the recirculation pattern is greatly influenced by the size and shape of the room, the positioning of furniture within the room, and the location and orientation of the fan. In general, the bulk of the air movement in the room occurs on the downstream side of the fan. This implies that the fan is fed principally by air that has returned to the inlet side of the fan by the medium of a converging stream of reverse flow just outside the diverging jet [D.1]. The buffer region between these two flows is one of relatively large-scale random turbulence except in

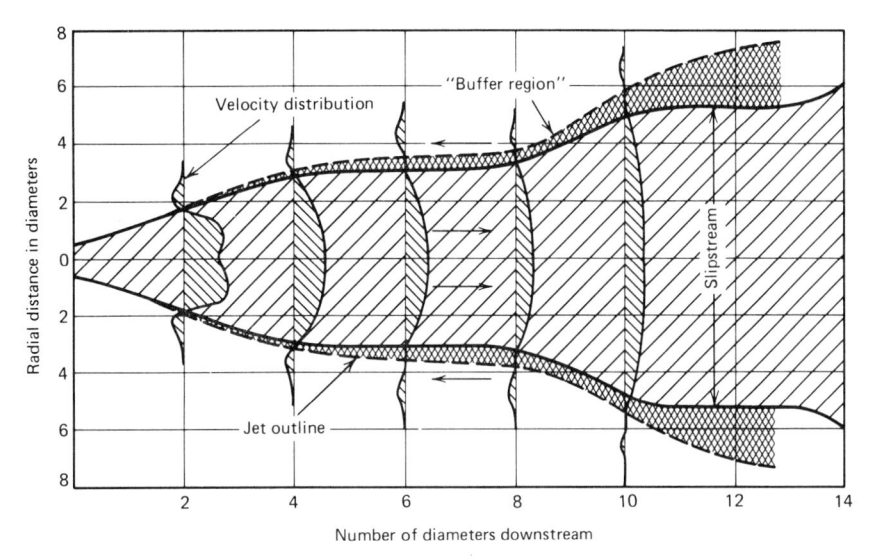

D.2 TEST REQUIREMENTS

Figure D.2 Typical jet features.

(Labels in figure: Velocity distribution; "Buffer region"; Slipstream; Jet outline; Radial distance in diameters; Number of diameters downstream; Test room not less than 30 diameters long × 20 diameters wide)

the close vicinity of the fan, where the flow reversal region is very narrow and well defined. The air entrained by the jet, through the action of fluid shear forces, is obtained from this buffer region. When the jet is powerful enough to reach an opposing surface, such as a wall, the stream is deflected, and this feature results in a relatively wide stream of reverse flow.

The local flow in the vicinity of the blade tips is of considerable interest. According to Smith [D.1] there are two possible flow regimes in this region (Fig. D.3). The one that he refers to as the "normal state" possesses no appreciable local recirculation of the air through the blade tips. In the second regime, known as the "vortex ring state," a stationary vortex ring of recirculating air is present; this phenomenon is often associated with local blade stalling.

To return to the problem of the general flow field created by the fan it is possible by design changes in spanwise blade loading to vary the overall properties of the jet. The air jet possessing the greatest penetration properties has a lesser rate of initial entrainment. Variations of this nature are inevitably associated with changes in the reverse flow and vortex conditions just downstream of the rotor blade tips.

D.2 TEST REQUIREMENTS

In the past, test codes have called for a single-velocity traverse at some specified distance downstream of the fan. On integration, the capacity of the

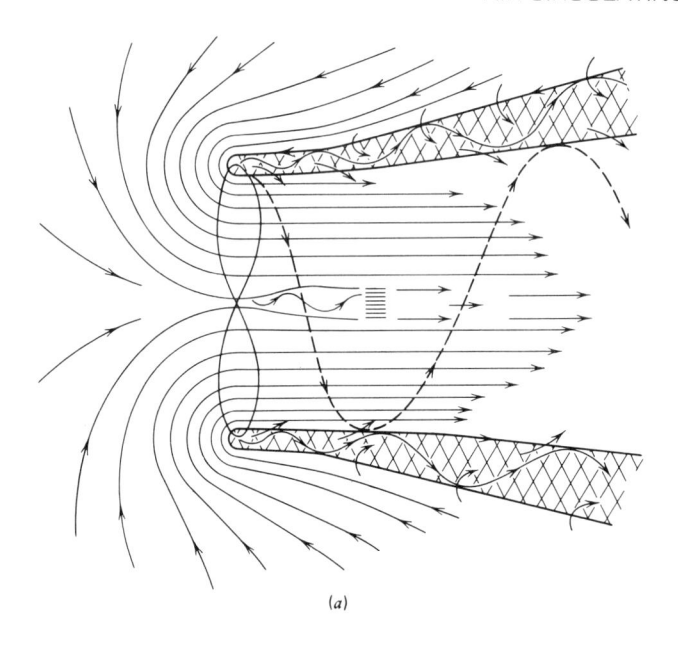

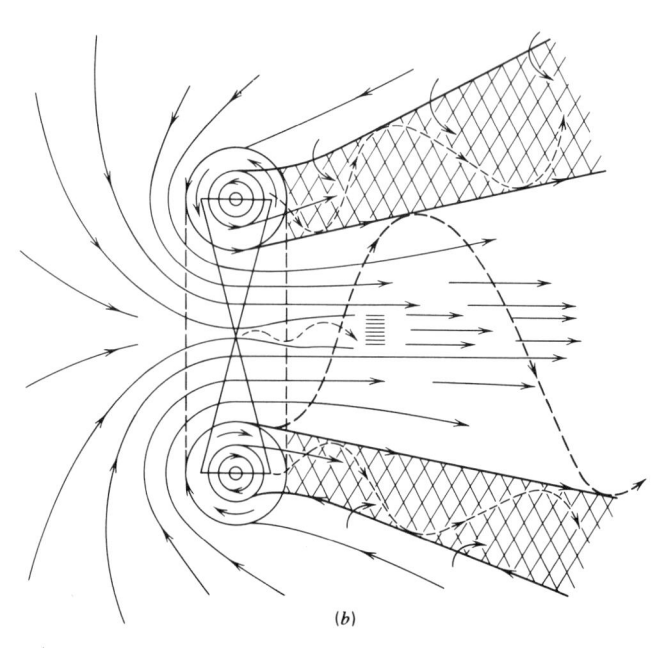

Figure D.3 Schematic illustrations of test states. (a) Normal flow state. (b) Vortex ring flow state.

D.3 DESIGN

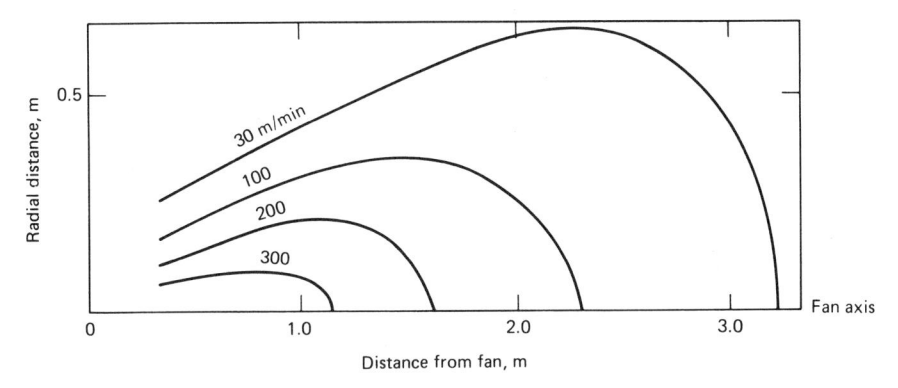

Figure D.4 Suggested form of test data presentation.

fan has been established at this station and this value divided by the power input has been used to specify the service factor of the fan. It is clear from Fig. D.2, however, that this test procedure can produce misleading data. Smith and Chambers [D.2] have therefore suggested the presentation of test data as in Fig. D.4. The combined use of data such as that contained in these latter two figures provides a more satisfactory specification of the fan duty.

Australian Standard AS 2071-1977 requires the contour for 100 m/min to be established during the test program. This approximately defines the personal comfort zone.

In the testing of such fans, the test room must be long enough to ensure that the end wall has no measurable influence on the jet penetration.

D.3 DESIGN

As stated previously, the design section will be limited to one specific set of conditions. Before the method is described, the group of assumptions on which design is based will be outlined.

D.3.1 Design Assumptions

The major design assumption concerns the spanwise distribution of axial velocity for which the fan will be designed. For the general-purpose fan, it has been found from design experience and test observations, including visual flow ones, that a velocity distribution such as that presented in Fig. D.5 gives good results. It is believed that the chances of ring vortex formation at the blade periphery are minimized by designing for a prescribed drop in velocity as the tip extremity is approached, as illustrated in Fig. D.5. The steady decrease in velocity from $x = 0.8$ inboard is dictated to some extent by aerodynamic design difficulties [D.1] and by practical difficulties associ-

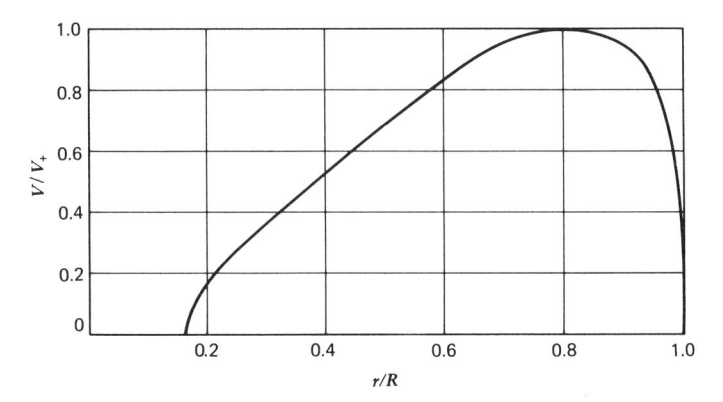

Figure D.5 A velocity distribution for design.

ated with blade shape and attachment. For example, if high velocity were designed for in the inner region, the blade chord and setting angle would both be excessively high. In addition, the curved inflow pattern of the air is such that a smaller axial velocity toward the blade root is, on fundamental grounds, almost inevitable. On the basis of the foregoing evidence, the distribution of Fig. D.5 has here been adopted for design.

In developing the design method, Smith was forced to make many assumptions and approximations. The flow at any given radius is assumed to be truly axial and independent of the adjacent flow. To each elementary annulus of width dr, the Rankine-Froude "actuator disk" theory has been applied in the process of determining the momentum added to the air; the momentum associated with the velocity pressure due to swirl is assumed to be negligible in computing the thrust on a blade element. Despite these rather sweeping assumptions, very good agreement has been obtained between the design and test values in all cases where the method has been applied.

D.3.2 Design Requirement

Design specifications may take two forms. Either (a) the design of a fan to make the best use of the power available from a specific electric motor may be called for, or (b) an air flow condition may be specified for which a fan suitably matched to a motor is required.

From the details to be given, either design can readily be carried out. Since the first specification is the one most commonly encountered, the design method will in the first instance be developed for this case.

D.3.3 Design Procedure

For case (a) the torque equation is the key to the design. From [D.1] this can be written

D.3 DESIGN

$$T = 4\pi\Omega R^4 V_+ \int_0^1 \frac{V}{V_+} a' x^3 \, dx \tag{D.1}$$

where V_+ is the peak axial velocity at $x = 0.8$ and a' is equal to $V_\theta/2\Omega r$.

The design values of T and Ω will be known from motor characteristics similar to those illustrated in Fig. D.6. The three remaining unknowns are R, V_+ and the distribution of a' with respect to x. The general design method is based on the relationships graphically presented in Fig. D.7. It will be seen that a', $C_L\sigma$, and Φ are all unique functions of $V/\Omega r$ for a given lift/drag ratio; the design, however, is not very sensitive to changes in this ratio. Since in the present instance V/V_+ is a unique function of x (see Fig. D.5), it can be shown that the preceding variables are also functions of $V_+/\Omega R$ and x. This being so, Eq. (D.1) can be rewritten for our particular case as

$$T = 4\pi\rho\Omega R^4 V_+ f\left(\frac{V_+}{\Omega R}\right) \tag{D.2}$$

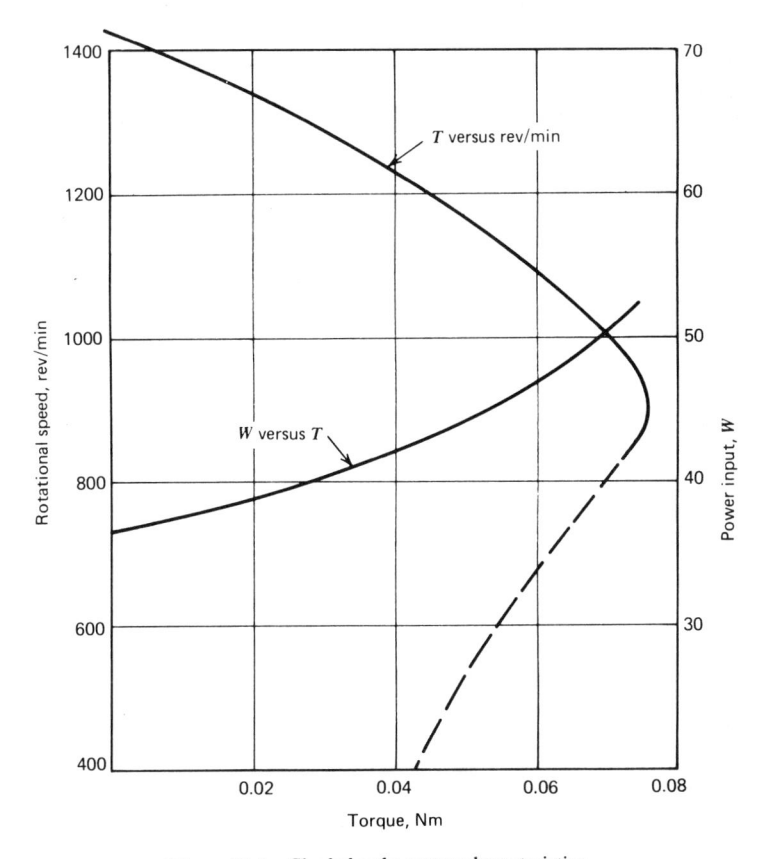

Figure D.6 Shaded-pole motor characteristics.

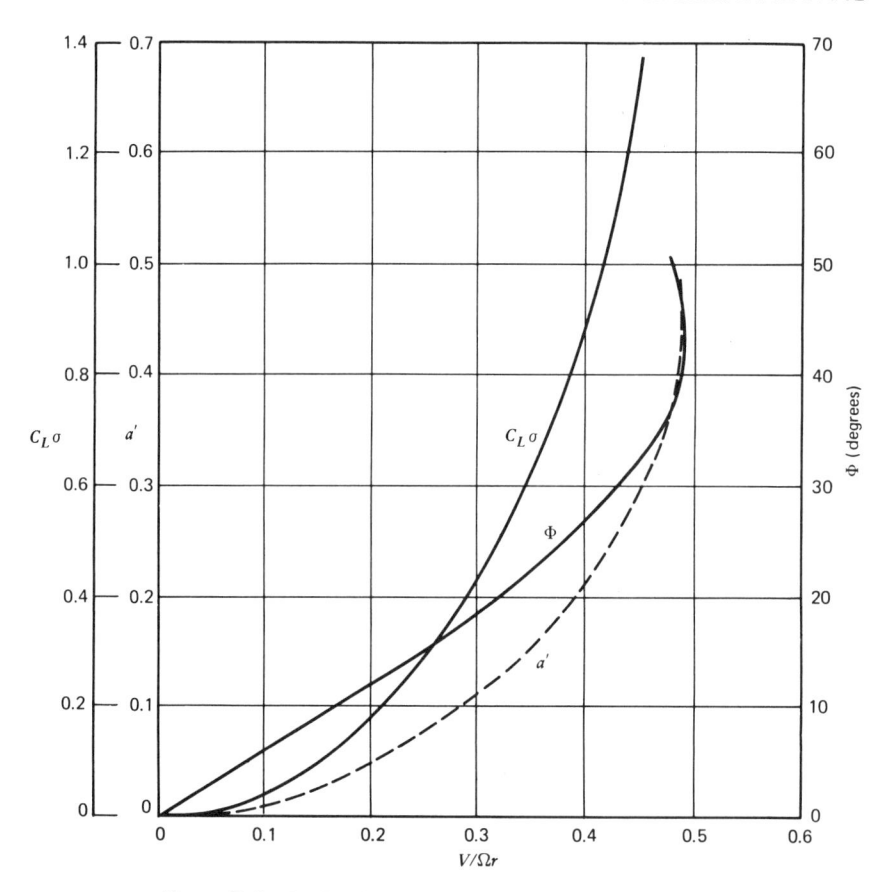

Figure D.7 Design data for general application, $C_L/C_{Dp} = 50$.

The function in this equation can be presented as in Fig. D.8. It will be noted that one variable, namely a', has been eliminated and individual integration avoided. It only remains to assume either R or V_+ and to solve the equation for the given values of T and Ω.

The volume flow at the fan can now be established from the general equation

$$\text{Volume flow rate} = 2\pi R^2 V_+ \int_0^1 \frac{V}{V_+} x\, dx \qquad (D.3)$$

which for the preceding specified velocity distribution reduces to

$$\text{Volume flow rate} = C_1 R^2 V_+$$

$$= 2.48 R^2 V_+ \qquad (D.4)$$

D.3 DESIGN

The thrust exerted by the fan is given in [D.1] by

$$\text{Thrust} = 4\pi\rho R^2 V_+^2 \int_0^1 \left(\frac{V}{V_+}\right)^2 x\, dx \tag{D.5}$$

or in our case by

$$\text{Thrust} = C_2 R^2 V_+^2$$

$$= 4.955 R^2 V_+^2 \tag{D.6}$$

for standard air conditions (see Section 2.1.2).

The remaining task is the computation of the blade properties at various radii, that is, values of x. Using the data previously presented in this subsection, values of Φ and $C_L\sigma$ are presented as functions of $(V_+/\Omega R)$ and x in Figs. D.9 and D.10. Before the design can proceed further, a choice of blade section and operating conditions for this airfoil section must be made.

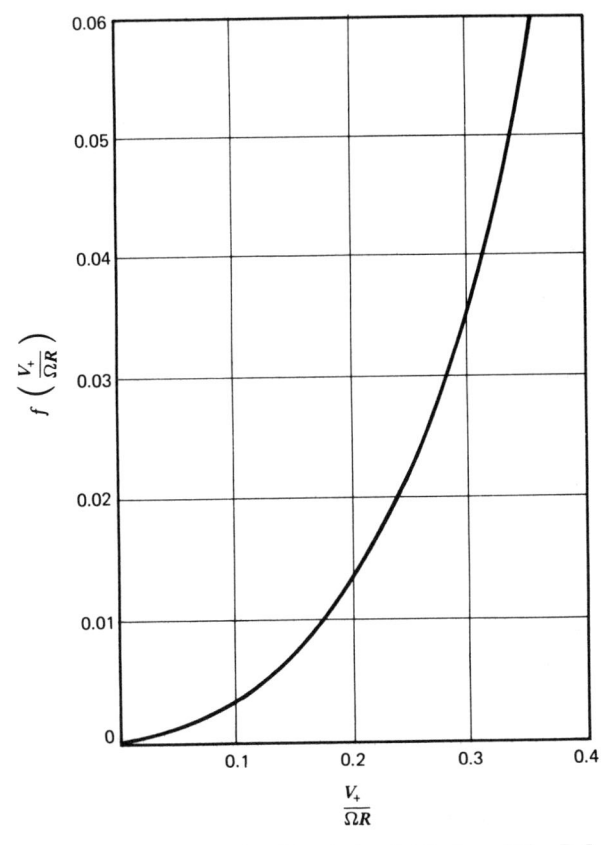

Figure D.8 Design function for velocity distribution of Fig. D.5

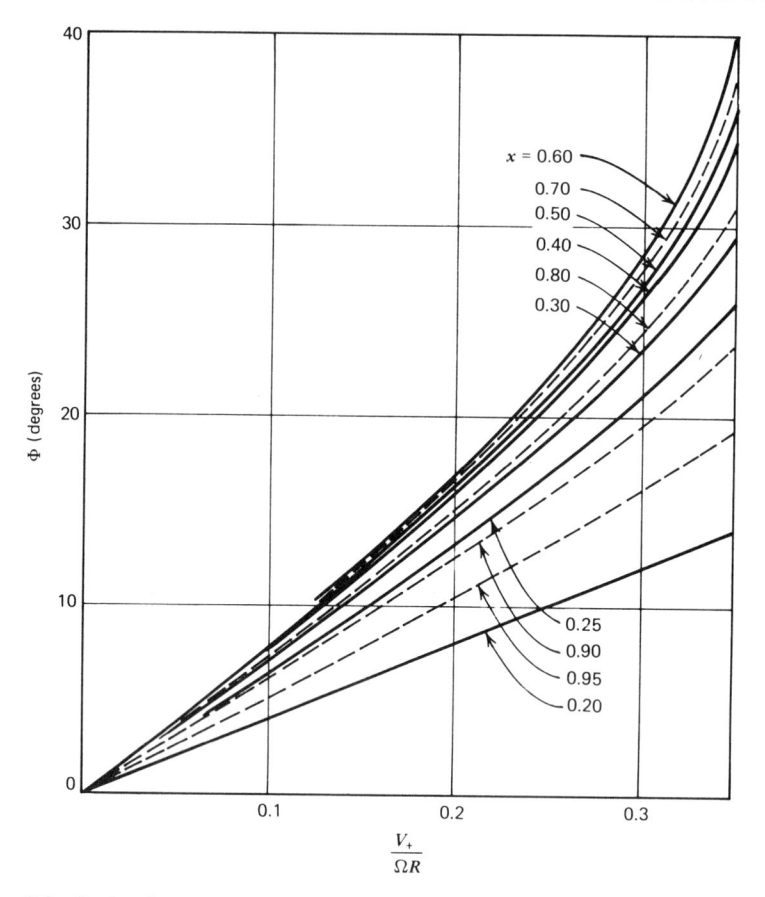

Figure D.9 Design flow angle relative to plane of rotation for velocity distribution of Fig. D.5.

In most cases, cambered plate blades are employed; an 8% camber is the selection of [D.1]. This airfoil at 3° incidence has a C_L of unity and a C_L/C_{D_P} of 50. The design parameters, as graphed, are related to this latter ratio value; a low value of 12 produced significant but not substantial changes in the design calculations. Hence the lift/drag ratio can be ignored as a design factor.

For the preceding set of conditions, σ can be established from Fig. D.10, for the constant spanwise value of $C_L = 1$. The chord follows from

$$c = \frac{2\pi x R \sigma}{n}$$

where n is the number of blades.

D.3 DESIGN

The angle made by the blade chord with the plane of rotation is

$$\Phi + \alpha = \Phi + 3°$$

where Φ is obtained from Fig. D.9.

The design is now complete. It will be seen that very little work is involved, particularly when full use is made of a tabulated design sheet and a graphical approach. The method can be applied to designs where the percentage camber and the spanwise distribution of C_L differ from the preceding selections. Modifications to the spanwise axial velocity distribution would involve the construction of new design graphs.

The second type of specification outlined in Section D.3.2 can be met by starting the design at Eq. (D.4). If the volume flow were specified at some given distance from the fan, then Fig. D.1 could be used in conjunction with Eq. (D.4) in the choice of values of R and V_+ to satisfy the requirements.

The next step requires the matching of the proposed fan to a suitable motor. Equation (D.2) is now solved by assuming either T or Ω and computing the other. The resulting relationship between T and Ω can be presented as a curve; when this line is superimposed on the T versus Ω curves for available shaded pole motors, a satisfactory matching of fan to motor may usually be achieved. In some instances it may be desirable to change the values of R and V_+ in which case a new solution of Eq. (D.4) is necessary. Nevertheless, by the preceding simple trial-and-error methods, a satisfactory solution of Eqs. (D.2) and (D.4) can readily be obtained for chosen values of R, V_+, T, and Ω.

The design procedure discussed here has centered on the shaded-pole type of motor, which possesses the characteristic of a relatively large variation of speed with torque. For split-phase or three-phase motors, which are comparatively insensitive to moderate torque changes, the design procedure can be shortened since Ω will be known initially.

D.3.4 Design Example

For a given shaded-pole motor it is assumed that a suitable operating point is given by

$$T = 0.088 \text{ Nm} \quad \text{and} \quad \Omega = 100 \text{ rad/s}$$

The fan, which is 0.305 m in diameter, can be computed after the selection of V_+ for appropriate T in Table D.1.

$$M = 4\pi\rho\Omega R^4 = 0.8134 \quad \text{for standard air}$$

438   AIR-CIRCULATING FANS

Table D.1 Selection of V_+

V_+	3.8	4.0	4.2
$V_+/\Omega R$	0.249	0.262	0.276
MV_+	3.091	3.254	3.416
$f(V_+/\Omega R)$	0.0225	0.0255	0.0285
T [Eq. (D.2)]	0.0695	0.0830	0.0974

From Table D.1 the design values of V_+ and $V_+/\Omega R$ are established as 4.07 m/s and 0.267, respectively.

$$\text{Volume flow rate} = 2.48 R^2 V_+ \qquad (D.3)$$

$$= 0.234 \text{m}^3/\text{s}$$

$$= 234 \text{ liter/s}$$

$$\text{Thrust} = 4.955 R^2 V_+^2 \qquad (D.6)$$

$$= 1.906 \text{ newtons}$$

The design details for a four-bladed rotor utilizing the 8% cambered plate airfoil are tabulated in Table D.2 for various radii.

A somewhat similar fan designed and tested by Smith [D.1] demonstrated close agreement between the theoretical and experimental values of motor speed, torque, and thrust. The velocity distribution at the Code test station, namely, 600 mm downstream of the rotor, is given in Fig. D.11.

In setting out the blade planform a measure of "industrial designing" may be employed, as suggested in [D.1]. The blade leading edge may be given an arbitrary curvature that then automatically fixes the contour of the trailing edge for the design chord distribution.

D.4 FAN ANALYSIS

For a given fan of unknown design, a performance estimate by analytical methods is often of great value. Such assessments may save testing time by showing the fan to be completely inadequate for the task at hand or they may assist in relating experimentally determined downstream circulation patterns with the spanwise blade loading.

Analysis in this instance differs from the ducted fan case in that the free rotor works always under a constant load condition. This eliminates the swirl coefficient as an independent variable. In analyzing a free fan the relevant task is the estimation of $V/\Omega r$ at each spanwise station.

Table D.2 Rotor Design

x	0.2	0.25	0.3	0.4	0.5	0.6	0.7	0.8	0.9	0.95
r (mm)	30	38	46	61	76	91	107	122	137	145
Φ	10.6	18.2	20.1	22.2	23.0	23.9	23.3	20.8	17.0	14.0
σ	0.134	0.416	0.512	0.623	0.672	0.713	0.694	0.548	0.364	0.247
c (mm)	6	25	37	60	80	102	117	105	78	56
$\Phi + \alpha$	13.6	21.2	23.1	25.2	26.0	26.9	26.3	23.8	20.0	17.0

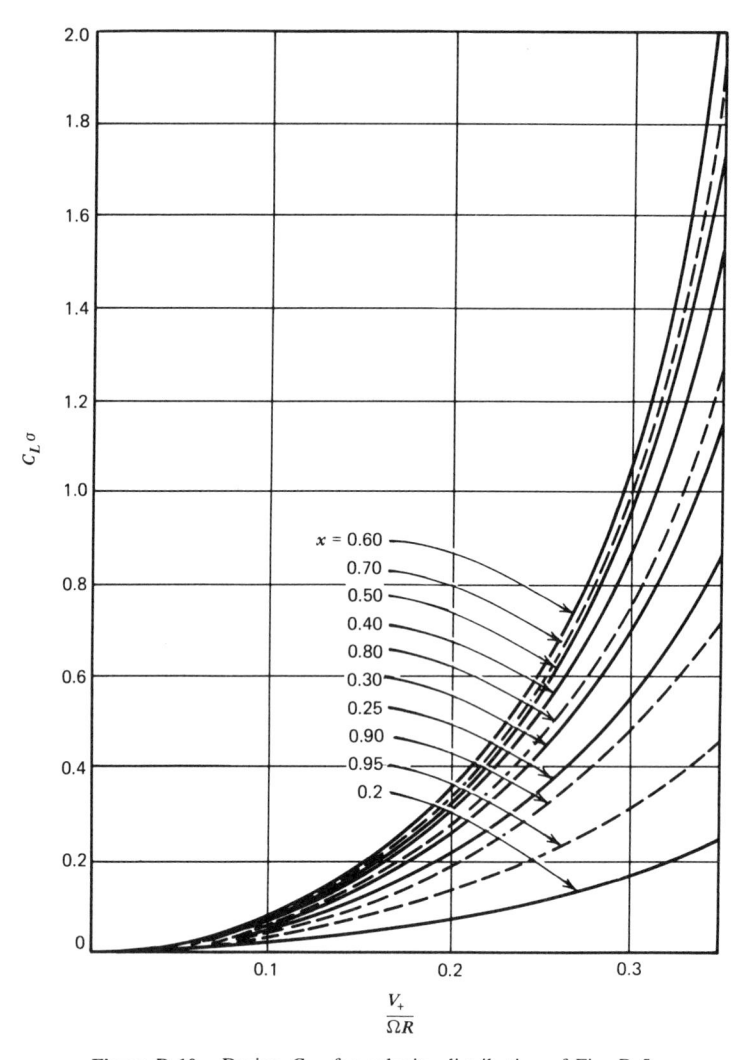

Figure D.10 Design $C_L\sigma$ for velocity distribution of Fig. D.5.

For a given station the values of σ and $(\Phi + \alpha)$ are ascertained by measurement. When the aerodynamic characteristics of the blade section are unknown, it is essential to choose a roughly equivalent section whose characteristics are known and to replace the original section by this in the computations. From an assumed value of C_L, a tentative value of α follows immediately (see Fig. 6.22 for cambered plates). The angle Φ is then obtained by subtraction and this leads to a tentative value of $V/\Omega r$ from the relationship of Fig. D.7. For this flow coefficient value the corresponding value of $C_L\sigma$ can be determined (Fig. D.7). A solution of the problem is achieved when

D.5 MISCELLANEOUS

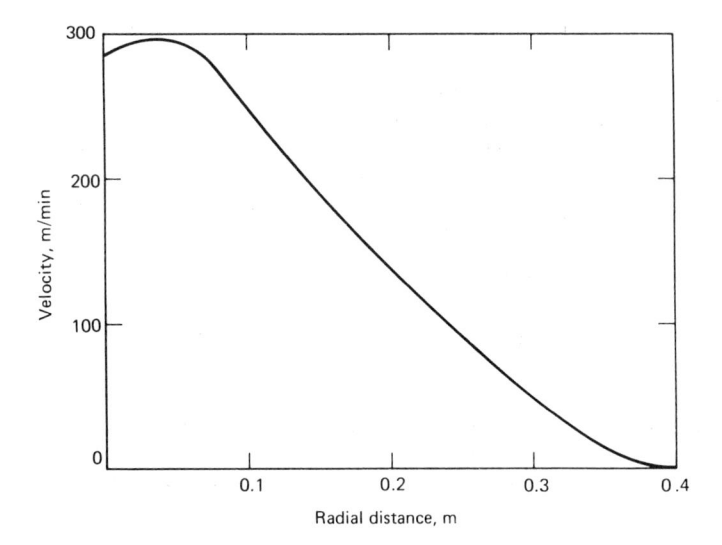

Figure D.11 Velocity distribution at Code test station [D.1].

this product is equal to that obtained by using the assumed value of C_L at the commencement of the process.

The distribution of V with x is tentatively established when a value of Ω is assumed. Equation (D.1) is now solved for T when the remaining variable a' is extracted from Fig. D.7 for the appropriate flow coefficient. A repetition of the above process permits the construction of a curve of T versus Ω; the fan operating point is given by the intersection of this line with the motor curve. With the rotational speed of the rotor thus determined, the velocity distribution follows immediately from the values of $V/\Omega r$ at each radius. The volume flow rate through the disk is given by Eq. (D.3).

Smith claims that an accuracy of prediction to within 5% can be obtained with the analysis procedure just outlined once experience has been gained.

D.5 MISCELLANEOUS

In the foregoing no attempt has been made to use or define an efficiency or figure of merit; these expressions are of doubtful value in the design of free fans. A better guide to the effectiveness of a particular fan is the dimensions of the downstream area within which the air movement exceeds a specified value (see Fig. D.4) for a given power input.

The influence of the axial flow velocity distribution on jet properties is not well understood and hence a more informed approach to this important design problem is not yet a possibility. Research in this field is obviously desirable.

Lastly, the question of noise is worthy of mention. The principles outlined in Chapter 16 are equally applicable to the free fan. Vortex noise can be minimized by eliminating regions of disturbed flow and local separation from the blade surfaces. In locating such regions stroboscopic observations of blade surface tufts are invaluable. A simple alternative method of isolating the spanwise blade region from which the noise is emanating involves the use of a stethoscope; if the detection or sensing piece is placed close to the upstream side of the blades, a radial sweep can be made. Any particular note can now be identified with the region from which it is emanating.

REFERENCES

D.1 V. J. Smith, Air circulator fans: a design method and experimental studies, Aust. Dept. Supply, Aero. Res. Labs, *Aero. Rep. 119,* 1960.

D.2 V. J. Smith and E. W. Chambers, The technique of testing air circulating fans, Aust. Dept. Supply, Aero. Res. Labs, *Aero. Tech. Memo 88,* 1952.

APPENDIX E

Airfoil Section Data

E.1 F-SERIES AIRFOILS

This airfoil series incorporates the C4 thickness form, possessing a circular-arc camber line with additional nose camber based on the NACA 230 camber line; the latter is variable. Hence the following equations apply equally well to the usual C4 blading sections.

The symmetrical C4 profile is given by the expression:

$$\pm y_t = \frac{t}{0.20}(0.3048x^{1/2} - 0.0914x - 0.8614x^2 + 2.1236x^3 - 2.9163x^4 + 1.9744x^5 - 0.5231x^6) \quad \text{(E.1)}$$

where x, y, and maximum thickness t are in percentage chord. Adding the NACA 230 camber line to a circular arc (Figs. E.1 and E.2) produces the camber line contour as for $x < 0.2025$

$$y_c = \left[\left(\frac{0.5}{\sin \theta/2}\right)^2 - (x - 0.5)^2\right]^{1/2} - \frac{0.5}{\tan \theta/2} + [120.5d(x^3 - 0.6075x^2 + 0.1147x)] \quad \text{(E.2)}$$

and for $x > 0.2025$

$$y_c = \left[\left(\frac{0.5}{\sin \theta/2}\right)^2 - (x - 0.5)^2\right]^{1/2} - \frac{0.5}{\tan \theta/2} + d(1 - x) \quad \text{(E.3)}$$

Clothing this camber line (Fig. E.3) with the profile shape of Eq. (E.1) in accordance with

$$\text{Upper surface} \begin{cases} x_U = x - y_t \sin \varphi \\ y_U = y_c + y_t \cos \varphi \end{cases} \quad \text{(E.4)}$$

444   AIRFOIL SECTION DATA

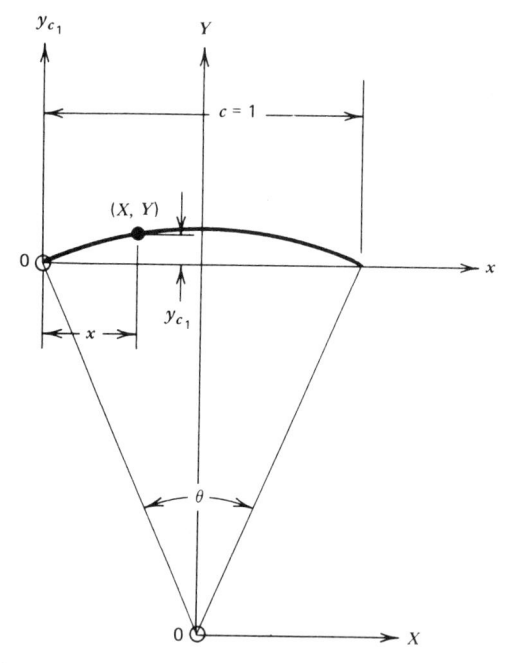

Figure E.1 Circular arc camber line co-ordinates.

Lower surface $\begin{cases} x_L = x + y_t \sin \varphi \\ y_L = y_c - y_t \cos \varphi \end{cases}$ (E.5)

where the angle φ is given by, (Fig. E.4)

$$\tan \varphi = \frac{dy_c}{dx} \quad (\text{E.6})$$

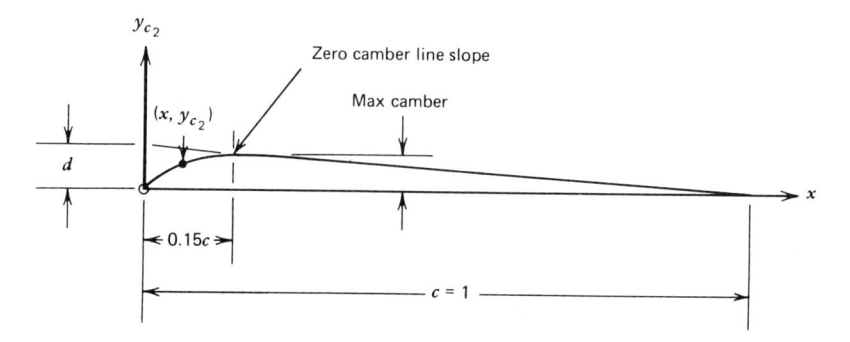

Figure E.2 NACA 230 camber line.

E.1 F-SERIES AIRFOILS

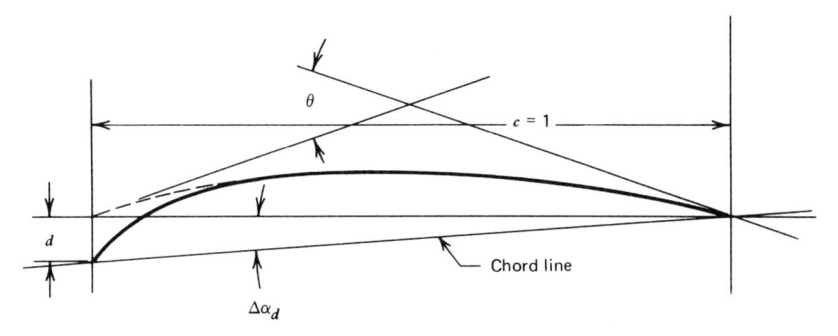

Figure E.3 Composite camber line.

Instead of determining the airfoil profile from Table E.1, a computer program incorporating the preceding equations will permit a direct printout of blade coordinates.

The C4 cambered airfoil coordinates are obtained when zero nose droop is selected.

A mathematical approach to profile shape also has advantages with respect to mechanical blade section properties such as area, centroid location, and moment-of-inertia determination.

The ensuing data for the F-series sections of solid content (defined in Fig. 6.5) are presented in Figs. E.5 to E.7 for 10% c thick sections, and in the following:

$$\bar{X} = 43.5 - 0.0036\theta - 0.048d \tag{E.7}$$

$$\bar{Y} = 0.164\theta + 53d \tag{E.8}$$

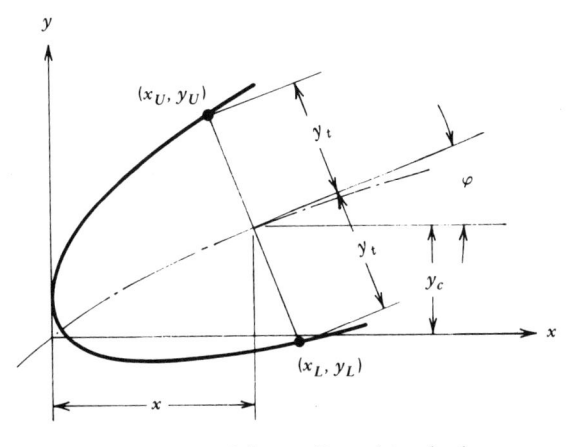

Figure E.4 Aerofoil co-ordinate determination.

AIRFOIL SECTION DATA

Table E.1 F-Series Airfoil Coordinates in Percentage Chord[a, b]

	$\theta = 10°$ $t = 10\%$ $d = 1\%$		Coefficients for Use in Eqs. (6.11) and (6.12)					
			Camber		Thickness		Droop	
x	Y'_U	Y'_L	k_U	k_L	l_U	l_L	m_U	m_L
0	0	0	0	0	0	0	0	0
1.25	2.161	−1.224	0.028	0.013	0.221	−0.128	0.372	0.194
2.5	2.979	−1.642	0.036	0.023	0.268	−0.197	0.448	0.324
5.0	4.165	−2.066	0.055	0.043	0.337	−0.286	0.616	0.537
7.5	5.010	−2.294	0.073	0.062	0.382	−0.347	0.726	0.685
10.0	5.647	−2.433	0.090	0.080	0.415	−0.392	0.793	0.776
15.0	6.514	−2.580	0.120	0.113	0.460	−0.450	0.832	0.834
20.0	7.035	−2.623	0.147	0.141	0.485	−0.481	0.797	0.800
30.0	7.542	−2.474	0.187	0.185	0.501	−0.501	0.700	0.701
40.0	7.581	−2.191	0.212	0.212	0.488	−0.489	0.602	0.601
50.0	7.244	−1.874	0.221	0.221	0.457	−0.456	0.503	0.501
60.0	6.555	−1.550	0.213	0.212	0.407	−0.404	0.404	0.401
70.0	5.521	−1.229	0.188	0.186	0.340	−0.335	0.305	0.301
80.0	4.158	−0.933	0.146	0.142	0.258	−0.251	0.205	0.201
90.0	2.493	−0.693	0.084	0.080	0.162	−0.156	0.103	0.100
95.0	1.552	−0.601	0.046	0.042	0.110	−0.105	0.053	0.050
100.0	0	0	0	0	0	0	0	0
LE rad.		1.2	NIL	NIL	0.12t		NIL	NIL
TE rad.		0.6	NIL	NIL	0.06t		NIL	NIL

[a] Accurate within the range of normal rotor blade design requirements, namely, $\theta = 10$ to $36°$, $t = 7$ to 13%, $d = 0$ to 3%.
[b] *Note:* All Y co-ordinates are measured normal to the chord line.

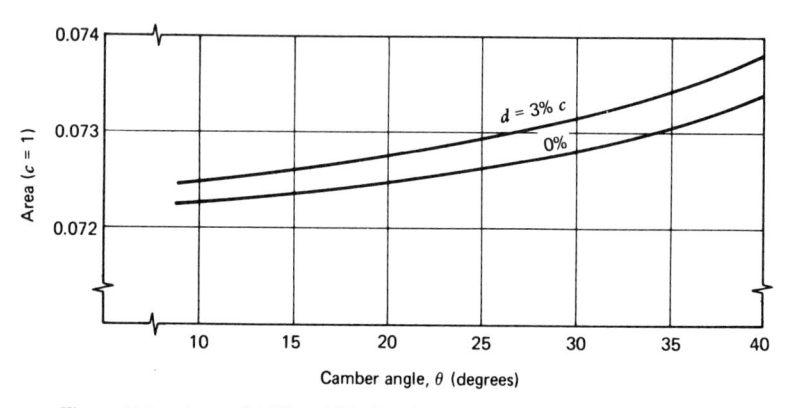

Figure E.5 Area of 10% c thick F-series sections versus camber angle.

E.2 65-SERIES

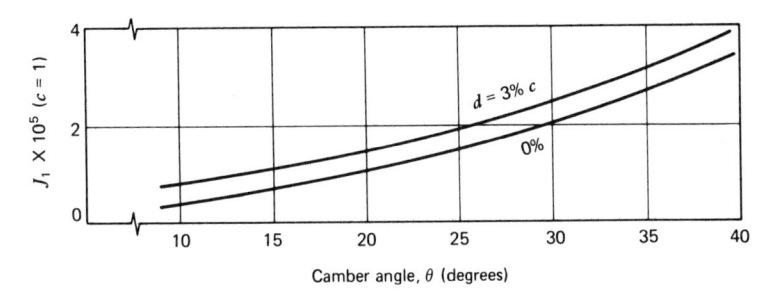

Figure E.6 Moment of inertia of 10% c thick F-series sections versus camber angle.

for centroid location where $\bar{X}$, $\bar{Y}$, and d are all in percentage chord. The moments of inertia, J_1 and J_2, about the principal axes I–I and II–II, respectively, are for unit chord length (see Fig. 16.2). The principal axis, I–I, is within a degree parallel to the airfoil chord line. Airfoil thickness is not a significant parameter in centroid determination. The moments of inertia and area are linear functions of the airfoil thickness to chord ratio and hence values for ratios other than 10% can be obtained from Figs. E.5 to E.7 and the multiplier, (thickness/chord)/0.10.

Finally, area and the moments of inertia are given when the preceding unit chord quantities are multiplied by the chord to the second and fourth powers, respectively.

E.2 65-SERIES AIRFOILS

As indicated in Section 6.8.2, 65-series airfoils are defined in terms of a mean camber line, designated by C_{L_0}. When the recommendations herein are accepted, the design will proceed in the general prescribed manner. On estab-

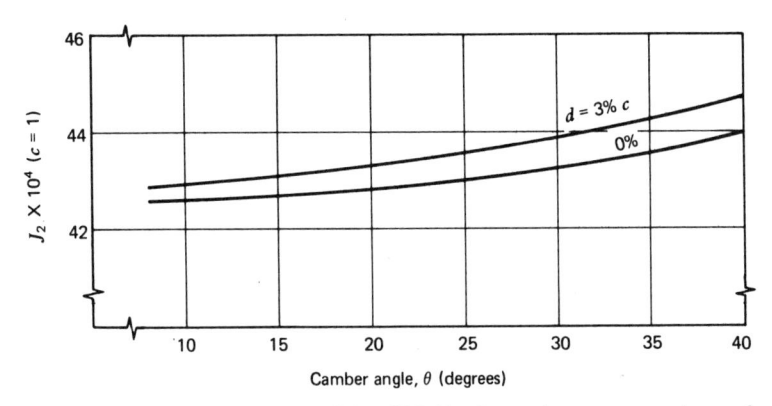

Figure E.7 Moment of inertia of 10% c thick F-series sections versus camber angle.

Table E.2 65-Series Airfoil Data in Percentage Chord

	$C_{L_0} = 1$		NACA 65-010
x	y_c	dy_c/dx	$\pm y_t$
0	0	-------	0
.5	.250	0.42120	.772
.75	.350	.38875	.932
1.25	.535	.34770	1.169
2.5	.930	.29155	1.574
5.0	1.580	.23430	2.177
7.5	2.120	.19995	2.647
10	2.585	.17485	3.040
15	3.365	.13805	3.666
20	3.980	.11030	4.143
25	4.475	.08745	4.503
30	4.860	.06745	4.760
35	5.150	.04925	4.924
40	5.355	.03225	4.996
45	5.475	.01595	4.963
50	5.515	0	4.812
55	5.475	−.01595	4.530
60	5.355	−.03225	4.146
65	5.150	−.04925	3.682
70	4.860	−.06745	3.156
75	4.475	−.08745	2.584
80	3.980	−.11030	1.987
85	3.365	−.13805	1.385
90	2.585	−.17485	.810
95	1.580	−.23430	.306
100	0	-------	0
		LE rad.	.687

lishing a design spanwise camber distribution, use can be made of Fig. 6.12 in obtaining equivalent C_{L_0} values. The related camber line is obtained when the values listed in Table E.2 are multiplied by the required C_{L_0}. The thickness form given in this table is then linearly adjusted to the desired blade section thickness and stretched along the camber line. The section coordinates are obtained using Eqs. (E.4) to (E.6) when the camber line slopes are linearly corrected for the desired C_{L_0} value.

E.3 OTHER AIRFOIL DATA

Table E.3

	Flat Undersurface Airfoils						Thickness Forms	
	Clark Y		Gö 436		Gö 623		RAF 6E	C4	NACA 0010
x	y_U	y_L	y_U	y_L	y_U	y_L	y	$\pm y_t$	$\pm y_t$
0	2.99	2.99	2.50	2.50	3.25	3.25	1.15	0	0
1.25	4.66	1.65	4.70	1.00	5.45	1.95	3.19	1.65	1.58
2.5	5.56	1.26	5.70	0.20	6.45	1.50	4.42	2.27	2.18
5.0	6.75	0.80	7.00	0.10	7.90	0.90	6.10	3.08	2.96
7.5	7.56	0.54	8.10	0.05	9.05	0.35	7.24	3.62	3.50
10	8.20	0.36	8.90		9.90	0.20	8.09	4.02	3.90
15	9.14	0.13	10.05		10.95	0.10	9.28	4.55	4.46
20	9.72	0.03	10.25		11.55	0.05	9.90	4.83	4.78
30	10.00		11.00		12.00		10.30	5.00	5.00
40	9.75		10.45		11.70		10.22	4.89	4.84
50	9.00		9.55		10.65		9.80	4.57	4.40
60	7.82		8.20		9.15		8.98	4.05	3.80
70	6.28		6.60		7.35		7.70	3.37	3.05
80	4.46		4.60		5.15		5.91	2.54	2.19
90	2.39		2.45		2.80		3.79	1.60	1.21
95	1.27		1.25		1.60		2.58	1.06	0.67
100	0.10		0		0.30		0.76	0	0.11
LE rad.							1.15	1.2	1.10
TE rad.							0.76	0.60	

E.3 OTHER AIRFOIL DATA

A selection has been made of airfoils possessing desirable properties with respect to fan and duct applications. The related airfoil section data are listed in Table E.3. Conversion to other thickness to chord ratios can be made assuming a linear change in section coordinates.

Notation

One of the problems associated with fan design theory is the lack of a universal notation system. This can be traced in part to a regional rather than international approach to the subject, in response to local demands for design guidance. In addition, the large number of variables to be specified, plus language spelling differences, have resulted in symbol nonuniformity.

Where possible, compressor notation has been adopted. However, the acceptance of radial equilibrium, and hence the free vortex flow concept, has enabled the replacement of flow angles by two velocity ratios as the main design parameters. Another difference is that all pressure gains and losses are expressed in terms of the mean axial dynamic pressure in the blading annulus. These developments follow from the design method developed by Patterson.

The symbols are presented in groups, facilitating an understanding of their derivation.

The notation of the first five chapters follows the normally accepted aerodynamic conventions.

1.1 VELOCITY VECTORS

1.1.1 Absolute Velocities

The resultant absolute velocity vector at any given point is not a normal requirement in fan practice. Instead only the axial and tangential components are required for design purposes.

V_a	Axial velocity component at any radial station within the fan annulus.
$\bar{V}_a$	Mean axial velocity component within the fan annulus.
V_θ	Tangential velocity component within the fan annulus.
Q	Volume flow rate.

SUFFIXES

p, s	These are used in conjunction with V_θ to denote conditions downstream of the prerotators and upstream of the straighteners, respectively.
N	Denotes value of V_{θ_p} for the no-lift rotor condition.

1.1.2 Relative Velocities

V_1	Velocity relative to rotor or stator blade at inlet.
V_2	Velocity relative to rotor or stator blade at outlet.
V_m	Mean velocity relative to rotor or stator blade.
β_1	Angle that V_1 makes with fan axis.
β_2	Angle that V_2 makes with fan axis.
β_m	Angle that V_m makes with fan axis.
β_N	Value of β_m for rotor no-lift condition.
$\beta_1 - \beta_2$	Flow turning angle.

SUFFIXES

r, p, s	Where necessary these suffixes may be used to differentiate between rotor, prerotator, and straightener conditions, respectively.

1.1.3 Rotational Velocities

Ω	Angular velocity of rotor ($2\pi N$).
Ωr	Blade velocity at radius r.
ΩR	Tip velocity.
N	Revolutions per second.

2.1 PRESSURES AND COEFFICIENTS

2.1.1 Pressures

H	Total pressure at a point.
P	Static pressure at a point.
Δ	Signifies a pressure differential.
Δh	Local total pressure change between two points.
ΔH	Mean total pressure change between two stations.
Δp	Local static pressure change between two points.
ITP	Mean inlet total pressure to fan unit.

3.1 FORCE COEFFICIENTS

OTP	Mean outlet total pressure from fan unit.
FTP	Fan total pressure (OTP − ITP), for in-line and blower units.
FITP	Fan inlet total pressure (−ITP), for exhaust units.

SUFFIXES

th	Signifies theoretical total pressure rise through rotor.
r, bl	Signifies actual total pressure rises through rotor and rotor plus stators, respectively.
R, P, S, D, DL	Denote total pressure losses, respectively, in rotor, prerotator, straightener, diffuser, and diffuser plus duct discharge (downstream loss).
T	Relates to the duct total pressure losses the fan unit must overcome, exclusive of internal losses.
IT	Relates to the duct total pressure losses upstream of exhaust fan inlet.

N.B.: Capital letter suffixes always denote total pressure losses.

2.1.2 Pressure Coefficients

k	Local total pressure coefficient $(\Delta h / \tfrac{1}{2} \rho \bar{V}_a^2)$
K	Mean total pressure coefficient $(\Delta H / \tfrac{1}{2} \rho \bar{V}_a^2)$

The first suffixes have meanings similar to those specified previously.

SECOND SUFFIXES (e.g., K_{R_P})

P, S, A	Signify losses due to profile, secondary, and annulus drag, respectively.

3.1 FORCE COEFFICIENTS

C_L	Lift coefficient with respect to mean velocity V_m.
C_L^*	Nominal lift coefficient.
C_{L_i}	Isolated airfoil lift coefficient.
C_D	Drag coefficient with respect to mean velocity V_m.
T_c	Torque coefficient
Th_c	Thrust coefficient
γ	Lift/drag ratio.

SECOND SUFFIXES

P, S, A	As in previous Section

4.1 FLOW AND SWIRL COEFFICIENTS (FREE VORTEX FLOW, $V_a = \overline{V}_a$)

λ	Local flow coefficient ($\overline{V}_a/\Omega r$)
Λ	Tip flow coefficient ($\overline{V}_a/\Omega R$)
ϵ_p	Local swirl coefficient upstream of rotor ($V_{\theta_p}/\overline{V}_a$)
ϵ_s	Local swirl coefficient downstream of rotor ($V_{\theta_s}/\overline{V}_a$)

SUFFIX

N	Denotes value of λ for no-lift condition

5.1 TORQUE, THRUST, AND POWER

T	Torque
Th	Thrust
W	Power input to rotor.
W_A	Air power (product of volume flow rate Q and total pressure differential ΔH_T or ΔH_{IT})

6.1 EFFICIENCIES

The loss of efficiency due to each component is expressed as a ratio of total pressure loss to the theoretical total pressure rise through the rotor (see Chapter 14). Hence efficiency is given by

η_R	Rotor efficiency $[1 - (K_R/K_{th})]$
η_{BL}	Blading efficiency $[1 - (K_R + K_P + K_S)/K_{th}]$
η_T	Total pressure efficiency of complete fan unit.
η_{IT}	Inlet total pressure efficiency of exhaust fan unit

7.1 GEOMETRIC PARAMETERS

r	Radius of elementary annulus.
r_b	Boss radius.
R	Radius of duct (approximately rotor tip radius).
x	Radius ratio (r/R).
x_b	Boss ratio (r_b/R)
c	Local blade chord of rotor or stator.

8.1 GREEK LETTERS USED

s	Circumferential spacing of adjacent rotor or stator blades at radius r.
σ	Blade solidity, c/s.
n	Number of rotor or stator blades.
θ	Blade camber angle.
ξ	Stagger angle, the angle between the airfoil chord line and the fan axis.
δ	Flow deviation angle, the angle between β_2 and the trailing-edge tangent.
i	Local flow incidence angle, the angle between β_1 and the leading-edge tangent.
α	Incidence angle, the angle between β_m and the airfoil chord line.
α_N	No-lift angle, the angle between ξ and β_N.
α_{od}	Incidence angle, the angle between β_m and the chord line of a circular arc airfoil with no additional nose camber.
ϕ	Angle between the airfoil chord line and the plane of rotation
Φ	Angle between the resultant blade velocity and the plane of rotation (Appendix D)

8.1 GREEK LETTERS USED

Letter	Small	Capital
Alpha	α	
Beta	β	
Gamma	γ	
Delta	δ	Δ
Epsilon	ϵ	
Eta	η	
Theta	θ	
Lambda	λ	Λ
Mu	μ	
Nu	ν	
Xi	ξ	
Pi	π	
Rho	ρ	
Sigma	σ	
Tau	τ	
Phi	ϕ, φ	Φ
Chi	χ	
Psi	ψ	
Omega	ω	Ω

Index

Abrasion, 280
Acoustics:
  attenuation, 295
  coupling, 293
  criteria, 295
  longitudinal modes, 293
  test chambers, 299
  transverse modes, 294
Aerodynamically smooth surface, 53
Airfoils:
  camber, 150, 158, 162, 197, 233, 235
  cascade, 143, 148, 167, 213, 321, 340
  center of lift, 171
  "clothing," 151, 449
  co-ordinates, 443
  definitions, 143
  effect of Reynolds number, 171, 220
  effect of roughness, 58, 154, 156, 220, 398
  isolated, 143, 151, 201
  multiplane interference, 167, 180
  nose droop, 151, 153, 200, 444
  profiles, 144, 152, 443
  stall, 147, 180, 201, 215
  thickness, 212, 215, 443
  types, 144
Airfoil forces:
  center of pressure, 171
  drag, 21, 146, 156, 160, 162, 166
  lift, 21, 145, 154, 159, 161, 165, 174
  pitching moment, 145, 284
Air straighteners, 349, 351
Analysis of blade performance:
  arbitrary vortex flow fans, 414
  cascade airfoils, rotors, 330, 340
  circulating fans, 438
  isolated airfoils, rotors, 326, 336
  prerotators, 332
  straighteners, 333
Angle of incidence, 22, 143, 153, 159, 180
Annulus flow:
  entrance lengths, 67
  hydraulic diameter, 46, 67
  skin friction, 38, 70
  velocity ratio, 45
Arbitrary vortex flow:
  alternate design methods, 25
  approximate design methods, 7, 179, 392
  design 404, 409
  linear swirl coefficient distributions, 25, 405
  radial equilibrium equations, 24
Aspect ratio:
  blade plan form, 222, 321
  rectangular diffuser, 88
  rectangular corner, 115
Atmospheric properties, 8–10
Axial gap between blade rows, effect on:
  noise, 240, 290
  pressure rise, 240
  vibration, 290
Axial velocity distributions:
  axial velocity ratio, 375, 391, 404
  ducted fans, 186
  ducts, 32, 40
  flow asymmetry, 393
  free fans, 430
  radial non-uniformity, 390

Balance weights, 282
Bernoulli's equation, 6, 11, 148, 186
Blade element, 179
Blades:
  air turning, 147, 192, 197
  aspect ratio, 222, 321
  camber, 150, 197
  design limits, 201, 214
  element, 179, 186
  forces, 21, 145, 189
  load distribution, 23, 25, 179, 193
  properties, 147
  sections and development, 152

457

Blades *(Continued)*
  solidity, 151, 180, 203
  thickness, 212, 215, 443
  types, 144
Bluff bodies:
  drag, 129
  duct blockage, 129, 212, 253
Boss ratio, effect on:
  fan efficiency, 265
  fan pressure, 180, 273
Boundary layer:
  equations of motion, 31
  growth, 13, 15, 34, 42
  laminar, 32
  "laminar sublayer," 52
  momentum transfer, 14
  separation, 16
  skin friction, 12, 35, 43
  thicknesses, 12
  transition, 14, 47
  turbulent, 39
Branched duct systems:
  general design, 65, 138
  pressure loss considerations, 126, 138

Cascade blading design, 213
Centroids, 282, 446
Characteristic curve:
  duct, 3
  fan, 3, 358
Codes of testing, 344, 349
  certification, 355
  laboratory, methods, 350
    equipment, 351
  on-site procedures, 356
Coefficients:
  drag, 21, 146
  flow, 177
  lift, 21, 145
  pitching moment, 145
  power, 263, 378
  pressure, 22, 177
  skin friction, 19, 35, 47
  swirl, 177
  thrust, 193, 262
  torque, 193, 262, 378
Compressibility, 4, 10
Conical bosses, 6, 416
Constant thickness blades:
  design example, 316
  lift and profile drag data, 162
  secondary drag, 224
Contractions:
  duct inlets, 79

  in ducts, 74, 249
  general types, 75
  inlet boxes, 76, 250
  resistance losses, 76
Contra-rotating rotors, 5, 187, 309
Corners:
  aspect ratio, 115
  compound bends, 124
  cross-sectional shape, 108
  diffusing, 119, 125
  flow features, 109
  "lobster-back," 117
  loss determination, 111
  mitered, 116
  outlet tangent loss, 109, 112
  radius ratio, 108
  secondary flow, 26, 109
  special types, 116
  splitters, 117
  turning, 108
  vaned, 119
Corrosion, 280
Cyclone separators, 131
Cylinders:
  drag, 129
  flow pattern, 128

Density, 8
Design development, ducts, 139
Deviation angle:
  prerotator vanes, 235
  rotor blades, 199
  straightener vanes, 213, 233
Diaphragm mounted fans, 1, 425
Diffusers:
  annular, 88, 95
  application, fan units, 253
  area ratio, 86
  aspect ratio, 88
  cross-sectional geometry, 88, 92
  efficiency or effectiveness, 85
  equivalent angle, 89
  flow control measures, 96, 101–105
  flow features, 87
  ideal pressure recovery, 86
  inlet flow influences, 92
  losses, 91
  optimum geometries, 88
  Reynolds number influences, 90
  swirl effects, 106
  testing, 382
  truncated, or cropped, 93
  vaned, 96
Dimensional analysis, 17

INDEX

Downstream loss coefficient, fans, 251, 265, 408
Drag components, fans:
  annulus, 221
  other, 227
  profile, 219, 243
  secondary, 221, 243
  tip clearance, 224
  total, 218
Ducts:
  arbitrary cross-sections, 72
  duct categories, 62
  general duct equations, 63
  general test procedures, 61, 139
  loss coefficients, 66
  summation of losses, 66
  system design, 65, 137
  system development, 137
Dust:
  blade fouling, 220
  erosion effects, 279

Efficiency:
  blading, 181, 260
  diffuser, 84, 251
  exhaust fan, 261, 265
  influence of:
    preswirl, 219, 265
    Reynolds number, 171, 220
    residual swirl, 260
    tip clearance, 226
  in-line fan, 260, 265
Ejectors, 131
Elliptical airfoils:
  applications, 161
  characteristics, 162
Entrance lengths, 67
Environmental matters:
  corrosion, 280–281
  erosion, 278
  heat, 277
  materials, 281
Erosion factors, 278–281
Excitation of blades:
  aerodynamic excitation, 284
  blade failures, 286, 290
  frequency parameter, 285
  natural blade frequencies, 285
  resonant frequencies, 284

Fairings:
  air cooling and service ducts, 304
  downstream centerbodies, 254
  nose cone, 248
  support struts, 243

Fan applications:
  cooling fans, 427
  mine ventilation, 307
  multistage units, 302, 309
  "off-the-shelf" units, 274, 303, 316, 319
  unitized equipment, 357
Fan diffusers:
  design data, 354–358
  effect of inlet velocity distribution, 253
Fan optimization:
  aerodynamic, 268
  design duty calculations, 270–273, 316–321
  general, 264
  optimum flow coefficients, 268
Fan pressures:
  fan inlet total, 4, 346
  fan total, 4, 346
Fan stall, 215, 397
Fan testing:
  rigs, 363
  types, 347
Fan unit:
  applications, 302
  components, 2
  ducted fan types, 5
  duty, 3
  recommended sizes, 302
  unducted types, 1
Flow angles, 149, 192
Flow asymmetry, 393
Flow categories, 28
Flow deflection, 147
Force coefficients, 19
Free fan:
  air recirculation properties, 429
  analysis, 435
  design, 431
  noise, 442
  testing, 429
Free vortex flow:
  effect on diffuser efficiency, 253
  equilibrium equation, 25
  fan design conditions, 186
Frequency parameter, 285
Friction velocity, 40
Fully rough surface, 53, 68

Geometry factors:
  laminar, 71
  turbulent, 72

Heat addition, 10, 62
Hot wire anemometer, 369

Humidity, 9
Hydraulic mean diameter, 44, 46, 73
Impulse fan, 6
Inlet boxes, 76, 250
Inlet flow quality, 393
Inlets:
    ducts, 79
    fans, 74, 249, 250
Interaction between duct components, 27, 65, 125, 138
Internal bodies in ducting:
    plates, rods, meshes, 129
    struts, 243
Intersecting ducts, 126

Job specification, 138, 356
Joint leaks, 73

Kinematic similarity, 17
Kinematic viscosity, 10

Leading edge, airfoils:
    camber, 151, 201
    local incidence, 150, 197
    "shock-free" entry, 279
    special nose shapes, 279
    stall, 163
Lift interference factor:
    definition, 167
    use in:
        blading analysis, 330
        design, 317, 320
Load factor:
    definition, 180
    design recommendations, 201–211
    limits, 211

Manufacturing techniques, 281
Materials:
    blades, 281
    erosion factors, 280
    strength, 282
Mean velocity location, 325, 404
Measurements:
    flow direction, 369
    mean value determination, 373
    power, 371
    pressures:
        static, 364, 367
        total, 367
        velocity, 367
    torque, 371, 374, 378
Mechanical considerations:
    balance, 287
    balance weights, 282
    blade centroid, 282, 445
    blade moments of inertia, 282, 446
    frequency parameter, 285
    reversed direction operation, 284
    stall condition, 286
    unsteady load factors, 284
Momentum theorem:
    boundary layer, 31
    fan blades, 184
Multistage units, 5, 302, 309

Noise, fan:
    control:
        aerodynamic design, 290
        silencers, 295
    measurement, 298
    noise induced, vibration, 294
    failure, 294
    ratings, 295
    sources, 291
Non-dimensional coefficients:
    derivation, 19
    dimensional analysis, 17

Obstructions to flow, 253
On-site testing:
    ducts, 141
    fans, 356
Operating point, 3
Optimum considerations, 264, 274

Packaged or unitized fans, 357, 425
Pipe flow:
    entry length, 67
    fully developed, 37, 45, 68
    laminar, 37
    losses, 68
    Reynolds number, 18
    roughness, 54
    transition, 67
    turbulent, 45
Plenums, 249
Power, 263
Pressure loading devices, testing, 253
Profile drag, 146, 219
Properties of air, various, 8–10

Radial equilibrium, 24
Radial velocity, 413
Resistance:
    components, 62

INDEX

nondimensional treatment, 65
Reynolds number considerations:
  airfoil lift and drag, 171
  contractions, 75, 81
  corners, 113, 123
  critical value, 19, 220
  diffusers, 90
  pipe flow, 18
  secondary drag, 221, 224
  turbulence level effects, 220
Rotor losses:
  profile drag, 219
  Reynolds number dependence, 22, 171, 220
  secondary drag, 221
  tip clearance losses, 224
  windage, 227
Roughness, surface:
  effects on:
    blade characteristics, 58
    duct losses, 54, 114
    skin friction, 54–57, 68
    transition, 51
  roughness values, 70
  uniform roughness, 53
  waviness, 51

Secondary flows:
  blade extremities, 26
  corners, 25
  effect on pipe losses, 109
Similarity, 17
Skin friction:
  flat plates, 36, 44
  "friction" velocity, 41
  laminar flow, 35–39
  in pipes, 37, 45
  with pressure gradient, 39, 42
  surface curvature, 42
  surface roughness, 54–59
  turbulent flow, 42–46
Specific speed, 178
Stator analysis, 332, 333
Stator design:
  axial blade spacing, 240, 291
  swirl coefficient dependence, 230
  types of stator vanes, 5
Stator losses:
  dependence on swirl and flow coefficients, 241
  drag coefficients, 243
Stressing of blades, 284
Surface curvature, effect on:
  flow separation, 87
  skin friction, 42

transition, 51
turbulence production, 42
Swirl:
  coefficients, 177
  distribution, 25, 405
  effect on efficiency:
    rotors, 219, 265
    stators, 245, 246
  limitations on, 183
  sign convention, 186
System control, 61
System design, 66, 137, 138
System effect factors, 344
System resistance, 65

Tail fairing:
  convergent, 253, 256
  divergent, 254
Test programs, ducts, 139, 141
Testing, fans:
  classes of, 347, 361
  commercial, 344
  developmental, 361, 412
Thrust:
  coefficient, 193, 262
  gradient, 193
Tip clearance, effects on:
  efficiency, 226
  pressure rise, 227
Torque:
  coefficient, 193, 262
  determination, 371, 378
  gradient, 193
Transition:
  effect of:
    adverse pressure gradient, 48
    instability, 48
    surface curvature, 51
    surface roughness, 51
    surface waviness, 51
  mechanisms, 47
Trouble shooting:
  ducts; 142
  fans, 387
Turning vanes in corners, 119
Two-dimensional channel flow, skin friction
  forces, 39, 70

Unducted fans, 1, 427

Variable pitch rotors, 307, 336
Vibration, 285

Viscosity, 10
Visual studies, 362, 387
Vortex flows, 23, 25
Vortex generators, 104
Wakes, 173
Wall stall, 214

Wire screens:
  diffuser flow control, 101
  resistance of, 129
"Work done" factor, 179, 310, 321, 390
Yaw meters, 348, 370